Handbook of Experimental Pharmacology

Volume 142

Editorial Board

G.V.R. Born, London
P. Cuatrecasas, Ann Arbor, MI
D. Ganten, Berlin
H. Herken, Berlin
K. Starke, Freiburg i. Br.
P. Taylor, La Jolla, CA

Springer
*Berlin
Heidelberg
New York
Barcelona
Hong Kong
London
Milan
Paris
Singapore
Tokyo*

Apoptosis and Its Modulation by Drugs

Contributors

G. Bauer, C. Bebb, M. Berard, J.-Y. Bonnefoy, R.G. Cameron,
M. Casamayor-Palleja, T. Defrance, Y. Delneste, L. Dini,
S. Dormann, M. Dragunow, I. Engelmann, D. Fau, G. Feldmann,
G. Feuer, M.-L. Gougeon, D. Haouzi, T. Ishii, T. Itoh, P. Jeannin,
T.D. Kim, G. Koopman, M. Lucas, K. Magyar, A. Moreau,
M. Nakamura, A. Neubauer, M. Neuman, D. Pessayre,
E. Roelandts, M. Saran, J. Savill, A. Schulz, N. Shear,
H.-U. Simon, H. Soga, B. Szende, C. Thiede, A.-M. Woodgate,
H. Yagi, L. Zhang

Editors

R.G. Cameron and G. Feuer

 Springer

Professor Ross G. CAMERON, M.D., Ph.D.
University of Toronto
Department of Pathology
Toronto Hospital, General Division
200 Elizabeth Street
Toronto, Ontario
CANADA M5G 2C4

Professor George FEUER, Ph.D., C. Med. Sci.
University of Toronto
Banting Institute
Department of Clinical Biochemistry
Toronto, Ontario,
CANADA M5G 2C4

With 73 Figures and 12 Tables

ISBN 3-540-66121-2 Springer-Verlag Berlin Heidelberg New York

Library of Congress Cataloging-in-Publication Data

Apoptosis and its modulation by drugs / contributors, G. Bauer ... [et al.} ; editors, R.G. Cameron and G. Feuer.
  p. cm. -- (Handbook of experimental pharmacology ; v. 142)
  Includes bibliographical references and index.
  ISBN 3540661212 (hc : acid-free paper)
  1. Apoptosis. 2. Pathology, Cellular. I. Bauer, G. (Georg) II. Cameron, R. G. III. Feuer, George, 1921 - IV. Series.

QH671 .A654 2000
616.07'8--dc21
                                                                99-052107

This work is subject to copyright. All rights are reserved, whether the whole or part of the material is concerned, specifically the rights of translation, reprinting, reuse of illustrations, recitation, broadcasting, reproduction on microfilm or in any other way, and storage in data banks. Duplication of this publication or parts thereof is permitted only under the provisions of the German Copyright Law of September 9, 1965, in its current version, and permission for use must always be obtained from Springer-Verlag. Violations are liable for prosecution under the German Copyright Law.

© Springer-Verlag Berlin Heidelberg 2000
Printed in Germany

The use of general descriptive names, registered names, trademarks, etc. in this publication does not imply, even in the absence of a specific statement, that such names are exempt from the relevant protective laws and regulations and therefore free for general use.

Product liability: The publishers cannot guarantee the accuracy of any information about dosage and application contained in this book. In every individual case the user must check such information by consulting the relevant literature.

Cover design: *design & production* GmbH, Heidelberg

Typesetting: Best-set Typesetter Ltd., Hong Kong

Production: Produserv, Berlin

SPIN: 10662498   27/3020 – 5 4 3 2 1 0 – Printed on acid-free paper

Foreword

Apoptosis is a fascinating concept for the basic scientist. This is not only because of the multifaceted variety of proposed and discovered mechanisms, but because apoptosis represents a fundamental pathway for cell renewal. The study of apoptosis has resulted in an array of discoveries on signal transduction and downstream effects that have facilitated and advanced many fields in biology, including research on cancer and other diseases. Thus, the apoptotic process can be viewed as the largest effort of the scientific community to understand how cells work and tissues assemble or remodel. The most direct consequence of this accumulated knowledge is a greater understanding of disease and pathological mechanisms. The end result of these efforts will be significant contributions to health and the adoption of new, never anticipated, therapeutic approaches.

This book represents the summation of considerable effort from a significant group of contributors from all over the world as well as from its editors. In this fashion, many viewpoints have been collected and subjected to thorough academic discussion. The concepts contained in this medically important volume will stimulate and renew the ideas of scientists and indeed, will generate additional work to advance biological knowledge even further. The emphasis of this volume cements what has been established, adds what has not been explored fully, and creates a fertile ground for further hypotheses that will lead to a more complete understanding of the apoptotic process.

Hence, the book is invaluable for both students and teachers of apoptosis and for practitioners of cell biology, while representing an exemplary source of reference for the medical-scientific community. The quality of the assembled data will serve as the underpinnings for moving apoptosis to the next frontier: that is, the exploration of the role of cell renewal as a continuous process from birth to death, precisely the life events that motivated scientists to study it with such fervor and dedication. Should this book achieve these goals, it will realize the aspirations of its editors, Ross Cameron and George Feuer.

The message contained in this book is well delineated and will serve as a comprehensive résumé of factors influencing apoptosis. The authors are sincerely congratulated for their efforts and the editors, whose character and energy are evident from the quality of the book, will return to their toils with the anticipation of further scholarly and new thematic synthesis in the not too distant future.

Felix de la Iglesia, M.D.
Department of Pathology and Experimental Toxicology,
Parke-Davis Pharmaceutical Research
Department of Pathology, School of Medicine,
University of Michigan, Ann Arbor, MI, USA

Preface

Apoptosis is a prevalent type of cell death involving many aspects of human biology and medicine. This text is a comprehensive study of what is known about the process of apoptosis, at the molecular, cellular, tissue, and organism levels of analysis. Molecules specifically involved in apoptosis – including p53, Bcl-2, CD95, interleukins and other cytokines, caspases, ROS and protein kinase C – are described and their roles in the cell death of thymocytes, T and B lymphocytes, cosinophils, nerve cells, skin cells, liver cells, and cancer cells are examined in detail. At the center of these analyses is the synthesis and integration of the molecular and cellular findings with the ultimate goal of the understanding and treatment of human diseases by means of modulation of apoptosis. For example, in Chap. 5, the extensive depletion of peripheral T lymphocytes by apoptosis as part of the chronic response to HIV infection is described. In Chap. 7, apoptosis of T cells is studied in the context of the imununosuppressive therapy used to prevent rejection following organ transplantation. The significance of apoptosis of various cell types is also explored in various chapters. Important human diseases in which apoptosis may play a role are discussed, including Alzheimer's disease, various malignancies, allergic diseases, psoriasis, autoimmune diseases, and chronic liver diseases.

Throughout the text, the theme of apoptosis and its modulation by drugs is discussed. A wide variety of drugs and chemicals is evaluated as to their ability to modify apoptosis. This includes chemotherapeutic drugs, glucocorticoids, p53, and deprenyl. Specific cell types are described with which these drugs and chemicals interact, and the nature of the interaction is specified, e.g., induction or inhibition of apoptosis.

The understanding of the process of apoptosis appears to be important for the care and treatment of patients with a variety of human diseases and, as such, represents an important field of human biology and medicine for clinicians and researchers alike.

We would like to thank the librarians of the University of Toronto Medical Library for their support and encouragement, and one of us (G.F.) would like to thank the Parke Davis Research Institute for its support of his work on this project. We would also like to express our gratitude to Mrs. Doris Walker from Springer-Verlag for her ongoing support of our efforts. In addition, we are both very pleased to have the opportunity to work closely together again on this important scientific and scholarly work.

Toronto, Ontario, Canada
Ross G. Cameron, George Feuer

List of Contributors

BAUER, GEORG, Abteilung Virologie, Institut für Medizinische Mikrobiologie und Hygiene, Universität Freiburg, Hermann-Herder-Str. 11, D-79104 Freiburg, Germany
e-mail: TGFB@sun1.ukl.uni-freiburg.de

BEBB, CHARLOTTE, Centre for Inflammation Research, Department of Clinical and Surgical Sciences (Internal Medicine), Royal Infirmary of Edinburgh, Lauriston Place, Edinburgh EH3 9YW, UK

BERARD, MARION, Inserm U 404, Institut Pasteur de Lyon, F-69365 Lyon Cedex 07, France

BONNEFOY, JEAN-YVES, Centre d'Immunologie Pierre Fabre, 5, avenue Napoléon III, BP97, F-74164 Saint Julien en Genevois, France

CAMERON, ROSS, Department of Pathology, University of Toronto and The Toronto Hospital, Toronto, Ontario, M5G 2C4, Canada
Fax: +1-416-586-9901

CASAMAYOR-PALLEJA, MONTSERRAT, Inserm U 404, Institut Pasteur de Lyon, F-69365 Lyon Cedex 07, France

DEFRANCE, THIERRY, Inserm U 404, Institut Pasteur de Lyon, Avenue Tony Garnier F-69365 Lyon Cedex 07, France
e-mail: defrance@albert.lyon151.inserm.fr

DELNESTE, YVES, Centre d'Immunologie Pierre Fabre, 5, avenue Napoléon III, BP97, F-74164 Saint Julien en Genevois, France
e-mail: yves.delneste@pierre-fabre.com

DINI, LUCIANA, Department of Biology, University of Lecce, via Prov.le Lecce-Monteroni, I-73100 Lecce, Italy
e-mail: ldini@ilenic.unile.it

DORMANN, SABINE, Abteilung Virologie, Institut für Medizinische
Mikrobiologie und Hygiene, Universität Freiburg,
Hermann-Herder-Str. 11, D-79104 Freiburg, Germany

DRAGUNOW, MIKE, Department of Pharmacology, University of Auckland,
Auckland, New Zealand
e-mail: m.dragunow@auckland.ac.nz

ENGELMANN, ILKA, Abteilung Virologie, Institut für Medizinische
Mikrobiologie und Hygiene, Universität Freiburg,
Hermann-Herder-Str. 11, D-79104 Freiburg, Germany

FAU, DANIEL, INSERM U481, Hôpital Beaujon, 100 Bd du Général Leclerc,
F-92118 Clichy, France

FELDMANN, GÉRARD, INSERM U327, Faculté de Médecine Xavier Bichat, 16
rue Henri Huchard, F-75870 Panis, France
e-mail: u327@bichat.inserm.fr

FEUER, GEORGE, Department of Clinical Biochemistry, Banting Institute,
University of Toronto, Toronto, Ontario, M5G 2C4, Canada
Fax: +1-416-512-9328

GOUGEON, MARIE-LISE, Unité d'Oncologie Virale/CNRS ERS 572,
Département SIDA et Rétrovirus, Institut Pasteur, 28 Rue du Dr. Roux,
F-75724 Paris Cedex 15, France
e-mail: mlgougeo@pasteur.fr

HAOUZI, DELPHINE, INSERM U481, Hôpital Beaujon, 100 Bd du Général
Leclerc, F-92118 Clichy, France

ISHII, TADASHI, Division of Immunology and Embryology, Department of
Cell Biology, Tohoku University School of Medicine, Seiryo-Machi 2-1,
Sendai 980-77, Japan

ITOH, TSUNETOSHI, Division of Immunology and Embryology, Department of
Cell Biology, Tohoku University School of Medicine, Seiryo-Machi 2-1,
Sendai 980-8575, Japan
e-mail: itoh@immem.med.tohoku.ac.jp

JEANNIN, PASCALE, Centre d'Immunologie Pierre Fabre, 5, avenue Napoléon
III, BP97, F-74164 Saint Julien en Genevois, France

KIM, THEO DANIEL, Abteilung für Innere Medizin, Schwerpunkt
Hämatologie Charite – Virchow Klinikum, Augustenburger Platz 1,
D-13353 Berlin, Germany
e-mail: labneub@charite.de

List of Contributors

KOOPMAN, GERRIT, Biomedical Primate Research Center, Department of Virology, Lange Kleiweg 157, NL-2288 GJ Rijswijk, The Netherlands
e-mail: koopman@bprc.nl

LUCAS, MIGUEL, Sevicio de Biología Molecular, Hospital Universitario Virgen Macarena, Facultad de Medicina, Avda. Sánchez Pizjuán 4, E-41009 Sevilla, Spain
e-mail: Lucas@cica.es

MAGYAR, KÁLMÁN, Department of Pharmacodynamics, Semmelweis University of Medicine, PO Box 370, H-1445 Budapest, Hungary
e-mail: Magykal@net.sote.hu

MOREAU, ALAIN, INSERM U327, Faculté de Médecine Xavier Bichat, 100 Bd du Général Leclerc, F-75018 Paris, France

NAKAMURA, MASANORI, Division of Immunology and Embryology, Department of Cell Biology, Tohoku University School of Medicine, Seiryo-Machi 2-1, Sendai 980-77, Japan

NEUBAUER, ANDREAS, Zentrum Innere Medizin, Klinik für Hämatologie, Onkologie und Immunologie, Philipps Universität Marburg, Baldinger Straße, D-35033 Marburg, Germany
e-mail: neubauer@mailer.uni-marburg.de

NEUMAN, MANUELA, Sunnybrook and Women's College Health Science Centre, Department of Pharmacology, Faculty of Medicine, University of Toronto, Toronto, Ontario, M4N 3M5, Canada
e-mail: manuela@fisher.sunnybrook.utoronto.ca

PESSAYRE, DOMINIQUE, INSERM U481, Hôpital Beaujon, 100 Bd du Général Leclerc, F-92118 Clichy Cedex, France
e-mail: Pessayre@bichat.inserm.fr

ROELANDTS, EDITH, Centre Hospital Regional, Pharmacie Centrale, 2, avenue Oscar Lambret, F-59000 Lille, France

SARAN, MANFRED, GSF-Forschungszentrum für Umwelt und Gesundheit Neuherberg, D-Neuherberg, Germany

SAVILL, JOHN, Centre for Inflammation Research, Department of Clinical and Surgical Sciences (Internal Medicine), Royal Infirmary of Edinburgh, Lauriston Place, Edinburgh EH3 9YW, UK
e-mail: j.savill@ed.ac.uk

SCHULZ, ANGELA, Abteilung Virologie, Institut für Medizinische
  Mikrobiologie und Hygiene, Universität Freiburg,
  Hermann-Herder-Str. 11, D-79104 Freiburg, Germany

SHEAR, NEIL, Sunnybrook Health Science Centre, Department of
  Pharmacology, Faculty of Medicine, University of Toronto, Toronto,
  Ontario, M4N 3M5, Canada

SIMON, HANS-UWE, Swiss Institute of Allergy and Asthma Research (SIAF),
  University of Zurich, Obere Strasse 22, CH-7270 Davos, Switzerland
  e-mail: hus@siaf.unizh.ch

SOGA, HIROYUKI, Division of Immunology and Embryology, Department of
  Cell Biology, Tohoku University School of Medicine, Seiryo-Machi 2-1,
  Sendai 980-77, Japan

SZENDE, BÉLA, 1st Institute of Pathology and Experimental Cancer
  Research, Semmelweis University of Medicine, PO Box 370,
  H-1445 Budapest, Hungary

THIEDE, CHRISTIAN, Medizinische Klinik und Poliklinik I,
  Universitätsklinikum Carl Gustav Carus, Technische Universität
  Dresden, Fetscherstr. 74, D-01307 Dresden, Germany
  e-mail: thiede@oncocenter.de

WOODGATE, ANN-MARIE, Departments of Pharmacology and Molecular
  Medicine, Faculty of Medicine and Health Science, The University of
  Auckland, New Zealand

YAGI, HIDEKI, Division of Immunology and Embryology, Department of
  Cell Biology, Tohoku University School of Medicine, Seiryo-Machi 2-1,
  Sendai 980-77, Japan

ZHANG, LI, Department of Pathology, The Toronto Hospital, General
  Division, CCRW 2-809, Toronto, Ontario M5G 2C4, Canada
  e-mail: lzhang@transplantunit.org

Contents

CHAPTER 1

Incidence of Apoptosis and Its Pathological and Biochemical Manifestations
R. Cameron and G. Feuer. With 6 Figures 1

A. Apoptosis: Characteristics and Scope 1
B. Cell Death 2
C. Features of Necrosis 4
    I. Occurrence 4
    II. Morphology 5
    III. Biochemistry 5
D. Features of Apoptosis 6
    I. Occurrence 6
    II. Morphology 6
    III. Biochemistry 7
E. Activation of Apoptosis 9
F. Incidence of Apoptosis 12
    I. Physiological Conditions 12
        1. Embryonic and Fetal Development 12
        2. Cell Turnover in Adult Tissues 12
        3. Involution of Adult Tissues 13
    II. Pathological Conditions 13
        1. Regression of Hyperplasia 13
        2. Pathological Atrophy 13
        3. Drugs 14
        4. Toxic Chemicals 14
        5. Chemical Carcinogens and Cancer Chemotherapy Agents 15
        6. Radiation and Hyperthermia 16
    III. Disease Conditions 16
        1. Cell-Mediated Immunity 16
        2. Ischemia 18
        3. Neurodegenerative Disorders 19
        4. Blood Cell Disorders 20
        5. Malignant Neoplasms 21

6. Viral Infection	21
7. Expression of Apoptosis in Other Diseases	22
G. Conclusions	23
References	24

CHAPTER 2

Molecular Cellular and Tissue Reactions of Apoptosis and Their Modulation by Drugs
R. CAMERON and G. FEUER 37

A. Introduction	37
B. Molecular Mediators of Apoptosis	37
I. TNF Receptor Family	37
II. Bcl-2 Gene Family	39
III. Caspases	41
IV. Cytokines	42
V. Co-Stimulatory Molecules	42
VI. Perforin and Granzyme B	43
VII. Protein Kinase C	43
VIII. Reactive Oxygen Species	43
IX. Glutathione	44
X. Inhibitor Polypeptides	44
C. Cell-Specific Pathways of Apoptosis	44
I. Immune System	44
1. T Cells	44
2. B Cells (and Plasma Cells)	45
3. Macrophages (and Dendritic Cells)	46
4. Eosinophils	46
5. Neutrophils	47
II. Nervous System	47
1. Neuronal Cells	47
III. Liver Cells	47
1. Hepatocytes	47
2. Kupffer Cells	48
IV. Malignant Cells	48
1. Leukemia (and Lymphoma)	48
2. Carcinoma	48
D. Tissue-Specific Reactions Involving Apoptosis	49
I. Inflammation and Hypersensitivity	49
II. Cancer (and Carcinogenesis)	51
III. Neurodegenerative Disorders	51
IV. Autoimmune Disorders	52
E. Potential of Modulation of Molecular, Cellular or Tissue Reactions by Drugs	52

I. Immunosuppressive Drugs	52
II. Chemotherapeutic Drugs	52
III. Natural Substances	52
References	53

CHAPTER 3

Hepatocyte Apoptosis Triggered by Natural Substances (Cytokines, Other Endogenous Molecules and Foreign Toxins)
D. PESSAYRE, G. FELDMANN, D. HAOUZI, D. FAU, A. MOREAU, and M. NEUMAN. With 4 Figures 59

A. Introduction	59
B. Fas-Mediated Apoptosis	60
I. Fas Ligand	60
II. Fas	61
III. Fas Signal Transduction in Lymphoid Cells	61
1. Caspase Activation	61
2. Permeabilization of Mitochondrial Membranes	63
3. Modulation by Caspase 8 Decoys (FLIPs), Cellular Inhibitors of Apoptosis (c-IAPs), Members of the Bcl-2 Family and Other Factors	65
4. Orientation of Cell Death Towards Apoptosis and/or Necrosis	67
5. Efflux of Reduced Glutathione and Other Effects	67
6. Fas Signaling Independent of Fas Ligand	67
IV. Role of Fas in the Control of the Immune System	68
V. Fas-Induced Hepatocyte Apoptosis	69
1. Agonistic Anti-Fas Antibodies	69
2. Activated Lymphocytes	71
3. Fratricidal Killing	72
4. Basal Hepatic Apoptosis	73
C. Tumor Necrosis Factor-α-Mediated Cell Death	74
I. TNF-α	74
II. TNF-α Receptors and Signal Transduction	75
III. Hepatotoxicity of TNF-α in Experimental Models	76
IV. Role of TNF-α in Human Liver Injury	78
D. Transforming Growth Factor-β and Activins	78
E. Small Endogenous Molecules	81
I. Ceramide, Sphingosine-1-phosphate, and Phosphatidylserine	81
II. Retinoic Acid	82
III. Bile Acids	83
IV. Extracellular ATP and Adenosine	86
V. Nitric Oxide	86

F. Foreign Toxins 86
G. Conclusions and Perspectives 90
References 91

CHAPTER 4

The Role of C-type Protein Kinases in Apoptosis
M. LUCAS. With 5 Figures 109

A. PKC Isozymes 109
B. PKC and Apoptosis 110
C. Caspases and PKC 112
D. Apoptosis Versus Mitosis 112
E. Cell Cycle, CDK and PKC Inhibitors 114
F. Capacitative Calcium Entry and Apoptosis 115
G. PKC Implication in the Sphingomyelin Pathway to Apoptosis ... 120
References 123

CHAPTER 5

How Does Programmed Cell Death Contribute to AIDS Pathogenesis?
M.-L. GOUGEON. With 4 Figures 127

A. Introduction 127
   I. The Pathogenesis of HIV Disease 127
   II. CD4 T Cell Homeostasis in HIV Infection 128
B. PCD in HIV Infection 129
   I. Influence of HIV-1 Genes on the Induction
     of Apoptosis 129
   II. Peripheral T Lymphocytes from HIV-Infected Subjects
     are Prematurely Primed for Apoptosis 130
   III. Relationship Between Apoptosis
     and Immune Activation 132
   IV. Relevance of PCD to Disease Progression
     and AIDS Pathogenesis 133
C. Molecular Control of HIV-Induced Apoptosis 134
   I. Negative Regulation of Bcl-2 Expression. Consequences
     on the Anti-Viral Cytotoxic Function 134
   II. Upregulation of the CD95 System 135
   III. Possible Effectors of CD95-Mediated Apoptosis.
     Consequences on CD4 T Cell Depletion 136
   IV. Other Cell Death Genes Involved in HIV-Induced
     Apoptosis 138
D. Interrelation of HIV-Induced Apoptosis and Cytokines 138

		I. Dysregulation of Cytokine Synthesis in HIV Infection	138
		II. The Disappearance of Th1 cells Is Related to Their Priming for Apoptosis	139
		III. Regulation of HIV-Induced Apoptosis by Cytokines	140
	E.	PCD and T Cell Renewal. Influence of HAART	142
	F.	Concluding Remarks	143
	References		144

CHAPTER 6

Apoptotic Cell Phagocytosis
J. SAVILL and C. BEBB. With 3 Figures 151

A.	Introduction	151
B.	Tissue Consequences of Cell Death	151
	I. Necrosis and Incitement of Inflammatory Injury	151
	II. Silent and Anti-Inflammatory Cell Clearance by Apoptosis	152
	III. Provocation of Immune Responses During Clearance of Apoptotic Cells	154
	IV. Resolving the Clearance Paradox for Apoptotic Cells	156
C.	Molecular Mechanisms by Which Phagocytes Recognise Cells Undergoing Apoptosis	158
	I. "Eat Me" Signals Displayed by Apoptotic Cells	158
	1. Exposure of Phosphatidylserine	158
	2. Sites Which Bind "Bridging" Proteins	160
	3. Carbohydrate Changes on Apoptotic Cells	161
	4. Intercellular Adhesion Molecule (ICAM)-3	162
	II. Phagocyte Receptors for Apoptotic Cells	162
	1. Thrombospondin Receptors: $\alpha v \beta 3$ and CD 36	163
	2. Scavenger Receptors	165
	3. CD14	165
	4. Phosphatidylserine Receptors (PSRs)	166
	5. Complement Receptors	166
	6. Murine ABC1 and C. Elegans CED-7 Proteins	167
	7. Intraphagocyte Signalling; CED-5 and CED-6	167
	III. Why So Many Recognition Mechanisms?	168
D.	Perturbations of Clearance in Disease	168
	I. C1q Deficiency	169
	II. Antiphospholipid Autoantibodies	170
E.	Promotion of Safe Clearance	170
	I. Glucocorticoids	171
	II. Other Factors	171
F.	Conclusions and Future Prospects	171
References		172

CHAPTER 7

T Cell Apoptosis and Its Role in Peripheral Tolerance
R. CAMERON and L. ZHANG 179

- A. Introduction 179
- B. Phenotypically Different Types of Apoptosis
  of T Lymphocytes 179
  - I. Activation Induced Cell Death 179
  - II. Veto Cell Phenomenon 179
  - III. Programmed Cell Death 180
  - IV. Activation Induced Cell Death of Human Peripheral
    T Cells 180
- C. Molecules Involved in T Cell Apoptosis 180
  - I. TNF Receptor Family 180
  - II. Bcl-2 Family 181
  - III. Caspases 182
- D. Regulators of T Cell Apoptosis 182
  - I. Cytokines (IL-2, IL-4, Interferon gamma, etc.) 182
  - II. Co-Stimulatory Molecules (B7, CD28, CTLA-4, etc.) 182
  - III. Effect of Viral Infection 183
- E. Mechanisms Involved in Peripheral Tolerance 183
  - I. Clonal Deletion 183
    1. Bacterial Superantigen-Induced AICD 183
    2. Alloantigen-Induced AICD 184
    3. Clonal Anergy 185
  - II. Suppression, Regulatory (Suppressor) T Cells 185
  - III. Immune Deviation (Th1 to Th2 Switching) 186
- F. Role of T Cell Apoptosis in Oral Tolerance
  and Autoimmunity 187
- G. Role of T Cell Apoptosis in Transplantation Tolerance 188
  - I. Mechanisms of Transplantation Tolerance 188
  - II. Potential of Immunosuppressive Drugs to Modulate T Cell
    Apoptosis and Induce Transplantation Tolerance 190
- H. Apoptosis and Immune Privilege 190
- I. Conclusions 191
- References 191

CHAPTER 8

Apoptosis of Nerve Cells
A.-M. WOODGATE and M. DRAGUNOW 197

- A. Introduction 197
  - I. Programmed Cell Death 199
- B. Apoptosis in the Brain 199

	I. Alzheimer's Disease	199
	II. Parkinson's Disease	200
	III. Cerebral Ischemia	201
	IV. Status Epilepticus	201
	V. Huntington's Disease	202
	VI. Other Brain Disorders	202
C. Models of Neuronal Apoptosis		202
	I. Developmental Nerve Cell Death	202
	II. Degenerative Nerve Cell Death	203
D. Biochemical Apoptosis Pathways in Neurons		204
	I. The Inducible Transcription Factors	204
	II. The Role of the ITFs in Apoptosis	204
	1. During CNS Development	204
	2. Degenerative Nerve Cell Death	205
	3. Evidence of a Role for c-Jun and c-Fos in Apoptotic Nerve Cell Death	206
	4. How Might c-Jun Mediate Neuronal Apoptosis?	206
	a. Upstream Mediators	206
	b. Downstream Mediators	207
	III. The Caspase Family	208
	1. Evidence of a Role for Caspases in Apoptotic Nerve Cell Death	209
	a. Developmental Nerve Cell Death	209
	b. Degenerative Nerve Cell Death	209
	2. Which Caspases Mediate Apoptotic Nerve Cell Death?	210
	3. Regulation of Apoptosis by the Caspases	211
	IV. The Bcl-2 Family	212
	1. Regulation of Bcl-2-Related Genes	212
	2. Evidence of a Role for Bcl-2-Related Genes in the Nervous System	213
	a. Developmental Nerve Cell Death	213
	b. Degenerative Nerve Cell Death	214
	3. How Does Bcl-2 Exert Its Neuroprotective Effects?	215
	V. Cell Cycle Regulators	215
	1. Evidence of a Role for p53 in Neuronal Apoptosis	216
	2. How Does p53 Mediate Neuronal Apoptosis?	217
	3. Cyclins and Cyclin-Dependent Kinases	217
E. Conclusion		218
References		218

CHAPTER 9

Use of p53 as Cancer Cell Target for Gene Therapy
C. THIEDE, T.D. KIM, and A. NEUBAUER. With 2 Figures 235

A. Introduction 235
B. Genetic Changes in Tumor Development 236
C. The p53 Tumor Suppressor Gene 236
    I. p53: From Structure to Function 236
    II. p53 and Induction of Apoptosis 239
    III. Alterations of p53 in Human Cancers 240
    IV. p53 Homologues 241
D. p53 and Gene Therapy 241
    I. Introduction 241
    II. Gene Therapy: General Remarks 242
    III. Rationale for p53-Targeting in Gene Therapy 242
    IV. Trials Reconstituting Wild-Type p53 242
        1. Adenoviruses 243
        2. Retroviruses 243
        3. Non-Viral Gene Delivery Systems 243
        4. In Vitro Data and Preclinical Trials 244
        5. Clinical Trials 245
    V. The ONYX-015 Virus 247
    VI. Future Directions 248
E. Summary 250
References 250

CHAPTER 10

Antioxidants: Protection Versus Apoptosis
Y. DELNESTE, E. ROELANDTS, J.-Y. BONNEFOY, and P. JEANNIN 257

A. Introduction 257
B. Apoptosis and the Cellular Redox Status 258
    I. Exogenous ROS or Oxidants can Trigger Apoptosis 259
    II. Apoptosis is Associated with an Alteration
        of the Redox Status 259
    III. The Antioxidant Activity of the Apoptosis Inhibitor
        Molecule Bcl-2 260
    IV. Are ROS Really Involved in Apoptosis? 261
C. Anti-Apoptotic Properties of Antioxidants: Mechanisms
   of Action 262
    I. ROS Scavenging and Reducing Activities
       of Antioxidants 262
    II. Replenishment of Intracellular GSH Levels 262
    III. Thiol Antioxidants Induce the Shedding
        of Membrane Fas 263

    IV. Thiol Antioxidants can Modulate the Generation
        of Second Messengers and the Expression-Activation
        of Transcription Factors 265
        1. Modulation of Signaling Molecules 265
        2. Modulation of Transcription Factors 266
D. Conclusions and Therapeutic Perspectives 267
References 267

CHAPTER 11

Reactive Oxygen Species and Apoptosis
G. BAUER, S. DORMANN, I. ENGELMANN, A. SCHULZ, and M. SARAN.
With 3 Figures 275

A. Introduction 275
B. Reactive Oxygen Species: Shotgun or Precision Tool? 276
C. Interdependencies of ROS 278
D. Physiological Sources of ROS 279
E. ROS and Apoptosis 281
    I. ROS-Dependent Apoptosis Under Physiological
       and Pathophysiological Conditions 281
    II. Evidence for the Role of ROS During Induction
        and Execution of Apoptosis 282
    III. Induction and Inhibition of Apoptosis by NO· 285
    IV. Peroxynitrite: An Efficient Apoptosis Inducer 287
    V. Glutathione: Key Element for the Regulation
       of Apoptosis 289
    VI. Mitochondria: Target and Source for ROS
        During Apoptosis Induction 290
    VII. Ceramides: First Class Second Messengers 291
F. Tumor Necrosis Factor: Apoptosis Induction Through Versatile
   Use of ROS 293
G. Apo/Fas-Mediated Apoptosis: ROS Involved
   in Synergistic Pathways 294
H. p53-Mediated Apoptosis: ROS Action Through Several
   Subsequent Steps 294
    I. TGF-Beta: Central Roles for ROS 295
I. ROS, Apoptosis and Tumorigenesis 296
    I. Intercellular Induction of Apoptosis: Elimination
       of Transformed Cells Through Diverse Extracellular
       and Intracellular ROS-Dependent Signaling Steps 296
    II. NO-Mediated Control of Tumorigenesis 299
    III. Sensitivity of Transformed Cells Against Natural
         Antitumor Mechanisms 300
References 302

CHAPTER 12

Clearance of Apoptotic Lymphocytes by Human Kupffer Cells. Phagocytosis of Apoptotic Cells in the Liver: Role of Lectin Receptors, and Therapeutic Advantages
L. Dini. With 4 Figures 319

A. Introduction 319
  I. Apoptotic Cells: Fast Food for Phagocytes 319
B. Recognizing Death: Phagocytosis of Apoptotic Cells
  in the Liver 321
  I. Liver Apoptosis 321
  II. Hepatic Lectin-Like Receptors 322
  III. Kupffer Cells Phagocytic Activity 325
C. Human Kupffer Cells Removal of Apoptotic Lymphocytes 330
  I. Lymphocyte Cell Surface Modifications 330
  II. Kupffer Cells Recognition and Phagocytosis
    of Apoptotic Lymphocytes 332
D. Concluding Remarks and Future Perspectives 333
References 336

CHAPTER 13

Drug-Induced Apoptosis of Skin Cells and Liver
M. Neuman, R. Cameron, N. Shear, and G. Feuer.
With 8 Figures 343

A. Prevalence of Drug-Induced Apoptosis 343
B. Methotrexate-Induced Apoptosis 343
  I. Apoptosis of Hepatocytes In Vivo 344
  II. Apoptosis of Hepatocytes In Vitro 344
    1. Initial Studies 344
    2. Effects of Co-exposures of Hepatocytes
      with Methotrexate and Ethanol 345
  III. Apoptosis of Skin Cells In Vitro 347
C. Acetaminophen-Induced Apoptosis of Hepatocytes
  and Skin Cells In Vitro 348
D. Valproic Acid-Induced Apoptosis of Hepatocytes In Vitro ... 350
E. Conclusion 353
References 353

CHAPTER 14

Apoptosis and Eosinophils
H.-U. Simon. With 6 Figures 357

A. Introduction 357

B. Characteristics and Measurements of Apoptotic Eosinophils 357
C. Role of Delayed Eosinophil Apoptosis for the Development
   of Eosinophilia in Allergic Tissues 360
D. Role of Tyrosine Kinases Activation in Cytokine-Mediated
   Antiapoptosis 361
E. The MEK-ERK MAP Kinase Pathway Does Not Mediate
   Antiapoptotic Signals Initiated Via the IL-5 Receptor 362
F. The Effects of Glucocorticoids on Eosinophil Apoptosis 363
G. Role of CD95 Ligand/CD95 Molecular Interactions
   in the Regulation of Eosinophil Apoptosis..................... 364
H. Nitric Oxide, but Not Eosinophil Hematopoietins,
   Mediates CD95 Resistance 365
I. Role of Sphingomyelinase-Mediated Pathways
   in CD95 Signaling 366
J. Role of Tyrosine Kinase Activation in CD95 Signaling 368
K. Concluding Remarks 369
References 370

CHAPTER 15

Thymocyte and B-Cell Death Without DNA Fragmentation
T. Itoh, M. Nakamura, H. Yagi, H. Soga, and T. Ishii.
With 19 Figures 375

A. Introduction 375
B. Functional and Structural Characteristics of the Thymus 375
    I. Differentiation of Thymocytes 375
   II. Thymocyte Selection 377
  III. Thymocyte Death 377
C. Functional and Structural Characteristics
   of the Germinal Center 386
    I. Affinity Maturation of B Cells 386
   II. B-Cell Selection 390
  III. B-Cell Death 390
D. Summary 392
E. Prospects 394
References 395

CHAPTER 16

Antigen Receptor-Induced Death of Mature B Lymphocytes
T. Defrance, M. Berard, and M. Casamayor-Palleja.
With 2 Figures 399

A. Introduction 399

B. Antigen Receptor-Induced Death and Maintenance
   of Peripheral B Cell Tolerance 400
   I. BCR-Induced Apoptosis of Germinal Center B Cells 400
   II. BCR-Induced Apoptosis of Virgin and Memory
       B Cells 405
C. Antigen Receptor-Induced Death and Homeostatic Regulation
   of the Mature B Cell Compartment 406
D. Positive and Negative Signaling Through the BCR 408
   I. Biochemical Events Associated with the Alternative BCR
      Signaling Pathways 408
   II. Parameters Affecting the Outcome of BCR Signaling 410
       1. Physical Properties of the Ag 410
       2. Costimulatory Signals 411
          a. Activated Complement Fractions 411
          b. T Cells and Microbial Factors 412
E. Molecular Control of the Apoptosis Sensitivity Threshold
   in Mature B Cells 412
   I. Developmental Regulation of the Survival Genes 413
   II. Activation-Induced Regulation of the Survival Genes 415
F. The Executioners of the BCR Apoptotic Pathway 416
   I. Early Transduction Events 416
   II. The Caspase Cascade 417
G. Concluding Remarks 419
References 420

CHAPTER 17

Modulation of Apoptosis and Maturation
of the B-Cell Immune Response
G. KOOPMAN. With 4 Figures 429

A. Introduction 429
B. Antigen Dependent B-Cell Maturation in Secondary
   Lymphoid Organs 430
   I. Anatomical Organization of the B-Cell Immune
      Response 430
   II. The Germinal Center Microenvironment 431
       1. Cellular Composition of the Germinal Center 431
       2. B-Cell Subpopulations 431
   III. Ig Switching and Somatic Hypermutation 434
C. Cell Surface Molecules Involved in Regulation of B-Cell
   Maturation and Apoptosis 435
   I. The TNF/NGF Receptor Family 435
   II. Adhesion Molecules 437
D. Regulation of B-Cell Maturation 438

I. The Initiation of the B-Cell Immune Response, Formation
      of Centroblasts 438
  II. Differentiation of Centroblasts into Centrocytes 440
 III. Differentiation of Centrocytes into Memory
      and Plasma Cells 440
E. Regulation of B-Cell Survival 442
   I. Apoptosis Regulation by Antigen and Follicular Dendritic
      Cells 442
  II. Apoptosis Regulation by T Cells 444
 III. Triple Check Hypothesis of B-Cell Selection 446
References 448

CHAPTER 18

The Neuroprotective and Neuronal Rescue Effect of (−)-Deprenyl
K. MAGYAR and B. SZENDE. With 3 Figures 457

A. Summary 457
B. Introduction 457
C. Clinical Benefits of (−)-Deprenyl Treatment 458
D. Effect of (−)-Deprenyl Against Oxidative Stress 459
E. Selegiline Induced Neuroprotection Against Toxic Insults ... 459
F. Apoptosis in Neurodegenerative Diseases 462
G. Effect of Deprenyl on Neuronal Apoptosis 463
H. Possible Mode of Action of (−)-Deprenyl on Apoptosis 466
References 467

Subject Index 473

CHAPTER 1
Incidence of Apoptosis and Its Pathological and Biochemical Manifestations

R. CAMERON and G. FEUER

A. Apoptosis: Characteristics and Scope

This volume is dedicated to the study of apoptosis and its modulation by drugs. The purpose of the book is to present (a) the molecular mechanisms of cell death by apoptosis in a comprehensive and stimulating manner, (b) the potential critical role of apoptosis in modifying selected diseases, and (c) a review of the effect of various drugs and chemicals on apoptosis. Studying apoptosis is a "hot" topic in research at present and it is applicable to many different areas of scientific and medical investigations. Contributors to this book are leading experts in this field, and the various papers attempt to synthesize views on basic mechanisms and molecular and genetic regulations. Several chapters present morphological changes through specific mediators, activators, and inhibitors, leading to final clinical end points of various diseases and to important diagnostic indicators of these conditions.

In a healthy state, cell degeneration and cell death are ongoing phenomena in multicellular organisms. These processes are balanced by cell renewal. From the normal tissue, the affected cells are removed by apoptosis. Apoptosis is connected with nuclear shrinkage and fragmentation (pyknosis) and condensation of cytoplasm. In the early stages, the cell membrane remains intact and the cytoplasmic and nuclear debris form granules termed apoptotic bodies.

Apoptosis plays an important role in development and in tissue homeostasis, and provides defense against oncogenesis and viral infection. The broad significance of this form of cell death is also related to the perception that it has an essential position in the onset of several illnesses. Apoptotic cells are often seen in many different disease conditions such as (a) autoimmune diseases, (b) HIV infections and acquired immunodeficiency diseases, (c) chronic viral hepatitis and recurrence of viral hepatitis in post-liver transplant, (d) organ transplant immunity and post-transplant rejection involving kidneys, lungs, heart, liver, and bone marrow, (e) cancer and chemotherapy of carcinomas and leukemias, (f) neurodegenerative disorders and development of Alzheimer's disease, and (g) several inflammatory conditions. Apoptotic cells are often seen in malignant tissues of many different types, during the course of chronic viral disease in the liver, and during degenerative processes in the nervous system. The process of apoptosis is also integral to the induction of tolerance in the immune system. Apoptosis caused by cytotoxins is considered as a defensive response that evolved to delete intracellular pathogens (VAUX et al. 1995).

B. Cell Death

The basis of all diseases is an injury to the cell. If the injury is too great or extensive, this results in irreparable changes in structure and function, leading to death of the cell. Cell death has fundamental importance in most pathological processes and it also plays an essential role in the regulation of normal tissue turnover by eliminating all debris formed from aged and dying cells. Ultrastructural abnormalities shown in cells dying in a variety of circumstances indicate two common patterns of morphological changes (WYLLIE 1981; SEARLE et al. 1982; Walker et al. 1988). In the first, the cell death is initiated through reactions to defined stimuli, followed by a sequence of intracellular changes. These morphological changes include marked swelling of mitochondria and the appearance of dense strictures in their matrix followed by progressive dissolution of the entire cell. This type of cell death is named necrosis (MCLEAN et al. 1965; KERR 1969, 1970, 1971). Necrosis refers to the progressive and complete degradation of cell structure that occurs after death. It represents an irreversible damage to cellular membranes associated with various injurious stimuli such as hypoxia, bacterial or viral infection, or corrosive chemicals, resulting in lysis (Fig. 1).

The second form of cell death named apoptosis is characterized by cell shrinkage, rapid condensation of the cytoplasm and nuclear chromatin accompanied by blebbing of the plasma membrane. This subsequently leads to the fragmentation of the cells into a cluster of membrane-bound structures, apoptotic granular bodies in which the integrity of various subcellular organelles is initially maintained. The apoptotic bodies are incorporated by phagocytes or neighboring cells, and DNA breaks up at the internucleosomal spaces into oligome fragments. This type of cell death is present in physiological conditions.

Naturally occurring cell death, unrelated to any causative agent, is also found in almost all tissues. Various terms have been used to describe natural death such as physiological cell death or programmed cell death, to distinguish it from pathological death brought about by disease. In physiological circumstances and during development, different sequences of events occur (KERR et al. 1972, 1987; WYLLIE et al. 1980). This involves prominent nuclear changes in response to hormonal stimuli and changes in other subcellular targets due to T cell or NK cell killing activities (Fig. 2).

These two distinct forms of cell death show major differences. Necrosis is a degenerative process that is associated with irreversible injury (TRUMP et al. 1981, 1982a). Apoptosis is connected with cellular self-destruction rather that degeneration (KERR 1971; WYLLIE et al. 1980; KERR et al. 1987) and requires

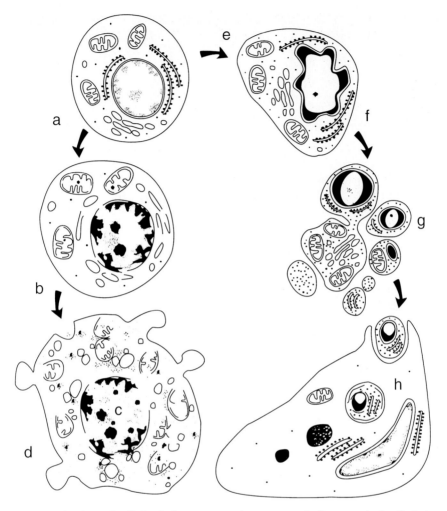

Fig. 1. Mechanism of cell death due to necrosis or apoptosis. In necrosis the first step is (a) an increase of intracellular volume, mitochondrial swelling (b) followed by vacuolization, dilatation of endoplasmic reticulum, blebbing, increased permeability, and condensed nuclei (pyknosis), (c) coagulation and karyolysis, (d) elimination of the cell by inflammation and phagocytosis. In apoptosis the first step is (e) shrinkage and pyknosis, (f) followed by budding and karyorrhexis and (g) break up and formation of apoptotic bodies which (h) may be destroyed by phagocytosis by macrophages. Adapted from WEEDON et al. (1973), and UEDA and SHAH (1994)

protein synthesis and fusion of subcellular components for its execution (LIEBERMAN et al. 1970; GALILI et al. 1982; COHEN and DUKE 1984). The phenomenon of apoptosis is also implicated in the physiological process of regulating organ size. Morphologically apoptosis involves fragmentation of the nucleus, and fusion of the nuclear chromatin and cytoplasm resulting in

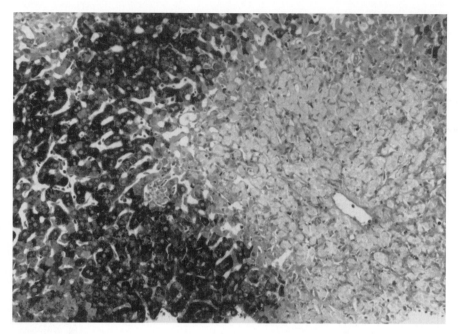

Fig. 2. Example of cell death by necrosis. This is a photomicrograph of liver tissue of a biopsy of a 36 year old man at about 65 h following an overdose of 30 g or 400 mg/kg of acetaminophen. There is cell death of all of the hepatocytes (*pale cells*) in the perivenous zone (or zone 3) with sparing of hepatocytes (*dark cells*) in the periportal and mid-zones. Periodic acid Schiff stain, ×200

membrane-encapsulated bodies: granules (Table 1). The presence of these bodies interferes with normal cell function and these granules are disposed of by neighboring cells without inflammation.

C. Features of Necrosis

I. Occurrence

Necrosis develops in various tissues due to severe hypoxia, ischemia, or during autolysis (JENNINGS and REIMER 1981; LAIHO et al. 1983; BORGES et al. 1987). It also occurs as the consequence of complement-mediated damage of the cell membrane, trauma, or exposure to several toxins (TRUMP et al. 1982a; LAIHO et al. 1983; GROMKOWSKI et al. 1986). Metabolic inhibitors such as fluorocitrate, iodoacetate, or cyanide, reactive oxygen metabolites, a variety of toxic chemicals, and the ionic pump inhibitor ouabain cause necrosis. Severe environmental conditions such as mild ischemia, hypoxia, and hyperthermia (BUCKLEY 1972; MCDOWELL 1973; HAWKINS et al. 1972; BISHOP et al. 1987) also provoke necrosis.

Table 1. Apoptotic bodies in physiological and pathological conditions

Condition	Apoptotic Bodies
Disease	
Viral hepatitis	Councilman bodies
	Yellow fever bodies
Alzheimer's disease	Amyloid β-protein granules
	Senile plaques
Parkinson's disease	Lewy bodies
Hashimoto's thyroiditis	Ashkenasy cells
	Hurthle bodies
Experimental	
Retinal degeneration	Apoptotic bodies in retina
Zinc deficiency	Paneth cells

II. Morphology

Reversible injury of the cell often leads to loss of specialized surface structures such as microvilli. These cells also show mild swelling of mitochondria. The glycogen stores are depleted, endoplasmic reticulum is dilated, ribosomes are detached, and chromatin is clumped irregularly at the nuclear membrane (TRUMP et al. 1981, 1982b). In the case of irreversible injury when necrosis sets in, the most characteristic effects occur in mitochondria (JENNINGS and REIMER 1981). These include gross swelling and granular changes in the matrix, representing the earliest ultrastructural changes associated with necrosis. Actually, mitochondria play a central role in the regulation of necrosis. These subcellular organelles can trigger cell death in a number of ways: by releasing and activating various proteins that mediate cell death, by the disruption of energy metabolism and electron transport, and by the alteration of cellular redox potential. Any or all of these mechanisms may give an explanation of how mitochondrial defects contribute to the pathogenesis of aging and of several human diseases (GREEN and REED 1998). During this process ribosomes are disintegrated and damage develops in the continuity of the plasma membrane and in the membranes of various subcellular organelles. Eventually the irregularly clumped chromatin disappears (WALKER et al. 1988).

By light microscopy the necrotic cells appear initially swollen, cytoplasm is eosinophilic, and nuclei show uniformly condensed chromatin (karyorrhexis), or pyknosis. Later, there is a dissolution of the chromatin masses (karyolysis). Ultimately, the necrotic cells are removed by phagocytes, and accompanied by an inflammatory reaction (WALKER et al. 1988).

III. Biochemistry

The biochemical changes are associated with the occurrence of a marked increase in the permeability of mitochondria and plasma membranes (TRUMP et al. 1982a). As a consequence, some components are leaving the cell and other electron-dense materials accumulate (HOFFSTEIN et al. 1975; WYLLIE et al. 1980). Trauma, toxic chemicals or a failure of the membrane pump connected with cellular energy depletion triggers off the move of cations from the cell and the accompanying fluid entry into the cell causes swelling (hydropic degeneration) (BUCKLEY 1972; SCHANNE et al. 1979; TRUMP et al. 1981; JENNINGS and REIMER 1981). When the electrolyte movement is severe, increased concentration of calcium activates membrane-bound phospholipases which metabolize phospholipid to lysophospholipid and fatty acids (TRUMP et al. 1982a; CHIEN et al. 1978; FARBER et al. 1981). This effect disrupts membrane continuity directly and indirectly by the detergent-like action of long chain fatty acid derivatives (CHIEN et al. 1978).

The accumulation of granular mitochondrial matrix residues is initiated by excess cytosolic calcium (TRUMP et al. 1982a). At the beginning of the injury this consists of inorganic calcium salt deposits and it also contains denatured matrix proteins in late stages of the evolution of necrosis. Following the loss of membrane integrity, cellular homeostasis is impaired and hydrolases are released from ruptured lysosomes. These potent enzymes cause a rapid acceleration of cellular disintegration. Consequently the concentrations of phospholipid, protein, RNA, DNA, and triglycerides rapidly decrease and the amounts of free amino acids, phosphates and fatty acids increase (TRUMP et al. 1981, 1982b).

D. Features of Apoptosis

I. Occurrence

Apoptosis is involved in the programmed elimination of cells in physiological conditions. This is an irreversible mechanism for the elimination of excess or damaged cells. Apoptosis also occurs during embryonic and fetal development. In adult life apoptosis regulates the size of organs and tissues. In pathological conditions apoptosis is responsible for the reduction of cells in different types of atrophy and in the regression of hyperplasia. It develops spontaneously in cancer cells and it is increased in both neoplasm and during normal cell proliferation triggered by a variety of agents applied in cancer chemotherapy. Apoptosis is enhanced by cell-mediated immune reactions and various toxins that also produce necrosis.

II. Morphology

Apoptosis manifests in single cells scattered in the affected organ in an "asynchronous" (apparently random) fashion and it is not associated with inflammation (WYLLIE et al. 1980; SEARLE et al. 1982; KERR et al. 1987; WALKER et al. 1988). Electron microscopic studies show, at the earliest stage, that nuclear chromatin is aggregated into dense masses attached to the nuclear membrane and that cytoplasm becomes concentrated. These changes are followed by further condensation of the cytoplasm and the nucleus breaks up into small fragments. The chromatin is segregated and some protuberances develop on the cell surface (blebbing). The pedunculated protuberances are separate and with bounded plasmalemmal sealing membrane apoptotic bodies are produced. These dense masses have a different texture from the chromatin and are sometimes present in the lucent part of nuclei or in their fragments. The condensation of the cytoplasm is often associated with the formation of vacuoles. The nuclear fragmentation and cellular budding usually characterize cells with a high nuclei-cytoplasm ratio such as in thymocytes (WYLLIE et al. 1980). In the acinar cells of salivary gland and pancreas the rough endoplasmic reticulum is rearranged into whorls before the cell becomes fragmented (WALKER 1987).

The apoptotic bodies are usually quickly phagocytosed by neighboring cells and degraded with phagolysosomes. In epithelial and tumor cells similar processes manifest and specialized mononuclear phagocytes also participate in the degradation (WYLLIE et al. 1980; KERR et al. 1987). In lining epithelia the apoptotic bodies are extruded from the surface (SEARLE et al. 1975; DON et al. 1977; WYLLIE 1981).

Light microscopic studies of apoptosis show diverse pictures. The shrinkage and budding of the cell is complete within a few minutes and discrete apoptotic bodies can be demonstrated at the end of the process (SANDERSON 1976; MATTER 1979). The size of the apoptotic bodies varies considerably. They are round or oval, some represent a single relatively large nuclear fragment surrounded by a thin cytoplasmic rim, others mostly consist of cytoplasm with a variable number of nuclear fragments.

III. Biochemistry

Early investigations of apoptosis revealed that it is an active process rather than simply degeneration of the cell (KERR 1971). It is connected with cytoplasmic and membrane surface changes, protein synthesis, and internucleosome cleavage of DNA.

The process of condensation observed by ultrastructural examinations and associated with an increased density suggest that the surface convolution and the removal of the apoptotic bodies are associated with redistribution of cytoplasmic microfilaments (CLOUSTON and KERR 1979; WYLLIE and MORRIS 1982). The rapid uptake of apoptotic bodies by neighboring cells probably depends on carbohydrate changes on the surface of these bodies. It may be that the carbohydrate changes represent the consequences of incorporation into the plasmalemma of membranes surrounding the cytoplasmic vacuoles that are formed during the development of apoptotic bodies. Actually, a discharge of the vacuole content has been described (KERR 1969, 1970; GALILI et al. 1982).

In the early stages of apoptosis, lysosomes are intact and it is unlikely that lysosomal enzymes are involved in triggering of this type of cell death (KERR 1967, 1971).

Protein synthesis seems to be a requirement in the formation of the apoptotic bodies. Inhibitors of protein synthesis suppress the occurrence of apoptosis of thymocytes and chronic lymphocytic leukemia cells treated with glucocorticoids (GALILI et al. 1982; COHEN et al. 1984; WYLLIE et al. 1984). Protein synthesis inhibitors also reduce the formation of apoptotic bodies in T lymphocytes deprived of interleukin-2 (WYLLIE 1981), in epithelial cells at the plane of fusion of the palliative processes in normal rat embryo (PRATT and GREEN 1976), and in various cells exposed to radiation or to cytotoxic drugs (LIEBERMAN et al. 1970; BEN-ISHAY and FARBER 1975; COHEN et al. 1985). All of these results indicate that protein synthesis is a required process in the development of apoptosis, but it is uncertain what the role of these proteins is. The synthesis of several proteins is increased following the treatment of thymocytes with glucocorticoids (VORIS and YOUNG 1981) but, in contrast, protein synthesis inhibitors do not block apoptosis induced by T lymphocytes (DUKE et al. 1983).

Among the biochemical events of apoptosis the double-strand cleavage of nuclear DNA at the regions between nucleosomes is reported for all cell types. This cleavage produces oligonucleosome fragments and it is catalyzed by endonuclease enzyme (WYLLIE 1980; SHALKA et al. 1981; COMPTON and CIDLOWSKI 1986). The endonuclease activity and DNA breakdown is inhibited by zinc (DUVALL and WYLLIE 1986). Some papers have reported that zinc deficiency enhances apoptosis in gut crypts (ELMES 1977; ELMES and JONES 1980).

Several recent studies have shown that the activation of the interleukin-1-beta-converting enzyme/Ced-3 family of proteases represents the end point in apoptotic cell death (FRASER and EVAN 1996). Other investigations have indicated that the loss of mitochondrial membrane potential is the critical step in cell death (ZAMZAMI et al. 1996; HENKART and GRINSTEIN 1996). Many members of the Bcl-2 family of genes play major roles in the regulation of the programmed cell death in many systems (YANG and KORSMEYER 1996). This family, including Bcl-x_l, are potent inhibitors that modulate cell death through inhibition of activation of caspases, a family of cysteine proteases (FRASER and EVAN 1996; CHENG et al. 1997; NAWA et al. 1998). In this way Bcl-x_l may facilitate protection against cell death (CLEM et al. 1998). Bcl-x_l can prevent apoptosis and maintain cell viability by averting loss of mitochondrial membrane potential that occurs as a consequence of the interleukin 1β-converting enzyme/Ced-3 protease activation (BOISE and THOMPSON 1998). The breakdown of Bcl-x_l during the execution phase of cell death converts it from a protective to a lethal protein (CLEM et al. 1998).

Apoptosis is involved in the death of hematopoietic progenitor cells after removal of the appropriate colony-stimulating factor. Pharmacological investigations indicated the role of protein kinase C in the suppression of apoptosis in interleukin-3 and granulocyte-macrophage-colony-stimulating factor dependent human myeloid cells (RAJOTTE et al. 1992; RINARDO et al. 1995). Overexpression of some protein kinase C isoform in factor-dependent human TF-1 cells enhances cell survival in the absence of cytokine. This affect is associated with an induction of Bcl-2 protein expression, an increase over the levels in empty vector transfections (GUBINA et al. 1998).

E. Activation of Apoptosis

The presence of apoptosis develops in four different phases. First, the presence of genes regulates the occurrence of programmed cell death. This prerequisite has been documented in developing organisms (ELLIS and HORVITZ 1986; SCHWARTZ et al. 1990) and in cell cultures (EVAN et al. 1992). Second, various signals trigger off the genetic program or an unbalanced signaling system can prevent the action of repressors. Specific signaling molecules include calcium ions, glucocorticoid hormones, and sphingomyelin. Initiation can also occur by imbalanced signaling such as lack of a growth factor (KYPRIANOU and ISAACS 1988) or due to a toxicant action the signaling pathway is inhibited (Aw et al. 1990). Third, the progression of the condition leads to the expression of genes manifesting in structural alterations such as cytoskeletal changes, cell shrinkage, nuclear pyknosis, chromatin changes, and DNA fragmentation (ARENDS and WYLLIE 1991). Fourth and finally, death and engulfment by phagocytosis of the whole cell or cell fragments terminates the apoptotic process (SAVILL et al. 1993) (Fig. 3).

Apoptotic signaling cascades are expressed in most if not all cells, and they are usually present in inactive forms (WYLLIE et al. 1980; RAFF et al. 1993). Apoptosis can be triggered by a variety of physiological and stress stimuli which initiate one or several distinct signaling pathways (Fig. 4). The activation of the specific pathway is dependent on the cell type and on the subcellular organelles, being the target of each type of stress. The various signaling pathways converge into a common final effector mechanism that disintegrates the dying cell (YUAN et al. 1993). The activation mechanism includes the ICE/Ced-3 family of cysteine proteases that reorganize subcellular structures in an orderly fashion. The integrity of plasma membrane is preserved and the disintegrated subcellular organelles are aggregated into membrane-bound vesicles called apoptotic bodies. Cellular fragments or dead cells are finally eliminated by neighboring cells or macrophages, by phagocytosis. The overall result of this process is that individual cells can be abolished without an inflammatory reaction producing tissue damage.

Intracellular Ca^{2+} signals activate apoptosis (NICOTERA et al. 1994). Calcium overload can trigger several lethal processes including disruption of the cytoskeletal organization, DNA damage, and mitochondrial dysfunction. When Ca^{2+} accumulates within the cytoplasm or other intracellular compartments, sudden increase of intracellular Ca^{2+} can quickly lead to cell necrosis,

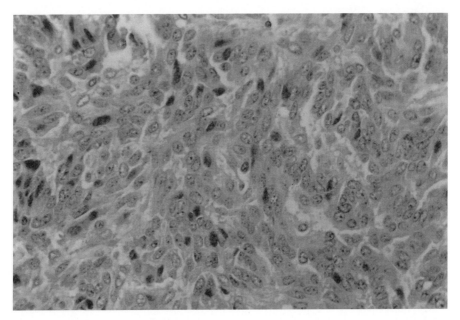

Fig. 3. Example of cell death by apoptosis. This is a photomicrograph of tumor tissue of a liver biopsy of a 58 year old man who has a large metastatic neuroendocrine carcinoma in the liver. This section shows only tumor cells and frequent apoptotic cells evident (*dark nuclei*) in all areas of the tumor tissue. Haematoxylin and eosin stain, ×300

and disturbances of Ca^{2+} signaling can also induce apoptosis (McConkey et al. 1990) (Fig. 5). Removal of extracellular Ca^{2+} can prevent nuclear changes manifest in apoptosis such as apoptotic body formation and DNA degradation, demonstrating Ca^{2+} requirement in apoptosis (Nicotera et al. 1994). Transfection of WEHI 7.2 thymoma cells with calbindin, a Ca^{2+}-binding protein, prevents apoptosis caused by calcium ionophore, cAMP, or glucocorticoids (Dowd et al. 1991). Several in vitro models of apoptosis are connected with a loss of the regulation of intracellular Ca^{2+} level and activation of Ca^{2+}-dependent endonuclease activity (McConkey et al. 1988). Ca^{2+}-mediated endonuclease activation is associated with the cytotoxicity of tributyltin and TCDD in thymocytes (McConkey et al. 1988; Aw et al. 1990). Ca^{2+} can induce endonuclease activity and initiate apoptosis in malignant cells and in cells infected with viruses (Nicotera et al. 1994).

Several studies described the sphingomyelin signal transduction pathway as an essential part in the mediation of apoptosis related to environmental stresses and to several cell surface receptors (Kolesnick and Golde 1994; Verhey et al. 1996). The sphingomyelin pathway is ubiquitous. Most, and probably all, mammalian cells are capable of signaling through the sphingomyelin system. The functioning sphingomyelin pathway is connected with the forma-

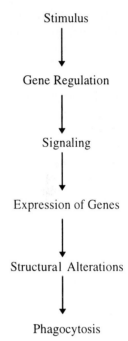

Fig. 4. Schematic illustration of various steps of apoptosis. Scheme represents cell death due to apoptosis. Various stimuli such as radiation, thermal actions, steroids, withdrawal of trophic hormones, cytokines and other growth factors, oxidants and other cytotoxic chemicals, anticancer agents, autoimmune disease, cell-mediated immunity, viral infections, activated signaling agents, and caspase cascade via gene regulation leading to structural damage, death, and elimination of cell debris by phagocytosis, or by macrophages originating from neighboring cells

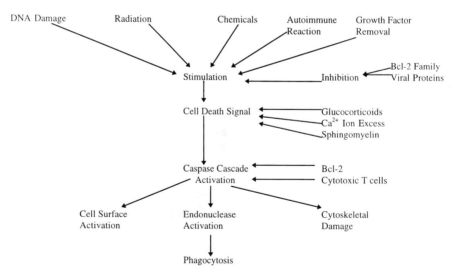

Fig. 5. Schematic illustration of various phases of apoptosis. The contribution of apoptosis to the pathogenesis of disease is rapid, leading to phagocytosis. Various stimuli affect the caspase cascade through signaling agents. The Bcl-2 family of proteins and some viral proteins such as crmA and p35 are known inhibitors of apoptosis. Although these events are important in the development of apoptosis, the mechanism of action of inhibition and the targets for the caspase cascade have not yet been identified

tion of ceramide that acts as a secondary messenger by activating a variety of cell functions (SPIEGEL et al. 1996; BALLON et al. 1996). Distinct receptors signal via the sphingomyelin pathway following ligand binding. Ceramide mediates apoptosis and several cellular functions, including differentiation of promyelocytes, proliferation of fibroblasts, and the survival of T9 glioma cells. The involvement of the sphingomyelin signaling system in apoptosis is associated with stress activation of acid sphingomyelinase to produce ceramide, and ceramide as a secondary messenger initiates apoptosis. Several environmental stresses that induce apoptosis such as ionizing radiation, heat shock, exposure to UV-C rays, and oxidative stress bring about rapid generation of ceramide through the activation of sphingomyelinase (HAIMOVITZ-FRIEDMAN et al. 1994; VERHIRJ et al. 1996). Understanding the role of pro- and antiapoptotic signaling involved in apoptosis mediated by ceramide, including their mode of action, may provide an opportunity to develop pharmacological means for intervention in the process of apoptosis (HAIMOVITZ-FRIEDMAN et al. 1997).

F. Incidence of Apoptosis
I. Physiological Conditions
1. Embryonic and Fetal Development
Controlled cell death is part of normal development. Several morphological studies reported that apoptosis is involved in the programmed elimination of cells during the embryonic and fetal period such as the deletion of the redundant epithelium at the plane of fusion of the palatine processes (HASSEL 1975), in the differentiation of the gut mucosa (HARMON et al. 1984; PIPON and STERLE 1986) and the retina (YOUNG 1984; PENFOLD and PROVIS 1986), and in the removal of interdigital webs (KERR et al. 1987).

2. Cell Turnover in Adult Tissues
Proliferating normal mammalian cells undergo spontaneous apoptosis, responsible for the continuous removal of the aged cells (KERR 1971; POTTER 1977; HUCKINS 1978; COLUMBANO et al. 1985; BURSCH et al. 1985; ALLAN et al. 1987). In the slowly proliferating cells apoptosis balances necrosis over a time period (WYLLIE et al. 1980), and the oscillation between these two processes may be regulated by soluble factors produced locally (LYNCH et al. 1986). In the rapidly proliferating cells the deletion of the cell is associated with movement from the site of production and apoptosis. These changes characterize the basal compartment of seminiferous tubules and gut crypts (POTTER 1977; ALLAN et al. 1987).

During the normal terminal differentiation of cells, the double-strand cleavage of DNA shows great similarity to processes occurring in apoptosis. This is exemplified by the differentiation in the lens of the eye (APPLEBY and MODAK 1977). Similarly, the residues of megakaryocytes remaining after platelet release in bone marrow greatly resemble the typical ultrastructural changes associated with apoptosis (RADLEY and HALLER 1983). Apoptotic bodies are found in lymphoid germinal centers of follicle cells due to apoptosis (SEARLE et al. 1982) and formed from macrophages in spleen (SWARTSENDHUBER and LONGDON 1963).

3. Involution of Adult Tissues

The growth of various cell populations is controlled by hormones and growth factors. Reduction or excess addition of these substances triggers off a rapid decrease of the cell number. In these circumstances, the fall of trophic hormone stimulation leading to cell deletion is connected with apoptosis. This occurs in the human premenstrual endometrium (HOPWOOD and LEVISON 1976), in the human breast towards the end of the menstrual cycle (FERGUSON and ANDERSON 1981), in the endometrium of the hamster at oestrus (SANDROW et al. 1979), in the ewe endometrium following parturition (O'SHEA and WRIGHT 1984), in the theca interna of sheep ovarian follicles during atresia (O'SHEA et al. 1978), and in the adrenal cortex of the neonatal rat (WYLLIE et al. 1973).

II. Pathological Conditions

1. Regression of Hyperplasia

In several cases in the processes of regression of hyperplasia, apoptosis is involved. This occurs after the removal of the proliferative stimulus producing hyperplasia in hepatic parenchymal cells by phenobarbital, lead nitrate, or cyproterone acetate (COLUMBANO et al. 1985; BURSCH et al. 1986), bile duct proliferation brought about by α-naphthyl isothiocyanate or ligation of the main bile duct (BHATHAL and GALL 1985), or pancreatic hyperplasia induced by trypsin inhibitor (OATES et al. 1986). In some cases hormone withdrawal is connected with the occurrence of apoptotic processes such as hormone-induced hyperplasia of the adrenal cortex (WYLLIE et al. 1980). Apoptosis is reported in renal parenchyma atrophy in hydronephrosis (KERR et al. 1984) and in hepatic atrophy brought about by mild ischemia (KERR et al. 1984). Apoptosis occurs in many tissue regressions and, in normal animals, apoptosis is involved in the catagen involution of hair follicles (WEEDON and STRATTIN 1981) and resorption of tissue around erupting teeth (SCHELLENS et al. 1982). Pancreas atrophy and salivary gland duct obstruction is associated with enhanced loss of secretory cells by apoptosis (POTTER 1977; MATHER 1979; WALKER 1987) and apoptotic changes in the vascular endothelial cells (POTTER 1977). Apoptosis is involved in normal regression of the corpus luteum (AZMI and O'SHEA 1982).

2. Pathological Atrophy

This is frequently associated with increased levels or withdrawal of hormones, or with the reduction of growth factor. Increased progesterone levels bring about apoptosis in cat oviduct lining (VERHAGE et al. 1984); increased glucocorticoids induce apoptosis in chronic lymphocytic leukemia cells (GALILI et al. 1982), in the cells of some lymphoid lines (BREWITT et al. 1983), and in thymocytes (LAPUSHIN and DE HARVEN 1971). Castration leading to pathological atrophy of the rat prostate or withdrawal of testosterone stimulation are connected with apoptosis of the epithelial cells (KERR et al. 1973; STIENS et al. 1981; STANFORD et al. 1984). Withdrawal of adrenocorticotropic hormone by excess prednisone administration significantly increases apoptosis in the adrenal cortex of rats (WYLLIE et al. 1973).

In T lymphocytes isolated from the blood of patients with infectious mononucleosis the withdrawal of the T lymphocyte growth factor, interleukin-2, induces apoptosis (Moss et al. 1985; BISHOP et al. 1985).

3. Drugs

Many drugs induce apoptosis in experimental condition or as side effects (UREN and VAUX 1996). Some of these actions are direct and affect the death pathway and some drugs interfere with biochemical mechanisms, the effect indirectly leading to apoptosis; for example, azide administration inhibits ATP synthesis and diphtheria toxin interferes with protein synthesis and subsequently apoptosis is induced. Since various pharmacological agents provoke the same reaction, it may be that the effect of drugs is associated with a nonspecific stress response leading to the formation of apoptotic bodies (Table 2).

4. Toxic Chemicals

Chronic copper administration is connected with an increased hepatic apoptosis in sheep (KING and BRENNER 1979). Acute lethal doses of copper or mercury in rainbow trout cause massive apoptosis in the gills (HOFFSTEIN et al. 1975). Various hepatotoxins such as 1,1-dichloroethylene, albitocin, and heliotrine given to experimental animals in high doses produce zonal necrosis, and administered in smaller doses they enhance apoptosis in less severely affected hepatic parenchyma (KERR 1967, 1969, 1970). Colchicine causes apoptosis in gut crypt (DUNCAN and HEDDLE 1984), interphase lymphocytes (BOMBASIREVIC et al. 1985), and affects microtubules. Toxic plant proteins, mycin, diphtheria toxin, and inhibitors of protein synthesis all induce apoptosis in the mouse colonic crypts (GRIFFITH et al. 1987). Apoptosis is also involved in the damage of the adrenal cortex of rats brought about by 9,10-dimethyl-1,2-benz(a)anthracene (KERR 1972). In acute mesodermal cell death, the apoptotic changes produced by the teratogenic compound 7-hydroxymethyl-12-methylbenz(a)anthracene in the developing rat are probably the consequence of the site-specific induction of this condition in the embryo (CRAWFORD et al. 1972). Shiga toxin formed from *Shigella dysenteriae* causes apoptosis in the absorptive epithelial cells of the rabbit small intestine (KEENAN et al. 1986).

Table 2. Inducers of Apoptosis

Physiological Factors	Damage Inducers	Cytotoxic Agents
Calcium	Antimetabolites	Actinomycin D
Glucocorticoids	Bacterial toxins	Aphidicolin
Growth factor withdrawal	Diphteria toxin	Bischlorethylnitrosourea
	Heliobacter pylori toxin	Colcemide
Loss of matrix attachment		Cycloheximide
	Shiga toxin	Cyclophosphamide
Neurotransmitters	Cytotoxic T cells	Dichlofenac sodium (Voltarol)
Dopamine	Free radicals	5-Fluorouracil
Glutamate	Heat shock	Hydroxyurea
N-Methyl-D-aspartate	Nutrient deprivation	Isopropyl-methane sulphonate
	Oncogenes	Mechlorethamine (nitrogen mustard)
Toxins	MYC, rel, EIA	
Abrin	Oxidants	Mefenamic acid
Albitocin	Viral infection	Mitomycin
β-Amyloid peptide		Triethylenethiophosphoramide
Aphidicolin	Chemotherapeutic Drugs	[³H] Thymidine
Azide		
Colcemid	Adriamycin	Cancer Causing Agents
Colchicine	Bleomycin	7,12-Dimethylbenz(a) anthracene
Copper salts	Cisplatin	9,10-Dimethylbenz(a) anthracene
1,1-Dichloroethylene	Cytosine arabinoside (Ara-C)	1,2-Dimethylhydrazine
Ethanol		7-Hydroxymethylbenz(a)anthracene
Heliotrine	Doxorubicin	
Mercury salts	Etoposide	Therapeutic Treatments
Mycin	Methotrexate	Hyperthermia
Raw soya flour	Myleran	Gamma radiation
Ricin	Taxol	Tritium beta particles
	Vincristine	UV radiation
		X-ray radiation

Treatment of several cultured mammalian cells with cell cycle phase specific antiproliferative drugs commonly results in apoptosis (BARRY et al. 1990). The cytotoxic outcome of low concentrations of colcemid, an antimitotic drug, on HeLa 53 cells is the induction of multipolar spindles and multipolar divisions. Aphidicolin, an inhibitor of DNA synthesis, causes apoptosis which varies as a function of aphidicolin concentration. It occurs later after the cells have progressed through the S phase (SHERWOOD and SCHIMKE 1994). These results indicate that the target of drug action in the cell cycle differs with colcemid and aphidicolin, which is of secondary importance in the induction of cytotoxicity and apoptosis.

5. Chemical Carcinogens and Cancer Chemotherapy Agents

Many different chemical carcinogens cause nuclear abnormalities associated with apoptotic body formation in proliferating epithelial cells in the gut of mice (MASKENS 1979; RONEN and HEDDLE 1984). Apoptosis is also involved in the action of several cancer-chemotherapeutic agents on normal proliferating cells and on neoplastic cell population (IRIJI and POTTER 1983; BENNETT et al.

1984). These substances included dimethylhydrazine, 1-β-d-arabinofurasonylcytosine (ara-C). The extent of apoptosis and subsequent cell death induced in proliferating tissues by these chemicals are not correlated with the rate of mitosis. Several cancer-chemotherapeutic agents produce more apoptotic bodies than occur in physiological conditions in normal highly proliferative tissues. The mechanism of action of these substances has not yet been established. It may be associated with their action on DNA, causing damage and affecting DNA turnover and repair.

Apoptosis induced by the cancer chemotherapy drugs ara-C, taxol, or etoposide in human acute myelogenous leukemic HL-60 cells is inhibited by the overexpression of Bcl-2 or Bcl-x_l. Taxol treatment brings about a molecular cascade of apoptosis, represented by an increase of cytochrome c and poly(ADP-ribose) polymerase or the DNA fragmentation factor cleavage activity of caspase-3. Taxol also raises phosphorylation of Bcl-2. This action and the mobility shift is associated with the 60 amino acid loop domain of Bcl-2 and Bcl-x_l which contains the phosphorylation sites and participates in the negative regulation of the antiapoptotic action of these gene proteins (FANG et al. 1998).

6. Radiation and Hyperthermia

Ionizing radiation induced by gamma ray, X-ray, or exposure to ultraviolet light significantly increases apoptosis in lymphocytes of the mouse intestinal epithelium (PRATT and SODILEFF 1972; GUNN et al. 1979; DUNCAN et al. 1983; SZEKELY and LOBREU 1985). Ionizing radiation generates reactive oxygen species and damages DNA, and its production of apoptosis is possibly related to these conditions. These treatments also greatly enhance the formation of apoptotic bodies in normal proliferating fetal and adult cell populations (POTTER 1977, 1985; HENDRY et al. 1982; ALLAN et al. 1987). Radiation of tumor cells causes both necrosis and apoptosis and necrosis is more advanced in certain cases (IRIJI and POTTER 1983, 1984). Mild hyperthermia brings about DNA damage and an inhibition of the DNA repair mechanism associated with the formation of apoptotic bodies in micronuclei (FORRITSMA and KONINGS 1986).

III. Disease Conditions

1. Cell-Mediated Immunity

Apoptosis plays an important role in the function of the immune system. This process is essential in the control of immune responses, cytotoxic killing, and in the elimination of immune cells recognizing self-antigens (EKERT and VAUX 1997). Several regulators of apoptosis have been identified such as the CD95 (also called Fas-ligand), and the Bcl-2 family of gene proteins. Malfunctioning of the immune system may be associated with reduced or enhanced cell death. Abnormality in the regulation of the apoptotic processes may lead to a variety of diseases including immunodeficiency and autoimmune disorders. The control of cell proliferation and the selection against autoreactive cells in the lymphoid system, i.e., the maintenance of homeostasis in the immune system is affected by the induction of apoptosis, e.g., autoreactive T cells are deleted by the process of apoptosis (OSBORNE 1996).

Cell mediated immune reactions are intrinsically associated with apoptosis. Apoptosis is the cause of programmed cell death (Fig. 6) in a great number of diseases where cell-mediated immune destruction of the tissues represents the underlying mechanism including acute and chronic hepatitis (KERR et al. 1984). In these cases acidophilic or Councilman bodies consist of apoptotic bodies. Apoptosis is also involved in primary biliary cirrhosis (BERNAVAN et al. 1981). The development of apoptosis has been reported following liver graft rejection (SEARLE et al. 1977), and in acute graft-versus-host disease in the human rectal epithelium (GALLUCCI et al. 1982). Features of apoptosis are present during cell death brought about by in vitro attachment of natural killer cells (BISHOP and WHITING 1983), K cells (SANDERSON and THOMAS 1977), or T cells (SANDERSON 1976a,b; LIEPINS et al. 1977; DON et al. 1977; MATTER 1979).

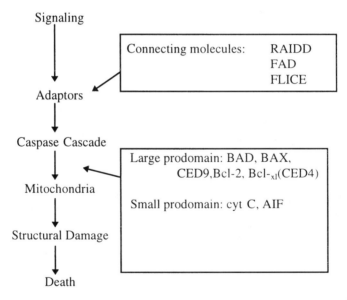

Fig. 6. Schematic illustration of the control and execution of programmed cell deaths. This scheme represents the control and execution stage of programmed death. Signalization occurs by the action of signals such as elevation of Ca^{2+}, glucocorticoids, and the sphingomyelin system, leading to the release of members of the caspase family through adaptors or connecting molecules such as RAIDD, FADD, or FLICE. The caspase cascade interacts with mitochondria on the large prodomain through the members of the Bcl-2 family such as Bcl-2, Bcl-x_l, CED-9 (Fas), BAD and Bax. This cascade is further amplified on the small prodomain through cytochrome C and AIF, leading to structural damage and death

The destruction of massive lymphoma cells in an in vitro culture by allogenic macrophages collected from the peritoneum of mice previously immunized with lymphoma is also associated with morphological changes characteristic of apoptosis. Biochemical investigations reported that in these circumstances the cleavage of nuclear DNA in the target cell is similar to apoptotic changes manifested in other conditions (COHEN et al. 1985; DUKE et al. 1983).

Some autoimmune diseases are correlated with gene abnormalities associated with the induction of apoptosis. A rare lymphoproliferative disorder of autoimmune origin in children is due to a mutation in the gene for CD95 (FISHER et al. 1995). An autoimmune disorder resembling systemic lupus erythematosus, reported in mice, is connected with the lack of functional CD95, or defects in Fas antigen that mediates apoptosis (WATANABE et al. 1992), or with the enforced expression of the Bcl-2 transgene in B-lymphoid cells (STRASSER et al. 1991).

The potent inflammatory cells, eosinophils, are involved in chronic allergic diseases (SIMON and BLASER 1995). Certain cytokines inhibit apoptosis of eosinophils causing tissue eosinophilia. This inhibition is expressed in the anti-apoptotic genes Bc-02, Bcl-x_l, and A_1, and spontaneous eosinophil apoptosis is connected with a decrease of protein and mRNA levels in Bcl-x_l (DIBBERT et al. 1998). In the regulation of eosinophil apoptosis Fas ligand/Fas receptor molecular interactions have been suggested. It was also shown that tyrosine phosphorylation is an important step in the development of the Fas receptor-linked transmembrane death signal in eosinophils (SIMON et al. 1998). A new signaling protein that specifically binds to the Fas death domain has been identified (YANG et al. 1997). Overexpression of this protein enhances Fas mediated apoptosis.

Fas-ligand induces apoptotic cell death in most cells that express its receptor (GREEN and WARE 1997). Fas-bearing cells include cells of the immune system and in this way tissues that naturally contain Fas-ligand kill infiltrating lymphocytes and inflammatory cells. Other roles of Fas in the body include the activation of cytotoxic T lymphocytes that often express high levels of Fas-ligand and hence the ability of Fas-ligand to kill cells bearing Fas accounts for some destructive effects mediated by these cells. Fas-ligand, therefore, not only protects tissues from immune assault but can also damage tissues expressing Fas.

2. Ischemia

Some investigators revealed that during an ischemic attack, due to the loss of blood supply, not only necrosis manifests but apoptotic cell death also occurs (BARR and TOMEI 1994). Outside the central ischemic zone cells die over a more protracted time period by apoptosis (SCHUMER et al. 1992). In a culture of both cardiac myocytes and neurons, ischemia induces apoptosis (GOTTLIEB et al. 1994). Inhibitors of apoptosis (Table 3) limit the infarct size in vitro (SCHUMER et al. 1992). Reperfusion of ischemic tissue is also associated with apoptosis

Table 3. Inhibitors of apoptosis

Physiological Factors	Pharmacological Agents	Viral Genes
Androgens	Calpain inhibitors	Adenovirus E1B
CD40 ligand	Cystein protease inhibitors	African swine fever virus
Estrogen	α-Hexochlorocyclohexane	Baculovirus IAP
Extracellular matrix	Phenobarbital	Baculovirus p35
Growth factors	PMA	Cowpox virus
Neutral amino acids	Miscellaneous drugs	Epstein-Barr virus
Zinc salts		Herpes virus

Table 4. Diseases associated with increase or inhibition of apoptosis

Increase	Inhibition
Neurodegenerative disorders	Autoimmune disorders
Alzheimer's disease	Immune medicated glomerulonephritis
Amyotrophic lateral sclerosis	Systemic lupus erythematosus
Cerebellar degeneration	Cancer
Parkinson's disease	Carcinomas with p53 mutations
Retinitis pigmentosa	Follicular lymphomas
Ischemic injury	Hormone-dependent tumours
Myocardial infarction	Breast cancer
Reperfusion injury	Ovarian cancer
Stroke	Prostate cancer
Viral infections	Viral infections
AIDS	Adenoviruses
Toxin-induced hepatic disease	Herpes viruses
Alcohol	Pox viruses

(SCHUMER et al. 1992; GOTTLIEB et al. 1994). This is connected with acute production of free radicals and flow of intracellular calcium, both potent inducers of apoptosis.

3. Neurodegenerative Disorders

A wide variety of neurodegenerative disorders are characterized by loss of neurons (ISACSON 1993). These include Parkinson's disease, Alzheimer's disease, spinal muscular atrophy, retinitis pigmentosa, amyotrophic lateral sclerosis, and various forms of cerebellar degeneration. In these diseases (Table 4) apoptotic cell death is suggested as the underlying mechanism. Many external and internal factors may contribute to the gradual loss of neurons such as calcium toxicity, excitatory toxicity, oxidative stress, mitochondrial lesions, deficiency of survival factors. Each of these factors contributes to the pathogenesis by predisposing the neurons to apoptosis (CHOI 1992; ZIV et al. 1993).

The presence of Lewy bodies, eosinophilic inclusions, are consistently observed in Parkinson's disease (LANG and LOZANO 1998). Lewy bodies have a dense spherical hyaline core and a variety of other constituents: structural filament, enzymes such as kinase and phosphatase, and cytosolic proteins are trapped in the granules during their formation. The Lewy body may represent a non-specific feature of Parkinson's disease and be unrelated to its pathogenesis, since these bodies are found in small numbers in other neurodegenerative diseases.

In Alzheimer's disease, extracellular deposits of amyloid β protein accumulates progressively in the plaques. Cerebral formation of amyloid fibers is probably the first event in the pathogenesis of Alzheimer's disease and the amyloid β protein may be responsible for the induction of apoptosis in neurons (LOO et al. 1993). Antioxidants can reverse the effect (LAFERLA et al. 1995). A recent experimental model of rat cerebellar granule neurons indicate that neuronal apoptosis is connected with an increase of metabolic products from amyloid β protein induced by β-secretase cleavage (GALLI et al. 1998).

Most cases of early-onset Alzheimer's disease are connected with mutations of genes encoding presenilin 1 and 2 proteins, which are processed by a regulated endoproteolysis. During apoptosis these proteins are cleaved by a caspase family protease, suggesting a potential role for apoptosis-associated breakdown of presenilins in the development of Alzheimer's disease (KIM et al. 1997).

In patients with amyotrophic lateral sclerosis having the form of copperzinc superoxide desmutase mutation, apoptosis is produced when they are exposed to free radicals. The superoxide-induced cell death can be inhibited by treatment with antioxidants or survival growth factors (TROY and SHELANSKI 1994).

Retinal degeneration is associated with the mutation of either rhodopsin, or peripherin, or β subunit of cyclic guanosine monophosphate phosphodiesterase gene. Any of these three mutations can lead to photoreceptor apoptosis (CHANG et al. 1993). This condition is initiated by the accumulation of mutant proteins or altered functional properties of these proteins.

Spinal muscular atrophies are characterized by progressive spinal cord motor neuron depletion. One of the genes associated with these disorders is the neuronal apoptosis inhibitory protein (ROY et al. 1995). In patients with spinal muscular atrophy, mutations in the gene may make motor neurons become more susceptible to apoptosis.

4. Blood Cell Disorders

The regulation of hematopoiesis is influenced by a number of growth factors including erythropoietin, thrombopoietin, stem cell factor, and colony-stimulating factors (FLEISCHMAN 1993). Hematopoietic growth factors are also required to support the survival of their target cells. If growth factors are absent or present only in low concentrations, apoptosis is developed. Hematopoietic growth factors are also essential in the survival of postmitotic blood cells such as neutrophils. Hematopoietic growth factors control blood cell production partly by inhibiting the occurrence of apoptosis (FAIRBAIRN et al. 1993).

Several hematological disorders are connected with reduced production of blood cells such as aplastic anemia, anemia associated with chronic disease, chronic neutropenia, and myelodysplastic syndromes. In some forms of these conditions, enhanced apoptotic cell death occurs in the bone marrow (BLACKWELL and CRAWFORD 1992). Probably one cause of this condition is an activation of genes that promote apoptosis.

5. Malignant Neoplasms

In a variety of malignant tumors apoptotic bodies are formed spontaneously, such as in squamous cell carcinoma (EL-LABBAN and OSORIS-HERRERA 1986), basal cell carcinoma (KERR and SEARLE 1972), breast cancer (MENDELSOHN 1960), leukemia (HUGGINS et al. 1974), and other malignancies (SEARLE et al. 1973, 1975), sometimes resulting in significant loss of cells. Apoptotic bodies are usually present in a scattered fashion throughout the whole tumor. Experimental carcinogenesis studies in rat liver have revealed that apoptotic bodies are present in preneoplastic cells and in the subsequently formed overt carcinoma (BURSCH et al. 1984; COLUMBANO et al. 1984). It seems that environmental factors participate in the occurrence of apoptosis. These factors include cell mediated immune changes (CURSON and WEEDON 1979), release of tumor necrosis factor (SARRAF and BOWEN 1986), and mild ischemia (PAULUS et al. 1979; SHERIDAN et al. 1984). But since apoptosis occurs at the early stages of cancer, it is likely this form of cell death is at least partly associated with intrinsic autoregulatory mechanisms.

6. Viral Infection

Cellular immune reactions participate in the elimination of damaged cells brought about by infection caused by certain viruses such as choriomeningitis virus (ZINKERNAGEL and DOHERTY 1966). This response is necessary to eliminate the infectious agent, but this process may also affect the transformed cells. Through these reactions the infection sometimes induces apoptosis. Apoptosis is enhanced in the epithelial cells of rectal crypts in patients with acquired immunodeficiency syndrome (KOTLER et al. 1986). The apoptotic bodies are associated with the elimination of the virus-infected cells by phagocytosis and thus prevent the dissemination of the virus particles (CLOUSTON and KERR 1985; ARIESEN and CAPRON 1991; MEYNARD et al. 1992). Apoptosis associated with the HIV virus not only affects the virus-infected cells but uninfected cells are also depleted (FINKEL et al. 1995). The mechanism of this action is connected with a cross-linkage of a glycoprotein in the viral envelope with a receptor in the cell (CD4), this sensitizes T-cells, and the apoptotic process is activated (BORDA et al. 1992). An HIV-encoded transactivator of viral and cellular genes may be involved in the induction of apoptosis (LI et al. 1995) and potentiate the killing of the cells (WESTENDORP et al. 1995). In the retina of scrapie virus infected hamsters apoptotic bodies have been found (HOGAN et al. 1981). Apoptosis of virus-infected cells can interrupt the replication of viruses and prevent the spread of infection to other cells (CLEM and MILLER 1993; LEVINE et al. 1993).

Transformation of fibroblasts, based on the activation of specific oncogenes and the functional inactivation of tumor suppressor genes, leads to the induction of apoptosis by transforming growth factor (TGF-β)-treated neighboring untransformed cells (WEINBERG 1989; FEARON and VOGELSTEIN 1990). However, fibroblasts transformed by bovine papillomavirus showed resistance against intercellular induction of apoptosis (MELCHINGER et al. 1996). In chemically transformed (methylcholanthrene) fibroblast cells apoptosis could be developed by TFG-β treatment. The bovine papillomavirus transformed cells were also resistant to the induction of apoptosis by reduction of the intracellular glutathione level.

7. Expression of Apoptosis in Other Diseases

A type II membrane protein, named Fas ligand, induces apoptotic cell death when bound to Fas antigen (SUDA et al. 1993). Fas ligand is expressed in the spleen, thymus, lung, small and large intestine, uterus, testis, prostate, seminal vesicle, and activated T cells (FRENCH et al. 1996). The membrane-bound Fas ligand is transformed by a metalloproteinase to a soluble form (KAYAGAKI et al. 1995). The soluble form of Fas ligand was found in the serum of patients with rheumatic diseases (NOSAWA et al. 1997), chronic congestive heart failure (NISHIGAKI et al. 1997), granular lymphocytic leukemia, and natural killer cell lymphoma (TANAKA et al. 1996). Human soluble Fas ligand binds to Fas-bearing cells such as cardiomyocytes and T cells, and produces apoptosis (TANAKA et al. 1995). It has been suggested that soluble Fas ligand may cause systemic tissue damage when released into circulation (NAGATA and GOLSTEIN 1995; TANAKA et al. 1995, 1996) and the ongoing loss of myocytes exerts an essential role in the pathogenesis of arrhythmogenic right ventricular dysplasia (MALLAT et al. 1996) and in end-stage heart failure (NARUDA et al. 1996). Recently another mechanism of T-cell mediated cytotoxicity was also published, based on the role of perforin (KAGI et al. 1994). In myocarditis, the development of myocardial cell damage is associated with perforin (SEKO et al. 1991). It seems, therefore, that in the destruction of myocytes the perforin-based pathway is essential, whereas further myocardial cell damage is associated with the Fas-based mechanism, because it is followed by apoptosis of myocytes (TOYOSAKI et al. 1998).

In ulcerative colitis, characterized by chronic inflammation, one form of epithelial injury is associated with apoptosis. In the active lesions, the proportion of apoptotic cells is enhanced in the epithelia of the colon (LEE 1993). Fas is expressed on the epithelia in normal colon as well as in the in the colon with ulcerative colitis lesion. Fas ligand mediated apoptosis probably takes part in the epithelial injury. Fas ligand transcripts are expressed only in the affected mucosa in patients with active ulcerative colitis and Fas ligand mRNA is strongly expressed in T lymphocytes that infiltrate into the lamina propria of the damaged mucosa (YEYAMA et al. 1998). The binding of Fas ligand on T lymphocytes induces apoptosis in the colon epithelia that express Fas, resulting in severe inflammation. Fas ligand mRNA positive cells infiltrating ulcerative colitis lesion are largely CD3 T lymphocytes. It seems that CD3 T lymphocytes with surface Fas ligand may be associated in the development of ulcerative colitis. Several publications revealed that the cross-linking of Fas by anti-Fas stimulate the production of interleukin 8 in colon epithelium (ABREU-MARTIN et al. 1995). Interleukin 8 is a potent inducer of neutrophil and lymphocyte migration (LARSEN et al. 1989; LINDLEY et al. 1988). Thus the high expression of Fas ligand in ulcerative colitis may trigger interleukin 8 synthesis, release from the epithelium of the colon, and promote the activation and migration of neutrophils and lymphocytes.

In Crohn's disease, another chronic inflammatory bowel disease, no Fas ligand transcripts were identified in the active phase. This finding suggests that Fas ligand is not involved in the inflammation associated with Crohn's disease. In ulcerative colitis, the disease is limited to the upper mucosal layer of the colon whereas in Crohn's disease the inflammatory lesion penetrates extensively transmurally through the digestive tract. Moreover, in Crohn's disease the skip lesions are not limited to the colon but spread segmentally to the ileum, jejunum, duodenum, and even further in the gastrointestinal tract, in contrast to ulcerative colitis which is limited to the colon. In active Crohn's condition, macrophages and not cytotoxic T lymphocytes have been thought to participate in the onset and progression of the lesions (ATTISON et al. 1988; MURCH et al. 1992). The cytotoxic T lymphocytes induced apoptosis is insignificant in Crohn's disease, indicating that the pathogenesis of Crohn's disease is different from that of ulcerative colitis. (YAYAMA et al. 1998).

G. Conclusions

Apoptosis is a well established process that plays an important role in a variety of physiological and pathological conditions. Apoptosis represents a process of cell death that manifests in all multicellular organisms. The phenomenon of apoptosis varies with cell type and stimuli. The unique character of apoptotic cell death is that it is regulated developmentally and thus it is also called programmed cell death (Fig. 6) (LOKSHIN and BEAULATON 1974). Cells dying during development undergo a unique and distinct set of structural changes which are similar or identical with changes occurring in cells dying in a wide variety of circumstances outside of development such as normal cell turnover in several tissues and in tumors, T-cell killing, atrophy induced by endocrine and other physiological stimuli, negative selection within the immune system and cell turnover following exposure to some toxic compounds, chemotherapy, hypoxia, or low doses of ionizing radiation. The process of cell death by apoptosis is clearly different from necrosis which is the consequence of extreme alterations of the cellular microenvironment.

The process of apoptosis can be divided into several steps – (a) the stimulus that initiates the cell death response, (b) the pathway by which the message is transferred to the cell, and (c) the effector mechanisms that carries out the death program (KRAMMER et al. 1994). The dying cell separates from its neighbors with a loss of specialized membrane structures and undergoes a period of distortion. Diverse stimuli may trigger a different death response in the cell, but the pathways converge into the same effector mechanisms with several identical key components, including a family of proteases called caspases. Following the activation of these proteases, they are directly or indirectly responsible for the varying morphological or biochemical changes characteristic of apoptosis. Finally, the neighboring cells are very competent in the phagocytosis of the apoptotic cells.

Apoptosis is a gene-regulated phenomenon, and great progress has been made to reveal the mechanism of this type of cell death. The occurrence of apoptotic cell death may provide a new insight into certain diseases. Further studies at the molecular level may lead to a clear view of the etiology and development of these diseases. A comprehensive understanding of the great variety of cellular processes undergone during apoptosis and further application of our knowledge concerning cell death can provide a solid basis for the development of novel therapeutic approaches and more effective ways of vaccination or gene therapy (THOMPSON 1995; UREN and VAUX 1996). They may also open new avenues to the application of pharmacological substances in diseases associated with apoptotic cell death.

Acknowledgements. We thank Marie Maguire for her excellent work in the preparation of this manuscript.

References

Abreu-Martin MT, Vidrich A, Lynch DH et al. (1995) Divergent induction of apoptosis and IL-8 secretion on HT-29 cells in response to TNF-alpha and ligation of Fas antigen. J Immunol 155:4147–4154

Adams JM, Cory S (1998) The Bcl-2 protein family: arbiters of cell survival. Science 281:1322–1326

Allan DJ, Harmon BV, Kerr JFR (1987) Cell death in spermatogenesis. In: Potten V (ed) Perspectives on mammalian cell death. Oxford University Press, Oxford, pp 229–258

Allison MC, Cornwall S, Poulter LW et al. (1988) Macrophage heterogeneity in normal colonic mucosa and in inflammatory bowel disease. Gut 29:1531–1538

Aneisen JC, Capron A (1991) Cell dysfunction and depletion in AIDS: the programmed cell death hypothesis. Immunol Today 12:102–105

Appleby DW, Modak SP (1977) DNA degeneration in terminally differentiating lens fiber cells from chick embryos. Proc Natl Acad Sci USA 74:5579–5583

Arch RH, Gedrich RW, Thompson CB (1998) Tumor necrosis factor receptor-associated factors (TRAFs) – a family of adapter proteins that regulates life and death. Genes Development 12:2821–2830

Arends MJ, Wyllie AH (1991) Apoptosis: mechanisms and roles in pathology. Int Rev Exper Pathol 32:223–254

Ashkenazi A, Dixit VM (1998) Death receptors: signaling and modulation. Science 281:1305–1308

Aw TV, Nicotera P, Manao L, Orrenius S (1990) Tributyltin stimulates apoptosis in rat thymocytes. Arch Biochem Biophys 283:46–50

Azmi TI, O'Shea JD. Mechanism of deletion of endothelial cells during regression of the corpus luteum. Lab Invest 51:206–217

Ballou LR, Laulederkind SK, Roslomiec E et al. (1996) Ceramid signaling and the immune response. Biochem Biophys Acta 1301:273–287

Banda NK, Bernier J, Kurahara DK, Kurrle R, Haigwood N, Sekaly RP, Finkel TH (1992) Crosslinking CD4 by human immunodeficiency virus gp120 primes T cell for activation-induced apoptosis. J Exp Med 176:1090–1106

Barr PJ, Tomei LD (1994) Apoptosis and its role in human disease. Biotechnology 12:487–493

Barry MA, Behnke CA, Eastman A (1990) Activation of programmed cell death (apoptosis) by cisplatinum, other anticancer drugs, toxins and hyperthermia. Biochem Pharmacol 40:2353–2362

Beewitt RW, Abbott AC, Bird CC (1983) Mode of cell death induced in human lymphoid cells by high and low doses of glucocorticoid. Br J Cancer 47:477–486

Ben-Ishay Z, Farber E (1975) Protective effects of an inhibitor of protein synthesis, cyclohexamide, on bone marrow damage induced by cytosine arabinoside or nitrogen mustard. Lab Invest 33:478–490

Bennett RE, Harrison MW, Bishop CT, Searle J, Kerr JFR (1984) The role of apoptosis in atrophy of the small gut mucosa produced by repeated administration of cytosine arabinoside. J Path 142:259–263

Bernuau D, Feldmann G, Degott C, Gisselbrecht C (1981) Ultrastructural lesions of bile ducts in primary biliary cirrhosis. A comparison with the lesions observed in graft versus host disease. Human Pathol 12:782–793

Bhathal PS, Gall JAM (1985) Deletion of hyperplastic biliary epithelial cells by apoptosis following removal of the proliferative stimulus. Liver 5:311–325

Bishop CJ, Moss DJ, Ryan JM, Burrows SR (1985) T lymphocytes in infectious mononucleosis II Response in vitro to interleukin 2 and establishment of T cell lines. Clin Exp Immunol 60:70–77

Bishop CJ, Rzepczyk CM, Stensel D, Anderson K (1987) The role of reactive oxygen metabolites in lymphocyte-mediated cytolysis. J Cell Sci 87:473–481

Bishop CJ, Whiting VA (1983) The role of natural killer cells in the intravascular death of intravenously injected murine tumor cells. Br J Cancer 48:441–444

Boise LH, Thompson CB (1998) Bcl-xl can inhibit apoptosis in cells that have undergone Fas-induced protease activation. Proc Natl Acad Sci USA 94:3759–3764

Borges M, Shu LG, Xhonneux R, Thoné F, Overloops P van (1987) Changes in ultrastructure and Ca^{2+} distribution in the isolated working rabbit heart after ischemia: a time-related study. Am J Path 126:97–102

Buckley IK (1972) A light and electron microscopic study of thermally injured cultured cells. Lab Invest 26:201–209

Bumbasirevic V, Lackovic V, Japundzic M (1985) Enhancement of apoptosis in lymphoid tissue by microtubule disrupting drugs. IRCS Med Sci 13:1257–1258

Bursch W, Düsterberg B, Schulte-Hermann R (1986) Growth, regression and cell death in rat liver as related to tissue levels of the hepatomitogen cyproterone acetate. Arch Toxicol 59:221–227

Bursch W, Lauer B, Timmermann-Trosiener I, Barthol G, Schuppler J, Schulte-Hermann R (1984) Controlled death (apoptosis) of normal and putative preneoplastic cells in rat liver following withdrawal of tumor promoters. Carcinogenesis 5:453–458

Bursch W, Taper HS, Lauer B, Schulte-Hermann R (1985) Quantitative histological and histochemical studies on the occurrence and stages of controlled cell death (apoptosis) during regression of rat liver hyperplasia. Virchows Arch Abt B Zellpath 50:153–166

Cheng EHY, Kirsch EG, Clem RJ, Ravi R, Kastan MB, Bedi A, Veno K, Hardwick JM (1997) Conversion of Bcl-2 to a Bax-like death effector by caspases. Science 274: 1966–1968

Chien KR, Abrams J, Serroni A, Martin JT, Farber JL (1978) Accelerated phospholipid degradation and associated membrane dysfunction in irreversible ischemic liver cell injury. J Biol Chem 253:4809–4817

Choi DW (1992) Excitotoxic cell death J Neurobiol 23:1261–1276

Clem RJ, Cheng EHY, Karp CL, Kirsch DG, Veno K, Takahashi A, Kastan MB, Griffin DE, Earnshaw WC, Veliuona MA, Hardwick JM (1998) Modulation of cell death by Bcl-xL through caspase interaction. Proc Natl Acad Sci USA 95:554–559

Clem RJ, Miller LK (1993) Apoptosis reduces both the in vitro replication and the in vivo infectivity of a baculovirus. J Virol 67:3730–3738

Clouston WM, Kerr JFR (1979) Microfilament distribution in cell death by apoptosis. Clin Exp Pharmacol Physiol 6:451–452

Clouston WM, Kerr JFR (1985) Apoptosis, lymphocytotoxicity and the containment of viral infections. Med Hypotheses 18:399–404

Cohen JJ, Duke PC (1984) Glucocortocoid activation of a calcium-dependent endonuclease in thymocyte nuclei leads to cell death. J Immun 132:38–42

Cohen JJ, Duke RC, Chervenak R, Sellins KS, Olson LK (1985) DNA fragmentation in targets of CTL: an example of programmed cell death in the immune system. Adv Exp Med Biol 184:493–508

Columbano A, Ledda-Columbano GM, Coni PP, Faa G, Liguori C, Santa Cruz G, Pani P (1985) Occurrence of cell death (apoptosis) during involution of liver hyperplasia. Lab Invest 52:670–675

Columbano A, Ledda-Columbano GM, Rao PM, Rajalakshmi S, Sarma DSR (1984) Occurrence of cell death (apoptosis) in preneoplastic and neoplastic liver cells. Am J Path 116:441–446

Compton MM, Cidlowski JA (1986) Rapid in vivo effects of glucocorticoids on the integrity of rat lymphocyte genomic deoxyribonucleic acid. Endocrinology 118:38–45

Crawford AM, Kerr, JFR, Currie AR (1972) The relationship of acute mesodermal cell death to the teratogenic effects of 7-OH-DMBA in the foetal rat. Br J Cancer 26:498–503

Curson C, Weedon D (1979) Spontaneous regression in basal cell carcinomas. J Cutaneous Pathol 6:432–437

Daoust PY, Wobeser G, Newstead JD (1984) Acute pathological effects of inorganic mercury and copper in gills of rainbow trout. Vet Pathol 21:93–101

Dibbert B, Daigle I, Braun D, Schranz C, Weber M, Blaser K, Zangemeister-Wittke I, Akbar AN, Simon HU (1998) Role of Bcl-xL in delayed eosinophil apoptosis mediated by granulocyte-macrophage colony-stimulating factor and interleukin-5. Blood 92:778–783

Don MM, Abbott G, Bishop CJ, Bundesen PG, Donald KJ, Searle J, Kerr JFR (1977) Death of cells by apoptosis following attachment of specifically allergized lymphocytes in vitro. Aust J Exp Biol Med Sci 55:407–417

Dowd DR, MacDonald PN, Komm BS, Maussler MR, Miesfeld R (1991) Evidence for early induction of calmudolin gene expression in lymphocytes undergoing glucocorticoid-mediated apoptosis. J Biol Chem 266:18423–18426

Duke RC, Chervenak R, Cohen JJ (1983) Endogenous endonuclease-induced DNA fragmentation: an early event in cell-mediated cytolysis. Proc Natl Acad Sci USA 80:6361–6365

Duncan AMV, Heddle JA (1984) The frequency and distribution of apoptosis induced by three non-carcinogenic agents in mouse colonic crypts. Cancer Lett 23:307–311

Duncan AMV, Ronen A, Beakey DH (1983) The frequency and distribution of apoptosis induced by three non-carcinogenic agents in mouse colonic crypts. Cancer Lett 23:307–311

Duvall E, Wyllie AH (1986) Death and the cell. Immunol Today 7:115–119

Ekert PG, Vaux DL (1997) Apoptosis and the immune system. Br Med Bull 53:591–603

El-Labban NG, Osorio-Herrera E (1986) Apoptotic bodies and abnormally dividing epithelial cells in squamous cell carcinoma. Histopathology 10:921–931

Ellis HM, Horvitz HR (1986) Genetic control of programmed cell death in the nematode C Elegans. Cell 44:817–829

Elmes M (1977) Apoptosis in the small intestine of zinc-deficient and fasted rats. J Path 123:219–223

Elmes M, Jones JG (1980) Ultrastructural studies of Paneth cell apoptosis in zinc deficient rats. Cell Tiss Res 208:57–63

Evan G, Littlewood T (1998) A matter of life and cell death. Science 281:1317–1321

Evan GI, Wyllie AH, Gilbert CS, Littlewood TD, Land H, Brooks M, Waters CM, Penn LZ, Hancock DC (1992) Induction of apoptosis in fibroblasts by c-myc protein. Cell 69:119–128

Fairbairn LJ, Cowling GJ, Reipert BM, Dexler TM (1993) Suppression of apoptosis Cell 74:823–832

Fang G, Chang BS, Kim CN, Perkins C, Thompson CB, Bhalla KN (1998) "Loop" domain is necessary for Taxol-induced mobility shift and phosphorylation of Bcl-2 as well as for inhibiting Taxol-induced cytosolic accumulation of cytochrome C and apoptosis. Cancer Res 58:3202–3208

Farber JL, Chien KR, Mittnacht S (1981) The pathogenesis of irreversible cell injury in ischemia. Am J Path 102:271–281

Fearon ER, Vogelstein B (1990) A genetic model for colorectal tumorigenesis. Cell 61:759–767

Ferguson DJP, Anderson TJ (1981) Morphological evaluation of cell turnover in relation to the menstrual cycle in the "resting" human breast. Br J Cancer 44:177–181

Finkel TH, Tudorwilliams G, Banda NK, Cotton MF, Curiel T, Monks C, Baba TW, Ruprecht RM, Kuppfer A (1995) Apoptosis occurs predominantly in bystander cells and not in productively infected cells of HIV-1 and SIV-infected lymph nodes. Nat Med 1:129–134

Fisher GH, Rosenberg FJ, Straus SE, Dale JK, Middleton LA, Lin AY, Strober W, Lenardo MJ, Puck JM (1995) Dominant interfering Fas gene mutations impair apoptosis in a human autoimmune lymphoproliferative syndrome. Cell 81:935–946

Forritsma JBM, Konings AWT (1986) DNA lesions in hyperthermic cell killing: effects of thermotolerance, procaine and erythritol. Radiat Res 106:89–97

French LE, Hahne M, Viard I, Radlgruber G, Zanone R, Becker K, Uruller C, Tschopp J (1996) Fas and Fas ligand in embryos and adult mice: ligand expression in several immune-privileged tissues and coexpression in adult tissues characterized by apoptotic cell turnover. J Cell Biol 133:335–343

Galili V, Leiserowitz R, Moreb J, Gamliel H, Gurfel D, Pollack A (1982) Metabolic and ultrastructural aspects of the in vitro lysis of chronic lymphocytic leukemia cells by glucocorticoids. Cancer Res 42:1433–1440

Galli C, Piccini A, Ciotti MT, Castellani L, Calisgano P, Zaccheo D, Tabaton M (1998) Increased amyloidogenic secretion in cerebellar granule cells undergoing apoptosis. Proc Natl Acad Sci USA 95:1247–1252

Gallucci BB, Sale GE, McDonald GB, Epstein R, Shulman HN, Thomas ED (1982) The fine structure of human rectal epithelium in acute graft-versus-host disease. Am J Surg Path 6:293–305

Goping IS, Gross A, Lavoie JN, Nguyen M, Jemmerson R, Roth K, Korsmeyer SJ, Shore GC (1998) Regulated targeting of Bax to mitochondria. J Cell Biol 143:207–215

Gottlieb RA, Buoleson KO, Kloner RA, Babior BM, Engler RL (1994) Reperfusion injury induces apoptosis in rabbit cardiomyocytes. J Clin Invest 94:1621–1628

Green DR, Reed JC (1998) Mitochondria and apoptosis. Science 281:1309–1312

Green DR, Ware CF (1997) Fas-ligand: privilege and peril. Proc Natl Acad Sci USA 94:5986–5990

Griffith GD, Leek MD, Gee DJ (1987) The toxic plant proteins ricin and abrin induce apoptotic changes in mammalian lymphoid tissues and intestine. J Path 151:221–229

Griffith TS, Lynch DH (1998) TRAIL: a molecule with multiple receptors and control mechanisms. Curr Op Immunol 10:559–563

Gromkowski SH, Brown TC, Cerutti PA, Cerottini JC (1986) DNA of human Raji target cells is damaged upon lymphocyte mediated lysis. J Immun 136:752–756

Gubina E, Rinaudo MS, Szallasi Z, Blumberg PM, Mufson RA (1998) Overexpression of protein kinase – isoform E but not E in human interleukin-3-dependent cells suppresses apoptosis and induces Bcl-2 expression. Blood 91:823–829

Gunn A, Scrimgeour D, Potts RC, Mackenzie LA, Brown RA, Swanson Beck J (1979) The destruction of peripheral blood lymphocytes by extracorporeal exposure to ultraviolet radiation. Immunology 50:477–485

Haimovitz-Friedman A, Kan CC, Ehleiter D et al. (1994) Ionizing radiation acts on cellular membranes to generate ceramide and induce apoptosis. J Exp Med 180:525–535

Haimovitz-Friedman A, Kolesnik R, Fuks Z (1997) Ceramide signaling in apoptosis. Br Med Bull 53:539–553

Harmon B, Bell L, Williams L (1984) An ultrastructural study on the "meconium corpuscles" in rat foetal epithelium with particular reference to apoptosis. Anat Embryol 169:119–124

Hassel KR (1975) The development of rat palatal shelves in vitro. An ultrastructural analysis of the inhibition of epithelial cell death and palate fusion by epidermal growth factor. Devl Biol 45:95–102

Hawkins HK, Ericsson JLE, Biberfeld P, Trump BF (1972) Lysosome and phagosome stability in lethal cell injury: morphologic tracer studies in cell injury due to inhibition of energy metabolism, immune cytolysis, and photosensitisation. Am J Path 68:255–288

Hendry JH, Potten CS, Chadwick C, Bianchi M (1982) Cell death (apoptosis) in the mouse small intestine after low doses: effects of dose rate, 147 MeV neutrons and 600 MeV (maximum energy) neutrons. Int J Radiat Biol 42:611–620

Henkart PA, Grinstein S (1996) J Exp Med 183:1293–1295

Hoffstein S, Gennaro DE, Fox AC, Hirsch J, Steuli F, Weissman G (1975) Colloidal lanthanum as a marker for impaired plasma membrane permeability in ischemic dog myocardium. Am J Path 79:207–218

Hogan RN, Baringer JR, Prusiner SB (1981) Progressive retinal degeneration in serapic-infected hamsters. A light and electron microscopic analysis. Lab Invest 44:34–42

Hopwood D, Levison DA (1976) Atrophy and apoptosis in the cyclical human endometrium. J Path 119:159–166

Huckins C (1978) The morphology and kinetics of spermatogenial degeneration in normal adult rats: an analysis using a simplified classification of the germinal epithelium: Anat Rec 190:905–926

Huggins CB, Yoshida H, Bird CC (1974) Hormone-dependent stem-cell rat leukemia evoked by a series of feedings of 7.12 dimethylbenz(a) anthracene. J Natl Cancer Inst 52:1301–1305

Iriji K, Potten CS (1983) Response of intestinal cells of differing topographical and hierarchical status to ten cytotoxic drugs and five sources of radiation. Br J Cancer 47:175–185

Iriji K, Potten CS (1984) Further studies on the response of intestinal crypt cells of different hierarchical status to eighteen different cytotoxic agents. Br J Cancer 55:113–123

Isaacson O (1993) On neuronal health Trends Neurosci 16:306–310

Jennings RB, Reimer KA (1981) Lethal myocardial ischemia injury. Am J Path 102:241–255

Kagi D, Vignaux F, Ledermann B, Burki K, Defractere V, Nagata S, Hengartner H, Golstein P (1994) Fas and perforin pathways as major mechanisms of T-cell mediated cytotoxicity. Science 265:528–530

Kawahara A, Kobayashi T, Nagata S (1998) Inhibition of Fas-induced apoptosis by Bcl-2. Oncogene 17:2549–2554

Keenan KP, Sharpnack DD, Collins H, Formal SB, O'Brien AD (1986) Morphological evaluation of the effects of Shiga toxin and E. coli Shiga-like toxin on the rabbit intestine. Am J Path 125:69–80

Kerr JFR (1967) Lysosome changes in acute liver injury due to heliotrine. J Path Bact 93:167–174

Kerr JFR (1969) An electron-microscope study of liver cell necrosis due to heliotrine. J Path 97:557–562

Kerr JFR (1970) An electron-microscope study of liver cell necrosis due to albitocin. Pathology 2:251–259

Kerr JFR (1971) Shrinkage necrosis: a distinct mode of cellular death. J Path 105:13–20

Kerr JFR (1972) Shrinkage necrosis of adrenal cortical cells. J Path 107:217–219

Kerr JFR, Bishop CJ, Searle J. Apoptosis in Anthony, MacSween (eds) (1984) Recent advances in histopathology, vol 12. Churchill, Livingstone, Edinburgh, pp 1–15

Kerr JFR, Searle J (1972) A suggested explanation for the paradoxically slow growth rate of basal-cell carcinomas that contain numerous mitotic figures. J Path 107:41–44

Kerr JFR, Searle J (1973) Deletion of cells by apoptosis during castration-induced involution of the rat prostate. Virchows Arch Abt B Zellpath 13:87–102

Kerr JFR, Searle L, Harmon BV, Bishop CJ (1987) Apoptosis. In: Patten (ed) Perspectives on mammalian cell death. Oxford University Press, Oxford, pp 93–128

Kerr JFR, Wyllie AH, Currie AR (1972) Apoptosis: a basic biological phenomenon with wide-ranging implications in tissue kinetics. Br J Cancer 26:239–257

Kim TW, Pettingell WH, Jung YK, Kovacs DM, Tanzi RE (1997) Alternative cleavage of Alzheimer-associated presenilins during apoptosis by a caspase-3 family protease. Science 277:373–375

King TP, Brenner I (1979) Autophagy and apoptosis in liver during the prehaemolytic phase of chronic copper poisoning in sheep. J comp Pathol 89:515–530

Kolesnick R, Golde DW (1994) The spingomyelin pathway in tumor necrosis factor and interleukin-7 signaling. Cell 77:325–328

Kotler DP, Weaver SC, Terzakis JA (1986) Ultrastructural features of epithelial cell degeneration in rectal crypts of patients with AIDS. Am J Surg Pathol 10:531–538

Koyagaki N, Kawasaki A, Ebata T, Ohmoto H, Ikeda S, Onoue S, Yoshino K, Okumura K, Tagita H (1995) Metalloproteinase-mediated release of human Fas ligand. J Exp Med 18:1777–1783

Krammer PH, Behrman I, Daniel P, Ohein J, Debatin KM (1994) Regulation of apoptosis in the immune system. Curr Opin Immunol 6:279–289

Kyprianou N, Isaacs JT (1988) Activation of programmed cell death in the rat ventral prostate after castration. Endocrinol 122:552–562

LaFerla FM, Tinkle BT, Bieberich CD, Haudenschild CC, Gay J (1995) The Alzheimer's A beta peptide induces neurodegeneration and apoptotic cell death in transgenic mice. Nat Genet 9:21–30

Laiho KV, Beresesky IK, Trump BF (1983) The role of calcium in cell injury: studies in Ehrlich ascites tumour cells following injury with anoxia and organic mercurials. Surv Synth Path Res 2:170–183

Lang AE, Lozano AM (1998) Parkinson's disease. N Engl J Med 339:1044–1053

LaPushin RW, Harven E de (1971) A study of glucocorticoid-induced pyknosis in the thymus and lymph node of the adrenalectomised rat. J Cell Biol 50:583–597

Larsen CG, Anderson AO, Appella E et al. (1989) The neutrophil-activating protein (NAP-1) is also chemotactic for lymphocytes. Science 243:1464–1466

Ledda-Columbano GM, Curto M, Riga R, Zedda AI, Menegazzi M, Sartori C, Shinozuka H, Bluethmann H, Poli V, Ciliberto G, Columbano A (1998) In vivo hepatocyte proliferation is inducible through TNF and IL-6 independent pathway. Oncogene 17:1039–1044

Lee FD (1993) Importance of apoptosis in the histopathology of drug related lesions in the large intestine. J Clin Pathol 46:118–122

Levine B, Huang Q, Isaacs JT, Reed JC, Griffen DE, Hardwick JM (1993) Conversion of lytic to persistent alphavirus infection by the Bcl-2 cellular oncogene. Nature 361:739–742

Lie CJ, Friedman DJ, Wang C, Metelev V, Pardee AB (1995) Induction of apoptosis in uninfected lymphocytes by HIV-1 Tat protein. Science 268:429–431

Lieberman MW, Verbin RS, Landay M, Liang H, Farber E, Lee TN, Starr R (1970) A probable role for protein synthesis in intestinal epithelial cell damage induced in vivo by cytosine arabinoside, nitrogen mustard, or X-irradiation. Cancer Res 30:942–951

Liepins A, Foanes RB, Lifter J, Choi YS, Harven E de (1977) Ultrastructural changes during T-lymphocyte-mediated cytolysis. Cell Immunol 28:109–124

Lie-Weber M, Lauer O, Hekele A, Coy J, Walczak H, Krammer PH (1998) A regulatory element in the CD95 (APO-1/Fas) ligand promoter is essential for responsiveness to TCR-mediated activation. Eur J Immunol 28:2373–2383

Lindley A, Aschauer H, Sufest JM (1988) Synthesis and expression in Escherichia coli of the gene encoding monocyte-derived neutrophil-activating factor: biological equivalence between natural and recombinant neutrophil-activating factor. Proc Natl Acad Sci USA 85:9199–9203

Lokshin RA, Beaulaton J (1974) Programmed cell death. Life Sci 15:1549–1565

Loo DJ, Copani A, Pike CJ, Whittemore ER, Walencervice AJ, Corman CW (1993) Apoptosis is induced by beta amyloid in cultured central nervous system neurons. Proc Natl Acad Sci USA 90:7951–7955

Loweth AC, Williams GT, James RFL, Scarpello JHB, Morgan NG (1998) Human islets of Langerhans express Fas ligand and undergo apoptosis in response to interleukin-1β and Fas ligation. Diabetes 47:727–732

Lynch MP, Nawas S, Gerchenson LE (1986) Evidence for soluble factors regulating cell death and cell proliferation in primary cultures of rabbit endometrial cells grown on collagen. Proc Natl Acad Sci USA 83:4784–4788

Mallat Z, Tedgui A, Fontaliran F, Frank R, Dungon M, Fontain G (1996) Evidence of apoptosis in antirhythmogenic right ventricular dysplasia. N Engl J Med 335:1190–1196

Maskens AP (1979) Significance of karyorrhetic index in 1,2-dimethylhydrasine carcinogenesis. Cancer Lett 8:77–86

Matter A (1979) Microcinematographic and electron microscopic analysis of target cell lysis induced by cytotoxic T lymphocytes. Immunology 36:179–190

McConkey DJ, Chow SC, Orrenius S, Fondal M (1990) NK cell-induced cytotoxicity is dependent on a Ca^{2+} increase in the target FASEB J 4:2661–2664

McConkey DJ, Hartzell P, Duddy SK, Hakansson H, Orrenius S (1988) 2,3,7,8-tetrachloro-di-benzo-p-dioxin kills immature thymocytes by Ca^{2+} mediated endonuclease activation. Science 242:256–258

McDowell EM (1972) Light- and electron-microscope studies of the rat kidney after administration of the citric acid cycle in vivo: changes in the proximal convoluted tubule during fluorocitrate poisoning. J Path 108:303–318

McLean AEM, McLean E, Judah JD (1965) Cellular necrosis in the liver induced and modified by drugs. Int Rev Exp Pathol 4:127–157

Melchinger W, Strauss S, Zucker B, Bauer G (1996) Antiapoptotic activity of bovine papilloma virus, implications for the control of oncogenesis. Int J Oncol 9:927–933

Mendelsohn ML (1960) Autoradiographic analysis of cell proliferation in spontaneous breast cancer of C3H mouse. II Growth and survival of cells labelled with tritiated thymidine. J Natl Cancer Inst 23:485–500

Meynard L, Otto SA, Jonker RR, Mijnster MJ, Keet RP, Miedema F (1992) Programmed death of T cells in HIV-1 infection. Science 257:217–219

Miwa K, Asano M, Horai R, Iwakura Y, Nagata S, Suda T (1998) Caspase 1-independent IL-1 beta release and inflammation induced by the apoptosis induces Fas ligand. Nature medicine 4:1287–1292

Moss DJ, Bishop CJ, Burrows SR, Ryan JM (1985) T lymphocytes in infectious mononucleosis I T cell death in vitro. Clin Exp Immunol 60:61–69

Murch SH, Braegger CP, Sessa WC (1992), High endothelin-1 immunoreactivity in Crohn's disease and ulcerative colitis. Lancet 339:381–385

Nagata S, Golstein P (1995) The Fas death factor. Science 267:1449–1456

Nasula J, Haider N, Virmani R, DiSalvo TG, Kolodgie FD, Hajjas RJ, Schmidt U, Semigran MJ, Dec GN, Khaw BA (1996) Apoptosis in myocytes in end-stage heart failure. N Engl J Med 335:1182–1189

Nicotera P, McConkey DJ, Jones DP, Orrenius S (1989) ATP stimulates Ca^{2+} uptake and increased the free Ca^{2+} concentration in isolated rat liver nuclei. Proc Natl Acad Sci USA 86:451–457

Nicotera P, Zhivotovsky B, Bellomo G, Orrenius S (1994) Low signaling in apoptosis. In Mihich E, Schimke RT (eds), Apoptosis. Plenum Press, New York, pp 97–108

Nishigaki K, Minatoguchi S, Seishima M, Asano K, Noda T, Yasuda N, H, Kumada H, Takemura M, Noma A, Tanaka T, Watanabe S, Fujiwara H (1997) Plasma Fas ligand, an inducer of apoptosis, and plasma soluble Fas an inhibitor of apoptosis, in patients with chronic congestive heart failure. J Am Cell Cardiol 29:1214–1220

Nosawa K, Kayagaki N, Tokano Y, Yagita H, Okumura K, Hasimoto H (1997) Soluble Fas (APO-1, CD-95) and soluble Fas ligand in rheumatic disease. Arthritis Rheum 40:1126–1129

O'Shea JD, Hay MF, Cran DG (1978) Ultrastructural changes in the theca interna during follicular atresia in the sheep. J Reprod Fertil 54:183–187

O'Shea JD, Wright PJ (1984) Involution and regeneration of the endometrium following parturition in the ewe. Cell Tiss Res 236:477–485

Oates PS, Morgan RGH, Light AM (1986) Cell death (apoptosis) during pancreatic involution after raw soya flour feeding in the rat. Am J Physiol 250:G4–G14

Osborne BA (1996) Apoptosis and the maintenance of homeostasis in the immune system. Current Opinion Immunol 8:248–254
Pails G, Lie Xiang Hong, Atassi G, Buyssens N (1982) Degree of differentiation and blood vessel proximity in B15 melanoma. Virchows Arch Abt B Zellpath 39:229–238
Penfold PL, Provis JM (1986) Cell death in the development of the human retina: phagocytosis of pyknotic and apoptotic bodies by retinal cells. Graefe's Arch Klin Exp Ophthal 224:549–553
Petak I, Mihalik R, Bauer PI, Suli-Vargha H, Sebestyen A, Koppur L (1998) BCNM is a caspase-mediated inhibitor of drug-induced apoptosis. Cancer Res 58:614–618
Peter ME, Krammer PH (1998) Mechanisms of CD95 (APO-1/Fas)-mediated apoptosis. Curr Op Immunol 10:545–551
Philips ES, Sternberg SS (1975) The lethal action of antitumor agents in proliferating cell system in vivo. Am J Path 81:205–218
Pipan N, Sterle M (1986) Cytochemical and scanning electron-microscopic analysis of apoptotic cells and their phagocytosis in mucoid epithelium of the mouse stomach. Cell Tiss Res 246:647–652
Potten CS (1977) Extreme sensitivity of some intestinal crypt cells to X and Y irradiation. Nature Lond 269:518–521
Potten CS (1985) Cell death (apoptosis) in hair follicles and consequent changes in the width of hairs after irradiation of growing follicles. Int J Radiat Biol 48:349–360
Pratt NE, Sodikoff M (1972) Ultrastructural injury following X-irradiation of rat parotid gland acinar cells. Arch oral Biol 17:1177–1186
Pratt RM, Green RM (1976) Inhibition of palatal epithelial cell death by altered protein synthesis. Devl Biol 54:135–145
Radley JM, Haller CJ (1983) Fate of senescent megakaryocytes in the bone marrow. Br J Haemat 53:277–287
Raff MC, Barres BA, Bume JF et al. (1993) Programmed cell death and the control of cell survival: lesions from the nervous system. Science 262:695–700
Rajotte D, Haddad P, Haman A, Crajoe EJ, Hoang T (1992) Role of protein kinase C and the Na+/H+ antiporter in suppression of apoptosis by granulocyte-macrophage colony stimulating factor and interleukin-3. J Biol Chem 267:9980–9987
Reynolds ES, Kanz MF, Chieco P. Moslen MT (1984) 1,1-Dichloroethylene: an apoptotic hepatotoxin ? Environ Health Perspect 57:313–320
Rinaudo M, Su K, Falk LA, Halder S, Mufson RA (1995) Human interleukin-3 receptor modulates Bcl-2 mRNA and protein levels through protein kinase C in TF-1 cells. Blood 86:80–88
Ronen A, Heddle JA (1984) Site-specific induction of nuclear abnormalities (apoptotic bodies and micronuclei) by carcinogens in mice. Cancer Res 44:1536–1540
Roy N, Meladevan MD, McLean M, Shutler G, Varaghi Z, Farshani R, Baird S et al. (1995) The gene for neuronal apoptosis inhibiting protein is partially deleted in individuals with spinal muscular atrophy. Cell 80:167–178
Rudi J, Kuck D, Strand S, Vonderbay A, Mariani SM, Krammer PH, Galle PR, Stremmel W (1998) Involvement of the CD95 (APO-1/Fas) receptor and ligand system in helicobacter pylori-induced gastric epithelial apoptosis. J Clin Invest 102:1506–1514
Sanderson CJ (1976a) The mechanism of T cell mediated cytotoxicity II. Morphological studies of cell death by time-lapse microcinematography. Proc R Soc Ser B 192:241–255
Sanderson CJ (1976b) The mechanism of T cell mediated cytotoxicity. V Morphological studies by electron microscopy. Proc R Soc Ser B 198:315–323
Sanderson CJ, Thomas JA (1977) The mechanism of K cell (antibody dependent) cell mediated cytotoxicity II. Characteristics of the effector cell and morphological changes in the target cell. Proc R Soc Ser B. 197:417–424

Sandford NL, Searle JW, Kerr JFR (1984) Successive waves of apoptosis in the rat prostate after repeated withdrawal of testosterone stimulation. Pathology 16:406–410

Sandrow BA, West NB, Norman RL, Brenner RM (1979) Hormonal control of apoptosis in hamster uterine luminal epithelium. Am J Anat 156:15–36

Sanejuva K, Tone S, Kottke TJ, Enari M, Sakahira H, Cooke CA, Durrieu F, Martins LM, Nagata S, Kaufmann SH, Earnshaw WC (1998) Transition from caspase-dependent to caspase-independent mechanisms at the onset of apoptotic execution. J Cell Biol 143:225–239

Sarraf GE, Bowen ID (1986) Kinetic studies on a murine sarcoma and an analysis of apoptosis. Br J Cancer 54:989–998

Savill J, Fadok V, Henson P, Haslett C (1993) Phagocyte recognition of cells undergoing apoptosis. Immunol Today 14:131–136

Schanne FAX, Kane AB, Young EE, Farber JL (1979) Calcium dependence of toxic cell death: a final common pathway. Science 206:700–702

Schellens JPM, Everts V, Beertsen W (1982) Quantitative analysis of connective tissue resorption in the supra-alveolar region of the mouse incisor ligament. J Periodont Res 17:407–422

Schumer M, Colombel MC, Sawesuk IS, Gobe G, Conner J, O'Toole KM, Olsson CA, Wise GJ, Buttyan R (1992) Morphologic, biochemical and molecular evidence of apoptosis during the reperfusion phase after brief periods of renal ischemia. Am J Pathol 140:831–838

Schwartz LM, Kisi L, Kay BK (1990) Gene activation is required for developmentally programmed cell death. Proc Natl Acad Sci USA 87:6594–6598

Searle J, Collins DJ, Harmon B, Kerr JFR (1973) The spontaneous occurrence of apoptosis in squamous carcinomas of the uterine cervix. Pathology 5:163–169

Searle J, Kerr JFR, Battersby C, Egerton WS, Balderson G, Burnett W (1977) An electron microscopic study of the mode of donor cell death in unmodified rejection of pig liver allografts. Aust J Exp Biol Med Sci 55:401–406

Searle J, Kerr JFR, Bishop CJ (1982) Necrosis and apoptosis: distinct mode of cell death with fundamentally different significance. Pathol A 17:229–259

Searle J, Lawson TA, Abbott PJ, Harmon B, Kerr JFR (1975) An electron-microscope study of the mode of cell death induced by cancer-chemotherapeutic agents in populations of normal and neoplastic cells. J Path 116:129–138

Seko Y, Shinkai Y, Kawasaki A, Yagita H, Okumura K, Tahaku F, Yasaki Y (1991) Expression of perforin in infiltrating cells in murine hearts with acute myocarditis caused by coxsackievirus B3. Circulation 84:788–795

Sheridan JW, Bishop CJ, Simmons RJ (1984) Effects of hypoxia in the kinetic and morphological characteristics of human melanoma cells grown as colonies in semi-solid agar medium. Br J Exp Path 65:171–180

Sherwood SW, Schimke RT (1994) Induction of apoptosis by cell-cycle phase specific drugs. In: Michich E, Schimke RT (eds) Apoptosis, Plenum, New York, pp 223–233

Simon HU, Blaser K (1995) Inhibition of programmed eosinophil death: A key pathogenic event for eosinophilia. Immunol Today 16:53–55

Simon HU, Yousefi S, Dibbert B, Hebestreit H, Weber M, Branch DR, Blaser K, Levi-Schaffer F, Anderson GP (1998) Role of tyrosine phosphorylation and L-tryosine kinase in Fas receptor-mediated apoptosis in eosinophils. Blood 92:547–551

Skalka M, Matyasova J, Cejkova M (1981) Post-radiation damage to thymus chromatin. Result of enzymic degradation of nucleosomes. Studies Biophys 85:71–72

Socini Y, Paakko P, Lehto V-P (1998) Histopathological evaluation of apoptosis in cancer. Am J Pathol 153:1041–1052

Spiegels S, Foster D, Kolesnik R (1996) Signal transduction through lipid second messengers. Curr Opin Cell Biol 8:159–167

Stiens R, Helpap B (1981) Regressive change in the rat prostate after castration. A study using histology, morphometrics and autoradiography with special reference to apoptosis. Pathol Res Pract 172:73–87

Strasser A, Whittingham, S, Vaux DL, Bath ML, Adams JM, Cory S, Harris AW (1991) Enforced BCL^2 expression in B-lymphoid cells prolongs antibody responses and elicits autoimmune disease. Proc Natl Acad Sci USA 88:8661–8665

Suda T, Takahashi T, Golstein P, Nagata S (1993) Molecular cloning and expression of the Fas ligand, a novel member of the tumor necrosis factor family. Cell 75:1169–1178

Swartsendhuber DC, Longdon CC (1963) Electron microscope observations on tangible body macrophages in mouse spleen. J Cell Biol 19:641–646

Takashima A, Naguchi K, Sato K, Hoshino T, Imahori K (1993) Proc Natl Acad Sci USA 90:7789–7793

Takasu T, Lyons JC, Park HJ, Song CW (1998) Apoptosis and perturbation of cell cycle progression in an acidic environment after hyperthermia. Cancer Res 58:2504–2508

Tanaka M, Suda Haze K, Nakamura N, Sato K, Kimura F, Motoyoshi K, Mizaki M, Tagawa S, Phga S, Hatake K, Drummond AH, Nagata S (1996) Fas ligand in human serum. Nature Med 2:317–322

Tanaka M, Suda T, Takahashi T, Nagata S (1995) Expression of the functional soluble form of human Fas ligand in activated lymphocytes. EMBOJ 14:1129–1135

Thompson CB (1995) Apoptosis in the pathogenesis and treatment of disease. Science 267:1456–1562

Thornberry NA, Lazebnik Y (1998) Caspases: enemies within. Science 281:1312–1316

Toyozaki T, Horol M, Tanaka M, Nagata S, Ohwada H, Marumo F (1998) Levels of soluble Fas ligand in myocarditis. Am J Cardiol 82:246–248

Troy CM, Shelanski ML (1994) Down-regulation of SOD Proc Natl Acad Sci USA 91:6384–6387

Trump BF, Beresesky IK, Cowley RA (1982) The cellular and subcellular characteristics of acute and chronic injury with emphasis on the role of calcium. In: Cowley RA, Trump BF (eds) Pathophysiology of shock anoxia and ischemia. Williams and Wilkins, Baltimore. pp 6–46

Trump BF, Beresesky IK, Osornio-Vargas AR (1981) Cell death and the disease process, the role of calcium. In Bowen, Lockshin (eds) Cell death in biology and pathology. Chapman & Hall, London, pp 209–242

Trump BF, Beresesky IK, Sato T, Laiko KV, Phelps PC, DeClaris N (1982) Cell calcium, cell injury and cell death. Environ Health Perspect 57:281–287

Tschopp J, Irmler M, Thome M (1998) Inhibition of Fas death signals by FLIPs. Curr Op Immunol 10:552–558

Uren AG, Vaux DL (1996) Molecular and clinical aspects of apoptosis. Pharmacol Ther 72:37–50

Vaux DL, Haecker G (1995) Hypothesis: that apoptosis caused by cytotoxins represents a defensive response that evolved to combat intracellular pathogens. Clin Exp Pharmacol Physiol 22:861–863

Verhage GH, Murray MK, Boomsma RA, Rehfeldt PA, Jaffe RC (1984) The postovulatory cat oviduct and uterus: correlation of morphological features and progesterone levels. Anat Rec 208:521–531

Verheij M, Bose R, Lin XH et al. (1996) Requirement for ceramid-initiated SAPL/JNK signaling in stress-induced apoptosis. Nature 380:75–79

Von Reyher M, Strater J, Kittstein W, Gschwendt M, Krammer PH, Moller P (1998) Colon carcinoma cells use different mechanisms to escape CD95-mediated apoptosis. Cancer Res 58:526–534

Voris BP, Young DA (1981) Glucocorticoid-induced proteins in rat thymus. J Biol Chem 256:11319–11329

Walker NI (1987) Ultrastructure of the rat pancreas after experimental duct ligation I the role of apoptosis and intraepithelial macrophages in acinar cell deletion. Am J Path 126:439–451

Walker NI, Harmon BV, Gobé GC, Kerr JFR (1988) Patterns of cell death. Meth Achiev Exp Pathol 13:18–54

Watanabe FR, Brennan CI, Copeland NG, Jenkins NA, Nagata S (1992) Lymphoproliferation disorder in mice explained by defects in Fas antigen that mediates apoptosis. Nature 356:314–317

Weedon D, Strutton G (1981) Apoptosis as the mechanism of the involution of follicles in catagen transformation. Acta Derm-Vener Stockh 61:335–339

Weinberg RA (1989) Oncogenes, antioncogenes and the molecular basis of multistep carcinogenesis. Cancer Res 49:3713–3721

Westendorp MD, Shatrov VA, Schulerosthoff K, Frank R, Kraft M, Los M, Krammer PH, Droge W, Lehmann V (1995) HIV-1 Tat potentiates TNF-induced NF-kappaB activation and cytotoxicity by altering the cellular redox state. EMBO J 14:546–554

Wyllie AH (1980) Glucocorticoid-induced thymocyte apoptosis is associated with endogenous endonuclease activation. Nature Lond 284:555–556

Wyllie AH (1981) Cell death: a new classification separating apoptosis from necrosis. In Bowen, Lockshin (eds) Cell death in biology and pathology. Chapman & Hall, London, pp 9–34

Wyllie AH, Kerr JFR, Currie AR (1973) Cell death in the normal neonatal rat adrenal cortex. J Path 111:255–261

Wyllie AH, Kerr JFR, Currie AR (1980) Cell death: the significance of apoptosis. Int Rev Cytol 68:251–306

Wyllie AH, Morris RG (1982) Hormone-induced cell death. Purification and properties of thymocytes undergoing apoptosis after glucocorticoid treatment. Am J Path 109:78–87

Wyllie AH, Morris RG, Smith AL, Dunlop D (1984) Chromatin cleavage in apoptosis: association with condensed chromatin morphology and dependence on macromolecular synthesis. J Path 142:67–77

Wyllie, AH, Kerr JFR, Macaskill IAM, Currie AR (1973) Adrenocortical cell deletion: the role of ACTH. J Path 111:85–94

Yang E, Korsmeyer SJ (1996) A discourse on the Bcl-2 family and cell death. Blood 88:386–401

Yang X, Khosavi-Far R, Chang HY, Baltimore D (1997) Daxx, a novel Fas-binding protein that activates JNK and apoptosis. Cell 89:1067–1076

Yeyama H, Kiyohara T, Sawada N, Isozaki K, Kitamura S, Kondo S, Niyagawa J, Kanayama S, Shinomura Y, Ishikawa H, Ohtani T, Nesu R, Nagada S, Matsusawa Y (1998) High Fas ligand expression on lymphocytes in lesions of ulcerative colitis. Gut 43:48–55

Young RW (1984) Cell death during differentiation of the retina in the mouse. J Comp Neurol 229:362–373

Yu R, Mondlekar S, Harvey KJ, Ucker DS, Kong A-NT (1998) Chemopreventive isothiocyanates induce apoptosis and caspase-3-like protease activity. Cancer Res 58: 402–408

Yuan J, Shahan S, Ledoux S et al. (1993) The C elegans cell death gene ced-3 encodes a protein similar to mammalian interleukin-1β-converting enzyme. Cell 75:641–652

Zanzami N, Susin SA, Marchetti P, Hirsch T, Gomez-Monterrey I, Castedo M, Kroemer G (1996) Mitochordrial control of nuclear apoptosis. J Exp Med 183:1522–1533

Zinkernagel RM, Doherty PC (1966) H-2 compatibility requirement for T-cell mediated lysis of target cells infected with choriomeningitis virus. J Exp Med 141:1427–1436

CHAPTER 2
Molecular Cellular and Tissue Reactions of Apoptosis and Their Modulation by Drugs

R. CAMERON and G. FEUER

A. Introduction

The process of cell death by apoptosis has been found to be regulated in a precise manner at the level of genes, and it is mediated by a complex series of molecules on the cell surface, in mitochondria, and in the cytosol.

The apoptotic process, is critical to the homeostasis of the immune system, particularly activation-induced cell death (AICD). There is a tremendous expansion of numbers of lymphocytes in response to antigens which is balanced by AICD to reduce the excess of lymphocytes and regain normalcy (CAMERON and ZHANG, Chap. 7, this volume). The pattern and extent of immune responses varies with different types of inflammatory or immune antigenic stimuli and so does the pattern and extent of apoptosis of immune cells which follows.

Apoptosis of immune cells is a constitutively (genetically) tightly regulated system that involves an interplay of different types of cells such as T and B lymphocytes, macrophages, eosinophils, and neutrophils. There is a network of specific receptors and ligands, costimulatory molecules, specific cytokines, inducers, and inhibitors that mediates the apoptotic process (SAVILL; DEFRANCE et al.; KOOPMAN, Chaps. 6, 16, and 17, respectively, this volume).

Apoptosis of cells in the nervous system such as neuronal cells is dependent on similar types of molecules as responsible for immune cell apoptosis. However, the sequence and timing of molecular events of apoptosis of nerve cells is quite different (WOODGATE and DRAGUNOW, Chap. 8, this volume). Apoptosis of various cell types also has an important role to play in various medical conditions including inflammation, hypersensitivity, neurodegenerative diseases, malignancy such as leukemia, autoimmune disorders, chronic viral diseases such as chronic hepatitis C viral infection, and during allograft rejection following organ transplantation. Recent results showing that various pharmacological agents could modulate, induce, or inhibit various aspects of the apoptotic process have received increasing attention.

B. Molecular Mediators of Apoptosis
I. TNF Receptor Family

The tumor necrosis factor receptor superfamily (TNFR) represents a family of proteins which share significant similarities in their extracellular ligand binding domains and in the intercellular effector or death domains. These receptors appear to transmit their signals by a protein to protein interaction resulting in either a death or a survival signal (BAKER and REDDY 1998). The CD95 molecule is a cell surface receptor of the TNFR superfamily that includes various molecules in immune regulation such as the TNF receptors I and II, CD27, CD30, and CD40 (ARCH et al. 1998; DUCKETT and THOMPSON 1997). The CD95 protein structure is characterized by three extracellular cystine rich domains found in all family members, a single transmembrane spanning region, highly homologous to the P55 TNFR. This intercellular death domain has been shown to transduce signals for apoptosis in the TNFR and the CD95 molecule (NAGATA 1997; PETER and KRAMMER 1998). The CD95 ligand is a type II transmembrane protein produced by the activated T cells and constituently expressed in a variety of tissues. The CD95 receptor is found on many activated immune cells whereas the CD95 is more restricted to CD8+ and CD4+ cytotoxic T cells, NK cells, and antigen presenting cells (SUDA et al. 1996).

Activation induced cell death of T lymphocytes has been shown to be mediated by CD95. This type of cell death can be neutralized by anti-CD95 antibodies (HARGREAVES et al. 1997). During the course of in vivo studies using transgenic mice which are FAS defective, namely the mlr/lpr mutant mice, their mature CD4+ T lymphocytes were resistant to activation induced cell death, i.e., they were dependent on FAS for apoptosis. The FAS gene in this model was shown to be essential for activation induced cell death in peripheral T lymphocytes (ROUVIER et al. 1993; SINGER and ABBAS 1994; NAGATA and GOLSTEIN 1995). Expression of the CD95 ligand simultaneously induced resistance to the apoptosis by means of CD95 ligand in naïve T cells. This resistance was induced in activated T cells but not in bystander cells. CD95 and TNFR1 mediated apoptosis occur in the presence of inhibitors of either RNA or protein synthesis and even enucleated cells undergo apoptosis upon CD95 activation, suggesting that all components necessary for apoptotic signal transduction are present de novo and that CD95 activation simply triggers this machinery (NAGATA and GOLSTEIN 1995). Apoptosis occurs in various cells and various tissues and CD95 is found abundantly in cells of the thymus, liver, heart, and kidneys. Mature T cells from lpr or gld mice do not die after activation and activated cells accumulate in the lymph nodes and spleens of these mice. When T cell hybridomas are activated in the presence of a CD95 neutralizing molecule they do not die. These results indicate that CD95 is involved in the activation induced cell death of T lymphocytes and it is part of a down-regulation of the immune system (SINGER and ABBAS 1994; NAGATA and GOLSTEIN 1995). CON-A activated mature mouse T lymphocytes shows a specific resistance to CD95 induced apoptosis during the S phase of their cell cycle (DAO et al. 1997).

A family of TNF associated proteins which bind to the TNF receptors and promote intracellular signal transduction from inside the cytoplasm have been designated as TRAF proteins. These TRAF proteins modulate the ability of receptors to trigger distinct signaling pathways that lead to phosphorylation and activation of protein kinases or to other transcription factors such as NF-κB (ARCH et al. 1998). One of these TNF type of receptors, CD30, serves as a binding site for TRAF proteins leading to the induction of NF-κB (DUCKETT and THOMPSON 1997). One of these TRAF proteins, TRAF2, appears to have a critical regulation effect on cell proliferation. When these proteins are expressed in abundance or activation is induced cell death there is depletion of TRAF2 (DUCKETT and THOMPSON 1997). One group of TNF related apoptosis inducing ligands or TRAIL has been found to induce apoptosis in tumor cells and some virally affected cells but has not as yet been found in normal cells (GOLSTEIN 1997; GRIFFITH and LYNCH 1998). The induction of NF-κB can lead to expression of genes that have an anti-apoptotic effect. Viruses that can induce NF-κB could therefore protect against apoptotic elimination of infected cells (BAEUERLE and BALTIMORE 1996; ASHKANAZI and DIXIT 1998; KASIBHATTA et al. 1999).

CD95 and the CD95 ligand have been shown to play an important role in three types of physiologic apoptosis: (a) peripheral deletion of activated mature T cells at the end of an immune response, (b) killing of targets such as virus infected cells or cancer cells by cytotoxic T cells and by NK cells, and (c) killing inflammatory cells at immune privileged sites such as the eye (ASHKANAZI and DIXIT 1998). Many viruses express anti-apoptotic proteins including caspase inhibitors, Bcl-2 homologues, and death effector domain containing proteins that are termed FLIPs (TSCHOPP et al. 1998). Cellular FLIPs structurally resemble caspase 8 except that they lack proteolytic activity.

Deficiencies in functional CD95 or its ligand manifest themselves in autoimmune syndromes. CD95-mediated apoptosis can be blocked by naturally occurring protein inhibitors which prevent apoptosis by serving as non-cleavable substrates for caspases (VARADHACHARY and SALGAME 1998).

The expression of CD95 ligands was studied in tissue sections (STRATER et al. 1999) using immunohistochemical staining and it was found that CD95 ligand was expressed in scattered lymphocytes, in lymphoid tissues of the thymus, lymph nodes, spleen, tonsil, and GI tract. A subset of plasma cells were prominent producers of CD95 ligand especially in the mucosa associated lymphoid tissue.

II. Bcl-2 Gene Family

Members of the Bcl-2 gene family encode proteins that function either to promote or inhibit apoptosis (TSUJIMOTO et al. 19985; VAUX et al. 1988; REED et al. 1990; REED 1998). Anti-apoptotic members such as Bcl-2 and Bcl-xl prevent programmed cell death with a wide variety of stimuli (CHENG et al. 1996; CHAO and KORSMEYER 1997). Conversely pro-apoptotic proteins exemplified by Bax and Bak can accelerate death and in some instances they are sufficient to cause apoptosis independent of additional signals. Bcl-2 related proteins are localized to the outer mitochondrial, outer nuclear, and endoplasmic reticular membranes (VAUX et al. 1988; REED et al. 1990; CHAO and KORSMEYER 1997). The ability of Bcl-2 to prevent apoptosis was clearly shown in experiments with knockout mice which show apoptosis of thymocytes and spleen cells (VEIS et al. 1993). Down regulation of the Bcl-2 gene product as in cytokine deprived activated T cells leads to apoptosis (AKBAR et al. 1996). Bcl-2 was shown to block cell-mediated cytotoxicity by allospecific cytotoxic lymphocytes when apoptosis was induced by degranulation as in the action of perforin and granzymes but not with apoptosis induced by cytotoxic lymphocytes by means of the CD95 pathway (CHIU et al. 1995). Bcl-2 has been documented to block apoptosis induced by chemotherapeutic drugs, ultraviolet radiation, free radicals, and some viruses such as the sindbis and baculoviruses (REED 1994, 1998). Bcl-2 inhibits CD95 induced apoptosis by preventing the event of cell death by inducing signaling complexes (KAWAHARA et al. 1998).

Experiments by KROEMER (1997) have shown that an important mechanism of the anti-apoptotic effect of Bcl-2 is the prevention of mitochondrial permeability transition which involves the opening of a larger channel in the inner mitochondrial membrane leading to free radical generation, release of calcium into the cytosol, and caspase activation either by direct or indirect control of the mitochondrial pore openings. By the prevention of this mitochondrial permeability transition, Bcl-2 also leads to free radical scavenging, ion efflux regulation, and caspase inhibition (KROEMER 1997; REED 1998). Other studies by KLUCK et al. (1997) and by YANG et al. (1997) had shown that overexpression of Bcl-2 prevented the efflux of cytochrome C from the mitochondria and also prevented the initiation of apoptosis. Both Bax and Bcl-2 were shown to insert into potassium chloride vesicles in a pH-dependent fashion and demonstrated microscopic ion efflux. Bcl-2 apoptotic regulators were shown by SCHLESINGER et al. 1997 to have the capacity to form ion channels in artificial lipid membranes. VANDER HEIDEN et al. (1997) also showed that Bcl-xl expressing cells adapted to growth factor withdrawal or staurosporine treatment by maintaining a decreased mitochondrial membrane potential. Bcl-xl expression also prevented mitochondrial swelling in response to agents that inhibited oxidative phosphorylation.

The antioxidant activity of Bcl-2 was documented in experiments by HOCKENBERY et al. (1993) and KANE et al. (1993) who showed that overexpression of Bcl-2 protected against H_2O_2 and menidione induced oxidative apoptosis. KANE et al. (1993) also found that overexpression of Bcl-2 in the GT1–7 neural cell line prevented necrosis resulting from glutathione depletion associated with the generation of reactive oxygen species. Neuronal cells prepared from mice deficient in the Bax gene were shown to be resistant to apoptosis induced by glutamate and kainate (XIANG et al. 1998). These results showed that Bax was required for neuronal cell death in response to some forms of cytotoxic injury. In contrast, the anti-apoptotic effect of Bcl-xl was shown when cytotoxic injury was induced in macrophage cell line (OKADA et al. 1998). In these studies, Bcl-xl but not Bcl-2 was highly inducible within 3h after stimulating macrophages with interferon gamma or LPS. Furthermore, Bcl-xl transfectants displayed substantial protection from toxic induced apoptosis by means of nitric oxide generation.

Structure-function analysis of Bcl-2 protein revealed conserved domains which were critical for homodimerization and heterodimerization between members of the Bcl-2 family of proteins (CHEN-LEAVY and CLEARY 1990; RADVANYI et al. 1990; HANADA et al. 1995; KELEKAR et al. 1997; SATTLER et al. 1997). For example, the structure and binding affinities of mutant Bak peptides indicate that the Bak peptide adopts an amphipathic alpha helix that interacts with Bcl-xl through hydrophobic and electrostatic interactions. Mutations in full length Bak that disrupt either type of interaction inhibit the ability of Bak to heterodimerase with Bcl-xl (SATTLER et al. 1997).

III. Caspases

Caspases (cystinoaspartic acid specific proteases) are a family of cysteine proteases that cleave their target proteins at aspartic acid residues in a defined cascade sequence (RAWLINGS and BARRETT 1994; ALNEMRI 1997). There are more than 12 caspases known to date which are expressed as precursors that are activated in a cascade-like cleavage parade (MEDEMA et al. 1997). This activation involves cleaving the molecule to 10 and 20 kilodalton subunits which then heterodimerase and became disassociated into tetramers that constitute the active enzyme (ENARI et al. 1995; NUNEZ et al. 1998). This activation was also shown by using specific inhibitors of caspases that block cell death (ENARI et al. 1995). Caspase-1 is the mammalian interleukin-1 β-converting enzyme which shows homology to the *C. elegans* cell death gene protein ced3, (YUAN et al. 1993; THORNBERRY et al. 1995; MARTIN et al. 1996; THORNBERRY and LAZEBNIK 1998). Caspase-3 or the apoptotic protease CPP32 is one of the caspases involved in cytotoxic T cell induced apoptosis which is mediated by granzyme B (DARMAN et al. 1995; ENARI et al. 1996; AMARANTE-MENDES et al. 1998; ZHENG et al. 1998). Caspase-8 or MACH is also involved in cytotoxic T lymphocyte induced apoptosis mediated by granzyme B (BOLDIN et al. 1996; MUZIO et al. 1996; MEDEMA et al. 1997).

Noncaspase target proteins which are inactivated by caspases include: (a) proteins of the DNA repair system (TAMURA et al. 1995), e.g., the poly ADP ribose polymerase which catalyzes the attachment of ADP ribose to nuclear proteins such as histones; (b) cytoskeletal or structural proteins such as nuclear lamins, phodren, cytokeratin 18, actin, and catinin B (VAUX et al. 1997; GROSS et al. 1999); (c) oncoproteins degraded by caspases including RB and MDM2 (VAUX et al. 1997); and (d) caspase activated DNA-ases such as DFF (LIU et al. 1997; SAMEJIMA et al. 1998) leading to chromosomal breakage (TAMURA et al. 1995).

Caspases have been shown to play a role in cytotoxic T cell induced apoptosis (DARMAN et al. 1995; TAMURA et al. 1995; MEDEMA et al. 1997), B lymphocyte induced cell death mediated by the B cell receptor (DEFRANCE et al., Chap. 16, this volume), and nerve cell death (WOODGATE and DRAGUNOW, Chap. 8, this volume). In some studies, the activation of caspases such as caspase-2 was found to be an early event in the apoptotic process (HARVEY et al. 1997).

IV. Cytokines

Cytokines such as IL-2 can increase or up-regulate Bcl-2 expression and prevent apoptosis in activated T cells. Using human IL-2 deprived activated T cells, it was possible to show that other cytokines such as IL-4, IL-7, and IL-15 could also prevent apoptosis of activated T cells in the absence of IL-2 (AKBAR et al. 1996). In contrast, sensitivity to the priming step for activation induced cell death was dependent on the cytokine interleukin-2 but not on cytokines IL-4, IL-7, or IL-15 (WANG et al. 1996). Furthermore, it was shown using transgenic mice which have a deficiency in the ability to use IL-2 that their T cells were resistant to CD95-mediated activation induced cell death, and that this defect could only be corrected by similar cytokines like IL-15 (VAN PARIJS et al. 1997). The kinetics of IL-2 production are as follows: messenger RNA is detectable within 3–5h and cytokine protein is also seen at this early time, cytokine mRNA is rapidly down-regulated shortly after it reaches a peak level at 6–12h, and the amount of cytokine produced is at least tenfold that seen in naïve cells with the same receptor (SWAIN et al. 1996). TCR stimulation of T lymphocytes that are activated in cycline in the presence of IL-2 leads to programmed cell death. This effect was shown to be mostly due to the ability of IL-2 to increase expression of mRNAs which encode ligands and receptors that mediate apoptosis (ZHENG et al. 1998). The pattern of cytokine production was shown to depend on the nature and dose of stimulation when T cell receptor complexes were used to elicit a diffuse array of effector activities (ITOH and GERMAIN 1997). For example, low concentrations of TCR ligand elicited only interferon gamma production. Increasing ligand recruits more cells into the interferon gamma pool and increases interferon gamma production per cell as well as inducing IL-2.

V. Co-Stimulatory Molecules

CD28 to B7 ligation provides co-stimulatory signals important for the development of T cell responses and CD28 is a principal co-stimulatory receptor for T cell activation. CD28 co-stimulation markedly enhances the production of lymphokines, especially of IL-2. In addition, CD28 sustains the late proliferative response of naïve T cell populations and enhances their long-term survival (SPERLING et al. 1996; TAI et al. 1997). CD28 deficient T cells were shown to be enhanced in their long term survival by cultures with IL-4 (STACK et al. 1998). In circulating T cells which express B7, a novel cell surface membrane protein was found; this is independent of co-stimulation by using anti CD28 antibodies (SOARES et al. 1997). Further studies showed that in fact cells expressing high levels of CD28 were entirely resistant to apoptosis by the CD95 pathway (McLeod et al. 1998). C28 co-stimulation was also shown to promote T cell survival by enhancing the expression of Bcl-x_l (Boise et al. 1995a,b; Radvanyi et al. 1996).

VI. Perforin and Granzyme B

Cytotoxic T cell induced apoptosis has been shown to be mediated by the molecular granzyme B (Darman et al. 1995; Enari et al. 1996; Bolden et al. 1996; Muzio et al. 1996; Medema et al. 1997; Amarante-Mendes et al. 1998; Zheng et al. 1998).

Lytic granules in cytotoxic T cells carry proteins such as granzyme B and also perforin. Perforin molecules are a family of proteins which induce pores in the membranes of cells and are often connected with molecules such as granzyme B which enter the target cell and induce apoptosis. A role for perforin was found by Spaner et al. (1998) in the activation induced cell death of T cells. The role of perforin in the control of T cell cytotoxicity was first clarified by studies in perforin deficient knockout mice (Kagi et al. 1994). Apoptotic cell death of allografted tumor cells by activated macrophages was shown to be independent of perforin by Yoshida et al. (1997).

VII. Protein Kinase C

Activation of protein kinase C blocks apoptosis and promotes cell survival of mature lymphocytes prone to apoptosis (Lucas et al. 1994; Lucas, Chap. 4, this volume). In addition, direct induction of cell apoptosis by ethanol is augmented by inhibiting protein kinase C which establishes a link between protein kinase C activity and ethanol toxicity and ethanol induced apoptosis (Aroor 1997; Lucas, Chap. 4, this volume). Inhibitors of protein kinase C such as storosporine have been shown to enhance the cytotoxic effects of various antitumor agents (Loch 1997; Lucas, Chap. 4, this volume). Cycloheximide causes apoptosis in sublethal doses in the liver by means of induction of oncogenes, and the accumulation of sphingosine and cycloheximide is also an endogenous modulator of protein kinase C activity (Alisenko 1997).

VIII. Reactive Oxygen Species

The generation of highly reactive oxygen species or ROS has been shown to induce apoptosis at different cell types (Delneste, Chap. 10, this volume; Bauer et al., Chap. 11, this volume). Hydrogen peroxide produced by monocytes-macrophages and neutrophils can trigger the death of bacterial cells as well as bystander cells. The antineoplastic drug deoxyrubicine reacts by generating reactive oxygen species. Bcl-2, the potent anti-apoptotic molecule, was shown to possess anti-oxidant activity by Hockenbery et al. (1993) and by Kane et al. (1993). The anti-apoptotic cell death gene CED-9 of the nematode worm has also been shown to have ROS regulatory activities (HENGARTNER and HORVITZ 1994). The apoptotic process was found to be regulated by the redox balance in a number of different cell types (DELNESTE, Chap. 10, this volume).

IX. Glutathione

Thiol antioxidants have been shown to protect cells against apoptosis. DELNESTE et al. (1996) showed that N-acetylcysteine was able to protect human T cells from CD95-mediated apoptosis. CHIBA et al. (1996) showed further that T cell sensitivity to CD95-mediated apoptosis was associated with low intracellular glutathione levels and that the cytoprotective effect of N-acetylcysteine was related to its ability to increase the intercellular glutathione levels. Antioxidants have also been found to modulate the generation of second messengers and the activation of transcription factors which are involved in the signaling pathways of apoptosis (DELNESTE, Chap. 10, this volume). As a consequence, anti-oxidants, especially the thiol anti-oxidants which have low pharmacological toxicity, have been proposed as treatments for patients shown to have diseases with altered redox status such as AIDS, cancer, or Alzheimer's disease (DELNESTE, Chap. 10, this volume).

X. Inhibitor Polypeptides

The inhibitor of apoptosis protein family (IAP) are widely expressed gene family of apoptotic inhibitors which appear to act to suppress apoptosis through direct caspase inhibition, primarily via caspase 3 and 7 and by modulation of the transcription NF-κB (LACASSE et al. 1998). One particular IAP type of protein named survivin was shown to inhibit caspase directly (DUCKETT et al. 1998; LACASSE et al. 1998). The inhibitory effects of the IAP family of proteins on apoptosis appear to involve a wide spectrum of cell types and triggering mechanisms of apoptosis compared to the Bcl-2 family of inhibitors of apoptosis, suggesting that the site of activity of IAP proteins is further downstream in the process than that of the Bcl-2 family (HARVEY et al. 1997; DUCKETT et al. 1997; LACASSE et al. 1998).

C. Cell-Specific Pathways of Apoptosis

I. Immune System

1. T Cells

Apoptosis of T lymphocytes has an essential role in developmental, physiologic, and pathological processes involving T cells including the deletion of T cell clones, the expression of self antigens in the thymus, elimination of T cells which are infected with viruses, and the homeostasis of T cell populations that have expanded following high dose antigen exposures. T cell apoptosis is a very precise and tightly regulated process which is coordinated by specific receptors and ligands such as CD95 and mediated by families of proteins such as caspases, interleukins, and various costimulatory molecules such as CD28 and B7 (CAMERON and ZHANG, Chap. 7, this volume). The precision and regulation of molecules involved in apoptosis of T cells is best exemplified by the patterns of cytokine production (SWAIN et al. 1996). Two distinct T helper cell clonal populations can be identified, each with a unique cytokine pattern. TH1 cells produce interleukin-2, interferon gamma, and GN-CFS whereas TH2 cells produce BSF1, a mast cell growth factor, and special T cell growth factor in addition to IL4, 5, and 6. In response to antigen stimulation, for example, there is a tenfold increase in IL2 production in TH1 cells within one to two days after exposure to antigen which is then rapidly down-regulated within hours (CAMERON and ZHANG, Chap. 7, this volume).

Recent studies of cell death of thymocytes in vivo have shown that the molecular pattern of regulation of cell death in thymocytes is quite different from peripheral T cells (ITOH et al., Chap. 15, this volume). In their studies, using TUNEL electron microscopy and TUNEL flow cytometry, it was evident that most thymocytes died by pyknosis either in situ or after exposure to injection of corticosteroids in vivo, and only showed DNA fragmentation following their phagocytosis by macrophages.

2. B Cells (and Plasma Cells)

The apoptosis of B cells in vivo is also a precise and tightly regulated process which most often takes place in the germinal centers of lymphoid tissues such as spleen and lymph nodes. Activated T cells express the ligand for CD95 which is involved in the apoptosis of B cells. Ligation of the B cell receptor in B cells which are not actively cycling protects them from CD95-mediated apoptosis. B cell proliferation takes place in germinal centers or extrafollicular foci in response to antigen stimulation and is coordinated with T cell proliferation. Prolonged or repeated exposure of cycling B cells to antigen and the concomitant decline of T helper cells leads to stimulation of apoptosis of B cells leading to the eventual downsizing of the responding B cell population (DEFRANCE et al. Chap. 16; KOOPMAN, Chap. 17, both this volume). B cells of the germinal center are able to interact with antigen and the immunoglobulin receptor and the immune response in this location is coordinated by interactions between follicular dendritic cells, germinal center B cells, and T cells (KOOPMAN et al. 1994; KOOPMAN, Chap. 17, this volume). These studies also showed that initially the presentation of antigen and interaction with adhesion molecules helped to maintain B cell survival and B cell activation. The ligand of CD40 which is located on T cells was also found to contribute to the stabilization of these germinal center B cell populations and to allow for the maturation of germinal center B cells (KOOPMAN, Chap. 17, this volume).

Morphologic studies of lymphoid cells within lymphoid tissues revealed that mature B cells were distinguished morphologically as plasma cells and showed strong positive immunostaining for the CD95 ligand responsible for apoptosis (STRATER 1996, 1999).

3. Macrophages (and Dendritic Cells)

In vivo, the normal fate of cells undergoing apoptosis is uptake and degradation of the intact dying cell by phagocytic cells such as macrophages, which serves to contain the toxic products of the dying cell and limit the extent of injury to surrounding tissues (SAVILL 1997; SAVILL, Chap. 6, this volume; BROWN and SAVILL 1999). A number of specific receptors have been identified on phagocytic cells involved in the uptake of apoptotic cells including phagocyte lectins, CD36, and the murine macrophage ABC-1 molecule (SAVILL 1997; SAVILL, Chap. 6, this volume). At sites where the numbers of apoptotic cells is abundant, such as in thymus, lymph node, bone marrow, liver, spleen, and inflammatory sites, it is common to find macrophages containing large numbers of apoptotic cells as part of their function in the degradation of large numbers of dying cells each day (SAVILL 1997). In addition, most tissues contain groups of resident macrophages such as in the kidney where glomerular mesangial cells can ingest apoptotic neutrophils as part of the resolution of glomerular inflammation (SAVILL et al. 1992; SAVILL, Chap. 6, this volume). In experiments studying the result of phagocytosis by monocyte macrophages, BROWN and SAVILL (1999) showed that, following the ingestion of opsonized zymosan, monocyte macrophages released CD95 ligand which triggered the CD95-mediated apoptosis of target neutrophils.

Dendritic cells of the myeloid lineage also ingest apoptotic cells and process them for presentation to MHC Class 1 and Class 2 restricted T cells (ROUVIER et al. 1998). In the presence of defects in the clearance of apoptotic cells, such as in animals with a genetic deficiency of C1Q molecules, the persistence of apoptotic cells without phagocytosis may be sufficient to stimulate an autoimmune response such as a systemic lupus erythematosus type of disease (ROUVIER et al. 1998). Donor tissue derived dendritic cells have been identified in recipients of kidney and liver tissue and may be responsible for a low level donor chimerism and an immunosuppressive effect post-transplantation (HART 1997).

4. Eosinophils

Apoptosis of eosinophils is an important process which decreases the numbers of eosinophils that have accumulated in tissues following inflammation, particularly of the "allergic" type. Eosinophil apoptosis is mediated by CD95 and CD95 ligand with the induction of sphingomyelinase and tyrosine kinase pathways, and involves the cascade of caspases. Glucocorticoids have a profound effect in the stimulation of apoptosis of eosinophils. In situations where the phagocytosis of apoptotic eosinophils is impaired or delayed, or when eosinophils develop a resistance to CD95 stimulated apoptosis, then an unlimited expansion of eosinophils can occur such as in nasal polyp tissue during allergic inflammation in chronic eosinophilic disorders (SIMON, Chap. 14, this volume).

5. Neutrophils

Granulocyte neutrophils are short-lived cells with half lives of less than 24 h. In the absence of appropriate stimuli, neutrophils undergo characteristic changes indicative of programmed cell death or apoptosis, including cell shrinkage, nuclear chromatin condensation, and DNA fragmentation into nucleosome length fragments. As a first line of defense, neutrophils are rapidly recruited to inflammatory sites, where the expression of their apoptotic program can be modified by a number of agents such as interleukin-2 and LPS which have been shown to inhibit neutrophil apoptosis and prolong their functional lifespan (GAMBORELLI et al. 1998). Apoptosis of neutrophils is followed by recognition and uptake and ingestion by macrophages and this is associated with a loss of neutrophil functions such as chemotaxis, phagocytosis, degranulation, and respiratory burst (GAMBORELLI et al. 1998; SAVILL, Chap. 6, this volume). From these studies it was apparent that the apoptosis of neutrophils was critical in the resolution of inflammation and the limiting of tissue injury by dying neutrophils. Molecules involved in apoptosis of neutrophils include caspases and also calpains and proteosomes (KNEPPER-NICOLAI et al. 1998). When apoptosis was accelerated by treatments with protein synthesis inhibitors in the studies of WHYTE et al. (1997) this was shown to promote an increased recognition and faster clearance by macrophages of the apoptotic neutrophils which had accumulated in human peripheral blood.

II. Nervous System

1. Neuronal Cells

Apoptosis of neuronal cells is an important mechanism of cell death in the nervous system during brain development and also in neurodegenerative diseases, which is mediated by the C-jun/jnk pathway and involves activation of caspases (WOODGATE and DRAGUNOW, Chap. 8, this volume). Studies by WHYTE et al. (1998) showed that the pro-apoptotic molecule Bax was required for the cell death of sympathetic and motor neurons in the setting of trophic factor deprivation. Neurons saved from apoptosis in Bax null mutant mice survived but did not develop normal functional capabilities and in fact the resultant supernumerary neurons and axons were atrophic.

III. Liver Cells

1. Hepatocytes

Studies of the apoptotic process in hepatocytes have been summarized by PESSAYRE et al. (Chap. 3, this volume). TNF alpha has been shown to have a strong pro-apoptotic action for hepatocytes in vitro and antibodies against

TNF alpha protect hepatocytes from apoptosis whereas glutathione depletion enhances apoptosis of hepatocytes.

Hepatocytes from patients chronically infected by hepatitis B virus (HBV) produce TNF alpha and this production seems to depend on HBX protein. In the human, however, in vivo there does not appear to be a strong association between expression of TNF alpha and apoptosis of hepatocytes in viral hepatitis. CD95 and CD95 ligand seems to be important in apoptosis of human hepatocytes in viral hepatitis.

2. Kupffer Cells

Kupffer cells represent the resident macrophages of the liver and are the critical cells for the phagocytosis of peripheral blood lymphocytes undergoing apoptosis such as after heat shock or cycloheximide treatments (FALASKA et al. 1996). Kupffer cells have specific lectin-like receptors involved in the recognition of apoptotic lymphocytes and the in vitro process of phagocytosis of apoptotic lymphocytes is a very rapid one, completed in only a few minutes of incubation (DINI, Chap. 12, this volume).

IV. Malignant Cells

1. Leukemia (and Lymphoma)

Hematopoietic cells require certain cytokines, including colony stimulating factors and interleukins, to maintain their viability and without these cytokines the program of apoptotic cell death is activated (LOTEM and SACHS 1996). Cells from many myeloid leukemias also require cytokines for viability and apoptosis is also activated in these leukemic cells after cytokine withdrawal, resulting in reduced leukemogenicity. This susceptibility of leukemic cells to the induction of apoptosis is regulated by the balance between apoptosis inducing genes such as the tumor suppresser wild type p53 and Bax and the apoptosis suppresser genes such as the oncogene mutant p53 and Bcl-2. Modulation of expression of apoptosis regulating genes could also be useful for the antileukemia therapy (LOTEM and SACHS 1996).

PAULLI et al. (1998) report evidence that apoptosis of normal and neoplastic lymphoid cells is regulated by a network of cytokines and that expression of CD95 is at high levels in all cutaneous CD30+ lymphomas which are significant due to their high rate of regression. Expression of Bcl-2 in lymphoproliferative conditions as shown by immunohistochemistry was common in nonregressing lesions, suggesting a protective effect on the lymphoid tumor cells from apoptosis.

2. Carcinoma

Studies of the extent of apoptosis in various types of carcinomas by means of quantitation of apoptosis associated proteins directly in tissue sections by

SOINI et al. (1998) revealed that the apoptotic index in most carcinomas was between 1 and 5% and the Bcl-2 expression as high as 50% in many carcinomas. A study of the expression of Bcl-2, Bax, and caspases in pancreatic carcinomas by VIRKAJARVI et al. (1998) revealed a strong correlation between the apoptotic index and the expression of caspases 3, 6, and 8 on immunostains. In vitro studies of the effect of Bcl-xl antisense oligonucleotides on a human gastric cancer cell line by KONDO et al. (1998) showed that overexpression of the Bak protein induced sensitization to apoptosis in gastric cancer cells, suggesting that the Bcl-2 gene family may be an important modulator of apoptosis for carcinoma cells. GORCZYCA et al. (1998) developed a technique which combined flow cytometry in conjunction with DNA strand break labeling assays and cell sorting for the study of solid tumors. They found that spontaneous apoptotic cells in solid tumors did not always show the typical features of apoptosis seen in treated cultured cells.

D. Tissue-Specific Reactions Involving Apoptosis
I. Inflammation and Hypersensitivity

Inflammatory and hypersensitivity reactions in tissues throughout the body are mediated by inflammatory and immune cells such as T and B lymphocytes and plasma cells, monocyte macrophages, eosinophils, and neutrophils. Apoptosis of large numbers of these inflammatory or immune cells and phagocytosis of apoptotic lymphocytes, eosinophils, and neutrophils are critical to the resolution of inflammatory and immune processes, and the restitution of tissue to normal conditions. Delays or deficiencies in this process can result in more severe and prolonged tissue injuries, fibrosis, and even more serious consequences (SAVILL 1997; SAVILL, Chap. 6, this volume).

Studies by SAVILL et al. (1992) of the glomerular cells of rats with experimental glomerulonephritis showed that apoptotic neutrophils were phagocytosed by inflammatory macrophages. In addition, they found that glomerular mesangial cells also had the ability to phagocytose apoptotic neutrophils. They also found that human mesangial cells in vitro ingested apparently intact human neutrophils which had been aged for 24h in culture but freshly isolated neutrophils were not ingested by the human mesangial cells in vitro. This phagocytic effect was inhibited by colchicine pretreatment confirming the active nature of the phagocytosis. This process would serve to limit the neutrophil mediated glomerular injury and potentially would play a role in determining whether there is resolution of glomerular inflammation. Failure of these mechanisms could lead to necrosis of neutrophils with direct release of toxic neutrophil contents such as lysosomes, further recruitment of leukocytes by chemotactic factors, and the development of fibrosis in tissue continually damaged prior to repair and restitution towards normal conditions (SAVILL, Chap. 6, this volume).

The importance of CD95 ligand induced apoptosis for the development of hepatitis was shown using various mouse models of hepatitis by KONDO et al. (1997). They report that cytotoxic T lymphocytes reactive against hepatitis B surface antigen caused an acute liver disease in hepatitis surface antigen positive transgenic mice. In the second model, mice were primed with *P. acnes* and challenged with lps which led to extensive apoptosis of hepatocytes which was prevented by the neutralization of CD95 ligand. CD95 null mice were resistant to lps induced mortality. MIWA et al. (1998) showed that CD95 ligand could induce the release of caspase 1 from peritoneal exudate cells and provoke the marked infiltration of neutrophils interperitoneally in Balb/c mice.

GOUGEON (Chap. 5, this volume) was able to show that during HIV infection CD4T cell death was mediated, not only directly by HIV replication as a consequence of viral gene expression but also indirectly through priming of uninfected cells to apoptosis when triggered by different agents. This phenomenon was shown by a number of related studies. A variety of blood cells taken from a large cohort of HIV positive patients was shown to be primed for apoptosis and showed increased fragility upon short-term culture and these cells included not only T cells but also monocytes, B cells, natural killer cells, and granulocytes. When lymph nodes of HIV infected patients were examined with immunostains there was evidence of apoptosis, not only in CD4T cells but also CD8T cells, B cells, and dendritic cells. The proportion of CD95 positive T lymphocytes in HIV infected patients increases as the disease progresses towards AIDS. In fact, serum concentrations of soluble CD95 and anti-CD95 autoantibodies were found to be predictive markers for the progression to AIDS. CD4T cells that express the HIV virus glycoprotein gp120 on their surface were shown to bind uninfected T cells leading to their apoptosis (GOUGEON and MONTAIGNER 1993; GOUGEON, Chap. 5, this volume). In addition, they showed that a rapid cell death apparently independent of known caspases and lacking DNA fragmentation was triggered by HIV gp120 interaction with the SDF-1 receptor, with the resultant cell death of normal CD4T cells. In fact, it was found that the majority of immune cells which die as a result of HIV infection undergo apoptosis by an indirect mechanism.

A series of studies have shown that viruses may act to promote their own intracellular persistence by enhancing the survival of virally infected cells by means of the anti-apoptotic properties of various viral gene products (LEVINE et al. 1993; YOUNG et al. 1997). For example, EB virus, adenovirus, and herpes virus produce gene products which are homologous to the anti-apoptotic proteins of Bcl-2. Baculovirus and cowpox virus produce proteins which inhibit caspases, and adenovirus and SV40 virus produce proteins which inactivate p53 leading to inhibition of the cell death of virally infected cells. These anti-apoptotic mechanisms would facilitate the maintenance of persistent virally infected cells and promote the continued production of progeny virus (YOUNG et al. 1997).

II. Cancer (and Carcinogenesis)

Since the initial discovery of the Bcl-2 gene in follicular lymphoma (SUJIMOTO et al. 1984), and the demonstration by VAUX et al. (1988) that Bcl-2 had anti-apoptotic properties, a large number of studies have demonstrated that human tumors often express Bcl-2 or alterations of the p53 gene. p53 has been shown to up-regulate the expression of the Bax protein which in turn forms heterodimers with Bcl-2 and blocks the action of Bcl-2, thereby inducing apoptosis (CHAO and KORSMEYER 1998). Alterations of the p53 gene are the most frequent genetic change in human cancer. In fact, it is estimated that about 50% of all human malignancies contain inactivating mutations of the p53 gene (BELLAMY 1997; THIEDE et al., Chap. 9, this volume). The ability of activation of p53 and the apoptosis of tumor cells has led to various strategies of cancer treatment such as gene therapy in which mutant p53s are targeted or wild type p53 genes are introduced (AMUNDSON et al. 1998; THIEDE et al., Chap. 9, this volume).

The balance between neoplastic and pre-neoplastic cell proliferation and apoptosis has been shown to be critical in the progression of cells during chemical carcinogenesis in rat liver by GRASL-KRAUPP et al. (1997). By modifying the growth rate of rat hepatocytes of neoplastic and pre-neoplastic lesions with the drug nafenopin, they were able to show that, with cessation of the drug, cell proliferation decreased and cell elimination by apoptosis increased. Similar studies reported by LYONS and CLARKE (1997) indicated that whenever cells showed an impaired apoptosis, there was a strong selection for further lesions in genes controlling cell proliferation and, conversely, in cell populations with increased proliferative capacity, there was strong selection for lesions conferring impaired apoptosis. These authors suggested that malignant conversion of a cell was associated with a synergy of mutations affecting both processes of cell proliferation and of apoptosis.

III. Neurodegenerative Disorders

Numerous studies have shown that toxins implicated in neurodegenerative diseases can trigger apoptotic death of neuronal cells in culture (WOODGATE and DRAGUNOW, Chap. 8, this volume). For example, beta amyloid, the major component of senile plaques in Alzheimer's disease, induces apoptosis in primary hippocampal and cortical cultures. Similar apoptotic neuronal cell death was induced by glutamate, hydrogen peroxide, and heavy metals. Inducible transcription factors such as fos and jun appear to be closely associated with neurodegenerative diseases such as Alzheimer's disease and also closely associated with apoptosis of neuronal cells. In addition, overexpression of a c-jun dominant negative mutant attenuates apoptosis triggered by nerve growth factor withdrawal in sympathetic neurons.

IV. Autoimmune Disorders

Autoimmunity in mice has been shown to be related to single gene defects. For example, lpr and gld mice have spontaneous mutations which are autosomal recessive on mouse chromosomes 19 and 1 respectively and lead to an autoimmune disease (NAGATA and GOLSTEIN 1995). These mice show mutations in CD95 and CD95 ligand on the surface of their CD4 and CD8 T cells with the resultant failure of activation induced cell death and subsequent lymphoproliferative disorder (NAGATA and GOLSTEIN 1995).

E. Potential of Modulation of Molecular, Cellular or Tissue Reactions by Drugs

I. Immunosuppressive Drugs

Immunosuppressive drugs can act to modulate T cell apoptosis and induce transplantation tolerance by a number of mechanisms: (a) inhibition of CD95 and CD95 ligand which is shown by 9-cisretinoic acid or glucocorticoids (YANG et al. 1995); (b) direct toxicity to specific cytotoxic T cells by the immunotoxin FN18-CRM9 (NEVILLE et al. 1996; FECHNER et al. 1997); (c) inhibition of IL-2 expression by cyclosporine (SIGAL and DUMONT 1992; ZHANG et al. 1998; WALDMANN and O'SHEA 1998); (d) shift of cytokine pattern from Th1 to Th2 expression by rapomycin, CTLA-4 immunoglobulin, anti-CD4 antibody, and cyclosporine (KABELITZ et al. 1998); and (e) induction of activation induced cell death in activated T cells by anti-CD3 antibody OKT3 or FK506 (SIGAL and DUMONT 1992; KABELITZ 1998).

II. Chemotherapeutic Drugs

Chemotherapeutic drugs can cause apoptosis of many kinds of cancer cells in vitro including drugs such as cisplatin, mitomycin, methotrexate, doxorubicin, and bleomycin at concentrations present in the sera of patients during therapy by means of up-regulation of the CD95 receptor and CD95 ligand (MUELLER et al. 1998). BCNU, an anti-cancer alkylating agent, could prevent apoptosis of human lymphoma cells by inhibiting caspases in vitro (PETAK et al. 1998). Daunorubicin was shown to cause apoptosis of leukemic cell lines in association with stimulation of sphingomyelin hydrolysis and ceramide generation (JAFFREZOU et al. 1996).

III. Natural Substances

A wide variety of naturally occurring substances of both plant and animal origin can induce apoptosis (PESSAYRE et al., Chap. 3, this volume).

Acknowledgement. Thanks to Marie Maguire for all of her excellent work in the preparation and formatting of this manuscript.

References

Akbar AN, Borthwick NJ, Wickremasinghe RG, Krajewski S, Reed JC, Salmon M (1996) Interleukin-2 receptor common γ-chain signaling cytokines regulate activated T cell apoptosis in response to growth factor withdrawal. Eur J Immunol 26:294–299

Alnemri ES (1997) Mammalian cell death proteases – a family of highly conserved aspartate specific cysteine proteases. J Cell Biochem 64:33–42

Amarante-Mendes GP, Kim CN, Liu L, Huang Y, Perkins CL, Green DR, Bhalla K (1998) Bc-Abl exerts its antiapoptotic effect against diverse apoptotic stimuli through blockage of mitochondrial release of cytochrome C and activation of caspase-3. Blood 91:1700–1705

Amundson SA, Myers TG, Fornace AJ Jr (1998) Roles for p53 in growth arrest and apoptosis: putting on the brakes after genotoxic stress. Oncogene 17:3287–3299

Arch RH, Gedrich RW, Thompson DB (1998) Tumor necrosis factor receptor-associated factors (TRAFs) – a family of adapter proteins that regulates life and death. Genes and Development 12:2821–2830

Ashkenazi A, Dixit VM (1998) Death receptors: signaling and modulation. Science 281:1305–1308

Baeuerle PA, Baltimore D (1996) NF-κB: ten years after. Cell 87:13–20

Baker SJ, Reddy EP (1998) Modulation of life and death by the TNF receptor superfamily. Oncogene 17:3261–3270

Baser S, Kolesnik R (1998) Stress signals for apoptosis: ceramide and c-Jun kinase. Oncogene 17:3277–3285

Bellamy CPC (1997) p53 and apoptosis. Brit Med Bull 53:522–538

Boldin MP, Goncharov TM, Goltsev YV, Wallach D (1996) Involvement of MACH, a novel MORT1/FADD-1 interacting protease, in Fas/APO-1 and TNF receptor-induced cell death. Cell 85:803–811

Boise, LH, Minn AJ, Noel PJ, June CH, Accavitti MA, Lindsten T, Thompson CB (1995) CD28 costimulation can promote T cell survival by enhancing the expression of Bcl-x_L. Immunity 3:87–98

Brown SB, Savill J (1999) Phagocytosis triggers macrophage rebase of Fas ligand and induces apoptosis of bystander leukocytes. J Immunol 162:480–485

Chao DT, Korsmeyer SJ (1997) Bcl-x_L-regulated apoptosis in T cell development. Int Immunol 9:1375–1384

Chen-Levy S, Cleary ML (1990) Membrane topology of the Bcl-2 proto-oncogenes protein demonstrated in vitro. J Biol Chem 265:4929–4933

Cheng EH-Y, Levine B, Boise LH, Thompson CB, Hardwick JM (1996) Bax-independent inhibition of apoptosis by Bcl-x_L. Nature 379:554–557

Chiba T, Takahashi S, Sato N, Ishii S, Kikuchi S (1996) Fas-mediated apoptosis is modulated by intracellular glutathione in human T cells. Eur J Immunol 26:1164–1169

Chiu VK, Walsh CM, Liu C-C, Reed JC, Clark WR (1995) Bcl-2 blocks degranulation but not Fas-based cell-mediated cytotoxicity. J Immunol 154:2023–2032

Cohen JJ, Duke RC (1984) Glucocorticoid activation of a calcium-dependent endonuclease in thymocyte nuclei leads to cell death. J Immunol 132:38–44

Dao T, Mehal WZ, Crispe IN (1998) IL-18 augments perforin-dependent cytotoxicity of liver NK-T cells. J Immunol 161:2217–2222

Darmon AJ, Nicholson DW, Bleackley RC (1995) Activation of the apoptotic protease CPP32 by cytotoxic T-cell-derived granzyme B. Nature 377:446–448

Deas O, Dumont C, MacFarlane M, Roubau M, Hebib C et al. (1998) Caspase-independent cell death induced by anti-CD2 or staurosporine in activated human peripheral T lymphocytes. J Immunol 161:3375–3383

Delneste Y, Jeannin P, Sebille E, Aubrey J-P, Bonnefoy J-Y (1996) Thiols prevent Fas (CD95)-mediated T cell apoptosis by down-regulating membrane Fas expression. Eur J Immunol 26:2981–2988

Duckett CS, Thompson CB (1997) CD30-dependent degradation of TRAF2: implications for negative regulation of TRAF signaling and the control of cell survival. Genes and Development 11:2810–2821

Enari M, Hug H, Nagata S (1995) Involvement of an ICE-like protease in Fas-mediated apoptosis. Nature 375:78–81

Enari M, Talamian RV, Wong WW, Nagata S (1996) Segmental activation of ICE-like and CPP-32-like proteases during Fas-mediated apoptosis. Nature 380:723–726

Falasca L, Bergamin A, Serafino A, Balabaud C, Dini L (1996) Human Kupffer cell recognition and phagocytosis of apoptotic peripheral blood lymphocytes. Exp Cell Res 224:152–162

Gamberale R, Giordano M, Trevani AS, Ardoregini G, Geffner JR (1998) Modulation of human neutrophil apoptosis by immune complexes. J Immunol 161:3666–3674

Golstein P (1997) Cell death: TRAIL and its receptors. Curr Biol 7:R750–R753

Gorczyca W, Bedner E, Burfeind P, Darzynkiewicz Z, Melamed MR (1998) Analysis of apoptosis in solid tumors by laser-scanning cytometry. Mod Pathol 11:1052–1058

Griffith TS, Lynch DH (1998) TRAIL: a molecule with multiple receptors and control mechanisms. Curr Op Immunol 10:559–563

Griffith TS, Chiu WA, Jackson GC, Lynch DH, Kubin MZ (1998) Intracellular regulation of TRAIL-induced apoptosis in human melanoma cells. J Immunol 161:2833–2840

Grasl-Kraupp B, Rattkay-Nedecky B, Mullauer L, Taper H, Huber W et al. (1997) Inherent increase of apoptosis in liver tumors: implications for carcinogenesis and tumor regression. Hepatology 25:906–912

Gross A, Yin X-M, Wong K, Wei MC, Jockel J, Millerman C et al. (1999) Caspase cleaved BID targets mitochondria and is required for cytochrome C release, while Bcl-x_L prevents this release but not tumor necrosis factor-R1/Fas death. J Biol Chem 274:1156–1163

Hanaida M, Aimé-Sempé C, Sato T, Reed JC (1995) Structure-function analysis of Bcl-2 protein. J Biol Chem 270:11962–11969

Hargreaves RG, Borthwick MJ, Montani MDS, Piccolella E, Carmichael P, Lechler RI et al. (1997) Dissociation of T cell anergy from apoptosis by blockade of Fas/APO-1 (CD95) signaling. J Immunol 158:3099–3107

Hart DNJ (1997) Dendritic cells: unique leukocyte populations which control the primary immune response. Blood 90:3245–3287

Harvey NL, Butt AJ, Kumar S (1997) Functional activation of Nedd 2/ICH-1 (caspase 2) is an early event in apoptosis. J Biol Chem 272:13134–13139

Hengartner MO, Horvitz HR (1994) C. elegans cell survival gene ced-9 encodes a functional homology of the mammalian proto-oncogene Bcl-2. Cell 76:665–676

Hockenberry D, Nunez G, Milliman RD, Schreiber RD, Korsmeyer SJ (1990) Bcl-2 is an inner mitochondrial membrane protein that blocks programmed cell death. Nature 348:334–336

Itoh Y, Germain RN (1997) Single cell analysis reveals regulated hierarchical T cell antigen receptor signaling thresholds. J Exp Med 186:757–766

Jacobson MD, Burne JF, King MP, Miyashita T, Reed JC, Raff MC (1993) Bcl-2 blocks apoptosis in cells lacking mitochondrial DNA. Nature 361:365–368

Jaffrezou J-P, Levade T, Bettarib A, Andrieu N, Bezombes C, Maestre N et al. (1996) Daunorubicin-induced apoptosis: triggering of ceramide generation through sphingomyelin hydrolysis. EMBO J 15:2417–2424

Kasibhatha S, Genestier L, Green DR (1999) Regulation of Fas-ligand expression during activation-induced cell death in T lymphocytes via nuclear factor κB. J Biol Chem 274:987–992

Kluck RM, Bossy-Wetzel E, Green DR, Neumeyer DD (1997) The release of cytochrome c from mitochondria: a primary site for Bcl-2 regulation of apoptosis. Science 275:1132–1136

Knepper-Nicolai B, Savill J, Brown SB (1998) Constitutive apoptosis in human neutrophils requires synergy between calpains and the proteasome downstream of caspases. J Biol Chem 273:30530–30536
Kondo S (1995) Apoptosis by antitumour agents and other factors in relation to cell cycle checkpoints. J Radiat Res 36:56–62
Kondo S, Shinomura Y, Kanayama S et al. (1998) Modulation of apoptosis by endogenous Bcl-x_L expression in MKN-45 human gastric cancer cells. Oncogene 17:2585–2591
Kondo T, Suda T, Fukuyama H, Adachi M, Nagata S (1997) Essential roles of the Fas ligand in the development of hepatitis. Nature Medicine 3:409–413
Kondo Y, Liu J, Haggi T, Barna BP, Kondo S (1998) Involvement of interleukin-1β-converting enzyme in apoptosis of irradiated retinoblastomas. Invest Ophthalmol Vis Sci 39:2769–2774
Kroemer G (1997) The protooncogene Bcl-2 and its role in regulating apoptosis. Nature Medicine 3:614–620
LaCasse EC, Baird S, Korneluk RG, MacKenzie AE (1998) The inhibitors of apoptosis (IAPs) and their emerging role in cancer. Oncogene 17:3247–3259
Levine B, Huang Q, Isaacs JT, Reed JC, Griffin DE, Hardwick JM (1993) Conversion of lytic to persistent alphavirus infection by the Bcl-2 cellular oncogene. Nature 361:739–742
Liu XS, Zou H, Slaughter C, Wang XD (1997) DFF, a heterodimeric protein that functions downstream of caspase-3 to trigger DNA fragmentation during apoptosis. Cell 89:175–184
Lotem J, Sachs L (1996) Control of apoptosis in hematopoiesis and leukemia by cytokines, tumor suppressor and oncogenes. Leukemia 10:925–931
Lyons SK, Clarke AR (1997) Apoptosis and carcinogenes. Br Med Bull 52:554–569
Martin SJ, Amarante-Mendes GP, Shi L, Chuang T-H, Casiano CA, O'Brien GA, Fitzgerald P, Ian EM, Bokoch GM, Greenberg AH, Green DR (1996) The cytotoxic protease granzyme B initiates apoptosis in a cell-free system by proteolytic processing and activation of the ICE/CED-3 family protease, CPP32, via a novel two-step mechanism. EMBO J 15:2407–2416
Medema JP, Toes REM, Scaffidi C, Zheng TS, Flavell RA, Melief CJM, Peter ME, Offringa R, Krammer PH (1997) Cleavage of FLICE (caspase-8) by granzyme B during cytotoxic T lymphocyte-induced apoptosis. Eur J Immunol 27:3492–3498
Miwa K, Asano M, Horai R, Iwakura Y, Nagata S, Suda T (1998) Caspase 1-independent IL-1β release and inflammation induced by apoptosis inducer Fas ligand. Nature Medicine 4:1287–1292
Mountz JD, Zhou T, Su X, Wu J, Cheng J (1996) The role of programmed cell death as an emerging new concept for the pathogenesis of autoimmune disease. Clin Immunol Immuno Path 80:50–514
Müller M, Wilder S, Bannasch D, Israeli D, Uhlback K, Li-Weber M et al. (1998) p53 activates the CD95 gene in response to DNA damage by anticancer drugs. J Exp Med 188:2033–2045
Muzio M, Chinnaiyan AM, Kischkel FC, O'Rourke K, Peter ME, Dixit VM (1996) FLICE, a novel FADD homologous ICE-CED-3-like protease, is recruited to the CD95 (Fas/APO-1) death-inducing signaling complex. Cell 85:817–827
Nagata S (1997) Apoptosis by death factor. Cell 88:355–365
Nagata S, Golstein P (1995) The Fas death factor. Science 267:1449–1455
Nunez G, Merino R, Grillot D, Gonzalez-Garcia M, (1994) Bcl-2 and Bcl-x: regulatory switches for lymphoid death and survival. Immunol Today 15:582–587
Okada S, Zhang H, Hatano M, Tokuhisa T (1998) A physiologic role of Bcl-x_L induced in activated macrophages. J Immunol 160:2590–2596
Paulli M, Berti E, Boven E, Kindl S, Bonoldi E et al. (1998) Cutaneous CD30+ lymphoproliferative disorders. Hum Pathol 29:1223–1230
Petak I, Mihalik R, Bauer PI, Suli-Vargha H, Sebestyen A, Kopper L (1998) BCNµ is a caspase-mediated inhibitor of drug-induced apoptosis. Cancer Res 58:614–618

Peter ME, Krammer PH (1998) Mechanisms of CD95 (APO-1/Fas)-mediated apoptosis. Curr Op Immunol 10:545–551

Radvanyi LG, Shi Y, Vaziri H, Sharma A, Dhala R, Mills GB, Miller RG (1996) CD28 costimulation inhibits TCR-induced apoptosis during a primary T cell response. J Immunol 156:1788–1798

Rawlings ND, Barrett AJ (1994) Families of cysteine peptidases. Methods Enzymol 244:461–486

Reed JC (1994) Bcl-2 and the regulation of programmed cell death. J Cell Biol 124:1–6

Reed JC (1998) Bcl-2 family proteins. Oncogene 17:3225–3236

Rouvier E, Luciani M-F, Golstein P (1993) Fas involvement in Ca^{2+}-independent T cell-mediated cytotoxicity. J Exp Med 177:195–200

Rovere P, Vallinoto C, Bondanza A, Crosti MC, Rescigno M et al. (1998) Bystander apoptosis triggers dendritic cell maturation and antigen-presenting function. J Immunol 161:4467–4471

Samijima K, Tone S, Kottke T, Enari M, Sakahira H, Cooke CA et al. (1998) Transition from caspase-dependent to caspase-independent mechanisms at the onset of apoptotic execution. J Cell Biol 143:225–239

Sattler M, Liang H, Nettesheim D, Meadows RP, Harlan JE et al. (1997) Structure of $Bcl-x_L$-Bak peptide complex: recognition between regulators of apoptosis. Science 275:983–986

Savill J (1997) Recognition and phagocytosis of cells undergoing apoptosis. Br Med Bull 53:491–508

Savill J, Smith J, Sarraf C, Ren Y, Abbott F, Rees A (1992) Glomerular mesangial cells and inflammatory macrophages ingest neutrophils undergoing apoptosis. Kidney Int 42:924–936

Schlesinger PH, Gross A, Yin X-M, Yamamoto K, Saito M et al. (1997) Comparison of the ion channel characteristics of proapoptotic Bax and antiapoptotic Bcl-2. Proc Natl Acad Sci USA 94:11357–11362

Simon H-U, Yousefi S, Dommann-Scherrer CC, Zimmermann DA, Bauer S, Barandum J, Blaser K (1996) Expansion of cytokine-producing CD4-CD8-T cells associated with abnormal Fas expression and hypereosinophilia. J Exp Med 183:1071–1082

Sigal NH, Dumont FJ (1992) Cyclosporin A, FK-506, and rapamycin: pharmocologic probes of lymphocyte signal transduction. Annu Rev Immunol 10:519–560

Soini Y, Paakko P, Lehto V-P (1998) Histopathological evaluation of apoptosis in cancer. Am J Pathol 153:1041–1052

Spaner D, Raju K, Radvanyi LG, Lin Y, Miller RG (1998) A role for perforin in activation-induced cell death. J Immunol 160:2655–2664

Sperling AI, Auger JA, Ehst BD, Rulifson IC, Thompson CB, Bluestone JA (1996) CD28/B7 interactions deliver a unique signal to naïve T cells that regulates cell survival but not early proliferation. J Immunol 157:3909–3917

Stack RM, Thompson CB, Fitch FW (1998) IL-4 enhances long-term survival of CD28-deficient T cells. J Immunol 160:2255–2262

Strater J, Mariani SM, Walczak H, Rucker FG, Leithauser F, Krammer PH, Moller P (1999) CD95 ligand (CD95L) in normal human lymphoid tissues. A subset of plasma cells are prominent producers of CD95L. Am J Pathol 154:193–201

Swain SL, Croft M, Daley C, Haynes L, Rogers P, Zhang X, Bradley LM (1996) From naïve to memory T cells. Immunol Rev 150:143–167

Tai X-G, Toyooka K, Yashiro Y, Abe R, Park C-S, Hamooka T et al. (1997) CD9-mediated costimulation of TCR-triggered naïve T cells leads to activation followed by apoptosis. J Immunol 159:3799–3807

Tamura T, Ishihara M, Lamphier MS, Tanaka N, Mak TW, Taki S, Taniguchi T (1995) An ARF-1-dependent pathway of DNA-damage-induced apoptosis in mitogen-activated T lymphocytes. Nature 376:596–599

Tanaka M, Itai T, Adachi M, Nagata S (1998) Downregulation of Fas ligand by shedding. Nature Medicine 4:31–36

Thornberry NA (1997) The caspase family of cysteine proteases. Br Med Bull 53:478–490
Thornberry NA, Lazebnik Y (1998) Caspases: enemies within. Science 281:1312–1316
Thornberry NA, Miller DK, Nicholson DW (1995) Interleukin-1β-converting enzyme and related proteases are potential targets in inflammation and apoptosis. Perspect Drug Discovery Design 2:389–399
Tschopp J, Irmler M, Thorne M (1998) Inhibition of Fas death signals by FLIPs. Curr Op Immunol 10:552–558
Van Parijs L, Abbas AK (1998) Homeostasis and self-tolerance in the immune system: turning lymphocytes off. Science 280:243–248
Vander Heiden MG, Chandel NS, Williamson EK, Schumacker PT, Thompson CB (1997) Bcl-x_L regulates the membrane potential and volume homeostasis of mitochondria. Cell 91:627–637
Varadhachary AS, Salgame P (1998) CD95 mediated T cell apoptosis and its relevance to immune deviation. Oncogene 17:3271–3276
Vaux DL, Cory S, Adams JM (1988) Bcl-2 gene promotes haemopoietic cell survival and cooperates with c-myc to immortalize pre-B cells. Nature 335:440–442
Vaux DL, Wilhelm S, Hacker G (1997) Requirements for proteolysis during apoptosis. Molec Cell Biol 17:6502–6507
Veis DT, Sorenson CM, Shutter JR, Korsmeyer SJ (1993) Bcl-2-deficient mice demonstrate fulminant lymphoid apoptosis, polycystic kidneys, and hypopigmented hair. Cell 75:229–240
Virkajarvi N, Paakko P, Soini Y (1998) Apoptotic index and apoptosis influencing proteins Bcl-2, mcl-1, Bax and caspases 3, 6 and 8 in pancreatic carcinoma. Histopathology 33:432–439
White FA, Keller-Peck CR, Knudson CM, Korsmeyer SJ, Snider WD (1998) Widespread elimination of naturally occurring neuronal death in Bax-deficient mice. J Neurosci 18:1428–1439
Whyte MKB, Savill J, Meagher LC, Lee A, Haslett C (1997) Coupling of neutrophil apoptosis to recognition by macrophages: coordinated acceleration by protein synthesis inhibitors. J Leukocyte Biol 62:195–202
Xiang H, Kinoshita Y, Knudson CM, Korsmeyer SJ et al. (1998) Bax involvement in p53-mediated neuronal cell death. J Neurosci 18:1363–1373
Yang JY, Liu X, Bhalla K, Kim CN, Ibrado AM, Cai J et al. (1997) Prevention of apoptosis by Bcl-2: release of cytochrome c from mitochondria blocked. Science 275:1129–1132
Yoshida R, Sanchez-Bueno A, Yamamoto N, Einaga-Naito K (1997) Ca^{2+}-dependent, Fas-and perforin-independent apoptotic death of allografted tumor cells by a type of activated macrophage. J Immunol 159:15–21
Young LS, Dawson CW, Eliopoulos AG (1997) Viruses and apoptosis. Br Med Bull 53:509–521
Yuan JY, Shaham S, Ledoux S, Ellis HM, Horvitz HR (1993) The C-elegans cell death gene ced3 encodes a protein similar to mammalian interleukin 1 beta converting enzyme. Cell 75:641–652
Zheng TS, Schlosser SF, Dao T, Hingorani R, Crispe IN, Boyer JL, Flavell RA (1998) Caspase-3 controls both cytoplasmic and nuclear events associated with Fas-mediated apoptosis in vivo. Proc Natl Acad Sci USA 95:13618–13623

CHAPTER 3
Hepatocyte Apoptosis Triggered by Natural Substances (Cytokines, Other Endogenous Molecules and Foreign Toxins)

D. PESSAYRE, G. FELDMANN, D. HAOUZI, D. FAU, A. MOREAU, and M. NEUMAN

A. Introduction

The necessity of maintaining a close balance between cell birth and cell death has caused multicellular organisms to evolve into unforgiving societies where cells that are no longer needed are requested to commit suicide.

For the doomed cells to disappear unobtrusively, a death programme is inserted in the genome of each cell (WHITE 1996). Cells constantly survey their external and internal milieus for survival signals (e.g., external growth factor, internal NF-κB nuclear translocation) and death signals (e.g., external Fas ligand, internal p53 overexpression). The cell then integrates these conflicting signals, particularly in the mitochondria/caspases system, to decide whether to live or commit suicide (GREEN and KROEMER 1998).

For this cell suicide to proceed discreetly (without inflammation), the apoptotic cell dissociates from neighboring cells, wraps its contents in a cross-linked protein scaffold (formed by tissue transglutaminase), condenses its chromatin beneath the nuclear membrane, and fragments its DNA first into large fragments, then between nucleosomes (OBERHAMMER et al. 1993c; PATEL and GORES 1995; SCHULTE-HERMANN et al. 1995). The cell then cuts both its cytoplasm and its nucleus into membrane-bound apoptotic bodies that express phospatidylserine on the outer leaflet of their plasma membrane and are phagocytized and digested mainly by macrophages and also by neighboring parenchymal cells (PATEL and GORES 1995; SCHULTE-HERMANN et al. 1995).

Like all other cells in the body, hepatocytes are subject to apoptosis (PATEL and GORES 1995; SCHULTE-HERMANN et al. 1995; FELDMANN 1997; GALLE 1997; JONES and GORES 1997). Apoptosis may play an important role in eliminating old hepatocytes, in decreasing liver mass when hepatic hyperplasia is no longer required, and in eliminating hepatocytes whose DNA has been damaged or which harbor viral proteins (COLUMBANO and SHINOZUKA 1996; JONES and GORES 1997).

Hepatocytes appear to be particularly susceptible to apoptosis, in particular apoptosis induced by Fas ligation (OGASAWARA et al. 1993). In many other cells, the expression of Bcl-2 exerts antiapoptotic effects as described below. It was thought, up to now, that Bcl-2 was not expressed in normal hepatocytes, explaining their sensitivity to Fas-mediated apoptosis (LACRONIQUE et al. 1996). However, a recent study suggests that Bcl-2 might be present in the inner membrane of rat liver mitochondria (MOTOYAMA et al. 1998). Furthermore, some hepatoma cells may re-express Bcl-2 (SAITO et al. 1998). In addition, hepatocytes may also slightly express Bcl-X_L, an antiapoptotic analogue of Bcl-2, and this expression may be increased after exposure to hepatocyte growth hormone (KOSAI et al. 1998), after treating hepatoma cell lines with dexamethasone (YAMAMOTO et al. 1998), or during liver regeneration (TZUNG et al. 1997).

Several endogenous or foreign compounds can trigger hepatocyte apoptosis. Man-made chemicals that trigger hepatocyte apoptosis in vitro are considered in Chap. 15. The aim of the present chapter is to review hepatocyte apoptosis induced by endogenous proteins and other natural substances. Particular emphasis is placed on the Fas/Fas ligand system, due to its major relevance in viral hepatitis (the most common cause of liver disease) and several other forms of hepatic apoptosis.

Due to the vast scope of this subject and space limitations, neither the topics covered by this review nor the list of references can be fully exhaustive.

B. Fas-Mediated Apoptosis

Cytolytic T lymphocytes cause apoptosis of target cells by several mechanisms, including the interaction of the Fas ligand expressed on the surface of T lymphocytes with the Fas (receptor) expressed on target cells (KÄGI et al. 1994).

I. Fas Ligand

Fas ligand (CD95 ligand) is a 40-kDa type II membrane glycoprotein which belongs to the tumor necrosis factor family (SUDA et al. 1993; SCHNEIDER et al. 1997). Membrane bound Fas ligand can be cut by a matrix metalloproteinase, releasing the extracellular region of the Fas ligand, a 26-kDa glycoprotein called soluble Fas ligand (TANAKA et al. 1995). Soluble Fas ligand has little apoptogenic activity (SCHNEIDER et al. 1998) and can work as an inhibitor of membrane-bound Fas ligand toxicity against hepatocytes (TANAKA et al. 1998).

Membrane-bound Fas ligand is mainly expressed by $CD8^+$ cytotoxic T lymphocytes and natural killer cells, and acts as a major effector of their cytotoxic effects (HANABUCHI et al. 1994; MONTEL et al. 1995). Fas ligand is also expressed by parenchymal cells in immunoprivileged sites such as eyes, testes, brain, and placenta, where it may destroy activated lymphocytes and avoid immune reactions (GRIFFITH et al. 1996; LEE et al. 1997; SAAS et al. 1997; BAMBERGER et al. 1997).

In the liver, rat Kupffer cells and hepatic sinusoidal endothelial cells slightly express Fas ligand mRNA in the basal state, and this expression is further increased when lipopolysaccharide is added to the culture medium (MÜSCHEN et al. 1998). In contrast, Fas ligand is not normally expressed in hepatocytes (GALLE et al. 1995). However, hepatocytes may express Fas ligand mRNA after exposure to dexamethasone (MÜSCHEN et al. 1998) and during several conditions causing oxidative stress, as reviewed later (GALLE et al. 1995; MÜLLER et al. 1997; HUG et al. 1997; STRAND et al. 1998). Hepatocarcinoma cells may escape immune surveillance either by downregulating Fas or by expressing Fas ligand (STRAND et al. 1996). Indeed, Fas ligand expression appears to be a frequent mechanism by which diverse cancer cells can kill cytotoxic T lymphocytes or natural killer cells and can thus escape immune control.

II. Fas

Fas (also termed APO-1 or CD95), a member of the tumor necrosis factor receptor superfamily, is a 45-kDa glycosylated type I-membrane protein (NAGATA 1997). The human Fas gene is composed of nine exons, all of which are conserved in the mRNA encoding the full-length Fas molecule (PAPOFF et al. 1996). Alternative splicing may produce several mRNA variants (PAPOFF et al. 1996). In the most abundant variant, exon 6 (which encodes the transmembrane fragment) is deleted (FERENBACH et al. 1997). This mRNA encodes for a soluble form of Fas, called FasTMDel (PAPOFF et al. 1996) or FasExo6Del (SCHUMANN et al. 1997). Extracellular FasTMDel acts as a decoy for the full length, membrane-bound Fas and decreases Fas-mediated apoptosis (PAPOFF et al. 1996).

Fas is abundantly expressed in the liver, thymus, lymphocytes, polynuclear cells, heart, lung, kidney, and ovary and is also weakly expressed in many other tissues (NAGATA and SUDA 1995). In the liver, Fas is present in hepatocytes (OGASAWARA et al. 1993; GALLE et al. 1995), Kupffer cells, and sinusoidal endothelial cells (MÜSCHEN et al. 1998). Fas may exhibit both cytoplasmic and plasma membrane expression in human hepatocytes (MOCHIZUKI et al. 1996) or human hepatocellular carcinoma cell lines (YANO et al. 1996). Human hepatocytes express both Fas, and, to a smaller extent, soluble Fas (KRAMS et al. 1998).

III. Fas Signal Transduction in Lymphoid Cells

Transduction of the Fas signal has been mainly studied in lymphoid cells (NAGATA 1997). Fas ligation causes caspase activation, permeabilization of mitochondrial membranes, glutathione efflux, and other effects.

1. Caspase Activation

Binding of the trimeric Fas ligand causes trimerization of Fas (SCHNEIDER et al. 1997; NAGATA 1997) (Fig. 1). The trimerized cytoplasmic region of Fas recruits a protein called FADD (Fas-associated protein with death domain) or Mort-1, through homotypic interaction of the death domain of Fas with the death domain of FADD (NAGATA 1997). FADD also possesses a death

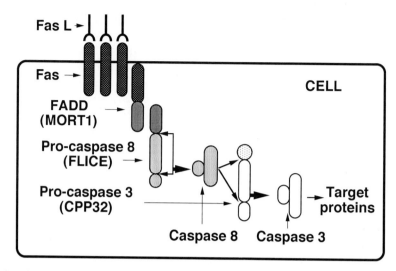

Fig. 1. Direct activation of caspases by Fas ligation. Binding of Fas ligand (Fas L) causes the trimerization of Fas. The adaptor molecule, FADD recruits procaspase 8 which autoactivates to caspase 8. The latter cuts pro-caspase 3 and other effector caspases. For simplicity, caspases are shown in their dimeric form, although they probably function as tetramers formed of two small and two large subunits

effector domain that serves as an adaptor molecule to recruit pro-caspase 8, also called FLICE (FADD-like interleukin-1 converting enzyme), or MACH-1 (Mort-1-associated CED-3) (NAGATA 1997). The prodomain of pro-caspase 8 contains two death effector domains that bind to the death effector domain of FADD.

Pro-caspase 8 is an initiator caspase that possesses intrinsic enzyme activity (MUZIO et al. 1998). When approximated (and probably oligomerized) by binding to FADD molecule(s), it may autoprocess to the active tetrameric species associating two large and two small catalytic subunits (MUZIO et al. 1998) (Fig. 1).

The activated caspase 8 may then activate effector caspases, including caspase 3 (also called CPP32/YAMA) and caspase 7 (NAGATA 1997). All caspases are cysteine proteases that cut after aspartate (THORNBERRY and LAZEBNIK 1998). They are synthesized as inactive prozymogens comprising a prodomain, a large subunit (about 20kDa), and a small subunit (about 10kDa) (THORNBERRY and LAZEBNIK 1998). After being cut after several aspartates by other caspases (or by autoprocessing, in the case of caspase 8), they then form the active enzyme which lacks the prodomain but associates the large and small subunits (Fig. 1), probably in a tetramer of two large and two small subunits (THORNBERRY and LAZEBNIK 1998).

In addition to the Fas/Fas ligand system, activated cytotoxic T lymphocytes can kill target cells through the perforin/granzyme B system (KÄGI et al. 1994).

Perforin makes holes in the cell membrane and, possibly, in post-endosomal intracellular vesicles, which may respectively allow the entry of extracellular and endocytosed granzyme B into the cytosol (NAGATA 1997; PINKOSKI et al. 1998). Although granzyme B is a serine protease, it also cuts proteins after aspartate (as do caspases) and can thus activate the caspase cascade (DARMON et al. 1995). Thus, the executioners of apoptosis are similar in the Fas/Fas ligand system and the perforin/granzyme B system (NAGATA 1997).

Activated caspases can then disassemble cell structures by cutting actin, β-catenin, and lamins (THORNBERRY and LAZEBNIK 1998). Lamins are intermediate filament proteins that form head-to-tail polymers under the nuclear membrane, forming a rigid structure (the nuclear lamina) that is involved in chromatin organization. The cleavage of lamins by caspases cause lamina to collapse, contributing to chromatin condensation (THORNBERRY and LAZEBNIK 1998). Activated caspase 3 also cuts ICAD (inhibitor of caspase-activated deoxyribonuclease) (SAKAHIRA et al. 1998). This inhibitory protein maintains CAD (caspase-activated deoxyribonuclease) in an inactive cytosolic complex (SAKAHIRA et al. 1998) and/or nuclear complex (SUMEJIMA and EARNSHAW 1998). Once ICAD has been cut, the liberated CAD may cause internucleosomal DNA fragmentation (SAKAHIRA et al. 1998). Caspases also cut, inactivate, or deregulate several proteins involved in DNA repair, mRNA splicing, and DNA replication (THORNBERRY and LAZEBNIK 1998). Finally, caspases alter the mitochondrial structure as described below.

2. Permeabilization of Mitochondrial Membranes

Mitochondria have two membranes, limiting the central mitochondrial matrix and the intermembranous space, respectively (FROMENTY and PESSAYRE 1995) (Fig. 2). The respiratory chain is located in the inner membrane. The transfer of electrons along the respiratory chain is associated with the extrusion of protons from the matrix into the intermembranous space. This creates a large membrane potential across the inner membrane, which is secondarily utilized to synthesize ATP. When ATP is needed, protons re-enter the matrix through Fo-ATPase, and a rotary motor in F1-ATPase synthesizes ATP (Fig. 2). Fas-mediated caspase activation increases the permeability of both the inner and the outer mitochondrial membranes (GREEN and KROEMER 1998).

Indeed, caspase activation may cause the mitochondrial permeability transition, a phenomenon due to the opening of a large pore in the inner mitochondrial membrane (Fig. 2). When human lymphoma cells are treated with an agonistic anti-Fas antibody, caspase activation precedes the disruption of the mitochondrial inner membrane potential (SUSIN et al. 1997). A synthetic caspase inhibitor prevents the collapse of the mitochondrial membrane potential (SUSIN et al. 1997). Recombinant caspase 1, also called ICE (interleukin-1β converting enzyme), causes a permeability transition-like swelling and disruption of the mitochondrial membrane potential in isolated rat liver mitochondria (SUSIN et al. 1997). These observations suggest that activated

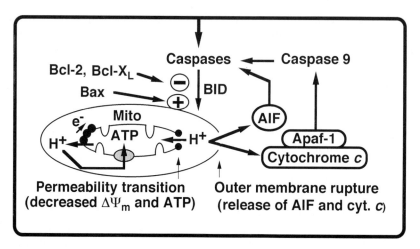

Fig. 2. Mitochondrial effects of Fas ligation. Activated caspase 8 cleaves BID. Truncated BID translocates to mitochondria and causes permeabilization of mitochondrial membrane(s). Bax has similar effects, whereas Bcl-2 and Bcl-X_L protect mitochondria. Outer membrane rupture releases apoptosis inducing factor (AIF) and cytochrome c, both of which cause caspase activation. The opening of the inner membrane permeability transition pore allows re-entry of protons into the mitochondrial matrix. This decreases the mitochondrial membrane potential ($\Delta\Psi_m$) and ATP synthesis

caspases may open the mitochondrial permeability transition pore (SUSIN et al. 1997) (Fig. 2).

The link between caspase activation and permeabilization of mitochondrial membranes is provided by BID, a proapoptotic member of the Bcl-2 family (LI et al. 1998; LUO et al. 1998). Activated caspase 8 cleaves BID, producing a truncated, C-terminal BID fragment that binds to mitochondria and induces release of cytochrome c (LI et al. 1998; LUO et al. 1998). Cytochrome c is normally present in the intermembranous space of mitochondria, where it is loosely associated with the inner membrane respiratory chain. The extrusion of cytochrome c in the cytoplasm may be due to rupture of the outer membrane (VANDER HEIDEN et al. 1997). Jurkat cells treated with an agonistic anti-Fas antibody exhibit swollen mitochondria with outer membrane discontinuities on electron microscopy. These damaged mitochondria release cytochrome c from the mitochondrial intermembranous space into the cytosol (VANDER HEIDEN et al. 1997).

It is not yet clear whether opening of the inner membrane mitochondrial transition pore causes outer membrane rupture or vice-versa. Due to the hyperosmolality of the mitochondrial matrix (GREEN and REED 1998), the opening of the mitochondrial permeability transition pore causes mitochondrial swelling in whole cells (HOEK et al. 1997). Whereas the inner mitochondrial membrane has many folds and can accommodate an increased matrix volume, the spherical outer membrane can burst when the mitochondrion swells (GREEN and KROEMER 1998; GREEN and REED 1998) (Fig. 2). The permeability transition pore may initially affect only a few mitochondria and the pore may open and close successively in a single mitochondrion (HÜSER et al. 1998). Therefore, previous swelling bursts in some mitochondria may cause the rupture of some outer membranes, although the overall inner membrane potential (averaging all mitochondria) may be subnormal (VANDER HEIDEN et al. 1997; GREEN and KROEMER 1998).

Alternatively, things may work the other way around (CAI and JONES 1998). Caspase activation may initially damage the outer membrane and release cytochrome c from mitochondria. The decreased availability of this component of the respiratory chain within mitochondria may hamper electron flow through the respiratory chain, and cause over-reduction of components located upstream, such as coenzyme Q. This over-reduction may increase the mitochondrial formation of superoxide anion and other reactive oxygen species (CAI and JONES 1998). The latter may then open the permeability transition pore (CAI and JONES 1998).

Whatever the mechanism, cytochrome c release further activates caspases in a circular loop (REED 1997; GREEN and KROEMER 1998) (Fig. 2). Cytosolic cytochrome c binds apaf-1 (apoptotic protease activating factor-1), the human analogue of CED-4 (*Cenorhabditis elegans* death-4) (REED 1997; PAN et al. 1998). In the presence of ATP, this may cause a conformational change in apaf-1 and allow it to bind to, and activate, caspase 9, which may then activate caspases 3 and 7, further increasing the apoptotic caspase (REED 1997; PAN et al. 1998) (Fig. 2). Fas activation also causes release of mitochondrial AIF (apoptosis-inducing factor) (SUSIN et al. 1997) (Fig. 2). This 57-kDa flavoprotein causes further dissipation of the mitochondrial membrane potential and secondary release of cytochrome c (SUSIN et al. 1999). AIF translocates to the nucleus, and it induces large scale (50-kb fragments) DNA fragmentation and chromatin condensation (SUSIN et al. 1997, 1999). Thus, the initial activation of caspases by transduction of the Fas signal may permeabilize mitochondria, release cytochrome c and AIF, and further activate caspases (Fig. 2).

Thus Fas ligation may activate caspases both directly (Fig. 1) and indirectly (Fig. 2). It has been suggested that different lymphoid cells may use these two pathways differently (SCAFFIDI et al. 1998). In some cells, Fas-mediated caspase activation may occur mainly by the direct (nonmitochondrial) pathway, while other cells may mainly activate caspases through the secondary (mitochondrial) pathway (SCAFFIDI et al. 1998).

3. Modulation by Caspase 8 Decoys (FLIPs), Cellular Inhibitors of Apoptosis (c-IAPs), Members of the Bcl-2 Family and Other Factors

The autopotentiating loop described in Fig. 2 (caspases cause the release of cytochrome c which further activates caspases) would imply that any minimal caspase activation could rapidly kill all cells in a catastrophic caspase/mitochondria/caspase reinforcing loop. This is not the case, thanks to several control mechanisms.

Different cellular FLIPs (Fas-associated death-domain-like interleukin-1 converting enzyme inhibitory proteins) resemble caspase 8, although they are themselves inactive (TSCHOPP et al. 1998). Through their death domain(s), these decoy proteins interact with FADD and caspase 8, preventing caspase 8 activation (KATAOKA et al. 1998; TSCHOPP et al. 1998).

Cellular IAPs (inhibitor of apoptosis), such as c-IAP-1, c-IAP-2, and ILP, directly bind to, and inhibit, several caspases (ROY et al. 1997; SUZUKI et al. 1998b).

The antiapoptotic, mitochondrial membrane-associated protein, Bcl-X_L, prevents the mitochondrial effects of Fas ligation in Jurkat cells (VANDER HEIDEN et al. 1997) (Fig. 2). Bcl-X_L may also have a more direct effect, as it completely inhibits apoptosis induced by microinjection of recombinant active caspase 8 in breast carcinoma cells (SRINIVASAN et al. 1998). Bcl-2, a close analogue of Bcl-X_L, also associates with mitochondria and prevents the collapse of the mitochondrial membrane potential induced by other agents (SHIMIZU et al. 1998). Bcl-2 might act by enhancing H^+ efflux from the mitochondrial matrix (SHIMIZU et al. 1998). In addition to its mitochondrial effect, Bcl-2 may also have several extramitochondrial effects. Indeed, Bcl-2 may increase the mRNA and protein of SERCA (sarcoplasmic/endoplasmic reticulum Ca^{2+}-ATPase), thus preserving the endoplasmic reticulum calcium store (KUO et al. 1998). In addition, Bcl-2 may also cause proteasomal degradation of IKB and nuclear translocation of NF-κB (DE MOISSAC et al. 1998), whose antiapoptotic effects are discussed later (in the context of tumor necrosis factor-α-induced cytotoxicity).

The overexpression of Bax (a proapoptotic analogue of the Bcl-2 family) has the opposite effects. Bax cooperates with the adenine nucleotide translocator to trigger the mitochondrial permeability transition (MARZO et al. 1998). Bax releases cytochrome c in isolated rat liver mitochondria (NARITA et al. 1998) and activates caspases in Jurkat cells (PASTORINO et al. 1998). Possibly because of the intrinsic toxicity of the mitochondrial permeability transition itself, Bax can kill mammalian cells whose caspases are inhibited (ADAMS and CORY 1998) or yeast cells which do not express caspases (GREEN and REED 1998).

In addition to caspases and members of the Bcl-2 family, several other factors modulate the opening of the mitochondrial permeability transition pore. Pore opening can be triggered by Ca^{2+}, electrophilic compounds, or reactive oxygen species, all of which may trigger or aggravate apoptosis in different models (GREEN and KROEMER 1998). In contrast, pore opening may be prevented by various anti-oxidants that prevent apoptosis in several models (GREEN and KROEMER 1998). Thus the combination of mitochondria and caspases can be considered as the site where antiapoptotic and proapoptotic signals are integrated before the cell makes its decision to live or die (GREEN and KROEMER 1998).

4. Orientation of Cell Death Towards Apoptosis and/or Necrosis

Mitochondria may also help decide whether the cell dies from necrosis, apoptosis, or both (GREEN and KROEMER 1998). Opening of the mitochondrial membrane transition pore may cause both caspase activation and ATP depletion (Fig. 2). Indeed, opening of this pore causes re-entry of protons into the matrix and collapse of the mitochondrial membrane potential, which is normally used to synthesize ATP (Fig. 2). Therefore, immediate opening of the pore in all mitochondria suppresses mitochondrial ATP synthesis (Fig. 2). If the cell cannot derive enough energy from anaerobic glycolysis, cell ATP decreases. Apoptosis is an active, ATP-requiring process (LEIST et al. 1997b). At low ATP levels, apoptosis cannot proceed, and cells die from necrosis instead (LEIST et al. 1997b). This occurs whenever cells are exposed to high concentrations of compounds that directly open the mitochondrial permeability transition pore in all mitochondria.

In contrast, if the pore only opens in some mitochondria, caspase activation may occur without an immediate decrease in cell ATP, so that apoptotic lesions develop. In the case of Fas ligation, the "race" between caspase activation (causing apoptosis) and ATP depletion (causing necrosis) (GREEN and KROEMER 1998) may be won by caspases because the Fas/FADD complex directly activates caspases (Fig. 1). However, secondary aggravation of mitochondrial lesions may then cause ATP depletion and secondary necrosis.

5. Efflux of Reduced Glutathione and Other Effects

Incubation of human Jurkat T lymphocytes with an agonistic anti-Fas antibody causes a rapid and specific cellular efflux of reduced glutathione (VAN DEN DOBBELSTEEN et al. 1996). GSH levels modulate Fas-mediated apoptosis (CHIBA et al. 1996). N-Acetylcysteine (a glutathione precursor) prevents both the depletion of glutathione and apoptosis in human T cells exposed to an agonistic anti-Fas antibody (CHIBA et al. 1996). Buthionine sulfoximine (an inhibitor of glutathione synthesis) has the opposite effects (CHIBA et al. 1996).

Although ceramide generation by cellular sphingomyelinases was initially proposed as an important mechanism for Fas-mediated apoptosis (CIFONE et al. 1995), more recent studies suggest that ceramide generation may only play a limited role in Fas-induced T cells apoptosis (WATTS et al. 1997; GAMEN et al. 1998). One of the ceramide metabolites may open the mitochondrial permeability transition pore and hasten apoptosis.

At least in human fibroblasts, Fas activation may also cause NF-κB (nuclear factor κB) activation, although less than tumor necrosis factor-α (RENSING-EHL et al. 1995).

6. Fas Signaling Independent of Fas Ligand

Through unknown mechanisms, UV light causes aggregation of both Fas and the tumor necrosis factor-α receptor (REHEMTULLA et al. 1997; ARAGANE et al. 1998). UV light-induced Fas oligomerization recruits FADD, activates caspases, and causes apoptosis, without requiring Fas ligand (REHEMTULLA et al. 1997; ARAGANE et al. 1998). Hydrophobic bile acids similarly cause Fas ligand-independent Fas aggregation, and trigger caspase activation and hepatocyte apoptosis in the absence of Fas ligand (FAUBION et al. 1999).

IV. Role of Fas in the Control of the Immune System

In lymphocytes, Fas ligand expression not only kills target cells but also activated lymphocytes, thus avoiding uncontrolled (auto)immune reactions. Indeed, lymphoproliferation and autoimmune manifestations are the main manifestations in mice with genetic defects in the Fas/Fas ligand system. In *lpr* (lymphoproliferation) or *gld* (generalized lymphoproliferative disease) mice, homozygous mutations of the Fas gene (*lpr* mice) or the Fas ligand gene (*gld* mice) affect Fas-mediated elimination of autoreactive B and T lymphocytes (WATANABE-FUKADA et al. 1992; TAKAHASHI et al. 1994a). This causes both hypergammaglobulinemia and accumulation of nonmalignant CD4/CD8 double negative T cells in lymphoid organs, leading to a generalized autoim-

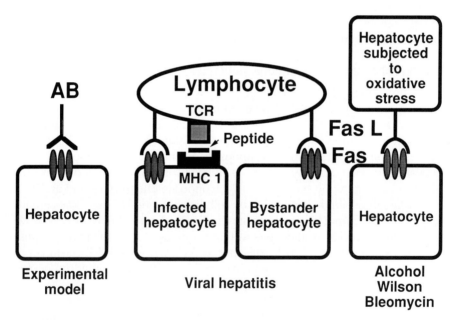

Fig. 3. Different models of Fas-mediated apoptosis. Agonistic anti-Fas antibodies (AB) cause the oligomerization and activation of Fas, reproducing the effects of Fas ligand (Fas L). Activated lymphocytes express Fas ligand and may kill both their specific immunologic targets (expressing viral peptides on major histocompatibility class 1 molecules) and also bystander (noninfected) hepatocytes. In alcohol abuse, Wilson's disease, or after exposure to some anticancer drugs, oxidative stress may cause the expression of Fas L by hepatocytes, which can kill each other through fratricidal killing

mune disease, with autoantibody production, glomerulonephritis, arthritis, and vasculitis (WATANABE-FUKADA et al. 1992; TAKAHASHI et al. 1994a). Patients with the autoimmune lymphoproliferative syndrome also have defects in the Fas gene and exhibit clinical features similar to those of *lpr* mice (FISHER et al. 1995; RIEUX-LAUCAT et al. 1995; KASAHARA et al. 1998).

V. Fas-Induced Hepatocyte Apoptosis

Fas-mediated hepatocyte apoptosis has been demonstrated in diverse experimental models and several human conditions. It can be triggered by agonistic anti-Fas antibodies, activated lymphocytes, or hepatocytes expressing Fas ligand (Fig. 3).

1. Agonistic Anti-Fas Antibodies

The use of agonistic anti-Fas antibodies (causing Fas aggregation) is an easy way to study the apoptotic effects of Fas activation in vitro or in vivo (Fig. 4).

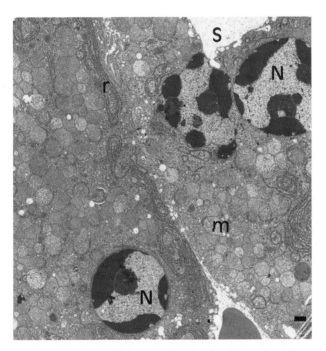

Fig. 4. Hepatic apoptosis induced by an anti-Fas antibody in mice. Mice were killed 4 h after the intraperitoneal administration of an anti-mouse Fas antibody (8 µg/mouse). Two apoptotic hepatocytes are visible. Nuclear chromatin is condensed beneath the nuclear membrane. Mitochondria (m) are tightly packed around the nucleus (N), whereas the rough endoplasmic reticulum (r) is shifted towards the cell periphery

Human hepatocytes rapidly underwent apoptosis when they were cultured with an agonistic anti-human Fas antibody alone (20 ng/ml) (GALLE et al. 1995). Apoptotic changes started after 5.5 h. All cells were dead at 7.5 h, and most were detached (GALLE et al. 1995).

In contrast, when mouse hepatocytes were cultured for 24 h with an agonistic anti-human Fas monoclonal antibody (1 μg/ml), apoptosis was not observed unless cycloheximide (10 μg/ml) or H7 (Seikagaku Kogyo Company, Tokyo, Japan), a serine/threonine kinase inhibitor, was also added (ROUQUET et al. 1996). With the anti-Fas/cycloheximide treatment, 55% of cells exhibited apoptotic changes (ROUQUET et al. 1996). The percentage of apoptotic cells was increased by genistein and herbimycin A, two tyrosine kinase inhibitors (ROUQUET et al. 1996). Apoptosis was prevented by caspase inhibitors (ROUQUET et al. 1996). Sphingomyelinase was not activated, and ceramide was not released; furthermore, exogenous ceramide did not cause mouse hepatocyte apoptosis (ROUQUET et al. 1996).

Suspended mouse hepatocytes exhibited apoptosis within hours following the addition of an anti-mouse Fas antibody (JO2, Pharmingen; 10 ng/ml) alone, without requiring cycloheximide or H7 (JONES et al. 1998). Apoptosis was associated with the processing of caspases 3 and 7, although poly(ADP-ribose) polymerase cleavage was not detected (JONES et al. 1998). In culture, however, these same hepatocytes also required cycloheximide to respond to the anti-Fas treatment (JONES et al. 1998).

In vivo, the seminal study of OGASAWARA et al. (1993) showed that the intraperitoneal administration of a monoclonal anti-Fas antibody (100 μg/mouse) rapidly killed wild-type mice but not *lpr* (lymphoproliferation) mice, which have Fas gene defects. In wild-type mice, few normal hepatocytes remained 2 h after injection of the anti-Fas antibody (OGASAWARA et al. 1993). Instead, most hepatocytes exhibited cytoplasm condensation and pyknosis of the nuclei, while condensed and fragmented nuclei were observed by electron microscopy (OGASAWARA et al. 1993).

Despite this morphological evidence of apoptosis, serum transaminase activity was considerably increased, as early as 2–3 h after injection (OGASAWARA et al. 1993). Transaminase release might be caused by secondary liver necrosis (possibly caused by secondary ATP depletion as discussed above). However, a second mechanism might also be involved. In several other apoptosis models, transglutaminase activation creates a cross-linked protein scaffold that wraps cell contents and may prevent plasma membrane rupture and release of cell content (FESUS et al. 1996). In contrast, Fas receptor stimulation did not cause transglutaminase activation in thymocytes (SZONDY et al. 1997). Hypothetically, failure of Fas to activate transglutaminase (if it is also true for hepatocytes) might permit liver enzyme extrusion in Fas-induced hepatic apoptosis.

Although Fas is expressed in many organs, administration of the anti-Fas antibody mainly damages the liver (OGASAWARA et al. 1993), possibly because hepatocytes do not express (or poorly express) Bcl-2. To determine whether forced hepatocellular expression of Bcl-2 would protect hepatocytes, transgenic mice were generated that expressed the human Bcl-2 gene product in their hepatocytes (LACRONIQUE et al. 1996). While administration of an anti-Fas antibody (10μg/mouse) caused hepatic apoptosis and death in nontransgenic mice, hepatic apoptosis was both delayed and reduced in the Bcl-2 transgenic mice, and 93% of them survived (LACRONIQUE et al. 1996).

Human recombinant hepatocyte growth factor (100μg) administered 6h and 0.5h prior to, and 3h after, an anti-Fas antibody (4μg or 8μg) increased Bcl-X_L expression in hepatocytes and prevented hepatic apoptosis and death in mice (KOSAI et al. 1998). IL-15 also prevented Fas-induced hepatic apoptosis, although the mechanism is unknown (BULFONE-PAUS et al. 1997).

2. Activated Lymphocytes

Activated cytotoxic lymphocytes are recruited to virus-infected hepatocytes, through interaction between the T cell receptor and viral peptides expressed on major histocompatibility complex class I molecules located on the surface of hepatocytes (Fig. 3). These virus-specific lymphocytes can kill the infected hepatocytes through Fas activation. A cytotoxic $CD8^+$ T lymphocyte clone, specific for the hepatitis B surface antigen, was injected intravenously into transgenic mice that expressed the hepatitis B surface antigen in the liver (KONDO et al. 1997). The T cell clone caused hepatocyte apoptosis and killed most animals within 3 days. Coadministration of a soluble form of Fas prevented apoptosis, and all mice survived (KONDO et al. 1997). Studies in lpr, gld, and control mice showed that the Fas system also played a role in the rapid elimination of hepatocytes transfected with the defective adenoviral vectors that are used in experimental gene therapy (OKUYAMA et al. 1998).

When present in large amounts, activated T cells may also kill uninfected hepatocytes in an antigen-independent manner, a phenomenon called "bystander killing" (Fig. 3). This phenomenon may involve both Fas ligand and tumor necrosis factor-α release. The role of Fas ligand was demonstrated in three models. In a first model, nontransgenic mice were primed with *Propionibacterium acnes*, which causes accumulation of macrophages and lymphocytes in the liver (KONDO et al. 1997). A subsequent challenge with lipopolysaccharide induced liver injury and killed the mice (KONDO et al. 1997). Neutralization of Fas ligand by administration of a soluble form of Fas decreased these effects (KONDO et al. 1997). In another model, concanavalin A, which activates T cells, caused severe hepatic injury in wild-type mice but mild injury in *lpr* and *gld* mice (SEINO et al. 1997; TAGAWA et al. 1998). A neutralizing antibody against Fas ligand reduced the aminotransferase increase (SEINO et al. 1997). In a third model, the transfer of *lpr* mouse spleen cells, which overexpress Fas ligand, to irradiated wild type mice caused hepatocyte apoptosis in vivo (BOBÉ et al. 1997).

Patients with chronic hepatitis B overexpress the Fas antigen (MOCHIZUKI et al. 1996). This increased expression mainly occurs in periportal hepatocytes that are close to the site of inflammatory cell infiltrates which are elective targets for immune destruction (Mochizuki et al. 1996). In patients with chronic hepatitis C, periportal hepatocytes also overexpress Fas (Hiramatsu et al. 1994), while liver-infiltrating mononuclear cells express Fas ligand (Mita et al. 1994). Increased serum concentrations of soluble Fas have been reported in patients with liver cirrhosis due to the hepatitis C virus (Seishima et al. 1997).

Hepatitis C virus-specific, human cytotoxic T lymphocyte lines were prepared from the peripheral blood lymphocytes of a patient who had cleared hepatitis C virus infection while on interferon therapy (Ando et al. 1997). These T cell clones killed Ag-bearing cells in vitro, by mechanisms involving Fas ligand, perforin, and tumor necrosis factor-α (Ando et al. 1997). The T cell clones also killed nonantigen bearing, bystander cells, although less efficiently (Ando et al. 1997).

3. Fratricidal Killing

Hepatocytes do not normally express Fas ligand, which prevents them from killing their neighbors. However, the Fas ligand promoter contains NF-κB binding sites (Takahashi et al. 1994b). Normally, NF-κB is maintained in the cytoplasm by IκB. However, reactive oxygen intermediates causes phosphorylation, ubiquitination, and proteasomal degradation of IκB, allowing nuclear translocation of NF-κB (Naumann and Scheidereit 1994). Conditions which increase reactive oxygen intermediates may thus cause Fas ligand expression by hepatocytes (Strand et al. 1998). At the same time, the increased formation of reactive oxygen intermediates might damage DNA, overexpress p53, and increase Fas expression by hepatocytes (Müller et al. 1997). The Fas ligand of a first hepatocyte may then interact with Fas on another hepatocyte, causing fratricidal killing. This form of cell death may occur under three conditions.

The first condition involves *alcoholism*, which increases the formation of reactive oxygen species and causes oxidative stress in the liver (Lettéron et al. 1993; Mansouri et al. 1997a). Fas ligand messenger RNA was detected by in situ hybridization in the hepatocytes of patients with alcoholic liver damage (Galle et al. 1995). At the same time, Fas was overexpressed in some hepatocytes.

The second condition involves *Wilson's disease*, which is due to mutations in a copper transporting P-type ATPase whose defects cause copper accumulation in the liver (Strand et al. 1998). Due to its ability to cycle between Cu^{2+} and Cu^+, copper is a powerful generator of reactive oxygen species and causes oxidative stress in the liver (Mansouri et al. 1997b). In patients with fulminant hepatic failure due to Wilson's disease, high Fas protein expression was observed on the hepatocyte plasma membrane in some areas, while Fas ligand mRNA became apparent in the cytoplasm of some hepatocytes located in the vicinity of apoptotic cells (Strand et al. 1998). When HepG2 hepatoma cells were treated with copper, cell surface Fas protein, Fas ligand mRNA, and Fas ligand protein were all increased, and these cells underwent apoptosis which was partially prevented by a neutralizing anti-Fas antibody or a caspase inhibitor (STRAND et al. 1998). This suggested that hepatocytes killed each other by fratricidal killing (Fig. 3).

The third condition involves the *anti-cancer drug, bleomycin*, which forms complexes with iron and other transition metals and produces reactive oxygen species (KANOFSKY 1986; HUG et al. 1997). Bleomycin increased the formation of reactive oxygen species and induced Fas ligand mRNA expression in HepG2 hepatoma cells (HUG et al. 1997). The latter effect was prevented by antioxidants and was reproduced by exposure to H_2O_2, suggesting that reactive oxygen intermediates were involved in the induction of Fas ligand (HUG et al. 1997). In another study, bleomycin and methotrexate (but not cisplatin) were shown to increase Fas ligand mRNA in HepG2 cells (MÜLLER et al. 1997). Bleomycin and methotrexate caused HepG2 apoptosis when present alone (MÜLLER et al. 1997). It was suspected that this drug-induced apoptosis may be due, at least in part, to Fas signaling and fratricidal killing. Indeed, the bleomycin-induced apoptosis was almost completely inhibited by an $F(ab)'_2$-anti-Fas antibody fragment known to interfere with Fas/Fas ligand interaction (MÜLLER et al. 1997).

Bleomycin, methotrexate, and cisplatin also overexpressed Fas mRNA and cell surface Fas protein in HepG2 cells (MÜLLER et al. 1997). This overexpression did not occur in hepatoma cells that did not express p53 or had a mutated p53, suggesting that wild-type p53 is somehow required for anticancer drug-induced enhancement of Fas expression (MÜLLER et al. 1997). To determine whether this enhanced Fas expression sensitizes hepatocytes to the effects of Fas ligation, HepG2 cells were first treated with bleomycin, methotrexate, or cisplatin alone for 48h, and then in combination with an anti-Fas antibody (0.1 μg/ml) for another 24h (MÜLLER et al. 1997). To differentiate between the specific effects of the anti-Fas antibody and cell death induced by the anticancer drug alone, surviving cells were expressed as the fraction of residual living cells with the anticancer drug only. The anti-Fas antibody alone caused specific cell death in only 10% of control HepG2 cells (MÜLLER et al. 1997). In contrast, in HepG2 cells treated with bleomycin, cisplatin, or methotrexate, the anti-Fas antibody caused specific apoptosis in 50%–75% of hepatocytes (MÜLLER et al. 1997).

Thus, some anti-cancer drugs can directly cause fratricidal apoptosis by inducing both Fas and Fas ligand expression in hepatocytes, and can also sensitize hepatocytes to exogenous causes of Fas ligation (MÜLLER et al. 1997).

4. Basal Hepatic Apoptosis

In the "streaming liver" hypothesis, hepatocytes are born in the periportal area and slowly pushed, along the sinusoids towards the centrilobular zone where they undergo apoptosis (ARBER et al. 1988; BENEDETTI et al. 1988). Interestingly, liver hyperplasia is found both in Fas-knockout mice (ADACHI et al. 1995) and in humans with mutated Fas (KASAHARA et al. 1998), suggesting a possible role of Fas in the apoptotic elimination of old hepatocytes.

C. Tumor Necrosis Factor-α-Mediated Cell Death

I. TNF-α

TNF-α belongs to the same superfamily as Fas ligand, and exhibits homologies with this ligand (NAGATA and GOLSTEIN 1995; SMITH et al. 1994). The TNF-α gene is located on human chromosome 6 in the human leucocyte antigen (HLA) region (BEUTLER and CERAMI 1988). Two biallelic, single base (guanine to adenosine) polymorphisms have been described in the human TNF-α promoter, at nucleotide −308 and nucleotide −238, respectively (WILSON et al. 1997). Homozygosity for the rare −308A allele ("TNF2") increases TNF-α synthesis and predisposes to lethal cerebral malaria and systemic lupus erythematosus (SULLIVAN et al. 1997). Although the functional significance of the rare −238A ("TNFA-A") allele is uncertain, an excess in this allele has been reported in patients with alcoholic steatohepatitis (GROVE et al. 1997).

The TNF-α gene promoter contains both NF-κB and AP-1 (activator protein-1) binding sites (ZWACKA et al. 1998). TNF-α is mainly produced by activated lymphocytes, monocytes, and macrophages, including Kupffer cells (TRACEY and CERAMI 1993). TNF-α production in macrophages or Kupffer cells is inhibited by dexamethasone and some prostaglandins (GONG et al. 1991) and is enhanced by viral infection (GREWE et al. 1994) or lipopolysaccharide (TRAN-THI et al. 1995). The latter increases the DNA-binding activity of both NF-κB and AP-1 (TRAN-THI et al. 1995).

TNF-α is also expressed by hepatocytes, particularly under conditions of oxidative stress such as ethanol exposure (NEUMAN et al. 1998). Hepatocyte TNF-α is also overexpressed during, and is involved in, liver regeneration (AKERMAN et al. 1992).

TNF-α is initially synthesized as a 26-kDa membrane-bound form of 233 amino acids, which is then proteolytically cleaved between Ala^{76} and Val^{77} to the 17-kDa secreted form of 137 amino acids (TRACEY and CERAMI 1993). Metalloproteinases, including TACE (TNF-α converting enzyme) and ADAM 17 (a disintegrin and metalloprotease 17) which are both members of the ADAM family of proteases, may cleave the membrane-bound cytokine at the cell surface (MOLHER et al. 1994; MCGEEHAN et al. 1994; GEARING et al. 1994; BLACK et al. 1997). Secreted TNF-α is an unglycosylated polypeptide which is active in its trimeric form (SMITH and BAGLIONI 1988; JONES et al. 1989). The cell surface, membrane-bound form of TNF-α is also active (TRACEY and CERAMI 1993). In this case, TNF-α-bearing cells kill target cells through cell-to-cell contacts (KRIEGLER et al. 1988; PEREZ et al. 1990; DECOSTER et al. 1995).

II. TNF-α Receptors and Signal Transduction

TNF-α exerts its functions through two cell surface receptors which are expressed on most cells, including hepatocytes (VOLPES et al. 1992; VANDENABEELE et al. 1995). TNFR1 has a molecular mass of 55kDa and is therefore also called TNFR-55, while the 75-kDa TNFR2 is also called TNFR75.

These two TNF-α receptors belong to the same superfamily as Fas and the NGF, CD40, and CD30 receptors (NAGATA and GOLSTEIN 1995). Both TNFR1 and TNFR2 are glycosylated transmembrane proteins and exhibit homologies in their extracellular domains (VANDENABEELE et al. 1995), which may explain why TNF-α binds to both receptors. This extracellular domain may be shed and may then act as a decoy receptor for TNF-α, thus decreasing the toxicity of circulating TNF-α (VAN ZEE et al. 1992). In contrast, the intracellular domains of TNFR1 and TNFR2 are completely different (VANDENABEELE et al. 1995), explaining why these receptors mediate different signals.

TNFR2 may signal for the proliferation of both thymocytes and cytotoxic T cells but may have no direct signaling effects for TNF-α-mediated cytotoxicity (TARTAGLIA et al. 1991). Nevertheless, TNFR2 might have some indirect effect by being able to recruit TNF-α and then "pass" this ligand to TNFR1 (ERICKSON et al. 1994). This might explain why TNFR2 –/– mice are less susceptible to TNF-α-mediated necrosis and death than wild type mice (ERICKSON et al. 1994).

TNFR1 signaling is thought to mediate mainly the cytotoxic effects of TNF-α (TARTAGLIA et al. 1993; ASHKENAZI and DIXIT 1998). Indeed, gene-targeted mice lacking TNFR1 do not develop TNF-α induced apoptosis (LEIST 1995b), while mice genetically deficient in Fas are susceptible (LEIST et al. 1996). Binding of the trimeric TNF-α to the extracellular domain of TNFR1 causes its trimerization. The trimerized intracellular domain then associates with the death domain of an adapter molecule, called TRADD (TNFR1-Associated Death Domain) (VANDENABEELE et al. 1995; NAGATA 1997; NATOLI et al. 1998). TRADD then recruits diverse signaling molecules that have both death-promoting and death-preventing effects (ASHKENAZI and DIXIT 1998).

Death-promoting effects are caused by the recruitment of FADD by TRADD. This causes the same proapoptotic effects as the recruitment of FADD by Fas. Briefly, FADD recruits procaspase 8 which autoactivates into the active caspase 8. The latter activates other caspases and also causes BID cleavage (GROSS et al. 1999). Translocation of truncated BID to mitochondria causes the release of mitochondrial cytochrome c into the cytosol. In the presence of ATP and apaf-1, cytosolic cytochrome c activates caspase 9 causing amplification of the caspase cascade (CHINNAIYAN et al. 1995; LI et al. 1988; GROSS et al. 1999). Indeed, two recent reports show that TNF-α induces apoptosis by acting on hepatic mitochondria (ANGERMULLER et al. 1998; BRADHAM et al. 1998), with depolarization of the inner membrane, release of cytochrome c in the cytosol, and caspase activation (BRADHAM et al. 1998). Ultrastructural alterations of hepatic mitochondria appear before any nuclear change

(ANGERMULLER et al. 1998). Alterations of the outer mitochondrial membrane occur first, followed by protrusions of the inner membrane through the outer membrane gaps (ANGERMULLER et al. 1998).

In addition to recruiting the proapoptotic FADD adaptor protein, TRADD also recruits RIP (receptor interacting protein) and TRAF2 (TNF Receptor-Associated Factor-2). These two polypeptides partly signal for anti-apoptotic effects (STANGER et al. 1995; HSU et al. 1996; ASHKENAZI and DIXIT 1998). Indeed, RIP and TRAF2 cause the phosphorylation and proteosomal degradation of IκB, an inhibitor of NF-κB that normally maintains NF-κB in an inactive, cytosolic complex (HSU et al. 1995; RÉGNIER et al. 1997). The resulting translocation of NF-κB into the nucleus may induce cell survival (BEG and BALTIMORE 1996; WANG et al. 1996) by suppressing some cell death signals (VAN ANTWERP et al. 1996), upregulating the expression of the c-IAPs (cellular inhibitors of apoptosis) (STEHLIK et al. 1998; WANG et al. 1998), and inhibiting caspase 8 activation (WANG et al. 1998). Second, TRAF2 activates the Jun NH$_2$-kinase/AP-1 system (ASHKENAZI and DIXIT 1998). Both NF-κB and AP-1 activation cause the upregulation of mitochondrial Mn-SOD (manganese-containing superoxide dismutase) (BORRELLO and DEMPLE 1997), which exerts protective effects by decreasing the toxicity of reactive oxygen species (WONG et al. 1989). In keeping with the antiapoptotic effects of NF-κB nuclear translocation through the enhanced transcription of antiapoptotic genes (MnSOD, cIAPs), NF-κB -knock-out mice exhibit liver apoptosis (BEG et al. 1995). Moreover, inhibiting the degradation of IκB and thus preventing the nuclear translocation of NF-κB also induces massive apoptosis in murine hepatocytes (BELLAS et al. 1997).

Thus, TNF-α signaling involves both the immediate activation of proapoptotic caspases, and the secondary, NF-κB-mediated transcription of antiapoptotic genes. This dual effect may explain why TNF-α-induced cytotoxicity often requires artificial inhibition of gene transcription by concomitant administration of transcriptional inhibitors. These inhibitors do not affect caspase activation by FADD, while they may prevent increased transcription of antiapoptotic genes by NF-κB nuclear translocation.

III. Hepatotoxicity of TNF-α in Experimental Models

The cytotoxicity of TNF-α has been studied in normal mouse hepatocytes (LEIST et al. 1994, 1995b, 1996, 1997a; SENALDI et al. 1998) and cultured rat hepatocytes (SHINAGAWA et al. 1991). Freshly isolated mouse hepatocytes are essentially insensitive to TNF-α cytotoxicity, unless transcriptional inhibitors are also added, such as actinomycin D, α-amanitin (an RNA polymerase II inhibitor), or D-galactosamine (whose hepatic metabolism decreases hepatic uridine nucleotide pool and causes liver-specific transcriptional arrest) (LEIST et al. 1994). In these sensitized cells, bleb formation, chromatin condensation, and oligonucleosomal DNA fragmentation preceding LDH release all indicate that cell death is initially caused by an apoptotic process. As in several (but not all) models of apoptosis, cycloheximide prevents cell death (LEIST et al. 1994), indicating that some ongoing synthesis of short-lived proteins is required. However, marked release of liver enzymes and necrosis may also occur (WANG et al. 1995). As discussed for Fas-mediated apoptosis, necrosis might be due to ATP depletion, and perhaps also to the noninduction of tissue transglutaminase during TNF-α-mediated apoptosis (LEIST et al. 1995a).

The proapoptotic action of TNF-α has been also investigated in HepG2 cells, a human hepatoma cell line (HILL et al. 1995; LEIST et al. 1997a). Translational inhibition by actinomycin D renders HepG2 cells susceptible to TNF-α-induced cytotoxicity. SV40-immortalized rat hepatocytes that have integrated the HBV genome are also sensitive to TNF-α when HBV expression is high, although the mechanism for this sensitization has not been clarified (GUILHOT et al. 1996).

Anti-TNF-α antibodies protect hepatocytes from apoptosis induced by the cytokine (LEIST et al. 1994), while preexisting glutathione depletion enhances toxicity (XU et al. 1998). Interferon-γ also potentiates TNF-α proapoptotic action (SHINAGAWA et al. 1991), while interleukin 1β (LEIST et al. 1995a), or keratinocyte growth factor (SENALDI et al. 1998) are protective. Prolonged cultures of rat hepatocytes were insensitive to TNF-α but became apoptotic when dimethylsulfoxide (an antioxidant) was removed from the culture medium (BOUR et al. 1996).

These in vitro observations have been confirmed by in vivo models (LEIST et al. 1994, 1995a, 1996, 1997a). Hepatocytes from mice receiving both TNF-α and actinomycin D exhibited chromatin condensation, apoptotic body formation, and significant oligonucleosomal DNA fragmentation that occurred before any increase in serum transaminase activity (LEIST et al. 1994; MORIKAWA et al. 1996). Necrosis, however, occurred at later times (LEIST et al. 1995a). Z-VAD-fluorometylketone, a caspase inhibitor, can protect mice from liver apoptosis induced by TNF-α (KÜNSTLE et al. 1997).

In another model, administration of a small dose of lipopolysaccharide (which causes the release of TNF-α) induced massive hepatic necrosis at 24h in galactosamine-sensitized mice, but not in control mice (HISHINUMA et al. 1990; LEIST et al. 1995a). Necrosis was prevented by an anti-TNF-α antibody (HISHINUMA et al. 1990; LEIST et al. 1995a). Similarly, administration of the *Pseudomonas aeruginosa* exotoxin A causes TNF-α release from T cells and hepatic apoptosis (SCHÜMANN et al. 1998).

A related in vivo experimental model is the *Corynebacterium parvum*/endotoxin model (HARBRECHT et al. 1994a). A single injection of killed *C. parvum* bacteria progressively causes macrophagic sinusoidal cell infiltrates and hepatic granulomas in rats (ARTHUR et al. 1985). When these rats or mice are then challenged, one week later, with a small dose of endotoxin, the hepatic macrophages release TNF-α, causing massive liver injury (ARTHUR et al. 1985), unless an anti-TNF-α antibody is also administered (HARBRECHT et al. 1994a).

Yet another in vivo model of TNF-α-mediated hepatocyte apoptosis is that of concanavalin A administration (TIEGS et al. 1992; LEIST et al. 1995a). This lectin activates CD4-positive lymphocytes (GANTNER et al. 1995). Their assembly in liver sinusoids increases the local production of several cytokines, particularly TNF-α. Although interferon-γ (TAGAWA et al. 1997) and Fas ligand (SEINO et al. 1997; KSONTINI et al. 1998; TAGAWA et al. 1998) are also involved, TNF-α also seems to play a role in mediating both the liver apoptosis that occurs 4h after concanavalin A administration, and the secondary necrosis that follows (GANTNER et al. 1995; KSONTINI et al. 1998; TRAUTWEIN et al. 1998). Indeed, anti-TNF-α antibodies decrease liver injury (TRAUTWEIN et al. 1998).

IV. Role of TNF-α in Human Liver Injury

Hepatocytes from patients chronically infected by HBV produce TNF-α (GONZALEZ-AMARO et al. 1996), and this production seems to depend on the HBX protein (LARA-PEZZI et al. 1998). Serum TNF-α is increased in chronic viral hepatitis (YOSHIOKA et al. 1989; SHERON et al. 1991). However, at least to our knowledge, no relationship has been reported between TNF-α serum levels and hepatocyte apoptosis in viral hepatitis. The current trend is to favor Fas-mediated rather than TNF-α-mediated apoptosis. In their model of HBV-transgenic mice, NAKAMOTO et al. (1997) demonstrated that hepatocytes were much less sensitive to destruction by TNF-α than by Fas ligand or interferon γ. Furthermore, the hepatitis C virus core protein may even inhibit TNF-α-mediated apoptosis in vitro (RAY et al. 1998).

Serum TNF-α is also increased in acute or chronic alcoholic liver disease (MCCLAIN and COHEN 1989; KHORUTS et al. 1991). Hypothetically, TNF-α might be involved in steatohepatitis (PESSAYRE et al. 1999). In rats, anti-TNF-α antibodies attenuate hepatic necrosis and inflammation caused by chronic exposure to ethanol (IIMURO et al. 1997).

D. Transforming Growth Factor-β and Activins

TGF-β is a member of a large superfamily including the activins, inhibins, bone morphogenic proteins, and several other growth and differentiation factors (GRANDE 1997). Although there are five known (highly homologous) TGF-β isoforms (named TGF-β1 to TGF-β5), only the first three are present in mammals. The most abundant, and the most extensively studied, isoform is TGF-β1.

TGF-β1 is initially synthesized as a 390-amino-acid precursor encompassing a signal peptide of 29 amino acids, then the latency associated peptide (LAP) of 249 amino acids, and then the C-terminal TGF-β1 peptide of 112 amino acids (GRANDE 1997). After cleavage of the signal peptide, the pro-TGF-β1 is again cut (between two arginines) into LAP and TGF-β. These polypeptides, however, remain associated by noncovalent interactions. Dimerization and disulfide formation eventually produces an inactive complex called the "small latent complex." This complex consists of the noncovalent association of a disulfide-linked TGF-β1 homodimer and a disulfide-linked LAP homodimer. A "large latent complex" can also be formed through additional disulfide bonding of LAP with LTBP (latent TGF-β binding protein). After secretion, these latent, inactive complexes may be dissociated by extracellular proteins (including plasmin and thrombospondin), releasing the active TGF-β1 homodimer.

TGF-β1 is synthesized in several cells and tissues (GRANDE 1997). In the liver, TGF-β1 is expressed in Kupffer cells, sinusoidal endothelial cells, and fat-storing, perisinusoidal cells (BEDOSSA and PARADIS 1995; DATE et al. 1998). Although normal hepatocytes in vivo, or freshly isolated hepatocytes in vitro may not express TGF-β1 message or protein, cultured hepatocytes and hepatoma cells may acquire the ability to express TGF-β mRNA and protein (BISSEL et al. 1995; CHUNFANG et al. 1996; GAO et al. 1996; GRESSNER et al. 1996; DATE et al. 1998). Similarly, in several diseases, hepatocytes may acquire the ability to synthesize TGF-β, as discussed later on. Finally, it has been suggested that hepatocytes might be able to internalize the TGF-β synthesized by other cells (ROTH-EICHHORN et al. 1998).

The cellular effects of TGF-βs are mediated by cell surface receptors. TGF-β receptor type I (TGF-β R-I) and TGF-β receptor type II (TGF-β R-II) are transmembrane proteins with serine/threonine kinase activity in their intracellular domains (HELDIN et al. 1997). The dimeric TGF-β protein first binds to a dimer of TGF-β R-II, which then associates with a TGF-β R-I dimer, and phosphorylates the cytoplasmic domains of the TGF-β R-I dimer. Thus, TGF-β, TGF-β R-II, and TGF-β R-I form a ligand-bound, phosphorylated tetrameric receptor complex responsible for signal transduction (HELDIN et al. 1997). Cytosolic Smad proteins seem to play an essential role in transducing the TGF-β signal into the nucleus (HELDIN et al. 1997; MASSAGUÉ et al. 1997).

TGF-β1 has a wide range of effects (GRANDE 1997). It acts on cell and tissue differentiation, has a potent immunosuppressive effect on the immune system, and plays a major role in extra-cellular matrix synthesis and remodeling (GRANDE 1997). TGF-β1 also inhibits the proliferation of several epithelial cells, including hepatocytes (NAKAMURA et al. 1985; RUSSELL et al. 1988); it may stimulate the production of p15, a nuclear protein which binds to, and inhibits, the cyclin D-cdk4,6 complex (GRANDE 1997). Finally, TGF-β1 triggers apoptosis in a large variety of normal or tumor cells, including hepatocytes (GRANDE 1997).

The first evidence that TGF-β1 can cause cell death in primary cultures of normal rat hepatocytes was reported by OBERHAMMER et al. (1991). In this first report, a small dose of TGF-β1 was used and there were no morphological signs of apoptosis or typical DNA fragmentation (OBERHAMMER et al. 1991). With a larger dose, LIN and CHOU (1992) provided convincing morphological and biochemical evidence that TGF-β1 can cause apoptosis in Hep3B cells (a human hepatoma cell line). OBERHAMMER and QIN (1995) and OBERHAMMER et al. (1992, 1993a,b) analyzed the proapoptotic role of TGF-β1 in hepatic cells. They found that apoptosis occurs in normal cultured rat hepatocytes exposed to sufficient amounts of TGF-β_1 (OBERHAMMER et al. 1992) and demonstrated that the intravenous administration of TGF-β1 could provoke programmed cell death during the regressive phase of liver hyperplasia caused by prior treatment with cyproterone acetate in rats (OBERHAMMER et al. 1992, 1993a). Nafenopin, a peroxisome proliferator, partly inhibited TGF-β1-induced apoptosis (OBERHAMMER and QIN 1995), an observation which had also been reported by BAYLY et al. (1994) in rat hepatoma cell lines.

The proapoptotic effects of TGF-β1 in normal adult or fetal, rat or mouse hepatocytes have been confirmed in several studies (BENEDETTI et al. 1995; OHNO et al. 1995; CAIN et al. 1996; FAN et al. 1996; SANCHEZ et al. 1996; GRESSNER et al. 1997; INAYAT-HUSSAIN et al. 1997; SANCHEZ et al. 1997; GILL et al. 1998). TGF-β1 also causes apoptosis in human hepatoma cell lines, such as Hep3B cells (CHUANG et al. 1994; PONCHEL et al. 1994) or HuH7 cells (FAN et al. 1996), as well as rat or mouse hepatoma cell lines, such as Morris cells (FUKUDA et al. 1993; YAMAMOTO et al. 1996, 1998), or FaO cells (ARSURA et al. 1997; LIM et al. 1997; CHOI et al. 1998). In general, TGF-β1-induced apoptosis requires incubation with relatively large doses or, as we have observed in our laboratory, with repeated doses of TGF-β1. Insulin receptor substrate 1 overexpression prevents transforming growth factor β1-induced apoptosis in human hepatocellular carcinoma cells (TANAKA and WANDS 1996).

The ability of TGF-β1 to cause hepatocyte apoptosis in vivo was demonstrated in transgenic mice overexpressing hepatic TGF-β1 (SANDERSON et al. 1995).

The initial mechanism(s) that trigger TGF-β1-induced hepatocyte apoptosis are not completely understood. Although Smad molecules are involved in the transduction of the TGF-β1 signal (MASSAGUÉ et al. 1997), their possible implication in TGF-β1-induced apoptosis has not been studied. TGF-β1 may inhibit the NFκB/ReL factors which are known to promote cell survival (ARSURA et al. 1997). TGF-β1 may also decrease Bcl-X_L, an antiapoptotic member of the Bcl-2 gene family, without changing the expression of proapoptotic Bax or Bad (YAMAMOTO et al. 1998).

More is known about later events. TGF-β1-induced apoptosis is associated with the activation of several caspases, including caspase 1 (CAIN et al. 1996), caspase 2 (CHOI et al. 1997), and caspase 3 (INAYAT-HUSSAIN et al. 1997). Caspase inhibitors, such as ZVAD-FMK or ZDEV-FMK, prevent caspase activation and apoptosis (CAIN et al. 1996; INAYAT-HUSSAIN et al. 1997). The mouse tissue transglutaminase promoter contains a TGF-β1 response element (RITTER and DAVIES 1998). Tissue transglutaminase is induced by TGF-β1 in a rat hepatoma cell line (FUKUDA et al. 1993), and causes extensive cross-linking of cytokeratin polypeptides (FUKUDA et al. 1991). TGF-β1-induced apoptosis in fetal rat hepatocytes is also associated with increased formation of reactive oxygen species and lowered glutathione contents (SANCHEZ et al. 1997), and apoptosis can be inhibited by radical scavengers (SANCHEZ et al. 1997). The nongenotoxic hepatocarcinogen, nafenopin, suppresses rodent hepatocyte apoptosis induced by TGF-β1 or Fas, by unknown mechanisms (GILL et al. 1998).

A possible pathogenic role of TGF-β is suspected (but not proven) in human liver disease. In rats, chronic ethanol administration markedly increases TGF-β1 mRNA and protein in the perivenular region of the liver (FANG et al. 1998). In humans, RT-PCR showed increased hepatic TGF-β transcripts in patients with alcohol-induced cirrhosis (LLORENTE et al. 1996). Hypothetically, TGF-β might be involved in several alcohol-induced steatohepatitis lesions (PESSAYRE et al. 1999). TGF-β can cause hepatocyte demise, and its ability to induce tissue transglutaminase and cross-link cytokeratins might be involved in the formation of Mallory bodies which are formed of cross-linked cytokeratin monomers (ZATLOUKAL et al. 1991). Finally, TGF-β1 stimulates collagen production by perisinusoidal Ito cells (CASINI et al. 1993), an effect which might contribute to the development of perisinusoidal fibrosis in alcoholic steatohepatitis (PESSAYRE et al. 1999).

Increased hepatic TGF-β transcripts are also found in patients with virus-induced cirrhosis (LLORENTE et al. 1996). Increased serum TGF-β1 levels and TGF-β1 immunostaining of both infiltrating cells and hepatocytes are found in patients with autoimmune hepatitis (BAYER et al. 1998).

It has also been suggested that hepatocarcinoma cells might cause the demise of surrounding normal hepatocytes by producing TGF-β. Although they were resistant to TGF-β1-induced apoptosis, HepG2 cells produced TGF-β and caused apoptosis in nontumoral hepatocytes (GRESSNER et al. 1997). Preliminary results have suggested that this might also occur in human hepatocarcinomas (LOTZ et al. 1998).

Activins are members of the same family as TGF-β, and also cause hepatocyte apoptosis in vitro and in vivo (SCHWALL et al. 1993; HULLY et al. 1994; DE BLESER et al. 1997).

E. Small Endogenous Molecules
I. Ceramide, Sphingosine-1-phosphate, and Phosphatidylserine

In several extrahepatic cells, apoptosis triggered by diverse stimuli (Fas ligation, TNF-α, X-rays, or diverse anticancer agents) may be accelerated by ceramide generation (JAFFRÉZOU et al. 1996; VERHEIJ et al. 1996). Neutral and acidic sphingomyelinases hydrolyse plasma membrane and endosomal sphingomyelin into phosphocholine and ceramide. The latter (and/or its metabolites) can reproduce many of the signaling events caused by Fas ligand or TNF-α. Ceramide causes apoptosis in several extrahepatic cell lines. However, as discussed above (see Fas section), ceramide does not appear to cause apoptosis in hepatocytes (ROUQUET et al. 1996), and it has been suggested that ceramide generation may play no, or a limited, role in Fas-mediated hepatocyte apoptosis (WATTS et al. 1997; GAMEN et al. 1998).

Phosphatidylserine has been reported to induce apoptosis in the CHO (Chinese hamster ovary) cell line (UCHIDA et al. 1998), and sphingosine-1-phosphate caused apoptosis in human hepatoma cells (HUNG and CHUANG 1996).

II. Retinoic Acid

Vitamin A (all-*trans*-retinol) is metabolized first into retinal and then all-*trans*-retinoic acid (tretinoin), which partly isomerizes to 13-*cis*-retinoic acid (isotretinoin) (CULLUM and ZILE 1985).

Retinoic acids exert gene-regulatory effects through three retinoic acid receptors (RARα, RARβ, and RARγ) and at least two retinoid X receptors (RXRα and RXRβ) (ZHANG et al. 1995). Liganded homodimers or heterodimers of these nuclear receptors then bind to responsive elements in the promoter regions of modulated genes (ZHANG et al. 1995). Retinoic acids exert profound effects on embryonic development, cell growth, and differentiation (CHAMBON et al. 1996) and can either inhibit (ORITANI et al. 1992) or induce (SU et al. 1994) apoptosis.

The proapoptotic effects of retinoic acid and other retinoids have been tested in cultured hepatocytes or hepatoma cells (NAKAMURA et al. 1995, 1996; KIM et al. 1996; FALASCA et al. 1998). All-*trans*-retinoic acid does not appear to induce apoptosis in normal adult rat hepatocytes (FALASCA et al. 1998), an observation also reported with acyclic retinoid (NAKAMURA et al. 1996). In contrast, 10μmol/l all-*trans*-retinoic acid provoked apoptosis in cultured, fetal rat hepatocytes (FALASCA et al. 1998), and 100μmol/l all-*trans*-retinoic acid caused 80% apoptosis in the Hep3B hepatoma cell line (KIM et al. 1996). A low concentration (5μmol/l) of all-*trans*-retinoic acid did not induce apoptosis in the HUH-7 hepatoma cell line, while the same concentration of acyclic retinoid caused apoptosis (NAKAMURA et al. 1995).

The mechanism for retinoic acid-induced apoptosis has not been thoroughly investigated. However, retinoic acid causes RAR and RXR-mediated increases in the mRNA and protein of tissue transglutaminase (ZHANG et al. 1995; JOSEPH et al. 1998). In rat tracheal epithelial cells, and human myeloma cell lines, tissue transglutaminase induction was associated with apoptosis (ZHANG et al. 1995; JOSEPH et al. 1998). In contrast, in rat hepatocytes, despite marked induction of tissue transglutaminase, the intracellular activity of this enzyme was somewhat decreased, and apoptosis did not occur (PIANCENTINI et al. 1992). This suggested that tissue transglutaminase was not activated in these hepatocytes. The intracellular activity of tissue transglutaminase is largely dependent on cytosolic calcium, and retinoic acid has been reported to decrease cell calcium in some epithelial cells (VARANI et al. 1991).

In humans, isotretinoin (13-*cis*-retinoic acid) is used for acne, while acitretin (an aromatic analog of all-*trans*-retinoic acid) and etretinate (the ethyl ester of acitretin) are used for psoriasis. All these retinoids are teratogenic in humans (LAMMER et al. 1985). The aromatized analogs (etretinate and acitretin) also cause cytolytic hepatitis in adults (FARRELL 1994). It is not known whether this adverse effect is related to the retinoid structure or, possibly, to metabolic activation of the phenyl ring.

Hypervitaminosis A causes a different type of liver damage in man (ZAFRANI et al. 1984). This effect is attributed to vitamin A itself and its storage in perisinusoidal lipocytes. These engorged cells compress the sinusoidal lumen and secrete collagen into the space of Disse, further closing the lumen. Portal hypertension, perisinusoidal fibrosis, and, sometimes, cirrhosis may develop (ZAFRANI et al. 1984).

III. Bile Acids

Cholestasis (failure of bile to reach the duodenum) is caused by obstruction of the biliary tree by cancer or stones, or by impairment of hepatocellular bile secretion or bile duct integrity due to genetic, autoimmune, or drug-induced diseases. Whatever the initial mechanism, the retention of hydrophobic bile acids within cholestatic hepatocytes may result in progressive liver injury.

Hydrophobic bile acids are toxic to mitochondria, where they uncouple state 4 respiration (LEE and WHITEHOUSE 1965), inhibit the respiratory chain (KRÄHENBÜHL et al. 1994b), increase the formation of reactive oxygen species (SOKOL et al. 1995), and open the mitochondrial membrane permeability transition pore (BOTLA et al. 1995). As explained above, pore opening causes either apoptosis (when enough ATP is maintained) or necrosis (when ATP is depleted). Indeed, hydrophobic bile acids cause ultrastructural apoptotic lesions and oligonucleosomal DNA fragmentation at low doses in vitro (PATEL et al. 1994) or in vivo (CHIECO et al. 1997), but induce ATP depletion and hepatocyte necrosis at higher doses (SPIVEY et al. 1993).

Under numerous circumstances, a major mechanism in the opening of the mitochondrial permeability transition pore is increased formation of reactive oxygen species (LEMASTERS et al. 1998). Oxidant injury to mitochondria seems to play a major role in triggering bile acid-induced liver lesions. Indeed, the antioxidant lazaroid, U83836E, inhibited lipid peroxidation and apoptosis in rat hepatocytes cultured with 50μmol/l glycochenodeoxycholate (PATEL and GORES 1997). Pretreatment with vitamin E reduced both oxidant injury to mitochondria and hepatocellular necrosis after intravenous administration of a high dose (100μmol/kg) of taurochenodeoxycholic acid to rats (SOKOL et al. 1998).

In addition to these mitochondrial effects, other bile acid-induced cell-destruction mechanisms have been demonstrated. In rat hepatocytes exposed to glycodeoxycholate (50μmol/l), cell Ca^{2+} did not change, whereas Mg^{2+} increased twofold (PATEL et al. 1994). Incubation of cells in an Mg^{2+}-free medium prevented this increase in Mg^{2+} and decreased nuclear DNA fragmentation. These observations suggest that the increase in cell Mg^{2+} activates

Ca^{2+}/Mg^{2+}-dependent endonucleases that contribute to DNA fragmentation (PATEL et al. 1994).

During glycochenodoxycholate-induced hepatocyte apoptosis, a decrease in nonnuclear serine-like protease activity coincided with an increase in nuclear activity, suggesting translocation of the protease from the cytosol to the nucleus (Kwo et al. 1995). A serine protease inhibitor indeed decreased DNA fragmentation and cell death (Kwo et al. 1995).

Another protease which is translocated to the nucleus is cathepsin B (ROBERTS et al. 1997). This cysteine protease is located not only in lysosomes, but also in several other cell fractions. Cathepsin B exhibits trypsin-like and other protease activities (ROBERTS et al. 1997). In bile duct-ligated rats, a threefold increase in apoptosis and a fourfold increase in trypsin-like nuclear protease activity were observed (ROBERTS et al. 1997). The purified nuclear protease activity was identified as cathepsin B. Inhibitors of cathepsin B blocked glycochenodeoxycholate-induced apoptosis in rat hepatocytes. Stable transfection of an antisense cathepsin B DNA reduced cathepsin B activity and glycochenodeoxycholate-induced apoptosis in McNtcp.24 cells expressing the Na/taurocholate cotransporting polypeptide involved in bile acid uptake. The cellular localization of cathepsin B during apoptosis was determined by nuclear immunoblots, immunocytochemistry, and by determining the location of fluorescence after expressing a cathepsin B fused to green fluorescent protein. All three approaches showed that cathepsin B was translocated from the cytoplasm to the nucleus during glycochenodeoxycholate-induced apoptosis (ROBERTS et al. 1997).

A recent report has establishes the role of Fas and caspase activation in bile acid-induced apoptosis (FAUBION et al. 1999). The toxic bile acid, glycochenodeoxycholate induced Fas oligomerization in the absence of Fas ligand (FAUBION et al. 1999). Fas oligomerization caused the activation of caspase 8 and effector caspases, followed by cathepsin B activation and apoptosis (FAUBION et al. 1999). These effects were prevented in Fas-deficient (*lpr*) mice or after addition of caspase inhibitors (FAUBION et al. 1999).

Other studies have determined the modulation of bile acid-induced apoptosis by diverse agents. Bile acids have been shown to activate protein kinase C (PKC) (STRAVITZ et al. 1996), which modulates cell death in other apoptotic models. JONES et al. (1997) provided evidence for a role of PKC activation in glycochenodeoxycholate-induced apoptosis. Membrane-associated total PKC activity was increased in bile acid-treated hepatocytes. Immunoblots demonstrated the translocation of PKC-α, PKC-δ, and PKC-ε to hepatocyte membranes. Direct activation of PKC-α and PKC-δ by the bile acid was also demonstrated. The PKC inhibitors chelerythrine and Go-6976 reduced glycochenodeoxycholate-induced hepatocyte apoptosis, whereas phorbol 12-myristate-13-acetate, a PKC agonist, had the opposite effects. Parallel changes occurred in cathepsin B activity, suggesting that PKC is somehow involved in cathepsin B activation (JONES et al. 1997).

Another intracellular mediator which can modulate cell death in other apoptotic models is cyclic AMP (cAMP). WEBSTER and ANWER (1998) demonstrated that cAMP is protective against glycochenodeoxycholate-induced apoptosis in rat hepatocytes. Indeed, cAMP analogs or agents that increase intracellular cAMP (glucagon and a combination of forskolin and 3-isobutyl-1-methylxanthine) inhibited apoptosis (WEBSTER and ANWER 1998). Evidence has also been provided that bile acids may activate mitogen activated protein kinase (MAPK) and that the cAMP-induced cytoprotection against bile acid-induced apoptosis may involve protein kinase A, MAPK, and phosphatidyl inositol-3-OH kinase (WEBSTER and ANWER 1998).

Most interestingly, cholestasis itself induces adaptive changes that limit the mitochondrial toxicity and proapoptotic effects of hydrophobic bile acids (LIESER et al. 1998). The mitochondrial permeability transition induced by glycochenodeoxycholate was reduced in hepatic mitochondria from bile duct-ligated rats. In these rat hepatocytes, glycochenodeoxycholate barely affected cell viability, although it markedly decreased the viability of control hepatocytes. Mitochondrial cardiolipin content was increased in bile duct-ligated rats. If these rats were fed a fatty acid-deficient diet, this cardiolipin increase was prevented and the susceptibility of mitochondria and hepatocytes to undergo bile acid-induced permeability transition and cell death was restored. Thus, under chronic cholestatic conditions, hepatocytes adapt to and resist the mitochondrial permeability transition (LIESER et al. 1998). This adaptive mechanism may explain the slow progression of liver injury under these conditions.

Whereas hydrophobic bile acids may cause cell death by inducing the mitochondrial permeability transition and by causing Fas ligand-independent Fas aggregation, in contrast the hydrophilic bile acid, ursodeoxycholic acid, seems to protect against the mitochondrial and apoptotic effects of hydrophobic bile acids (KRÄHENBÜHL et al. 1994a; RODRIGUES et al. 1998). Co-incubation with tauroursodeoxycholic acid reduced apoptosis caused by glycochenodeoxycholic acid in rat hepatocytes (BENZ et al. 1998a,b). S-Adenosylmethionine also reduced bile acid-induced apoptosis, and the combination of tauroursodeoxycholate and S-adenosyl methionine had additive protective effects (BENZ et al. 1998a).

In humans, administration of ursodeoxycholic acid slows disease progression in primary biliary cirrhosis (POUPON et al. 1987) and several other chronic cholestatic diseases (BEURS et al. 1998). Ursodeoxycholic acid may act by slightly reducing the ileal absorption of toxic, hydrophobic bile acids, by causing Ca^{2+}-stimulated insertion of transport proteins in the canalicular membrane, and by exerting the direct cytoprotective effects described above (BEURS et al. 1998).

Primary biliary cirrhosis is an autoimmune liver disease causing slowly progressive bile duct injury and rarefaction. Biliary cells exhibit both necrosis and apoptosis (BERNUAU et al. 1981). The infiltrating lymphocytes that surround bile ducts cause the apoptotic death of biliary cells through Fas- and perforin/granzyme B-mediated apoptosis (HARADA et al. 1997). Because ursodeoxycholic acid slows disease progression, it is tempting to speculate that hydrophobic bile acids might aggravate bile duct lesions (BEURS et al. 1998). Indeed, ursodeoxycholate administration significantly decreased DNA fragmentation in the biliary epithelial cells in patients with primary biliary cirrhosis (KOGA et al. 1997). Similarly, ursodeoxycholate feeding impedes the development of chronic cholestatic liver disease in mdr2-knock-out mice which lack the ability to secrete protective phospholipids into bile (VAN NIEUWKERK et al. 1996).

IV. Extracellular ATP and Adenosine

Extracellular ATP is toxic to hepatocytes through P2 purinoceptors (CHOW et al. 1997). Death occurs either by necrosis or apoptosis.

External adenosine causes apoptosis through P1 purinoceptors mostly in nonhepatic cells (CHOW et al. 1997).

V. Nitric Oxide

Nitric oxide has dual effects on hepatic apoptosis. On the one hand, nitric oxide can de-energize hepatic mitochondria, open the mitochondrial permeability transition pore, release mitochondrial calcium, and cause hepatocyte cell death, unless cytosolic Ca^{2+} is chelated (RICHTER et al. 1994). Macrophage-derived nitric oxide induces apoptosis of rat hepatoma cells in vivo (NISHIKAWA et al. 1998), and nitric oxide may be involved in the hepatic apoptosis caused by the concomitant administration of lipopolysaccharide, TNF-α, and antioxidants (WANG et al. 1998).

On the other hand, a first small dose of nitric oxide may prevent hepatocyte death caused by a second, high dose of nitric oxide (KIM et al. 1995). Induction of resistance is prevented by cycloheximide, suggesting upregulation of protective protein(s) (KIM et al. 1995). Nitric oxide induced heat shock protein 70 expression and prevented apoptosis in hepatocytes cultured with tumor necrosis factor-α and actinomycin D (KIM et al. 1997a). Furthermore, nitric oxide caused S-nitrosylation and inactivation of caspases (KIM et al. 1997b). This prevented the hepatocyte apoptosis caused by removal of growth factor or exposure to tumor necrosis factor-α or an anti-Fas antibody (KIM et al. 1997b). In vivo, delivery of nitric oxide to the liver blocks tumor necrosis factor-α-induced apoptosis and fulminant hepatic failure (SAAVEDRA et al. 1997), while inhibition of nitric oxide production aggravates liver injury caused by endotoxin or hemorrhagic shock (HARBRECHT et al. 1994b, 1995; SZABO et al. 1994).

F. Foreign Toxins

Actractyloside is a toxic glucosidic component of *Actractylis gummifera* L., a plant that causes hypoglycemia, liver failure, and renal failure in North African children who use it as chewing-gum (LARREY 1997). Actractyloside binds to the mitochondrial adenine nucleotide translocator (which is part of the mitochondrial permeability transition pore), opens this pore, and can trigger apoptosis (ZAMZAMI et al. 1996).

Apoptin is a small protein synthesized by the genome of the chicken anemia virus, which causes aplastic anemia and thymocyte destruction in young chickens (NOTEBORN and VAN DER EB 1998). Apoptin induces apoptosis in diverse cancer cell lines, including hepatoma cells, but not in normal cells (although thymocytes and erythroblasts were not tested) (NOTEBORN and VAN DER EB 1998). The apoptotic effect is independent of p53 and is enhanced, rather than suppressed, by Bcl-2 expression, suggesting possible applications in cancer therapy (NOTEBORN and VAN DER EB 1998).

Cocaine, an alcaloid from *Erythroxylon coca*, may cause liver damage in drug users (WANLESS et al. 1990). Cocaine is transformed by cytochrome P450 and flavin adenine nucleotide-dependent monoxygenases into electrophilic metabolites that covalently bind to proteins and/or undergo redox cycling, producing the superoxide anion and causing lipid peroxidation (BOELSTERLI and GÖDLIN 1991). Cocaine also impairs mitochondrial respiration, both in isolated rat liver mitochondria exposed to cocaine in vitro and in mitochondria from rats treated with cocaine in vivo (DEVI and CHAN 1997). In mice, cocaine administration causes early ultrastructural mitochondrial membrane discontinuities and late mitochondrial swelling (GOTTFRIED et al. 1986) and a combination of early apoptosis and late necrosis (CASCALES et al. 1994). In humans, the liver lesions induced by cocaine include centrilobular necrosis, and microvesicular and macrovacuolar steatosis (WANLESS et al. 1990).

Curcumin, a component of the plant *Curcuma Ionga*, which is used as a spice and food preservative, elevates p53 and c-Myc proteins and causes apoptosis in HepG2 cells (JIANG et al. 1996).

Etoposide and its analog, GL331 (Genelabs, California) are two semisynthetic derivatives from the plant toxin, podophyllotoxin. These topoisomerase II inhibitors cause DNA strand breaks (CLARKE et al. 1993) and induce apoptosis in Hep3B, HepG2, and other cell lines (HUANG et al. 1996). Whereas wild type thymocytes undergo etoposide-induced apoptosis, in contrast, homozygous null p53 thymocytes are resistant (CLARKE et al. 1993). This might suggest that DNA damage may cause p53 overexpression, Bax upregulation, and caspase activation as described later in relation to germander.

Fumonisin B, a mycotoxin product of *Fusarium moniliforme*, caused apoptosis in mouse liver and kidney after repeated exposure (SHARMA et al. 1997).

Germander (Teucrium chamaedrys L.) is a medicinal plant which has been used since ancient times for its alleged choleretic and antiseptic properties. Germander was generally considered safe, until germander capsules were marketed for use in weight control diets. This popular indication and the fad for natural medicine led to large scale utilization and to an epidemic of hepatitis in France (LARREY et al. 1992).

Germander contains saponins, glycosides, flavonoids, and furano *neo*-clerodane diterpenoids (LARREY et al. 1992). These diterpenoids were shown to be responsible for the in vivo hepatotoxicity of germander in mice (LOEPER et al. 1994). Hepatotoxicity was prevented by preadministration of a single dose of troleandomycin, a specific inhibitor of cytochrome P450 3A, and was enhanced by pretreatment with either dexamethasone or clotrimazole, two cytochrome P4503A inducers (LOEPER et al. 1994).

In vitro, the furano *neo*-clerodane diterpenoids of germander were activated by cytochrome P4503A into electrophilic metabolites that covalently bound to proteins, depleted cellular glutathione and protein thiols, increased cytosolic [Ca^{2+}], activated Ca^{2+}-dependent tissue transglutaminase (forming a cross-linked protein scaffold), and caused both internucleosomal DNA fragmentation and typical ultrastructural apoptotic lesions in isolated rat hepatocytes (LEKEHAL et al. 1996; FAU et al. 1997). Although the germander diterpenoids also inhibited mitochondrial respiration, the loss of cell ATP was moderate (FAU et al. 1997).

Formation of reactive metabolites may damage not only proteins but also DNA. DNA lesions activate protein kinases, such as DNA-PK (DNA-dependent protein kinase) and ATM (mutated in ataxia telangectasia) (EVAN and LITTLEWOOD 1998). These kinases may phosphorylate both p53 and Mdm-2. Normally Mdm-2 interacts with p53 and signals its degradation. DNA-damage-induced phosphorylation of either p53 or Mdm-2 prevents the two proteins from interacting and thus stabilizes p53 (EVAN and LITTLEWOOD 1998). The overexpression of p53 upregulates Bax (CANMAN and KASTAN 1997). As explained above, Bax localizes in mitochondria, releases mitochondrial cytochrome *c*, activates caspases, and causes apoptosis (ROSSÉ et al. 1998). Germander diterpenoids caused marked overexpression of p53 in hepatocytes from rats treated with dexamethasone, a cytochrome P450 3A inducer which increases the formation of electrophilic metabolites (FAU et al. 1997). However, only mild p53 overexpression occurred in nonpretreated rat hepatocytes, although these hepatocytes also underwent apoptosis. This suggested that p53 overexpression was not the main mechanism of germander-induced apoptosis. Instead, it was concluded that electrophilic metabolites may stimulate apoptosis by decreasing cellular thiols, increasing [Ca^{2+}], and activating Ca^2-dependent transglutaminase and endonucleases (FAU et al. 1997). In keeping with this hypothesis, apoptotic cell death was prevented by decreasing metabolic activation (with troleandomycin), preventing depletion of glutathione (with cystine), blocking activation of Ca^{2+}-modulated enzymes (with calmidazolium), or inhibiting internucleosomal DNA fragmentation (with aurintricarboxylic acid) (LEKEHAL et al. 1996; FAU et al. 1997).

Related calcium-activated mechanisms may also cause liver cell necrosis (BELLOMO and ORRENIUS 1985). Whereas germander diterpenoids caused hepatocyte apoptosis in vitro, they mainly caused necrosis, with only a few apoptotic hepatocytes in vivo (FAU et al. 1997). The reasons for these in vitro/in vivo differences have not been elucidated (FAU et al. 1997).

Administration of an aqueous extract of *Teucrium stocksanium* caused occasional hepatic apoptosis and cerebral neuron loss in rats (TANIRA et al. 1996).

Microcystin-LR is a cyclic heptapeptide produced by the blue-green alga, *Microcystis aeruginosa* (SOLTER et al. 1998). This toxic alga proliferates during the algae blooms caused by sewer and fertilizer runoffs. Microcystin-LR causes diverse liver lesions, including hepatic apoptosis in rats (SOLTER et al. 1998). Administration of microcystin-LR increases serum bile acid concentrations in rats. Microcystin-LR is also a potent inhibitor of serine/threonine protein phosphatases and causes the hyperphosphorylation of several hepatic proteins (reviewed in SOLTER et al. 1998). It would be interesting to know whether Bcl-2 and Bcl-X_L are also phosphorylated.

Paclitaxel (Taxol) and *taxotere* are microtubule-stabilizing, antineoplastic agents derived from the bark of the yew tree, *Taxus brevifolia*. Paclitaxel inhibits mitochondrial respiration and is toxic to hepatocytes (MANZANO et al. 1996). Paclitaxel and taxotere were shown to activate caspases (SUZUKI et al. 1998a) and cause apoptosis in other cell lines, possibly due to decreased expression of Bcl-2 (LIU et al. 1994) and increased phosphorylation of both Bcl-2 (HALDAR et al. 1996) and Bcl-X_L (PORUCHYNSKY et al. 1998). However, the significance of this increased phosphorylation is not clear, since Bcl-2 phosphorylation has been shown to prevent apoptosis in a recent study (RUVOLO et al. 1998 and references therein).

Perillyl alcohol, a monoterpene derived from lavender, was found to increase the apoptotic index, and decrease tumor weight, in rat liver tumors caused by previous diethylnitrosamine exposure (MILLS et al. 1995). This apoptosis-enhancing effect was tentatively ascribed to an increased expression of transforming growth factor β receptors caused by perillyl alcohol (MILLS et al. 1995).

Prostaglandins PGA_2 and Δ^{12}-PGJ_2 induce apoptosis in human hepatocarcinoma cell lines (LEE et al. 1995; AHN et al. 1998).

Solamargine, a compound purified from the Chinese herb, *Solanum incanum*, initiated the apoptosis of hepatoma cells, possibly by triggering the expression of tumor necrosis factor-α receptors (KUO et al. 1997).

Thapsigargin is a guaianolide component of the Mediterranean plant, *Thapsia garganica* L. (Linnaeus) (TREIMAN et al. 1998). Thapsigargin inhibits endoplasmic reticulum Ca^{2+}-ATPases, empties endoplasmic reticulum Ca^{2+} stores, increases cytosolic and mitochondrial Ca^{2+}, opens the mitochondrial permeability transition pore (HOEK et al. 1997), and causes apoptosis in several cell types (TREIMAN et al. 1998), including hepatoma cells (TSUKAMATO and KANEKO 1993).

Staurosporine, a bacterial alkaloid, is a potent inhibitor of protein kinase C and other cell cycle-dependent protein kinases (SWE and SIT 1997). It induces telophase arrest and apoptosis in Chang liver cells (SWE and SIT 1997). In addition, p53 and c-Myc proteins are increased (JIANG et al. 1996).

Vespa orientalis (Oriental hornet) venom, a complex of endonuclease and phospholipase, produces apoptosis in normal human hepatocytes (NEUMAN et al. 1991) by inducing the mitochondrial permeability transition.

The hepatic apoptosis induced by administration of the plant lectin, *concanavalin A*, bacterial *lipopolyssacharides*, or the *Pseudomona aeruginosa exotoxin A* is mediated by TNF-α release and has been considered in the section Tumor Necrosis Factor-α-Mediated Cell Death.

Many other natural substances have been shown to induce apoptosis in nonhepatic cells, although their effects remain to be tested in hepatocytes.

G. Conclusions and Perspectives

The progress made in the last ten years has revolutionized our earlier conceptions of cell life and death. Due to the rapid disappearance of apoptotic cells and the long persistence of necrotic cells, past descriptions of human liver lesions mainly reported necrosis, while apoptosis-like lesions were rarely mentioned. The programmed cell death occurring during embryogenesis or cellular turnover was opposed to the necrotic cell death caused by immune reactions, man-made chemicals, or foreign toxins. In the last ten years, however, we have learned that viral hepatitis and several forms of immune-mediated hepatitis may initially involve an apoptotic process (often associated with secondary or concomitant necrosis), whereas foreign molecules can induce apoptosis, necrosis, or both. Indeed, several plants are smart enough to use the apoptotic machinery of the cell to kill the animals that consume them.

An ever increasing number of endogenous substances are being recognized as signaling proapoptotic or antiapoptotic messages in diverse, paracrine, autocrine, or intracellular pathways. We are also beginning to learn how the cell integrates these opposite signals. Whereas the mitochondrial permeability transition was initially considered to be an in vitro oddity with little in vivo significance, the seminal works of Guido Kroemer and others have placed mitochondria at the center of the cell's decision either to live or to die, and to orient cell death either towards necrosis (through ATP depletion) or apoptosis (through the caspase/mitochondria/caspase reinforcing loop). We are also beginning to understand the several inhibitory molecules (FLICE, c-IAPs, Bcl-2, Bcl-X_L) that prevent this auto-potentiating loop from killing all cells as soon as any caspase is activated.

The therapeutic applications that can be foreseen in the next ten years are even more fascinating. We already know how to prevent hepatocyte apoptosis in several animal models, so that clinical applications should be forthcoming. Whereas prolonged inhibition of apoptosis would be dangerous (due to its beneficial role in tissue homeostasis, viral eradication, and cancer prevention), short-term inhibition of immune-mediated apoptosis might be life saving in immune-mediated, drug-induced fulminant hepatitis. In patients with viral-induced fulminant hepatitis, antiapoptotic strategies might be combined with anti-viral agents to avoid chronicity.

Indeed, whereas there may be too much apoptosis in fulminant viral hepatitis, there may not be enough in chronic viral hepatitis. Insufficient immunologic destruction of infected hepatocytes may allow viral persistence in these cases. After a period of anti-viral therapy alone, agents that would increase the apoptosis of infected hepatocytes might complete viral eradication, without exposing the patient to the risk of fulminant hepatitis.

In cancer, finally, some agents seem to kill neoplastic hepatocytes selectively without killing normal cells. Hopefully these agents will improve the presently disappointing management of unresectable hepatocarcinomas.

References

Adachi M, Suematsu S, Kondo T, Ogasawara J, Tanaka T, Yoshida N, Nagata S (1995) Targeted mutation in the Fas gene causes hyperplasia in peripheral lymphoid organs and liver. Nature Genetics 11:294–300

Adams JM, Cory S (1998) The Bcl-2 protein family: arbiters of cell survival. Science 281:1322–1326

Ahn SG, Jeong SY, Rhim H, Kim IK (1998) The role of c-Myc and heat shock protein 70 in human hepatocarcinoma Hep3B cells during apoptosis induced by prostaglandin A_2/Δ^{12}-prostaglandin J_2. Biochem Biophys Acta 1448:115–125

Akerman P, Cote P, Yang SQ, McClain C, Nelson S, Bagby GJ, Diehl AM (1992) Antibodies to tumor necrosis factor-α inhibit liver regeneration after partial hepatectomy. Am J Physiol 263:G579-G585

Ando K, Hiroishi K, Kaneko T, Moriyama T, Muto Y, Kayagaki N, Yagita H, Okumura K, Imawari M (1997) Perforin, Fas/Fas ligand, and TNF-α pathways as specific and bystander killing mechanisms of hepatitis C virus-specific human CTL. J Immunol 158:5283–5291

Angermüller S, Künstle G, Tiegs G (1998) Pre-apoptotic alterations in hepatocytes of TNFα-treated galactosamine-sensitized mice. J Histochem Cytochem 46:1175–1183

Aragane Y, Kulms D, Metze D, Wilkes G, Pöppelmann B, Luger TA, Schwarz T (1998) Ultraviolet light induces apoptosis via direct activation of CD95(Fas/APO-1) independently of its ligand CD95L. J Cell Biol 140:171–182

Arber M, Zajicek G, Ariel I (1988) The streaming liver. II. Hepatocyte life history. Liver 8:80–87

Arsura M, FitzGerald MJ, Fausto N, Sonenshein GE (1997) Nuclear factor-κB/Rel blocks transforming growth factor β1-induced apoptosis of murine hepatocyte cell lines. Cell Growth Differentiation 8:1049–1059

Arthur MJP, Bentley IS, Tanner AR, Kowalski Saunders P, Millward-Sadler GH, Wright R (1985) Oxygen-derived radicals promote hepatic injury in the rat. Gastroenterology 89:1114–1122

Ashkenazi A, Dixit VM (1998) Death receptors: signaling and modulation. Science 281:1305–1308

Bamberger AM, Schulte HM, Thuneke I, Erdmann I, Bamberger CM, Asa SL (1997) Expression of the apoptosis-inducing Fas ligand (FASL) in human first and third trimester placenta and choriocarcinoma cells. J Clin Endocrin Metab 82:3173–3175

Bayer EM, Herr W, Kanzler S, Waldmann C, Meyer Zum Büschenfelde KH, Dienes HP, Lohse AW (1998) Transforming growth factor-β1 in autoimmune hepatitis: correlation of liver tissue expression and serum levels with disease activity. J Hepatol 28:803–811

Bayly AC, Roberts RA, Dive C (1994) Suppression of liver cell apoptosis in vitro by the non-genotoxic hepatocarcinogen and peroxisome proliferator nafenopin. J Cell Biol 125:197–203

Bedossa P, Paradis V (1995) Transforming growth factor-beta (TGF-beta): a key role in liver fibrogenesis. J Hepatol 22:37–42

Beg AA, Baltimore D (1996) An essential role for NF-κB in preventing TNF-α-induced cell death. Science 274:782–784

Beg AA, Sha WC, Bronson RT, Ghosh S, Baltimore D (1995) Embryonic lethality and liver degeneration in mice lacking the RelA component of NF-κB. Nature 376: 167–170

Bellas RE, FitzGerald MJ, Fausto N, Sonenshein GE (1997) Inhibition of NF-κB activity induces apoptosis in murine hepatocytes. Am J Pathol 151:891–896

Bellomo G, Orrenius S (1985) Altered thiol and calcium homeostasis in oxidative hepatocellular injury. Hepatology 5:876–882

Benedetti A, Di Sario A, Baroni GS, Jezequel AM (1995) Transforming growth factor β1 increases the number of apoptotic bodies and decreases intracellular pH in isolated periportal and perivenular rat hepatocytes. Hepatology 22:1488–1498

Benedetti A, Jezequel AM, Orlandi F (1988) A quantitative evaluation of apoptotic bodies in rat liver. Liver 8:172–177

Benz C, Angermüller S, Klöters-Plachky P, Sauer P, Stremmel W, Stiehl A (1998a) Effect of S-adenosylmethionine versus tauroursodeoxycholic acid on bile acid-induced apoptosis and cytolysis in rat hepatocytes. Eur J Clin Invest 28:577–583

Benz C, Angermüller S, Tox U, Klöters-Plachky P, Riedel HD, Sauer P, Stremmel W, Stiehl A (1998b) Effect of tauroursodeoxycholic acid on bile-acid-induced apoptosis and cytolysis in rat hepatocytes. J Hepatol 28:99–106

Bernuau D, Feldmann G, Degott C, Gisselbrecht C (1981) Ultrastructural lesions of bile ducts in primary biliary cirrhosis. A comparison with the lesions observed in graft versus host disease. Hum Pathol 12:782–793

Beuers U, Boyer JL, Paumgartner G (1998) Ursodeoxycholic acid in cholestasis: potential mechanisms of action and therapeutic applications. Hepatology 28:1449–1453

Beutler B, Cerami A (1988) Tumor necrosis, cachexia, shock, and inflammation: a common mediator. Ann Rev Biochem 57:505–518

Bissel DM, Jarnagin WR, Roll FJ (1995) Cell specific expression of transforming growth factor-beta in rat liver: evidence for autocrine regulation of hepatocyte proliferation. J Clin Invest 96:447–455

Black RA, Rauch CT, Kozlosky CJ, Peschon JJ, Slack JL, Wolfson MF, Castner BJ, Stocking KL, Reddy P, Srinivasan S, Nelson N, Boiani N, Schooley KA, Gerhart M, Davis R, Fitzner JN, Johnson RS, Paxton RJ, March CJ, Cerretti DP (1997) A metalloproteinase disintegrin that releases tumour necrosis factor-α from cells. Nature 385:729–733

Bobé P, Bonardelle D, Reynès M, Godeau F, Mahiou J, Joulin V, Kiger N (1997) Fas-mediated damage in MRL hemopoietic chimeras undergoing *lpr*-mediated graft versus host disease. J Immunol 159:4197–4204

Boelsterli UA, Göldlin C (1991) Biomechanisms of cocaine-induced hepatocyte injury mediated by the formation of reactive metabolites. Arch Toxicol 65:351–360

Botla R, Spivey JR, Bronk SF, Gores GJ (1995) Ursodeoxycholic acid (UDCA) inhibits the mitochondrial membrane permeability transition induced by glycochenodeoxycholate: a mechanism of UDCA cytoprotection. J Pharmacol Exp Ther 272:930–938

Borello S, Demple B (1997) NF-κB-independent transcriptional induction of the human manganous superoxide dismutase gene. Arch Biochem Biophys 348:289–294

Bour ES, Ward LK, Cornman GA, Isom HC (1996) Tumor necrosis factor-α-induced apoptosis in hepatocytes in long-term culture. Am J Pathol 148:485–495

Bradham CA, Qian T, Streetz K, Trautwein C, Brenner DA, Lemasters JJ (1998) The mitochondrial permeability transition is required for tumor necrosis factor alpha-mediated apoptosis and cytochrome c release. 18:6353–6364

Bulfone-Paus S, Ungureanu D, Pohl T, Lindner G, Paus R, Rückert R, Krause H, Kunzendorf U (1997) Interleukin-15 protects from lethal apoptosis in vivo. Nature Medicine 3:1124–1128

Cai J, Jones DP (1998) Superoxide in apoptosis. Mitochondrial generation triggered by cytochrome c loss. J Biol Chem 273:11401–11404

Cain K, Inayat-Hussain SH, Couet C, Cohen GM (1996) A cleavage-site-directed inhibitor of interleukin-1β-converting enzyme-like proteases inhibits apoptosis in primary cultures of rat hepatocytes. Biochem J 314:27–32

Canman CE, Kastan MB (1997) Role of p53 in apoptosis. Adv Pharmacol 41:429–460

Cascales M, Alvarez A, Gasco P, Fernandez-Simon L, Sanz N, Bosca L (1994) Cocaine-induced liver injury in mice exhibits specific changes in DNA ploidy and induces programmed death of hepatocytes. Hepatology 20:992–1001

Casini A, Pinzani M, Milani S, Grappone C, Galli G, Jezequel AM, Schuppan D, Rotella CM, Surrenti C (1993) Regulation of extracellular matrix synthesis by transforming growth factor β1 in human fat storing cells. Gastroenterology 105: 245–253

Chambon P, Olson JA, Ross AC (1996) The retinoid revolution. FASEB J 10:939–1040

Chiba T, Takahashi S, Sato N, Ishii S, Kikuchi K (1996) Fas-mediated apoptosis is modulated by intracellular glutathione in human T cells. Eur J Immunol 26: 1164–1169

Chieco P, Romagnoli E, Aicardi G, Suozzi A, Forti GC, Roda A (1997) Apoptosis induced in rat hepatocytes by in vivo exposure to taurochenodeoxycholate. Histochem J 29:875–883

Chinnaiyan AM, O'Rourke K, Tewari M, Dixit VM (1995) FADD, a novel death domain-containing protein, interacts with the death domain of Fas and initiates apoptosis. Cell 81:505–512

Choi KS, Lim IK, Brady JN, Kim SJ (1998) ICE-like protease (caspase) is involved in transforming growth factor β1-mediated apoptosis in FaO rat hepatoma cell line. Hepatology 27:415–421

Chow SC, Kass GEN, Orrenius S (1997) Purines and their role in apoptosis. Neuropharmacology 36:1149–1156

Chuang LY, Hung WC, Chang CC, Tsai JH (1994) Characterization of apoptosis induced by transforming growth factor-β1 in human hepatoma cells. Anticancer Res 14:147–152

Chunfang G, Gressner G, Zoremba M, Gressner AM (1996) Transforming growth factor β (TGF-β) expression in isolated and cultured rat hepatocytes. J Cell Physiol 167:394–405

Cifone MG, Roncaioli P, De Maria R, Camarda G, Santoni A, Ruberti G, Testi R (1995) Multiple pathways originate at the Fas/APO-1(CD95) receptor: sequential involvement of phosphatidylcholine-specific phospholipase C and acidic sphingomyelinase in the propagation of the apoptotic signal. Embo J 14:5859–5868

Clarke AR, Purdie CA, Harrison DJ, Morris RG, Bird CC, Hooper ML, Wyllie AH (1993) Thymocyte apoptosis induced by p53-dependent and independent pathways. Nature 362:849–852

Columbano A, Shinozuka H (1996) Liver regeneration versus direct hyperplasia. FASEB J 10:1118–1128

Cullum ME, Zile MH (1985) Metabolism of all-trans-retinoic acid and all-trans-retinyl acetate. Demonstration of common physiological metabolites in rat small intestinal mucosa and circulation. J Biol Chem 260:10590–10596

Darmon AJ, Nicholson DW, Bleakley RC (1995) Activation of the apoptotic protease CPP32 by cytosolic T-cell-derived granzyme B. Nature 377:446–448

Date M, Matsuraki K, Matsushita M, Sakitani K, Shibano K (1998) Differential expression of transforming growth factor-β and its receptors in hepatocytes and non-parenchymal cells of rat liver after CCl_4 administration. J Hepatol 28:572–581

De Bleser PJ, Niki T, Xu G, Rogiers V, Geerts A (1997) Localization and cellular sources of activins in normal and fibrotic rat liver. Hepatology 26:905–912

De Moissac D, Mustapha S, Greenberg AH, Kirshenbaum LA (1998) Bcl-2 activates the transcription factor NF-κB through the degradation of the cytoplasmic inhibitor IκBα. J Biol Chem 273:23946–23951

Decoster E, Vanhaesebroeck B, Vandenabeele P, Grooten J, Fiers W (1995) Generation and biological characterization of membrane-bound, uncleavable murine tumor necrosis factor. J Biol Chem 270:18473–18478

Devi BG, Chan AWK (1997) Impairment of mitochondrial respiration and electron transport chain enzymes during cocaine-induced hepatic injury. Life Sci 60:849–855

Erickson SL, de Sauvage FJ, Kikly K, Carver-Moore K, Pitts-Meek S, Gillet N, Sheehan KCF, Schreiber RD, Goeddel DV, Moore MW (1994) Decreased sensitivity to tumour-necrosis factor but normal T-cell development in TNF receptor-2-deficient mice. Nature 372:560–563

Evan G, Littlewood T (1998) A matter of life and death. Science 281:1317–1322

Falasca L, Favale A, Gualandi G, Majetta G, Devirgilis LC (1998) Retinoic acid treatment induces apoptosis or expression of more differentiated phenotype on different fractions of cultured fetal rat hepatocytes. Hepatology 28:727–737

Fan G, Ma X, Kren BT, Steer CS (1996) The retinoblasma gene product inhibits TGF-β1 induced apoptosis in primary rat hepatocytes and human HuH-7 hepatoma cells. Oncogene 12:1909–1919

Fang C, Lindros KO, Badger TM, Ronis MJJ, Ingelman-Sundberg M (1998) Zonated expression of cytokines in rat liver: effects of chronic ethanol and the cytochrome P450 2E1 inhibitor, chlormethiazole. Hepatology 27:1304–1310

Farrell GC (1994) Drug-induced liver disease. Churchill Livingstone: Melbourne

Fau D, Lekehal M, Farrell G, Moreau A, Moulis C, Feldmann G, Haouzi D, Pessayre D (1997) Diterpenoids from germander, a herbal medicine, induce apoptosis in isolated rat hepatocytes. Gastroenterology 113:1334–1346

Faubion WA, Guicciardi ME, Miyoshi H, Bronk SF, Roberts PJ, Svingen PA, Kaufmann SH, Gores GJ (1999) Toxic bile salts induce rodent hepatocyte apoptosis via direct activation of Fas. J Clin Invest 103:137–145

Feldmann G (1997) Liver apoptosis. J Hepatol 26 (Suppl. 2):1–11

Ferenbach DA, Haydon GH, Rae F, Malcomson RDG, Harrison DJ (1997) Alteration in mRNA levels of Fas splice variants in hepatitis C-infected liver. J Pathol 183:299–304

Fesus L, Madi A, Balajthy Z, Nemes Z, Szondy Z (1996) Transglutaminase induction by various cell death and apoptosis pathways. Experientia 52:942–949

Fisher GN, Rosenberg FJ, Strauss SE, Dale JK, Middleton LA, Lin AY, Strober W, Lenardo MJ, Puck JM (1995) Dominant interfering Fas gene mutations impair apoptosis in a human lymphoproliferative syndrome. Cell 81:935–946

Fromenty B, Pessayre D (1995) Inhibition of mitochondrial β-oxidation as a mechanism of hepatotoxicity. Pharmacol Ther 67:101–154

Fukuda K, Xie R, Chiu JF (1991) Demonstration of cross-linked cytokeratin polypeptides in transplantable rat hepatoma cells. Biochem Biophys Res Commun 176:441–446

Fukuda K, Kojiro M, Chiu JF (1993) Induction of apoptosis by transforming growth factor-β1 in the rat hepatoma cell line McA-RH7777: a possible association with tissue transglutaminase expression. Hepatology 18:945–953

Galle PR (1997) Apoptosis in liver disease. J Hepatol 27:405–412

Galle PR, Hofmann WJ, Walczak H, Schaller H, Otto G, Stremmel W, Krammer PH, Runkell L (1995) Involvement of the CD95 (APO-1/Fas) receptor and ligand in liver damage. J Exp Med 182:1223–1230

Gamen S, Anel A, Pineiro A, Naval J (1998) Caspases are the main executioners of Fas-mediated apoptosis, irrespective of the ceramide signalling pathway. Cell Death Differ 5:241–249

Gantner F, Leist M, Lohse AW, Germann PG, Tiegs G (1995) Concanavalin A-induced T-cell mediated hepatic injury in mice: the role of tumor necrosis factor. Hepatology 21:190–198

Gao C, Gressner G, Zoremba M, Gressner AM (1996) Transforming growth factor β (TGF-β) expression in isolated and cultured rat hepatocytes. J Cell Physiol 167: 394–405

Gearing AJH, Beckett P, Christodoulou M, Churchill M, Clements J, Davidson AH, Drummond AH, Galloway WA, Gilbert R, Gordon JL, Leber TM, Mangan M, Miller K, Nayee P, Owen K, Patel S, Thomas W, Wells G, Wood LM, Wooley K (1994) Processing of tumour necrosis factor-α precursor by metalloproteinases. Nature 370:555–557

Gill JH, James NH, Roberts RA, Dive C (1998) The non-genotoxic hepatocarcinogen nafenopin suppresses rodent hepatocyte apoptosis induced by TGF-β_1, DNA damage and Fas. Carcinogenesis 19:299–304

Gong JH, Sprenger H, Hinder F, Bender A, Schmidt A, Horch S, Nain M, Gemsa D (1991) Influenza virus infection of macrophages. Enhanced tumor necrosis factor-α (TNF-α) gene expression and lipopolysaccharide-triggered TNF-α release. J Immunol 147:3507–3513

Gonzalez-Amaro R, Garcia-Monzón C, Garcia-Buey L, Moreno-Otero R, Alonso JL, Yagüe E, Pivel JP, López-Cabrera M, Fernández-Ruiz E, Sánchez-Madrid F (1994) Induction of tumor necrosis factor α production by human hepatocytes in chronic viral hepatitis. J Exp Med 179:841–848

Gottfried MR, Kloss MW, Graham D, Rauckman EJ, Rosen GM (1986) Ultrastructure of experimental cocaine hepatotoxicity. Hepatology 6:299–304

Grande JP (1997) Role of transforming growth factor-β in tissue injury and repair. Proc Soc Exp Biol Med 214:27–40

Green DR, Kroemer G (1998) The central executioners of apoptosis: caspases or mitochondria? Trends Cell Biol 8:267–271

Green DR, Reed JC (1998) Mitochondria and apoptosis. Science 281:1309–1312

Gressner AM, Polzar B, Lahme B, Mannhertz HG (1996) Induction of rat liver parenchymal cell apoptosis by hepatic myofibroblasts via transforming growth factor β. Hepatology 23:571–581

Gressner AM, Lahme B, Mannhertz HG, Polzar B (1997) TGF-β-mediated hepatocellular apoptosis by rat and human hepatoma cells and primary rat hepatocytes. J Hepatol 26:1079–1092

Grewe M, Gausling R, Gyufko R, Hoffmann R, Decker K (1994) Regulation of the mRNA expression for tumor necrosis factor-α in rat liver macrophages. J Hepatol 20:811–818

Griffith TS, Yu X, Herndon JM, Green DR, Ferguson TA (1996) CD95-induced apoptosis of lymphocytes in an immune privileged site induces immunological tolerance. Immunity 5:7–16

Gross A, Yin XM, Wang K, Wei MC, Jockel J, Milliman C, Erdjument-Bromage H, Tempst P, Korsmeyer SJ (1999) Caspase cleaved BID targets mitochondria and is required for cytochrome c release, while BCL-X_L prevents this release but not tumor necrosis factor-R1/Fas death. J Biol Chem 274:1156–1163

Grove J, Daly AK, Bassendine MF, Day CP (1997) Association of a tumor necrosis factor promoter polymorphism with susceptibility to alcoholic steatohepatitis. Hepatology 26:143–146

Guilhot S, Miller T, Cornman G, Isom HC (1996) Apoptosis induced by tumor necrosis factor-α in rat hepatocyte cell lines expressing hepatitis B virus. Am J Pathol 148:801–814

Haldar S, Chintapalli J, Croce CM (1996) Taxol induces Bcl-2 phosphorylation and death of prostate cancer cells. Cancer Res 56:1253–1255

Hanabuchi S, Koyanagi M, Kawasaki A, Shinohara N, Matsuzawa A, Nishimura Y, Kobayashi Y, Yonehara S, Yagita H, Okumura K (1994) Fas and its ligand in a general mechanism of T-cell-mediated cytotoxicity. Proc Natl Acad Sci USA 91: 4930–4934

Harada K, Ozaki S, Gershwin ME, Nakamura Y (1997) Enhanced apoptosis relates to bile duct loss in primary biliary cirrhosis. Hepatology 26:1399–1405

Harbrecht BG, Di Silvio M, Demetris AJ, Simmons RL, Billiar TR (1994a) Tumor necrosis factor-α regulates in vivo nitric oxide synthesis and induces liver injury during endotoxemia. Hepatology 20:1055–1060

Harbrecht BG, Stadler J, Demetris AJ, Simmons RL, Billiar TR (1994b) Nitric oxide and prostaglandin interact to prevent hepatic damage during murine endotoxemia. Am J Physiol 266:G1004–G1010

Harbrecht BG, Wu B, Watkins DL, Marshall HP Jr, Peitzman AB, Billiar TR (1995) Inhibition of nitric oxide synthase during hemorrhagic shock increases hepatic injury. Shock 4:332–337

Heldin CH, Miyazono K, Dijke PT (1997) TGF-β signalling from cell membrane to nucleus trough SMAD proteins. Nature 390:465–471

Hill DB, Schmidt J, Shedlofsky SI, Cohen DA, McClain CJ (1995) In vitro tumor necrosis factor cytotoxicity in HepG2 liver cells. Hepatology 21:1114–1119

Hiramatsu N, Hayashi N, Katayama K, Mochizuki K, Kawanishi Y, Kasahara A, Fusamoto H, Kamada T (1994) Immunohistochemical detection of Fas antigen in liver tissue of patients with chronic hepatitis C. Hepatology 19:1354–1359

Hishinuma I, Nagakawa JI, Hirota K, Miyamoto K, Tsukidate K, Yamanaka T, Katayama KI, Yamatsu I (1990) Involvement of tumor necrosis factor-α in development of hepatic injury in galactosamine-sensitized mice. Hepatology 12:1187–1191

Hoek JB, Walajtys-Rode E, Wang X (1997) Hormonal stimulation, mitochondrial Ca^{2+} accumulation, and the control of the mitochondrial permeability transition in intact hepatocytes. Mol Cell Biochem 174:173–179

Hsu H, Xiong J, Goeddel V (1995) The TNF receptor 1-associated protein TRADD signals cell death and NF-κB activation. Cell 81:495–504

Hsu H, Shu H-B, Pan M-G, Goeddel DV (1996) TRADD-TRAF2 and TRADD-FADD interactions define two distinct TNF receptor 1 signal transduction pathways. Cell 84:299–308

Huang TS, Shu CH, Shih YL, Huang HC, Su YC, Chao Y, Yang WK, Whang-Peng J (1996) Protein tyrosine phosphatase activities are involved in apoptotic cancer cell death induced by GL331, a new homolog of etoposide. Cancer Lett 110:77–85

Hug H, Strand S, Grambihler A, Galle J, Hack V, Stremmel W, Krammer PH, Galle PR (1997) Reactive oxygen intermediates are involved in the induction of CD95 ligand mRNA expression by cytostatic drugs in hepatoma cells. J Biol Chem 272:28191–28193

Hully JR, Chang L, Schwall RH, Widmer HR, Terrell TG, Gillett NA (1994) Induction of apoptosis in the murine liver with recombinant human activin A. Hepatology 20:854–861

Hung WC, Chuang LY (1996) Induction of apoptosis by sphingosine-1-phosphate in human hepatoma cells is associated with enhanced expression of Bax gene product. Biochem Biophys Res Commun 229:11–15

Hüser J, Rechenmacher CE, Blatter LA (1998) Imaging the permeability pore transition in single mitochondria. Biophys J 74:2129–2137

Iimuro Y, Gallucci RM, Luster MI, Kono H, Thurman RG (1997) Antibodies to tumor necrosis factor alpha attenuate hepatic necrosis and inflammation caused by chronic exposure to ethanol in the rat. Hepatology 26:1530–1537

Inayat-Hussain SH, Couet C, Cohen GM, Cain K (1997) Processing/activation of CPP32-like proteases is involved in transforming growth factor β1-induced apoptosis in rat hepatocytes. Hepatology 25:1516–1526

Jaffrézou JP, Levade T, Bettaïeb A, Andrieu N, Bezombes C, Maestre N, Vermeersch S, Rousse A, Laurent G (1996) Daunorubicin-induced apoptosis: triggering of ceramide generation through sphingomyelin hydrolysis. EMBO J 15:2417–2424

Jiang MC, Yang-Yen HF, Lin JK, Yen JJY (1996) Differential regulation of p53, c-Myc, Bcl-2 and Bax protein expression during apoptosis induced by widely different stimuli in human hepatoblastoma cells. Oncogene 13:609–616

Jones BA, Gores GJ (1997) Physiology and pathophysiology of apoptosis in epithelial cells of the liver, pancreas, and intestine. Am J Physiol 273:G1174–G1188

Jones BA, Rao YP, Stravitz RT, Gores GJ (1997) Bile salt-induced apoptosis of hepatocytes involves activation of protein kinase C. Am J Physiol 272:G1109–G1115

Jones EY, Stuart DI, Walker NPC (1989) Structure of tumor necrosis factor. Nature 338:225–228

Jones RA, Johnson VL, Buck NR, Dobrota M, Hinton RH, Chow SC, Kass GEN (1998) Fas-mediated apoptosis in mouse hepatocytes involves the processing and activation of caspases. Hepatology 27:1632–1642

Joseph B, Lefebre O, Méreau-Richard C, Danzé PM, Belin-Plancot MT, Formstecher P (1998) Evidence for the involvement of both retinoic acid receptor- and retinoic X receptor-dependent signaling in the induction of tissue transglutaminase in the human myeloma cell line RPMI 8226. Blood 91:2423–2432

Kägi D, Vignaux F, Ledermann B, Bürki K, Depraetere V, Nagata S, Hengartner H, Goldstein P (1994) Fas and perforin as molecular mechanisms of T cell-mediated cytotoxicity. Science 265:528–530

Kanofsky JR (1986) Single oxygen production by bleomycin. A comparison with heme-containing compounds. J Biol Chem 261:13546–13550

Kasahara Y, Wada T, Niida Y, Yachie A, Seki H, Ishida Y, Sakai T, Koizumi F, Koizumi S, Miyawaki T, Tanigushi N (1998) Novel Fas (CD95/APO-1) mutations in infants with a lymphoproliferative disorder. Int Immunol 10:195–202

Kataoka T, Schröter M, Hahne M, Shneider P, Irmler M, Thome M, Froelich CJ, Tschopp J (1998) FLIP prevents apoptosis induced by death receptors but not by perforin/granzyme B, chemotherapeutic drugs, and gamma irradiation. J Immunol 161:3936–3942

Khoruts A, Stahnke L, McClain CJ, Logan G, Allen JI (1991) Circulating tumor necrosis factor, interleukin-1 and interleukin-6 concentration in chronic alcoholic patients. Hepatology 13:267–276

Kim YM, Bergonia H, Lancaster JR Jr (1995) Nitrogen oxide-induced autoprotection in isolated rat hepatocytes. FEBS Lett 374:228–232

Kim DG, Jo BH, You KY, Ahn DS (1996) Apoptosis induced by retinoic acid in Hep3B cells in vitro. Cancer Lett 107:149–159

Kim YM, de Vera ME, Watkins SC, Billiar TR (1997a) Nitric oxide protects rat hepatocytes from tumor necrosis factor-α-induced apoptosis by inducing heat shock protein 70 expression. J Biol Chem 272:1402–1411

Kim YM, Talanian RV, Billiar TR (1997b) Nitric oxide inhibits apoptosis by preventing increases in caspase-3-like activity via two distinct mechanisms. J Biol Chem 272:31138–31148

Koga H, Sakisaka S, Ohishi M, Sata M, Tanikawa K (1997) Nuclear DNA fragmentation and expression of Bcl-2 in primary biliary cirrhosis. Hepatology 25:1077–1084

Kondo T, Suda T, Fukuyama H, Adachi M, Nagata S (1997) Essential roles of the Fas ligand in the development of hepatitis. Nature Medicine 3:409–413

Kosai K, Matsumoto K, Nagata S, Tsujimoto Y, Nakamura T (1998) Abrogation of Fas-induced fulminant hepatic failure in mice by hepatocyte growth factor. Biochem Biophys Res Commun 244:683–690

Krähenbühl S, Fischer S, Talos C, Reichen J (1994a) Ursodeoxycholate protects oxidative mitochondrial metabolism from bile acid toxicity: dose-response study in isolated rat liver mitochondria. Hepatology 20:1595–1601

Krähenbühl S, Talos C, Fischer S, Reichen J (1994b) Toxicity of bile acids on the electron transport chain of isolated rat liver mitochondria. Hepatology 19:471–479

Krams SM, Fox CK, Beatty R, Cao S, Viallueva JC, Esquivel CO, Martinez OM (1998) Human hepatocytes produce an isoform of Fas that inhibits apoptosis. Transplantation 65:713–721

Kriegler M, Perez C, DeFay K, Albert I, Lu SD (1988) A novel form of TNF/cachectin is a cell surface cytotoxic transmembrane protein: ramifications for the complex physiology of TNF. Cell 53:45–53

Ksontini R, Colagiovanni DB, Josephs MD, Edwards III CK, Tannahill CL, Solorzano CC, Norman J, Denham W, Clare-Salzler M, MacKay SLD, Moldawer LL (1998) Disparate roles for TNF-α and Fas ligand in concanavalin A-induced hepatitis. J Immunol 160:4082–4089

Künstle G, Leist M, Uhlig S, Revesz L, Feifel R, MacKenzie A, Wendel A (1997) ICE-protease inhibitors block murine liver injury and apoptosis caused by CD95 or by TNF-α. Immunol Lett 55:5–10

Kuo KW, Hsu SH, Sheu HM, Chai CY, Lin CN (1997) Solamargine purified from *Solanum incanum* Chinese herb initiates apoptosis of hepatoma cell by triggering the gene expression of human TNFR1 and 2. FASEB J 11:A1114

Kuo TH, Kim HRC, Zhu L, Yu Y, Lin HM, Tsang W (1998) Modulation of endoplasmic reticulum calcium pump by Bcl-2. Oncogene 17:1903–1910

Kwo P, Patel T, Bronk SF, Gores GJ (1995) Nuclear serine protease activity contributes to bile acid-induced apoptosis in hepatocytes. Am J Physiol 268:G613–G621

Lacronique V, Mignon A, Fabre B, Viollet B, Rouquet N, Molina T, Porteu A, Henrion A, Bouscary D, Varlet P, Joulin V, Kahn A (1996) Bcl-2 protects from lethal hepatic apoptosis induced by an anti-Fas antibody in mice. Nature Medicine 2:80–86

Lammer EJ, Chen, DT, Hoar RM, Agnish ND, Benke PJ, Braun JT, Curry CJ, Fernhoff PM, Grix AW, Lott IT, Richard JM, Sun SC (1985) Retinoic acid embryopathy. N Engl J Med 313:837–841

Lara-Pezzi E, Majano PL, Gómez-Gonzalo, Garcóa-Monzón, Moreno-Otero R, Levrero M, López-Cabrera M (1998) The hepatitis B virus X protein up-regulates tumor necrosis factor α gene expression in hepatocytes. Hepatology 28:1013–1021

Larrey D (1997) Hepatotoxicity of herbal remedies. J Hepatol 26 (Suppl. 1): 47–51

Larrey D, Vial T, Pauwels A, Castot A, Biour M, David M, Michel H (1992) Hepatitis after germander (*Teucrium chamaedrys*) administration: another instance of herbal medicine hepatotoxicity. Ann Intern Med 117:129–132

Lee JH, Kim HS, Jeoung SY, Kim IK (1995) Induction of p53 and apoptosis by Δ^{12}-PGJ$_2$ in human hepatocarcinoma SK-Hep-1 cells. FEBS Lett 368:348–352

Lee J, Richburg JH, Younkin SC, Boekelheide K (1997) The Fas system is a key regulator of germ cell apoptosis in the testis. Endocrinology 138:2081–2088

Lee MJ, Whitehouse MW (1965) Inhibition of electron transport and coupled phosphorylation in liver mitochondria by cholanic (bile) acids and their conjugates. Biochim Biophys Acta 100:317–328

Leist M, Gantner F, Bohlinger I, Germann PG, Tiegs G, Wendel A (1994) Murine hepatocyte apoptosis induced in vitro and in vivo by TNF-α requires transcriptional arrest. J Immunol 153:1778–1788

Leist M, Gantner F, Bohlinger I, Tiegs G, Germann PG, Wendel A (1995a) Tumor necrosis factor-induced hepatocyte apoptosis precedes liver failure in experimental murine shock models. Am J Pathol 146:1220–1234

Leist M, Gantner F, Jilg S, Wendel A (1995b) Activation of the 55 kDa TNF receptor is necessary and sufficient for TNF-induced liver failure, hepatocyte apoptosis, and nitrite release. J Immunol 154:1307–1316

Leist M, Gantner F, Künstle G, Bohlinger I, Tiegs G, Bluethmann H, Wendel A (1996) The 55-kD tumor necrosis factor receptor and CD95 independently signal murine hepatocyte apoptosis and subsequent liver failure. Mol Medicine 2:109–124

Leist M, Gantner F, Naumann H, Bluethmann H, Vogt K, Brigelius-Flohé R, Nicotera P, Volk H-D, Wendel A (1997a) Tumor necrosis factor-induced apoptosis during the poisoning of mice with hepatotoxins. Gastroenterology 112:923–934

Leist M, Single B, Castoldi AF, Kühnle S, Nicotera P (1997b) Intracellular adenosine triphosphate (ATP) concentration: a switch in the decision between apoptosis and necrosis. J Exp Med 185:1481–1486

Lekehal M, Pessayre D, Lereau JM, Moulis C, Fourasté I, Fau D (1996) Hepatotoxicity of the herbal medicine, germander. Metabolic activation of its furano diterpenoids by cytochrome P450 3A depletes cytoskeleton-associated protein thiols and forms plasma membrane blebs in rat hepatocytes. Hepatology 24:212–218

Lemasters JL, Nieminen AL, Qian T, Trost LC, Elmore SP, Nishimura Y, Crowe RA, Cascio WA, Bradham CA, Brenner DA, Herman B (1998) The mitochondrial permeability transition in cell death: a common mechanism in necrosis, apoptosis and autophagy. Biochim Biophys Acta 1366:177–196

Lettéron P, Duchatelle V, Berson A, Fromenty B, Fisch C, Degott C, Benhamou JP, Pessayre D (1993) Increased ethane exhalation, an in vivo index of lipid peroxidation, in alcohol-abusers. Gut 34:409–414

Li H, Zhu H, Xu C, Yuan J (1998) Cleavage of BID by caspase 8 mediates the mitochondrial damage in the Fas pathway of apoptosis. Cell 94:491–501

Lieser MJ, Park J, Natori S, Jones BA, Bronk SF, Gores GJ (1998) Cholestasis confers resistance to the rat liver mitochondrial permeability transition. Gastroenterology 115:693–701

Lim IK, Joo HJ, Choi KS, Sueoka E, Lee MS, Ryu MS, Fujiki H (1997) Protection of 5α-dihydrotestosterone against TGF-β-induced apoptosis in FaO cells and induction of mitosis in HepG$_2$ cells. Int J Cancer 72:351–355

Lin JK, Chou CK (1992) In vitro apoptosis in the human hepatoma cell line induced by transforming growth factor β1. Cancer Res 5:385–388

Liu Y, Bhalla K, Hill C, Priest DG (1994) Evidence for involvement of tyrosine phosphorylation in taxol-induced apoptosis in a human ovarian tumor cell line. Biochem Pharmacol 48:1265–1272

Llorente L, Richaud-Patin Y, Alcocer-Castillejos N, Ruiz-Soto R, Mercado MA, Orozco H, Gamboa-Dominguez A, Alcocer-Varela J (1996) Cytokine gene expression in cirrhotic and non-cirrhotic human liver. J Hepatol 24:555–563

Loeper J, Descatoire V, Lettéron P, Moulis C, Degott C, Dansette P, Fau D, Pessayre D (1994) Hepatotoxicity of germander in mice. Gastroenterology 106:464–472

Lotz G, Nagy P, Patonai A, Kiss A, Nemes B, Szalay F, Schaff (1998) TGF-β1 and apoptosis in human hepatocellular carcinoma and focal nodular hyperplasia. J Hepatol 28:175A

Luo X, Budihardjo I, Zou H, Slaughter C, Wang X (1998) Bid, a Bcl2 interacting protein, mediates cytochrome c release from mitochondria in response to activation of cell surface death receptors. Cell 94:481–490

McClain CJ, Cohen DA (1989) Increased tumor necrosis factor production by monocytes in alcoholic hepatitis. Hepatology 9:349–351

McGeehan GM, Becherer JD, Bast RC Jr, Boyer CM, Champion B, Connolly KM, Conway JG, Furdon P, Karp S, Kidao S, McElroy AB, Nichols J, Ptyzwansky KM, Schoenen F, Sekut L, Truesdale A, Verghese M, Warner J, Ways JP (1994) Regulation of tumour necrosis factor-α processing by a metalloproteinase inhibitor. Nature 370:558–561

Mansouri A, Fromenty B, Berson A, Robin MA, Grimbert S, Beaugrand M, Erlinger S, Pessayre D (1997a) Multiple hepatic mitochondrial DNA deletions suggest premature oxidative aging in alcoholics. J Hepatol 27:96–102

Mansouri A, Gaou I, Fromenty B, Berson A, Lettéron P, Degott C, Erlinger S, Pessayre D (1997b) Premature oxidative aging of mitochondrial DNA in Wilson's disease. Gastroenterology 113:599–605

Manzano A, Roig T, Bermudez J, Bartrons R (1996) Effects of taxol on isolated rat hepatocyte metabolism. Am J Physiol 40:C1957–C1962

Marzo I, Brenner C, Zamzami N, Jürgensmeier JM, Susin SA, Vieira HLA, Prévost MC, Xie Z, Matsuyama S, Reed JC, Kroemer G (1998) Bax and adenine nucleotide translocator cooperate in the mitochondrial control of apoptosis. Science 281: 2027–2031

Massagué J, Hata A, Lui F (1997) TGF-β signalling through the Smad pathway. Trends Cell Biol 7:187–192

Mills JJ, Chari RS, Boyer IJ, Gould MN, Jirtle RL (1995) Induction of apoptosis in liver tumors by the monoterpene perillyl alcohol. Cancer Res 55:979–983

Mita E, Hayashi N, Lio S, Takehara T, Hijioka T, Kasahara A, Fusamoto H, Kamada T (1994) Role of Fas ligand in apoptosis induced by hepatitis C virus infection. Biochem Biophys Res Commun 204:468–474

Mochizuki K, Hayashi N, Hiramatsu N, Katayama K, Kawanishi Y, Kasahara A, Fusamoto H, Kamada T (1996) Fas antigen expression in liver tissue of patients with chronic hepatitis B. J Hepatol 24:1–7

Mohler KM, Sleath PR, Fitzner JN, Cerretti DP, Alderson M, Kerwar SS, Torrance DS, Otten-Evans C, Greenstreet T, Weerawarna K, Kronheim SR, Petersen M, Gerhart M, Kozlosky CJ, March CJ, Black RA (1994) Protection against a lethal dose of endotoxin by an inhibitor of tumour necrosis factor processing. Nature 370:218–220

Montel AH, Bochan MR, Goebel WS, Brahmi Z (1995) Fas-mediated cytotoxicity remains intact in perforin and granzyme B antisense transfectants of a human NK-like cell line. Cell Immunol 165:312–317

Morikawa A, Sugiyama T, Kato Y, Koide N, Jiang G-Z, Takahashi K, Tamada Y, Yokochi T (1996) Apoptotic cell death in the response of D-galactosamine-sensitized mice to lipopolysaccharide as an experimental endotoxic shock model. Infect Immun 64:734–738

Motoyama S, Kitamura M, Saito S, Minamiya Y, Suzuki H, Saito R, Terada K, Ogawa J, Inaba H (1998) Bcl-2 is located predominantly in the inner membrane and crista of mitochondria in rat liver. Biochem Biophys Res Commun 249:628–636

Müller M, Strand S, Hug H, Heninemann EA, Walczak H, Hoffmann WJ, Stremmel W, Krammer PH, Galle PR (1997) Drug-induced apoptosis in hepatoma cells is mediated by the CD95 (APO-1/Fas) receptor/ligand system and involves activation of wild-type p53. J Clin Invest 99:403–413

Müschen M, Warskulat U, Douillard P, Gilbert E, Häussinger D (1998) Regulation of CD95 (APO-1/Fas) receptor and ligand expression by lipopolysaccharide and dexamethasone in parenchymal and non-parenchymal rat liver cells. Hepatology 27: 200–208

Muzio M, Stockwell Br, Stennicke HR, Salvesen GS, Dixit VM (1998) An induced proximity model for caspase-8 activation. J Biol Chem 273:2926–2930

Nagata S (1997) Apoptosis by death factor. Cell 88:355–365

Nagata S, Goldstein P (1995) The Fas death factor. Science 267:1449–1455

Nagata S, Suda T (1995) Fas and Fas ligand: lpr and gld mutations. Immunology Today 16:39–43

Nakamoto Y, Guidotti LG, Pasquetto V, Schreiber RD, Chisari FV (1997) Differential target cell sensitivity to CTL-activated death pathways in hepatitis B virus transgenic mice. J Immunol 158:5692–5697

Nakamura T, Tomita Y, Hirai R, Yamaoka K, Kaji K, Ichihara A (1985) Inhibitory effects of TGF-β on DNA synthesis of adult rat hepatocytes in primary culture. Biochem Biophys Res Commun 133:1042–1050

Nakamura N, Shidoji Y, Yamada Y, Hatakeyama H, Moriwaki H, Muto Y (1995) Induction of apoptosis by acyclic retinoid in the human hepatoma-derived cell line, HUH-7. Biochem Biophys Res Commun 207:382–388

Nakamura N, Shidoji Y, Moriwaki H, Muto Y (1996) Apoptosis in human hepatoma cell line induced by 4,5-didehydro geranylgeranoic acid (acyclic retinoid) via down-regulation of transforming growth factor-alpha. Biochem Biophys Res Commun 219:100–104

Narita M, Shimizu S, Ito T, Chittenden T, Lutz RJ, Matsuda H, Tsujimoto Y (1998) Bax interacts with the permeability transition pore to induce permeability transition and cytochrome c release in isolated mitochondria. Proc Natl Acad Sci USA 95: 14681–14686

Natoli G, Costanzo A, Guido F, Moretti F, Levrero M (1998) Apoptotic, non-apoptotic, and anti-apoptotic pathways of tumor necrosis factor signalling. Biochem Pharmacol 56:915–920

Naumann M, Scheidereit C (1994) Activation of NF-κB in vivo is regulated by multiple phosphorylations. EMBO J 13:4597–4607

Neuman MG, Ishay J, Zimmerman HY, Eshchar J (1991) Hepatotoxicity of Vespa orientalis venom sac extract. Pharmacol Toxicol 69 [Suppl 1]:1–36

Neuman MG, Shear NH, Bellentani S, Tiribelli C (1998) Role of cytokines in ethanol-induced cytotoxicity in vitro in HepG2 cells. Gastroenterology 115:157–166

Nishikawa M, Sato EF, Kuroki T, Utsumi K, Inoue M (1998) Macrophage-derived nitric oxide induces apoptosis of rat hepatoma cells in vivo. Hepatology 28:1474–1480

Noteborn MHM, Van der Eb AJ (1998) Apoptin-induced apoptosis: potential for antitumor therapy. Drug Resistance Updates 1:99–103

Oberhammer F, Qin HM (1995) Effect of three tumour promoters on the stability of hepatocyte cultures and apoptosis after transforming growth factor-β1. Carcinogenesis 16:1363–1371

Oberhammer F, Bursch W, Parzefall W, Breit P, Erber E, Stadler M, Schulte-Hermann R (1991) Effect of transforming growth factor β on cell death of cultured rat hepatocytes. Cancer Res 51:2478–2485

Oberhammer FA, Pavelka M, Sharma S, Tiefenbacher R, Purchio AF, Bursch W, Schulte-Hermann R (1992) Induction of apoptosis in cultured hepatocytes and in regressing liver by transforming growth factor β1. Proc Natl Acad Sci USA 89:5408–5412

Oberhammer F, Bursch W, Tiefenbacher R, Fröschl G, Pavelka M, Purchio T, Schulte-Hermann R (1993a) Apoptosis is induced by transforming growth factor-β1 within 5 hours in regressing liver without significant fragmentation of DNA. Hepatology 18:1238–1246

Oberhammer F, Fritsch G, Schmied M, Pavelka M, Printz D, Purchio Y, Lassmann H, Schulte-Hermann R (1993b) Condensation of the chromatin at the membrane of an apoptotic nucleus is not associated with activation of an endonuclease. J Cell Sci 104:317–326

Oberhammer F, Wilson JW, Dive C, Morris ID, Hickman JA, Wakeling AE, Walker PR, Sikorska M (1993c) Apoptotic death in epithelial cells: cleavage of DNA to 300 and/or 50kb fragments prior to or in the absence of internucleosomal fragmentation. EMBO J 12:3679–3684

Ogasawara J, Watanabe-Fukunaga R, Adachi M, Matsuzawa A, Kasugai T, Kitamura Y, Itoh N, Suda T, Nagata S (1993) Lethal effects of the anti-Fas antibody in mice. Nature 364:806–809

Ohno K, Ammann P, Fasciati R, Maier P (1995) Transforming growth factor β1 preferentially induces apoptotic cell death in rat hepatocytes cultured under pericentral-equivalent conditions. Toxicol Appl Pharmacol 132:227–236

Okuyama T, Li XK, Funeshima N, Fujino M, Sasaki K, Kita Y, Kosuga M, Takahashi M, Saito H, Suzuki S, Yamada M (1998) Fas-mediated apoptosis is involved in the elimination of gene-transduced hepatocytes with E1/E3-deleted adenoviral vectors. J Gastroenterol Hepatol 13 (Suppl.):S113–118

Oritani K, Kaisho T, Nakajima K, Hirano T (1992) Retinoic acid inhibits interleukin-6-induced macrophage differentiation and apoptosis in a murine hematopoietic cell line, Y6. Blood 80:2298–2305

Pan G, Humke EW, Dixit VM (1998) Activation of caspases triggered by cytochrome c in vitro. FEBS Lett 426:151–154

Papoff G, Cascino I, Eramo A, Starace G, Lynch DH, Ruberti G (1996) An N-terminal domain shared by Fas/Apo-1 (CD95) soluble variants prevents cell death in vitro. J Immunol 156:4622–4630

Pastorino JG, Chen ST, Tafani M, Snyder JW, Farber JL (1998) The overexpression of Bax produces cell death upon induction of the mitochondrial permeability transition. J Biol Chem 273:7770–7775

Patel T, Bronk SF, Gores GJ (1994) Increases of intracellular magnesium promote glycodeoxycholate-induced apoptosis in rat hepatocytes. J Clin Invest 94:2183–2192

Patel T, Gores GJ (1995) Apoptosis and hepatobiliary disease. Hepatology 21:1725–1741

Patel T, Gores GJ (1997) Inhibition of bile-salt-induced hepatocyte apoptosis by the antioxidant lazaroid U83836E. Toxicol Appl Pharmacol 142:116–122

Perez C, Albert I, DeFay K, Zachariades N, Gooding L, Kriegler M (1990) A nonsecretable cell surface mutant of tumor necrosis factor (TNF) kills by cell-to-cell contact. Cell 63:251–258

Pessayre D, Fromenty B, Mansouri A (1999) Drug-induced steatosis and steatohepatitis. In: John J Lemasters and Anna-Liisa Niemenen, editors, Mitochondria in Pathogenesis, Plenum Press, New York (in press)

Piacentini M, Cerù MP, Dini L, Di Rao M, Piredda L, Thomazy V, Davies PJA, Fesus L (1992) In vivo and in vitro induction of "tissue" transglutaminase in rat hepatocytes by retinoic acid. Biochim Biophys Acta 1135:171–179

Pinkoski MJ, Hobman M, Heiben JA, Tomaselli K, Li F, Seth P, Froelich CJ, Bleachley RC (1998) Entry and trafficking of granzyme B in target cells during granzyme B-perforin-mediated apoptosis. Blood 92:1044–1054

Ponchel F, Puisieux A, Tabone E, Michot JP, Fröschl G, Morel AP, Frébourg T, Fontanière B, Oberhammer F, Ozturk M (1994) Hepatocarcinoma-specific mutant p53–249ser induces mitotic activity but has no effect on transforming growth factor β1-mediated apoptosis. Cancer Res 54:2064–2068

Poruchynsky MS, Wang EE, Rudin CM, Blagosklonny MV, Fojo T (1998) Bcl-X_L is phosphorylated in malignant cells following microtubule disruption. Cancer Res 58:3331–3338

Poupon R, Poupon E, Calmus Y, Chrétien Y, Ballet F, Darnis F (1987) Is ursodeoxycholic acid an effective treatment for primary biliary cirrhosis? Lancet i:834–836

Ray RB, Meyer K, Steele R, Shrivastava A, Aggarwal BB, Ray R (1998) Inhibition of tumor necrosis factor (TNF-α)-mediated apoptosis by hepatitis C virus core protein. J Biol Chem 273:2256–2259

Reed JC (1997) Cytochrome c: can't live with it – can't live without it. Cell 91:559–562

Régnier CH, Song HY, Gao X, Goeddel DV, Cao Z, Rothe M (1997) Identification and characterization of an IκB kinase. Cell 90:373–383

Rehemtulla A, Hamilton CA, Chinnaiyan AM, Dixit VM (1997) Ultraviolet radiation-induced apoptosis is mediated by activation of CD-95 (Fas/APO-1). J Biol Chem 41:25783–25786

Rensing-Ehl A, Hess S, Ziegler-Heitbrock HWL, Riethmüller G, Engelmann H (1995) Fas/apo-1 activates nuclear factor κB and induces interleukin-6 production. J Inflammation 45:161–174

Richter C, Gogvadze V, Schalpbach R, Schweitzer M, Schlegel J (1994) Nitric oxide kills hepatocytes by mobilizing mitochondrial calcium. Biochem Biophys Res Commun 205:1143–1150

Rieux-Laucat F, Le Deist F, Hivroz C, Roberts IAG, Debatin KM, Fisher A, de Villartay JP (1995) Mutations in Fas associated with human lymphoproliferative syndrome and autoimmunity. Science 268:1347–1349

Ritter SJ, Davies PJA (1998) Identification of a transforming growth factor-β1/bone morphogenic protein 4 (TGF-β1/BMP4) response element within the mouse tissue transglutaminase gene promoter. J Biol Chem 273:12798–12806

Roberts LR, Kurosawa H, Bronk SF, Fesmier PJ, Agellon LB, Leung WY, Mao F, Gores GJ (1997) Cathepsin B contributes to bile salt-induced apoptosis of rat hepatocytes. Gastroenterology 113:1714–1726

Rodrigues CM, Fan G, Ma X, Kren BT, Steer CJ (1998) A novel role for ursodeoxycholic acid in inhibiting apoptosis by modulating mitochondrial membrane perturbation. J Clin Invest 101:2790–2799

Rossé T, Olivier R, Monney L, Rager M, Conus S, Fellay I, Jansen B, Borner C (1998) Bcl-2 prolongs cell survival after Bax-induced release of cytochrome c. Nature 391:496–499

Roth-Eichhorn S, Kühl K, Gressner AM (1998) Subcellular localization of (latent) transforming growth factor β and the latent TGF-β binding protein in rat hepatocytes and hepatic stellate cells. Hepatology 28:1588–1596

Rouquet N, Carlier K, Briand P, Wiels J, Joulin V (1996) Multiple pathways of Fas-induced apoptosis in primary culture of hepatocytes. Biochem Biophys Res Commun 229:27–35

Roy N, Deveraux QL, Takahashi R, Salvesen GS, Reed JC (1997) The c-IAP-1 and c-IAP-2 proteins are direct inhibitors of several caspases. EMBO J 16:6914–6925

Russell WE, Coffey RJ Jr, Quelette AJ, Moses HL (1988) TGF-β reversibly inhibits the early proliferation response to partial hepatectomy. Proc Natl Acad Sci USA 85:5126–5130

Ruvolo PP, Deng X, Carr BK, May WS (1998) A functional role for mitochondrial protein kinase Cα in Bcl2 phosphorylation and suppression of apoptosis. J Biol Chem 273:25436–25442

Saas P, Walker PR, Hahne M, Quiquerez AL, Schnuriger V, Perrin G, French L, Van Meir EG, de Tribolet N, Tschopp J, Dietrich PY (1997) Fas ligand expression by astrocytoma in vivo: maintaining immune privilege in the brain? J Clin Invest 99:1173–1178

Saavedra JE, Billiar TR, Williams DL, Kim YM, Watkins SC, Keefer LK (1997) Targeting nitric oxide (NO) delivery in vivo. Design of a liver-specific NO donor prodrug that blocks tumor necrosis factor-α-induced apoptosis and toxicity in the liver. J Med Chem 40:1947–1954

Saito H, Ebinuma H, Takahashi M, Kaneko F, Wakabayashi K, Nakamura M, Ishii H (1998) Loss of butyrate-induced apoptosis in human hepatoma cell lines HCC-M and HCC-T having substantial Bcl-2 expression. Hepatology 27:1233–1240

Sakahira H, Enari M, Nagata S (1998) Cleavage of CAD inhibitor in CAD activation and DNA degradation during apoptosis. Nature 391:96–99

Samejima K, Earnshaw WC (1998) ICAD/DFF regulator of apoptotic nuclease is nuclear. Exp Cell Res 243:453–459

Sanchez A, Alvarez AM, Benito M, Fabregat I (1996) Apoptosis induced by transforming growth factor-β in fetal hepatocyte primary cultures. J Biol Chem 271:7416–7422

Sanchez A, Alvarez AM, Benito M, Fabregat I (1997) Cycloheximide prevents apoptosis, reactive oxygen species production, and glutathione depletion induced by transforming growth factor β in fetal rat hepatocytes in primary culture. Hepatology 26:935–943

Sanderson N, Factor V, Nagy P, Kopp J, Kondaiah P, Wakefield L, Roberts AB, Sporn MB, Thorgeirsson SS (1995) Hepatic expression of mature transforming growth factor β1 in transgenic mice results in multiple tissue lesions. Proc Natl Acad Sci USA 92:2572–2576

Scaffidi C, Fulda S, Srinivasan A, Friesen C, Li F, Tomaselli KJ, Debatin KM, Krammer PH, Peter ME (1998) Two CD95 (APO-1/Fas) signaling pathways. Embo J 17:1675–1687

Schneider P, Bodmer JL, Holler N, Mattmann C, Scuderi P, Terskikh A, Peitsch MC, Tschopp J (1997) Characterization of Fas (Apo-1, CD95)-Fas ligand interaction. J Biol Chem 272:18827–18833

Schneider P, Holler N, Bodmer JL, Hahne M, Frei K, Fontana A, Tschopp J (1998) Conversion of membrane-bound Fas(CD95) ligand to its soluble form is associated with downregulation of its proapoptotic activity and loss of liver toxicity. J Exp Med 187:1205–1213

Schulte-Hermann R, Bursch W, Grasl-Kraupp B (1995) Active cell death (apoptosis) in liver biology and disease. Progress Liver Dis 13:1–35

Schümann J, Angermüller S, Bang R, Lohoff M, Tiegs G (1998) Acute hepatotoxicity of *pseudomonas aeruginosa* exotoxin A in mice depends on T cells and TNF. J Immunol 161:5745–5754

Schumann H, Morawietz H, Hakim K, Zerkowski HR, Eschenhagen T, Holtz J, Darmer D (1997) Alternative splicing of the primary Fas transcript generating soluble Fas antagonists is suppressed in the human ventricular myocardium. Biochem Biophys Res Commun 239:794–798

Schwall RH, Robbins K, Jardieu P, Chang L, Lai C, Terrell TG (1993) Activin induces cell death in hepatocytes in vivo and in vitro. Hepatology 18:347–356

Seino KI, Kayagaki N, Takeda K, Fukao K, Okumura K, Yagita H (1997) Contribution of Fas ligand to T cell-mediated hepatic injury. Gastroenterology 113:1315–1322

Seishima M, Takemura M, Saito K, Ando K, Noma A (1997) Increased soluble Fas (sFas) concentrations in HCV-positive patients with liver cirrhosis. J Hepatol 27: 424–427

Senaldi G, Shaklee CL, Simon B, Rowan CG, Lacey DL, Hartung T (1998) Keratinocyte growth factor protects murine hepatocytes from tumor necrosis factor-induced apoptosis *in vivo* and *in vitro*. Hepatology 27:1584–1591

Sharma RP, Dugyala RR, Voss KA (1997) Demonstration of in-situ apoptosis in mouse liver and kidney after short-term repeated exposure to fumonisin B1. J Comp Pathol 117:371–381

Sheron N, Lau J, Daniels HM, Goka J, Eddleston A, Alexander GJM, Williams R (1991) Increased production of tumor necrosis factor-alpha in chronic hepatitis B virus infection. J Hepatol 12:241–245

Shimizu S, Eguchi Y, Kamiike W, Funahashi Y, Mignon A, Lacronique V, Matsuda H, Tsujimoto Y (1998) Bcl-2 prevents apoptotic mitochondrial dysfunction by regulating proton flux. Proc Natl Acad Sci USA 95:1455–1459

Shinagawa T, Yoshioka K, Kakumu S, Wakita T, Ishikawa T, Itoh Y, Takayanagi (1991) Apoptosis in cultured rat hepatocytes: the effects of tumour necrosis factor α and interferon γ. J Pathol 165:247–253

Smith CA, Farrah T, Goodwin RG (1994) The TNF receptor superfamily of cellular and viral proteins: activation, costimulation, and death. Cell 76:959–962

Smith RA, Baglioni C (1987) The active form of tumor necrosis factor is a trimer. J Biol Chem 262:6951–6954

Sokol RJ, Winklhofer-Roob BM, Devereaux MW, McKim JM (1995) Generation of hydroperoxides in isolated rat hepatocytes and hepatic mitochondria exposed to hydrophobic bile acids. Gastroenterology 109:1249–1256

Sokol RJ, McKim JM, Goff MC, Ruyle SZ, Devereaux MW, Han D, Packer L, Everson G (1998) Vitamin E reduces oxidant injury to mitochondria and the hepatotoxicity of taurochenodeoxycholic acid in the rat. Gastroenterology 114:164–174

Solter PF, Wollenberg GK, Huang X, Chu FS, Runnegar MT (1998) Prolonged sublethal exposure to the protein phosphatase inhibitor microcystin-LR results in multiple dose-dependent hepatotoxic effects. Toxicol Sci 44:87–96

Spivey JR, Bronk SF, Gores GJ (1993) Glycochenodeoxycholate-induced lethal hepatocellular injury in rat hepatocytes. Role of ATP depletion and cytosolic free calcium. J Clin Invest 92:17–24

Srinivasan A, Li F, Wong A, Kodandapani L, Smidt R, Krebs JF, Fritz LC, Wu JC, Tomaselli KJ (1998) Bcl-XL functions downstream of caspase-8 to inhibit Fas- and tumor necrosis factor receptor 1-induced apoptosis of MCF7 breast carcinoma cells. J Biol Chem 273:4523–4529

Stanger BZ, Leder P, Lee T-H, Kim E, Seed B (1995) RIP: a novel protein containing a death domain that interacts with Fas/APO-1 (CD95) in yeast and causes cell death. Cell 81:513–523

Stehlik C, de Martin R, Binder BR, Lipp J (1998) Cytokine induced expression of porcine inhibitor of apoptosis (iap) family member is regulated by NF-κB. Biochem Biophys Res Commun 243:827–832

Strand S, Hofmann WJ, Hug H, Müller M, Otto G, Strand D, Mariani SM, Stremmel W, Krammer W, Krammer PH, Galle PR (1996) Lymphocyte apoptosis induced by CD95 (APO-1/Fas) ligand-expressing tumor cells – A mechanism of immune evasion? Nature Medicine 2:1361–1366

Strand S, Hofmann WJ, Grambihler A, Hug H, Volkmann M, Otto G, Wesch H, Mariani SM, Hack V, Stremmel W, Krammer PH, Galle PR (1998) Hepatic failure and liver cell damage in acute Wilson's disease involve CD95 (APO-1/Fas) mediated apoptosis. Nature Medicine 4:588–593

Stravitz RT, Rao YP, Vlahcevic ZR, Gurley EC, Jarvis WD, Hylemon PB (1996) Hepatocellular protein kinase C activation by bile acids: implications for regulation of cholesterol 7α-hydroxylase. Am J Physiol 271:G293–G303

Su IJ, Lay ZD, Cheng AL, Chang YC (1994) Modulation of retinoic acid receptor alpha, growth factors and proto-oncogenes in retinoic acid-induced apoptosis of Ki-1 lymphoma cell line. Int J Oncol 4:1089–1095

Suda T, Takahashi T, Goldstein P, Nagata S (1993) Molecular cloning and expression of the Fas ligand, a novel member of the tumor necrosis factor family. Cell 75:1169–1178

Sullivan KE, Wooten C, Schmeckpeper BJ, Goldman D, Petri MA (1997) A promoter polymorphism of tumor necrosis factor α associated with systemic lupus erythematosus in African-Americans. Arthitis Rheum 40:2207–2211

Susin SA, Zamzami N, Castedo M, Daugas E, Wang HG, Geley S, Fassy F, Reed JC, Kroemer G (1997) The central executioner of apoptosis: multiple connections between protease activation and mitochondria in Fas/APO-1/CD95- and ceramide-induced apoptosis. J Exp Med 186:25–37

Susin SA, Lorenzo HK, Zamzami N, Marzo I, Snow BE, Brothers GM, Mangion J, Jacotot E, Costantini P, Loeffler M, Larochette N, Goodlett DR, Aebersold R, Siderovski DP, Penninger JM, Kroemer G (1999) Molecular characterization of mitochondrial apoptosis-inducing factor. Nature 397:441–445

Suzuki A, Kawabata T, Kato M (1998a) Necessity of interleukin-1β converting enzyme cascade in taxotere-initiated death signaling. Eur J Pharmacol 343:87–92

Suzuki A, Tsutomi Y, Akahane K, Araki T, Miura M (1998b) Resistance to Fas-mediated apoptosis: activation of caspase 3 is regulated by cell cycle regulator p21^{WAF1} and IAP gene family ILP. Oncogene 17:931–939

Swe M, Sit KH (1997) Staurosporine induces telophase arrest and apoptosis blocking mitosis exit in human Chang liver cells. Biochem Biophys Res Commun 236:594–598

Szabo C, Southan GJ, Thiemermann C (1994) Beneficial effects and improved survival in rodent models of septic shock with S-methylisothiourea sulfate, a potent and selective inhibitor of inducible nitric oxide synthase. Proc Natl Acad Sci USA 91:12472–12476

Szondy Z, Molnar P, Nemes Z, Boyiadzis M, Kedei N, Toth R, Fésüs L (1997) Differential expression of tissue transglutaminase during in vivo apoptosis of thymocytes induced via distinct signalling pathways. FEBS Lett 404:307–313

Tagawa Y-I, Sekikawa K, Iwakura Y (1997) Suppression of concanavalin A-induced hepatitis in IFN-$\gamma^{-/-}$ mice, but not in TNF-$\alpha^{-/-}$ mice. J Immunol 159:1418–1428

Tagawa Y-I, Kakuta S, Iwakura Y (1998) Involvement of Fas/Fas ligand system-mediated apoptosis in the development of concanavalin A-induced apoptosis. Eur J Immunol 28:4105–4113

Takahashi T, Tanaka M, Brannan CI, Jenkins NA, Copeland NG, Suda T, Nagata S (1994a) Generalized lymphoproliferative disease in mice, caused by a point mutation in the Fas ligand. Cell 76:969–976

Takahachi T, Tanaka M, Inazawa J, Abe T, Suda T, Nagata S (1994b) Human Fas ligand: gene structure, chromosoamal location and species specificity. Int Immunol 6:1567–1574

Tanaka S, Wands JR (1996) Insulin receptor substrate 1 overexpression in human hepatocellular carcinoma cells prevents transforming growth factor β1-induced apoptosis. Cancer Res 56:3391–3394

Tanaka M, Suda T, Takahachi T, Nagata S (1995) Expression of the functional soluble form of human Fas ligand in activated lymphocytes. EMBO J 14:1129–1135

Tanaka M, Itai T, Adachi M, Nagata S (1998) Downregulation of Fas ligand by shedding. Nature Medicine 4:31–36

Tanira MOM, Wasfi IA, Al Homsi M, Bashir AK (1996) Toxicological effects of *Teucrium stocksianum* after acute and chronic administration in rats. J Pharm Pharmacol 48:1098–1102

Tartaglia LA, Weber RF, Figari IS, Reynolds C, Palladino MA, Goeddel DV (1991) The two different receptors for tumor necrosis factor mediate distinct cellular responses. Proc Natl Acad Sci USA 88:9292–9296

Tartaglia LA, Ayres TM, Wong GH, Goeddel DV (1993) A novel domain within the 55 kd TNF receptor signals cell death. Cell 74:845–853

Tiegs G, Hentschel J, Wendel A (1992) A T-cell-dependent experimental liver injury in mice inducible by concanavalin A. J Clin Invest 90:196–203

Thornberry NA, Lazebnik Y (1998) Caspases: enemies within. Science 281:1312–1316

Tracey KJ, Cerami A (1993) Tumor necrosis factor, other cytokines and disease. Annu Rev Cell Biol 9:317–343

Tran-Thi TA, Decker K, Baeuerle PA (1995) Differential activity of transcription factors NF-κB and AP-1 in rat liver macrophages. Hepatology 22:613–619

Trautwein C, Rakemann T, Brenner DA, Streetz K, Licato L, Manns MP, Tiegs G (1998) Concanavalin A-induced liver cell damage: activation of intracellular pathways triggered by tumor necrosis factor in mice. Gastroenterology 114:1035–1045

Treiman M, Caspersen C, Chritensen SB (1998) A tool coming of age: thapsigargin as an inhibitor of sarco-endoplasmic reticulum Ca^{2+}-ATPases. Trends Pharmacol Sci 19:131–135

Tschopp J, Irmler M, Thome M (1998) Inhibition of Fas death signal by FLIPs. Current Opinion Immunol 10:552–558

Tsukamoto A, Kaneko Y (1993) Thapsigargin, a Ca^{2+}-ATPase inhibitor, depletes the intracellular Ca^{2+} pool and induces apoptosis in human hepatoma cells. Cell Biol Int 17:969–970

Tzung SP, Fausto N, Hockenbery DM (1997) Expression of Bcl-2 family during liver regeneration and identification of Bcl-X as a delayed response gene. Am J Pathol 150:1985–1995

Uchida K, Emoto K, Daleke DL, Inoue K, Umeda M (1998) Induction of apoptosis by phosphatidylserine. J Biochem 123:1073–1078

Van Antwerp DJ, Martin SJ, Kafri T, Green DR, Verma IM (1996) Suppression of TNF-α-induced apoptosis by NF-κB. Science 274:787–789

van den Dobbelsteen DJ, Nobel SI, Schlegel J, Cotgreave IA, Orrenius S, Slater AFG (1996) Rapid and specific efflux of reduced glutathione during apoptosis induced by anti-Fas/APO-1 antibody. J Biol Chem 271:15420–15427

Van Nieuwerk CM, Elferinck RP, Groen AK, Ottenhoff R, Tytgat GN, Dingemans KP, Van Den Bergh Weerman MA, Offerhaus GJ (1996) Effects of ursodeoxycholate and cholate feeding in liver disease in FVB mice with a disrupted mdr2 P-glycoprotein gene. Gastroenterology 111:165–171

Van Zee KJ, Kohno T, Fisher E, Rock CS, Moldawer LL, Lowry SF (1992) Tumor necrosis soluble receptors circulate during experimental and clinical inflammation and can protect against excessive tumor necrosis factor α in vitro and in vivo. Proc Natl Acad Sci USA 89:4845–4849

Vandenabeele P, Declercq W, Beyaert R, Fiers W (1995) Two tumour necrosis factor receptors: structure and function. Trends Cell Biol 5:392–399

Vander Heiden MG, Chandel NS, Williamson EK, Schumacker PT, Thompson CG (1997) Bcl-X_L regulates the membrane potential and volume homeostasis of mitochondria. Cell 91:627–637

Varani J, Gibbs DF, Inman DR, Shah B, Fligiel SEG, Voorhees JJ (1991) Inhibition of epithelial cell adhesion by retinoic acid. Relationship to reduced extracellular matrix production and alterations in Ca^{2+} levels. Am J Pathol 138:887–895

Verheij M, Bose R, Lin XH, Yao B, Jarvis WD, Grant S, Birrer MJ, Szabo E, Zon LI, Kyriakis JM, Haimowitz-Friedman A, Kuks Z, Kolesnick RN (1996) Requirement for ceramide-initiated SPAK/JNK signalling in stress-induced apoptosis. Nature 380:75–79

Volpes R, Van Den Cord JJ, de Voo R, Desmet VJ (1992) Hepatic expression of type A and type B receptors for tumor necrosis factor. J Hepatol 14:361–369

Wang C-Y, Mayo MW, Baldwin AS Jr (1996) TNF- and cancer therapy-induced apoptosis: potentiation by inhibition of NF-κB. Science 274:784–787

Wang C-Y, Mayo MW, Korneluk RG, Goeddel DV, Baldwin AS Jr (1998) NF-κB antiapoptosis: induction of TRAF1 and TRAF2 and c-IAP1 and c-IAP2 to suppress caspase-8 activation. Science 281:1680–1683

Wang JH, Redmond HP, Watson RWG, Bouchier-Hayes D (1995) Role of lipopolysaccharide and tumor necrosis factor-α in induction of hepatocyte necrosis. Am J Physiol 269:G297–G304

Wang JH, Redmond HP, Wu QD, Bouchier-Hayes D (1998) Nitric oxide mediates hepatic injury. Am J Physiol 275:G1117–G1126

Wanless IR, Dore S, Gopinath N, Tan J, Cameron R, Heathcote EJ, Blendis LM, Levy G (1996) Histopathology of cocaine hepatotoxicity. Report of four patients. Gastroenterology 98:497–501

Watanabe-Fukunada R, Brannan CI, Copeland NG, Jenkins NA, Nagata S (1992) Lymphoproliferation disorder in mice explained by defects in Fas antigen that mediates apoptosis. Nature 356:314–317

Watts JD, Gu M, Polverino AJ, Patterson SD, Aerbersold R (1997) Fas-induced apoptosis of T cells occurs independently of ceramide generation. Proc Natl Acad Sci USA 94:7292–7296

Webster CR, Anwer MS (1998) Cyclic adenosine monophosphate-mediated protection against bile acid-induced apoptosis in cultured rat hepatocytes. Hepatology 27:1324–1331

White E (1996) Life, death and the pursuit of apoptosis. Genes Develop 10:1–15

Wilson AG, Symons JA, McDowell TL, McDevitt HO (1997) Effects of a polymorphism in the human necrosis factor α promoter on transcriptional activation. Proc Natl Acad Sci USA 94:3195–3199

Wong GHW, Elwell JH, Oberley LW, Goeddel DV (1989) Manganous superoxide dismutase is essential for cellular resistance to cytotoxicity of tumor necrosis factor. Cell 58:923–932

Xu Y, Jones BE, Neufeld DS, Czaja MJ (1998) Glutathione modulates rat and mouse hepatocyte sensitivity to tumor necrosis factor α toxicity. Gastroenterology 115:1229–1237

Yamamoto M, Ogawa K, Morita M, Fukuda K, Komatsu Y (1996) The herbal medicine inchin-ko-to inhibits liver cell apoptosis induced by transforming growth factor β1. Hepatology 23:552–559

Yamamoto M, Fukuda K, Miura N, Suzuki R, Kido T, Komatsu Y (1998) Inhibition by dexamethasone of transforming growth factor β1-induced apoptosis in rat hepatoma cells: a possible association with Bcl-XL induction. Hepatology 27: 959–966

Yano H, Fukuda K, Haramaki M, Momosaki S, Ogasawara S, Higachi K, Kojiro M (1996) Expression of Fas and anti-Fas-mediated apoptosis in human hepatocellular carcinoma cell lines. J Hepatol 25:454–464

Yoshioka KS, Kakumu S, Arao M, Tsutsumi Y, Inoue M (1989) Tumor necrosis factor alpha production by peripheral blood mononuclear cells of patients with chronic liver disease. Hepatology 10:769–773

Zafrani ES, Bernuau D, Feldmann G (1984) Peliosis-like ultrastructural changes of the hepatic sinusoids in human chronic hypervitaminosis A: report of three cases. Hum Pathol 15:1166–1170

Zamzami N Susin SA, Marchetti P, Hirsh T, Gomez-Monterrey I, Castedo M, Kroemer G (1996) Mitochondrial control of nuclear apoptosis. J Exp Med 183:1533–1544

Zatloukal K, Böck G, Rainer I, Denk H, Weber H (1991) High molecular weight components are main constituents of Mallory bodies isolated with a fluorescence activated cell sorter. Lab Invest 64:200–206

Zhang LX, Mills KJ, Dawson MI, Collins SJ, Jetten AM (1995) Evidence for the involvement of retinoic acid receptor RARα-dependent signaling pathway in the induction of tissue transglutaminase and apoptosis by retinoids. J Biol Chem 270:6022–6029

Zwacka RM, Zhang Y, Zhou W, Halldorson J, Engelhardt JF (1998) Ischemia/reperfusion injury in the liver of BALB/c mice activates AP-1 and nuclear factor κB independently of IκB degradation. Hepatology 28:1022–1030

CHAPTER 4
The Role of C-type Protein Kinases in Apoptosis

M. Lucas

A. PKC Isozymes

C-type protein kinases (PKCs) mediate a multitude of signal transduction pathways triggered by phospholipid hydrolysis. Diacylglycerol is the main activator of PKCs, in addition to the modulation by Ca^{2+} of conventional PKC isozymes. The hydrolysis of phosphatidylinositol biphosphate by its specific phosphodiesterase provides both activators diacylglycerol and Ca^{2+} since inositol triphosphate releases Ca^{2+} from an intracellular, non-mitochondrial, calcium pool.

PKC isozymes contain an amino terminal regulatory peptide and a carboxy terminal catalytic domain. Three subclasses of PKC can be differentiated:

1. Conventional PKCs α, βI, βII and γ, which are modulated by diacylglycerol, phosphatidylserine and Ca^{2+}
2. Novel PKCs δ, ε, η and θ, which are regulated by diacylglycerol and phosphatidylserine
3. Atypical PKCs ζ, ι and λ, which are stimulated by phosphatidylserine, but its regulation is poorly documented

Given the diversity of PKC isozymes, the differential tissue expression and the relatively poor specificity for in vitro substrates, the subcellular distribution and membrane targeting emerge as the main determinants of in vivo activity. In the presence of diacylglycerol, PKC binds membranes containing phosphatidylserine with high affinity. This changes the conformation of the protein and releases the auto-inhibitory substrate from the active site of the PKCs.

The regulatory domain of PKCs contains:

1. The pseudosubstrate, an autoinhibitory domain
2. A cysteine-rich sequence that binds diacylglycerol and its functional analogue, phorbol esters
3. A β-sheet domain that binds acidic phospholipids and a Ca^{2+} binding pocket.

The overall cycle of PKC activation includes (see Newton 1997): the association of newly synthesized protein kinases with the cytoskeleton;

phosphorylation by a PKC kinase, autophosphorylation and release into the cytosol; anchoring of PKCs to membrane-bound isozyme-specific proteins; binding of diacylglycerol, phosphatidylserine and Ca^{2+}, which increases the affinity for phosphatidylserine; and interaction of the active form of PKC with targeting proteins that lead the enzyme to its substrate.

In addition to the regulation by second messenger binding, a fine-tune mechanism regulates PKCs by phosphorylation, subcellular localization and interaction with specific targeting proteins. An interesting feature of the activation process is that the treatment of cells with tumor-promoting phorbol esters results in the activation but then depletion of phorbol ester-responsive PKC isozymes. These data are consistent with a suicide model whereby activation of PKC triggers its own degradation via the ubiquitin-proteasome pathway (Lu et al. 1998).

Protein kinase C isozymes play distinct roles in cellular function in the balance proliferation/apoptosis/survival. Low PKC activity is associated with apoptosis (Sánchez et al. 1992) and the selective role of PKC isozymes in apoptosis has been documented in leukemia cells (Murray and Fields 1997). PKCα is important for cellular differentiation and PKCβII is required for proliferation. PKCι has been described to have a role in cell survival (Murray and Fields 1997) and protects K562 cells against drug-induced apoptosis. K562 cells, which are resistant to most apoptotic agents, undergo apoptosis when treated with the protein phosphatase inhibitor okadaic acid. Overexpression of PKCι leads to increased resistance to, whereas inhibition of PKCι expression sensitizes cells to okadaic acid-induced apoptosis. Overexpression of the related atypical PKCζ has no protective effect, demonstrating that the effect is isozyme-specific. PKCι also protects K562 cells against taxol-induced apoptosis, indicating that it plays a general protective role against apoptotic stimuli.

B. PKC and Apoptosis

The death of cells in normal tissue turnover is called apoptosis or programmed cell death (Kerr et al. 1972). Apoptosis occurs during fundamental physiological processes such as embryo morphogenesis, the development of immune tolerance, aging and tissue degeneration, as well as cell proliferation and tumorigenesis (McConkey et al. 1990; Fesus et al. 1991; Golstein et al. 1991; Green and Scott 1994; Wright et al. 1994). Morphological and molecular events include chromatin condensation, formation of the apoptotic bodies, shrinkage, fragmentation of DNA into oligonucleosome-sized fragments and, at a later state, progressive cell degradation, swelling and membrane rupture (Wyllie et al. 1980). Oncogenes and tumor suppressor genes are clearly involved. In fact, p53-dependent and independent pathways have been described (Lowe et al. 1993; Clarke et al. 1993), as well as an altered expression of oncogenes c-fos and c-myc (Buttyan et al. 1988; Clark and Gillespie

1997), whereas a protein encoded by the oncogene Bcl-2 was shown to block programmed cell death (HOCKENBERRY et al. 1990).

Agents or conditions inducing apoptosis show a variable degree of dependency on different pathways depending on the cell type, the state of the cell and the apoptosis-inducing agent (GOLSTEIN et al. 1991). The interpretation of the role of PKC in the apoptotic pathways was complicated by conflicting reports. It is conceivable that conflicting observations on the role of PKCs in the regulation of apoptosis reflect cell type-specific responses to triggering agents (GUBINA et al. 1998), as well as the tissue-specific expression of PKC isozymes. Most experiments supporting the role of PKCs in apoptosis can be classified into two groups: 1) the apoptotic effect of PKC inhibitors and 2) the protection against apoptosis and promotion of cell survival by activation of PKCs.

1. PKC inhibitors trigger the apoptotic death in a number of cell types under several conditions: in mouse natural killer cells and cytotoxic T lymphocytes (MIGLIORATI et al. 1994); in B cells where apoptosis is triggered by the PKC inhibitor chelerytrine (BONNEFOYBERARD et al. 1994); inhibition of PKC by staurosporine triggers apoptosis of insulin-secreting RIN m5F cells without raising cytosolic free calcium (SÁNCHEZ et al. 1993); PKC inhibitors induce apoptosis in malignant glioma cells (COULDWELL et al. 1994); selective PKC inhibitors block IL-2-mediated proliferation of murine T cells and cause apoptosis (GÓMEZ et al. 1994); inhibitors of PKC block the prolongation of cell survival and induce DNA fragmentation in neutrophils (ADACHI et al. 1993); direct induction of cell apoptosis by ethanol is augmented by inhibiting protein kinase C and establishes a link between protein kinase C activity, ethanol toxicity and ethanol-induced apoptosis (AROOR and BAKER 1997).
2. Data supporting the assertion that PKC activation blocks apoptosis are well documented: the activation of PKC promotes cell survival of mature lymphocytes prone to apoptosis (LUCAS et al. 1991; LUCAS et al. 1994); the combination of a calcium ionophore and a protein kinase activator (PMA) inhibits corticosterone-induced apoptosis in lymphocytes (ISEKI et al. 1993); apoptosis of B cells in germinal centers can be arrested by protein kinase C-activating phorbol esters (KNOX et al. 1993); translocation of PKC from the cytosol mediates phosphatidyl inositol-dependent pathway of rescue germinal center B cells from apoptosis (KNOX and GORDON 1994); phorbol esters protect endothelial cells (HAIMOVITZ-FRIEDMAN et al. 1994a) and pre-T cells (RADFORD 1994) against radiation-induced apoptosis; activation of tyrosine kinase by basic fibroblast growth factor causes the translocation of the PKCα isozyme into the membrane and arrests apoptosis (HAIMOVITZ-FRIEDMAN et al. 1994a).

The selective dependency of some cell lines on specific PKC isozymes has been applied to the targeted apoptosis of tumor cells. The androgen-independent cells of prostate cancer have been proposed as a target for the therapy (O'BRIAN 1998). PKCα allows the cells in androgen-independent prostate cancer to acquire a selective growth advantage through the overexpression of PKCα and this adaptive response renders the cells dependent on constitutively active PKCα for their survival.

C. Caspases and PKC

An intriguing feature of the apoptotic pathway is that the caspase type of cysteine proteases, which drive the terminal effector events (THORNBERRY 1996), regulate the activity of c-type protein kinases. Indeed, 7-hydroxystaurosporine, a protein kinase C inhibitor, is a potent inducer of apoptosis in cell lines that lack p53 and are usually resistant to apoptosis. Caspases, triggered during 7-hydroxystaurosporine-induced apoptosis (SHAO et al. 1997a), in turn regulate PKC in two ways: hyperphosphorylation of PKCα and proteolytic activation of PKC δ and βI (SHAO et al. 1997b). PKC α, βI, βII, δ, and ζ activities have been studied in HL60 cells challenged with 7-hydroxystaurosporine or the topoisomerase inhibitors, camptothecin and etoposide. 7-hydroxystaurosporine has no effect on PKCζ and inhibits the kinase activity of PKC βI, βII, and δ. PKCα activity is initially inhibited and subsequently increases as cells undergo apoptosis with 7-hydroxystaurosporine treatment. Camptothecin and etoposide also markedly enhance PKCα activity during apoptosis in HL60 cells. Another target for specific proteolysis is PRK2, a protein kinase C-related kinase, which is cleaved by caspase during Fas- and staurosporine-induced apoptosis. The major apoptotic cleavage sites of PRK2 lie within its regulatory domain, suggesting that its activity may be deregulated by proteolysis (Cryns et al. 1997).

D. Apoptosis Versus Mitosis

Cyclin dependent kinases (CDKs) are key regulators in the cell cycle. CDKs control the major steps between different phases of the cell cycle through the phosphorylation of target proteins like histones, cytoskeletal proteins, tumor suppressors, transcription factors and others. CDKs are faced with two main tasks: 1) the completion of cell cycle steps before others can start, and 2) the alternation of steps of the cell cycle in the proper sequence. Cyclins are the regulatory subunits of CDKs. While CDKs are synthesized at relatively constant rates, the level of cyclins varies significantly throughout the phases of the cell cycle (ELLEDGE 1996). The transition of cells through mid/late G1 is mediated by D-type cyclins in complex with CDK4 and CDK6. CDK2 and E-cyclin carry the cell to the end of the G1 phase. CDK2/cyclin A drives the entry into S phase. During S phase, B-cyclin switches partners and associates with CDC2 in late G2 phase. B-type cyclins associate with CDC2 kinases and program the passage of the cell through M phase. D-type cyclins are expressed throughout the cell cycle in response to mitogen activation.

The regulation of the cell cycle takes place at different levels:
1. Modulation of the transcriptional activity of genes encoding cyclins.
2. Direct covalent modification of CDK by CDK-activating kinases and phosphatases. The holoenzymes can be negatively regulated by phosphorylation, so that even though the CDC2/cyclin B complexes are progressively formed as cyclin B accumulates, the kinase remains inactive and its catalytic activity is restricted to mitosis.
3. Regulation by CDK inhibitors that bind CDK/cyclin complexes and block the kinase activity. The cyclin-kinase inhibitors (CKI) are a group of proteins acting as inhibitory subunits by binding CDK/cyclin complexes. Two main groups of CKI have been characterized: the INK4 group of p15, p16, p18 and p19 are quite specific for G1 CDKs; the group of p21, p27 and p57 has a wider action and associates with most CDK/cyclin complexes.
4. Proteolysis-driven progression from G1 to S (via CDC34) and triggering of anaphase and exit from mitosis (via APC, anaphase promoting complex). Both CDC34 and APC encode ubiquitin-conjugating enzymes that degrade cyclins and inhibitors of the cell cycle transition (KING et al. 1996).

During apoptosis, certain cell cycle regulatory proteins are inappropriately expressed, such as cyclin-dependent-kinase 4/cyclin D, and alterations in specific phosphorylation events, mediated by protein kinases and phosphatases, have been described (DAVIS et al. 1997). Apoptosis is morphologically related to premature mitosis, an aberrant form of mitosis. The uncoupling of timing for p34^{cdc2} activation and the completion of DNA replication causes the so-called "mitotic catastrophe" or premature mitosis that apparently results from mitosis during DNA replication (NURSE 1990; HEALD et al. 1993). p34^{cdc2} is a highly regulated serine-threonine kinase that controls entry into mitosis. The regulation of p34^{cdc2} is known to involve a network of kinases and phosphatases that may respond to the state of DNA replication, as well as forming complexes with cyclins (NURSE 1990). Entry into M phase is determined by activation of p34^{cdc2} that requires p34^{cdc2} dephosphorylation of phosphotyrosine and phosphothreonine and association with cyclin B. The wee1 tyrosine kinase maintains mitotic timing and coordinates the transition between DNA replication and mitosis by protecting the nucleus from the cytoplasmically activated cdc2 kinase (HEALD et al. 1993). The active form of the kinase leads to the phosphorylation of key substrates: H1 histone, p60src, lamins, centrosomal proteins, and other proteins that need to be displaced from chromatin to allow chromosome condensation. The complex p34^{cdc2}/cyclin B initiates the dissolution of the nuclear membrane and promotes chromatin condensation, events that take place during both mitosis and apoptosis (MEIKRANTZ et al. 1994). Premature p34^{cdc2} activation may be a general mechanism by which cells, induced to undergo apoptosis, initiate the disruption of the nucleus. This was deduced from experiments with fragmentin and with staurosporine, which induces dephosphorylation of p34^{cdc2} and apoptosis in lymphoma and mammary carcinoma cell lines (SHI et al. 1994). This hypothesis has been questioned since OBERHAMMER et al. (1994) showed that chromatin condensation during apoptosis appears to be due to a rapid proteolysis of nuclear matrix proteins which does not involve the $p34^{cdc2}$ kinase; in contrast to mitosis, dephosphorylation and activation of $p34^{cdc2}$ does not occur in apoptotic cells. Nonetheless, different observations support the hypothesis that apoptosis may be due, in part, to an uncoordinated attempt by a nondividing cell to reenter and progress through the cell cycle (DAVIS et al. 1997).

E. Cell Cycle, CDK and PKC Inhibitors

Several protein kinase inhibitors have demonstrated a potential for use in the therapy of human cancers. Staurosporine, a potent PKC inhibitor with broad specificity, enhances the cytotoxic effects of various antitumor agents with different modes of action. Staurosporine potentiates apoptosis through events that occur downstream of DNA damage, and implicates the unscheduled activation of cyclin A-dependent kinase during the inhibition of DNA synthesis as a possible cause (LOCK et al. 1997). Staurosporine induces not only apoptotic cell death in a wide variety of mammalian cells, but also premature initiation of mitosis in cells arrested in S phase by DNA inhibitors. Chromosome condensation occurs in both staurosporine-induced apoptosis and premature mitosis. However, neither formation of mitotic spindles nor mitosis-specific phosphorylation of MPM-2 antigens is observed in apoptosis, unlike premature mitosis. The $p34^{cdc2}$ kinase activated in normal and prematurely mitotic cells remains inactive in the apoptotic cells, probably because the active cyclin B/$p34^{cdc2}$ complex is almost absent in the S phase-arrested cells. Phosphorylation of histones, which is associated with mitotic chromosome condensation, does not occur in the apoptotic cells. Therefore, staurosporine-induced apoptosis and premature mitosis are different in their requirements for $p34^{cdc2}$ kinase activation and histone phosphorylation (YOSHIDA et al. 1997). The role of protein kinases in the staurosporine-mediated events during the progression of the cell cycle remains to be studied.

The inhibition of CDKs has raised considerable interest in apoptosis research in view of their essential role in the regulation of the cell cycle. Olomoucine (6-(benzylamino)-2-[(2-hydroxyethyl)amino]-9-methylpurine), roscovitine (6-(benzylamino)-2(R)-{[1-(hydroxymethyl)propyl]amino}-9-isopropylpurine), and other N6,2,9-trisubstituted adenines exert a strong inhibitory effect on the $p34^{cdc2}$/cyclin B kinase. Inhibition of CDK with olomoucine and related compounds clearly arrests cell proliferation of many tumor cell lines at G1/S and G2/M transitions and also triggers apoptosis in the target tumor cells in vitro and in vivo. Thus, from a pharmacological point of view, olomoucine may represent a model compound for a new class of antimitotic and antitumor drugs (HAVLICEK et al. 1997). The kinase specificity of roscovitine has been investigated using highly purified kinases (including protein kinase A, G and C isozymes, myosin light-chain kinase, casein kinase 2, insulin receptor tyrosine kinase, c-src, v-abl and CDKs). The high selectivity of roscovitine for some cyclin-dependent kinases provides a useful antimitotic reagent for cell cycle studies and may prove interesting for the control of cells with deregulated cdc2, cdk2 or cdk5 kinase activities (MEIJER et al. 1997).

Experiments on taxol-induced activation of $p34^{cdc2}$ kinase and subsequent apoptosis (SHEN et al. 1998) have shown the protective effect of PMA on taxol-induced apoptosis. The blocking effect of PMA appears to be mediated by preventing the dephosphorylation of the Tyr-15 residue of $p34^{cdc2}$. Although the degree of specificity of the PMA effect was not established, this study focused interest on the possible relation of cell signals mediated by PKCs to cell cycle progression.

Gliobastoma cells, whose proliferation is highly dependent on PKCα, are very resistant to drug induced apoptosis by an undefined pathway. The inhibition of PKC by a novel specific inhibitor, Ro 31–82–220 involves the accumulation of p53 and of insulin-like growth factor-1 binding protein-3 (a pro-apoptotic protein), as well as the conversion of the retinoblastoma tumor suppressor protein to the hypophosphorylated and activated form (SHEN and GLACER 1998). These cells express PKCα at a high level and it is associated with a decreased synthesis of p53 protein, suggesting the regulation by PKCα of the apoptotic p53-dependent pathway.

Many signals from DNA damage are funneled through the p53 protein which, in turn, shuts down the cell cycle in the early G1 phase (see Fig. 1). p53 is known to induce the synthesis of p21 CDK inhibitor, which affects a variety of cyclin/CDK complexes and, therefore, can provoke at any point exit from the cell cycle (EL DEIRY et al. 1993). The retinoblastoma protein, pRB, in its hypophosphorylated form, constrains the advance of the cell cycle, while the formerly phosphorylated pRB losses its growth-suppressing power (BARTEK et al. 1996). The connection of PKC with p53, pRB and, therefore, with cell cycle regulation provides a cross-talk between signals mediating proliferation and apoptosis.

F. Capacitative Calcium Entry and Apoptosis

Bcl-2, first described as an inner mitochondrial membrane protein that blocks programmed cell death (HOCKENBERRY et al. 1990), is associated with the nuclear envelope and the endoplasmic reticulum, as well as the mitochondrial membrane (JACOBSON et al. 1993). The inhibition, by the oncoprotein Bcl-2, of the apoptosis induced by withdrawal of interleukins was clearly associated with the repartitioning of intracellular calcium (BAFFY et al. 1993). These observations were reinforced by experiments with thapsigargin, an inhibitor of the calcium pumping ATPase of the endoplasmic reticulum (Fig. 2) that causes persistent depletion of intracellular calcium stores and produces apop-

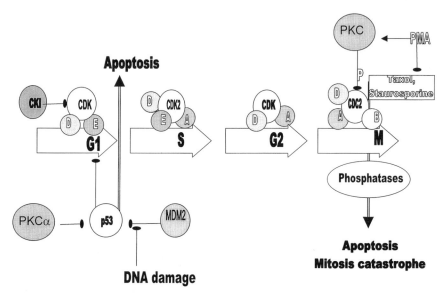

Fig. 1. Cell cycle, apoptosis and premature mitosis. Cyclin dependent kinases (*CDK*) complex with cyclins (*A, B, D, E*) and drive cell cycle phases from G1 to mitosis (*M*). p53 functions as a transcription factor. MDM2 (the human homologue of mouse double minute 2) neutralizes p53 by binding to its DNA-binding domain. CDK inhibitors (*CKI*) block CDK/cyclin complexes; the gene encoding the CKI is a target for p53-mediated regulation and is responsible for p53-mediated G1 arrest and apoptosis. Many signals from DNA damage are funneled through the p53 protein. PKCα and PKC inhibitors have been shown to regulate p53 action. The entry into the mitosis phase requires the activation of p34^{cdc2} (*CDC2*), following dephosphorylation and association with cyclins. The uncoupling of these events by p34^{cdc2} dephosphorylation via phosphatases (*P*), before DNA replication is completed, causes mitosis catastrophe and apoptosis. p34^{cdc2} dephosphorylation and activation mediate staurosporine- and taxol-induced apoptosis and both are blocked by phorbol-myristate acetate (PMA)

tosis of hepatoma cell lines (KANEKO and Tsukamoto 1994). This apparent paradox (the association of calcium depletion and apoptosis) can be explained by taking into account the so-called "capacitative" model of calcium entry, which proposes that calcium concentration is regulated by the degree of depletion of the endoplasmic reticulum calcium pool. Interestingly, this store-operated calcium entry mechanism is inhibited by stimulants of protein kinase C, the phorbol esters (MONTERO et al. 1993). The inhibition of calcium entry should block the activation of calcium-dependent enzymes associated with the apoptotic reactions. In fact, LAM et al. (1994) explained the role of Bcl-2 in the repression of apoptosis as mediated through the regulation of endoplasmic reticulum-associated calcium fluxes. The induction of apoptosis by thapsigargin is blocked by Bcl-2 and may be explained by assuming that the oncoprotein, by inhibiting calcium leaks from the endoplasmic reticulum, hinders the thapsigargin-induced "capacitative" calcium entry. This could also be a general

The Role of C-type Protein Kinases in Apoptosis

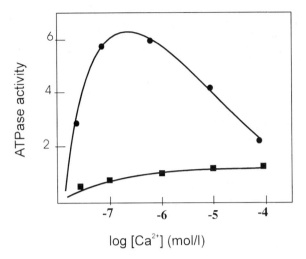

Fig. 2. ATPase activity in membranes of neutrophils. A crude membrane preparation was made by differential centrifugation of ultrasonically lysed neutrophils. ATPase activity was assayed by the release of [^{32}P] from γ[^{32}P]ATP in the absence (*circles*) and in the presence (*squares*) of 0.5 μM thapsigargin. Free calcium concentrations, given as pCa values, were buffered in EGTA-containing medium

mechanism of the abrogation of apoptosis by phorbol esters, since they too inhibit the capacitative calcium entry into the cytosol (MONTERO et al. 1993).

We have recently addressed the role of calcium and PKC in the activity of endonucleases and apoptosis (unpublished results). In human neutrophils thapsigargin produced a rapid rise of [Ca^{2+}]$_i$ with a sustained second phase and activated the endonuclease leading to the breakdown of 60–80% of the DNA in 24h and apoptosis (Fig. 3). PMA inhibited the second phase of calcium entry and completely blocked the activation of the endonuclease induced by thapsigargin. A similar profile of DNA breakdown can be reproduced in RIN m5F cells (Fig. 4).

The regulation by calcium of the neutrophilic endonuclease could be achieved either directly, as a cofactor, or through the expression of an endonuclease-encoding gene. In addition, it is worth pointing out the autoregulation by calcium ions of [Ca^{2+}]$_i$ via the store-regulated capacitative calcium entry. Studies on calcium fluxes and phosphorylation experiments have shown that two plasma membrane proteins close to 50 and 64 kDa are phosphorylated in PMA-challenged neutrophils. Calcium entry by the capacitative mechanism is sensitive to the depletion of the intracellular calcium pool by thapsigargin. In resting neutrophils, the non-phosphorylated form of the protein allows basal calcium entry and in thapsigargin-challenged neutrophils, the depletion of the non-mitochondrial calcium pool, enhances the capacitative calcium entry. The phosphorylation of membrane-associated proteins by PMA inhibits calcium

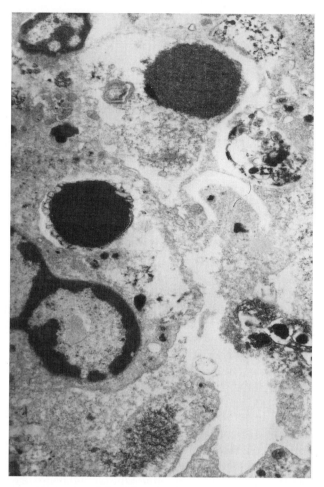

Fig. 3. Electron microscopy of thapsigargin-challenged neutrophils. Cells were incubated for 24 h in the presence of 0.5 μM thapsigargin, centrifuged and fixed for microscopic study. The photo shows nuclear chromatin condensation, apoptotic bodies and membrane alterations

uptake by the neutrophils in both the resting and thapsigargin-activated cells and, therefore, blocks the triggering of the Ca^{2+}-dependent endonuclease (Fig. 5).

The relationship of the anti-apoptotic Bcl-2 family of proteins with the regulation of cytosolic calcium can be deduced directly from their conformation and structural domains. The anti-apoptotic protein Bcl-XL has three domains in close spatial proximity which form an extended hydrophobic cleft. X-ray and NMR studies (MUNCHMORE et al. 1996) have demonstrated that, in addition to the three domains, there are seven alpha helices in Bcl-XL which align in a conformation similar to the membrane insertion structure of bacte-

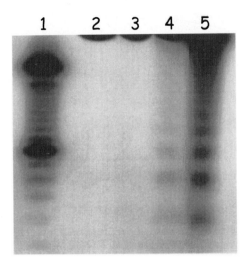

Fig. 4. Internucleosomal breakdown of DNA in RINm5F cells. *Lane 1*, 100 bp molecular size marker; *lane 2*, control cells incubated for 24 h; *lane 3*, cells incubated for 24 h in medium containing 100 nM phorbol-myristate-acetate and 0.5 µM thapsigargin; *lane 4*, cells incubated for 8 h with 0.5 µM thapsigargin; *lane 5*, cells incubated for 24 h with 0.5 µM thapsigargin. RINm5F cells were incubated for the indicated periods at 37°C. DNA was extracted with phenol/chloroform, labeled with $\alpha[^{32}P]dCTP$ using the Klenow fragment of the polymerase and the molecular size was analyzed by electrophoresis on a 1% agarose gel followed by autoradiography

rial toxins, raising the possibility for the formation of a pore or membrane channel. The formation of a regulatable ion pore in the endoplasmic reticulum and nuclear membrane supports the hypothesis that the regulation of intracellular calcium is the one of the main activities of the Bcl-2 protein.

The phosphorylation of Bcl-2 has been suggested as a direct mechanism by which PKC might regulate apoptosis. It is worth noting that a direct effect of PKC on Bcl-2 has been described, indicating that hematopoietic growth factors inhibit apoptosis by phosphorylation of Bcl-2 (May et al. 1993). Indeed, Bcl-2 function is partly regulated by phosphorylation/dephosphorylation mechanisms via the PKC system, and phosphorylated Bcl-2 prevents the apoptosis of lymphoma cells (Murata et al. 1997). Bryostatin 1, which downregulates PKC, as well as staurosporine and its 7-hydroxy derivative, which directly inhibit the enzyme, circumvent the resistance of Bcl-2-overexpressing leukemic cells to ara-C-induced apoptosis and activation of the protease cascade. These results highlighted the mediation by PKC of the anti-apoptotic effect of Bcl-2 and raised the possibility that modulation of the Bcl-2 phosphorylation status contributes to this effect (Wang et al. 1997).

PKC appears to regulate the expression of the *Bcl-2* gene. Suppression of apoptosis by v-abl PTK is associated with PKC signaling and the upregulation of Bcl-XL (Chen et al. 1997). Along this line, Gubina et al. (1998) have recently reported that the epsilon isoform of PKC allows the survival of inter-

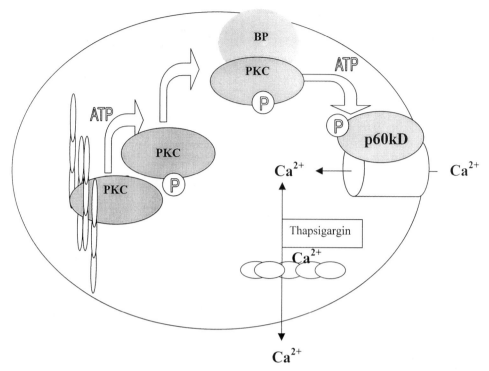

Fig. 5. Regulation by PKC of plasma membrane phosphorylation and capacitative calcium entry. Thapsigargin, by inhibiting the calcium pumping ATPase of the endoplasmic reticulum, provokes the depletion of this intracellular calcium pool. Store-operated mechanisms regulate capacitative calcium entry in resting cells. Following the maturation, phosphorylation and cellular activation, PKC is released into the cytosol and moves to the plasma membrane. A PKC-binding protein (*BP*) facilitates the activation by diacylglycerol and links PKC to a protein (p60kDa) that, upon phosphorylation, inhibits calcium entry. By activating PKC, PMA (not shown) inhibits calcium entry and, therefore, the apoptosis-associated calcium-activated reactions such as the activation of endonucleases

leukin-3 dependent cells in the absence of the cytokine. Overexpression of PKCε persists during all phases of the cell cycle, induces the expression of Bcl-2 and suppresses apoptosis. Moreover, experiments in human erythropoietin cell lines showed that the cytokine receptor increases Bcl-XL by a PKC-dependent pathway (Tsushima et al. 1997).

G. PKC Implication in the Sphingomyelin Pathway to Apoptosis

The sphingomyelin pathway is a ubiquitous, evolutionarily conserved signaling system which is initiated through the hydrolysis of the plasma membrane phospholipid, sphingomyelin, to generate ceramide. The generation of ceramide takes place via the action of sphingomyelinases, sphingomyelin-specific phospholipases that split sphingomyelin into ceramide and phosphorylcholine. Sphingolipid breakdown products have anti-proliferative and tumor-suppressor properties (HANNUN and LINARDIC 1993), and the hydrolysis of sphingomyelin and ceramide mediates the effects of tumor necrosis factor α (TNFα). A great variety of receptors, CD28 and CD95, and the receptors for TNFα, IL-1β, progesterone, γ-interferon and glucocorticoids trigger the sphingomyelin pathway, generating ceramide and activating a variety of cellular functions (SPIEGEL et al. 1996). Most mammalian cells are sensitive to agents acting through the sphingomyelin pathway.

Ceramide acts as a second messenger in activating the apoptotic cascade. Diverse cytokine receptors and environmental stresses utilize ceramide to signal apoptosis. In several cell systems ceramide is linked to the stress-activated protein kinase (SAPK)/c-jun kinase (JNK) cascade to signal apoptosis. Coordinated regulation of stress- and mitogen-activated protein kinases (SAPK and MAPK) are associated with the influence of ceramide and sphingosine because ceramide-mediated lethality is primarily associated with the strong stimulation of SAPK and weak inhibition of MAPK; and because sphingosine-mediated lethality is primarily associated with a weak stimulation of SAPK and strong inhibition of MAPK. The dominant basal influence of the MAPK cascade allows sustained proliferation, whereas redirection of this balance toward the SAPK cascade initiates apoptotic cell death (JARVIS et al. 1997).

The generation of ceramide and induction of apoptosis by environmental stresses, such as UV and ionizing radiation, may occur via different models. Ionizing radiation acts on cellular membranes of bovine and aortic endothelial cells to generate ceramide and initiate apoptosis, suggesting an alternative to the hypothesis that direct DNA damage mediates radiation-induced cell killing. HAIMOVITZ-FRIEDMAN et al. (1994b) indicated that PKC activation blocked both radiation-induced sphingomyelin hydrolysis and apoptosis. Radiation appears to utilize caspases which are downstream of ceramide generation to execute apoptosis. Nonetheless, additional studies will be required to further define the mechanism of radiation-induced apoptosis since UV radiation activates a variety of cytokine receptors (LIU et al. 1996).

One of the most striking features of apoptosis is that dying cells disappear from the tissue without generation of any inflammatory reaction. This contrasts with necrosis where internal materials (mainly mitochondrial proteins) reach the extracellular space and cause an inflammatory reaction (WYLLIE 1997). The mediation by ceramide of TNF signals is particularly interesting since it causes both pro-inflammatory and apoptotic processes. The 55kDa TNF receptor initiates apoptosis via a death-domain adaptor protein complex downstream of acid sphingomyelinase. The proliferative and pro-inflammatory effects of TNF are mediated by the following events: (1) generation of ceramide by the neutral sphingomyelinase; (2) activation of ceramide-activated protein kinases (CAPK) and; (3) triggering of the extracellular regulated kinase (ERK) (see HAIMOVITZ-FRIEDMAN et al. 1997). Ceramide acts on different targets including: ceramide-activated protein kinase and phosphatase, the guanine nucleotide exchange protein Vav, and the atypical protein kinase C isoform ζ.

Ceramide induces programmed cell death (OBEID et al. 1993) as well as the activation of the sphingomyelin pathway (JARVIS et al. 1994), processes that are inhibitable by the protein kinase C activators, phorbol-myristate acetate and synthetic diglycerides, suggesting opposite roles for diglyceride- and ceramide-mediated signals in the regulation of apoptosis. Ceramide is a possible mediator of apoptosis in response to a number of agents, including interferon and hypoxia, that cause sphingomyelin hydrolysis. In addition, a ceramide-activated protein phosphatase can mediate the effects of ceramide (DOBROWSKY and HANNUN 1992). On the other hand, sphingosine, a breakdown product of sphingolipids, is well known for its pharmacological inhibition of PKC. The translocation of PKC to the plasma membrane is central to the accessibility for second messengers and substrates and to the regulation by ceramide that inactivates PKCα, probably by dephosphorylation. PKCs, in turn, inhibit ceramide-mediated apoptosis by activating sphingosine kinase (CUVILLIER et al. 1996). The coincidence of both complementary events, through inhibition of the phosphorylation and activation of the phosphatase of target proteins, may argue in favor of PKC-regulated mechanisms in the sphingomyelin apoptotic pathway.

The engagement of the sphingomyelin pathway in signaling apoptosis is tightly regulated by anti-apoptotic control mechanisms, and the balance between pro- and anti-apoptotic systems determines the magnitude of the apoptotic response in vitro and in vivo. Understanding both pro- and anti-apoptotic signaling involved in ceramide-mediated apoptosis and the mode of their coordinated functions may yield opportunities for pharmacological interventions with potential for clinical applications (HAIMOVITZ-FRIEDMAN et al. 1997).

The activation of caspase commits most cells to apoptosis and, therefore, resting cells may be equipped with appropriate suppressors of the proteolytic attack. The apoptotic pathway is poised between suppression and activation controlled by agents acting through a great variety of signals, both transcriptional and non-transcriptional. WYLLIE (1997) suggested that a variation of the suppression level might be very effective in turning on apoptosis. In resting cells, continuous synthesis of labile protective proteins seems to be required to restrain apoptosis. In fact, cycloheximide in sublethal doses causes apoptosis in liver and induces expression of the c-myc, c-fos, c-jun and p53 genes and the accumulation of sphingosine, which might be important in mediating cycloheximide-induced apoptosis as an endogenous modulator of protein kinase C activity (ALESSENKO et al. 1997). The interactions of different apoptotic pathways depend on PKC activity in determining the fate of the cells in the balance of apoptosis, survival and proliferation. The targets of the kinase activity seem to be widely distributed in the apoptotic pathways and work, in many instances, downstream of the main apoptotic steps.

Acknowledgement. This work was partially supported by grant number 96/278 from the Fondo de Investigaciones Sanitarias.

References

Adachi S, Kubota M, Matsubara K, Wakazono Y, Hirota H, Kuwakado K, Akiyama Y, Mikawa H (1993) Role of protein kinase c in neutrophil survival enhanced by granulocyte colony–stimulating factor. Exp Hematol 21:1709–1713

Alessenko AV, Boikov PY, Filippova GN, Khrenov AV, Loginov AS, Makarieva ED (1997) Mechanisms of cycloheximide induced apoptosis in liver cells. FEBS Lett 416:113–116

Aroor AR, Baker RC (1997) Ethanol-induced apoptosis in human HL-60 cells. Life Sci 61:2345–2350

Baffy G, Miyashita T, Williamson JR, Reed JC (1993) Apoptosis induced by withdrawal of interleukin-3 (IL-3) from an IL-3-dependent hematopoietic cell line is associated with repartitioning of intracellular calcium and is blocked by enforced Bcl-2 oncoprotein production. J Biol Chem 268:6511–6519

Bartek J, Bartova J, Lukas J (1996) The retinoblastoma protein and the restriction point. Curr Opin Cell Biol 8:805–814

Bonnefoyberard N, Genestier L, Flacher M, Revillard JP (1994) The phosphoprotein phosphatase calcineurin controls calcium-dependent apoptosis in B cell lines. Eur J Immunol 24:325–329

Buttyan R, Zakeri Z, Lockshin R, Wogelmuth D (1988) Cascade induction of c-fos, c-myc and heat shock 70 K transcripts during the regression of the ventral prostate glan. Mol Endocrinol 2:650–657

Chen Q, Turner J, Watson AJ, Dive C (1997) v-Abl protein tyrosine kinase (PTK) mediated suppression of apoptosis is associated with the up-regulation of Bcl-XL. Oncogene 15:2249–2254

Clark W, Gillespie DA (1997) Transformation by v-Jun prevents cell cycle exit and promotes apoptosis in the absence of serum growth factors. Cell Growth Differ 8:371–380

Clarke AR, Purdie CA, Harrison DJ, Morris RG, Bird CC, Hooper ML, Wyllie AH (1993) Thymocyte apoptosis is induced by p53-dependent and independent pathways Nature 362:849–852

Couldwell WT, Hinton DR, He SK, Chen TC, Sebat I, Weiss MH, Law RE (1994) Protein kinase C inhibitors induce apoptosis in human malignant glioma cell lines FEBS Lett 345:43–46

Cuvillier O, Pirianov G, Kleuser B, Vanek PG, Coso OA, Gutkind JS, Spiegel S (1996) Suppression of ceramide-mediated programmed cell death by sphingosine-1-phosphate. Nature 381:800–803

Cryns VL, Byun Y, Rana A, Mellor H, Lustig KD, Ghanem L, Parker PJ, Kirschner MW, Yuan J (1997) Specific proteolysis of the kinase protein kinase C-related kinase 2 by caspase-3 during apoptosis. Identification by a novel, small pool expression cloning strategy. J Biol Chem 272:29449–29453

Davis PK, Dudek SM, Johnson VW (1997) Select alterations in protein kinases and phosphatases during apoptosis of differentiated PC12 cells. J Neurochem 68:2338–2347

Dobrowsky RT, Hannun YA (1992) Ceramide stimulates a cytosolic protein phosphatase. J Biol Chem 267:5048–5051

El Deiry WS, Tokino T, Velculescu VE et al. (1993) WAF1, a potential mediator of p53 tumor suppression. Cell 75:817–825

Elledge SJ (1996) Cell cycle check points: preventing an identity crisis. Science 274:1664–1672

Fesus L, Davies PJA, Piacentini M (1991) Apoptosis: molecular mechanisms in programmed cell death. Eur J Cell Biol 56:170–177

Golstein P, Ojcius DM, Young JDE (1991) Cell death mechanisms and the immune system. Immunol Rev 121:29–65

Gómez J, Delahera A, Silva A, Pitton C, Garcia A, Rebollo A (1994) Implication of protein kinase C in IL-2-mediated proliferation and apoptosis in a murine T cell clone. Exp Cell Res 213:178–182

Green DR, Scott DW (1994) Activation-induced apoptosis in lymphocytes. Curr Opin Immunol 6:476–487

Gubina E, Rinaudo MS, Szallasi Z, Blumberg PM, Mufson RA (1998) Overexpression of protein kinase C isoform epsilon but not delta in human interleukin-3-dependent cells suppresses apoptosis and induces bcl-2 expression. Blood 91:823–829

Haimovitz-Friedman A, Balaban N, Mcloughlin M, Ehleiter D, Michaeli J, Vlodavsky I, Fuks Z (1994a) Protein kinase C mediates basic fibroblast growth factor protection of endothelial cells against radiation–induced apoptosis. Cancer Res 54:2591–2597

Haimovitz-Friedman A, Kan CC, Ehleiter D, Persaud RS, Mcloughlin M, Fuks Z, Kolesnick RN (1994b) Ionizing radiation acts on cellular membranes to generate ceramide and initiate apoptosis. J Exp Med 180:525–535

Haimovitz-Friedman A, Kolesnick RN, Fuks Z (1997) Ceramide signaling in apoptosis. Br Med Bull 53:539–553

Havlicek L, Hanus J, Vesely J, Leclerc S, Meijer L, Shaw G, Strand M (1997) Cytokinin-derived cyclin-dependent kinase inhibitors: synthesis and cdc2 inhibitory activity of olomoucine and related compounds. J Med Chem 40:408–412

Hannun YA, Linardic CM (1993) Sphingolipid breakdown products: anti–proliferative and tumor-suppressor lipids. Biochim Biophys Acta 1154:223–236

Heald R, MaLoughlin M, McKeon F (1993) Human Wee1 maintains mitotic timing by protecting the nucleus from cytoplasmically activated Cdc2 kinase. Cell 74:463–474

Hockenberry D, Núñez G, Milliman C, Schreiber RD, Korsmeyer SJ (1990) Bcl-2 is an inner mitochondrial membrane protein that blocks programmed cell death. Nature 348:334–336

Iseki R, Kudo Y, Iwata M (1993) Early mobilization of Ca^{2+} is not required for glucocorticoid-induced apoptosis in thymocytes. J Immunol 151:5198–5207

Jacobson MD, Burne JF, King MP, Miyashita T, Reed JC, Raff MC (1993) Bcl-2 blocks apoptosis in cells lacking mitochondrial DNA. Nature 361:365–369

Jarvis WD, Kolesnick RN, Fornari FA, Traylor RS, Gerwitz DA, Grant S (1994) Induction of apoptotic DNA damage and cell death by activation of the sphingomyelin pathway. Proc Natl Acad Sci USA 91:73–77

Jarvis WD, Fornari FA Jr, Auer KL, Fremerman AJ, Szabo E, Birrer MJ, Johnson CR, Barbour SE, Dent P, Grant S (1997) Coordinate regulation of stress-and mitogen-activated protein kinases in the apoptotic actions of ceramide and sphingosine. Mol Pharmacol 52:935–947

Kaneko Y, Tsukamoto A (1994) Thapsigargin–induced persistent intracellular calcium pool depletion and apoptosis in human hepatoma cells. Cancer Lett 79:147–155

Kerr JFR, Wyllie AH, Currie AR (1972) Apoptosis: A basic biological phenomenon with wide-ranging implications in tissue kinetics. Br J Cancer 26:239–257

King RW, Deshaies RJ, Peters JM, Kirschner MW (1996) How proteolysis drives cell cycle. Science 274:1652–1659

Knox KA, Honson GD, Gordon J (1993) Distribution of cAMP in secondary follicles and its expression in B cell apoptosis and CD40-mediated survival. Int Immunol 5:1085–1091

Knox KA, Gordon J (1994) Protein tyrosine kinases couple the surface immunoglobulin of germinal center b cells to phosphatidyl inositol-dependent and independent pathways of rescue from apoptosis. Cell Immunol 155:62–76

Lam M, Dubyak G, Chen L, Nuñez G, Miesfeld RL, Distelhorst CW (1994) Evidence that BCL-2 represses apoptosis by regulating endoplasmic reticulum–associated Ca^{2+} fluxes. Proc Natl Acad Sci USA 91:6569–6573

Liu ZG, Hsu H, Goeddel DV et al. (1996) Dissection of TNF receptor 1 effector functions: JNK activation is not linked to apoptosis while NF-κB activation prevents cell death. Cell 87:565–576

Lock RB, Thompson BS, Sullivan DM, Stribinskiene L (1997) Potentiation of etoposide-induced apoptosis by staurosporine in human tumor cells is associated with events downstream of DNA-protein complex formation. Cancer Chemother Pharmacol 39:399–409

Lowe SW, Schmitt EM, Smith SW, Osborne BA, Jacks T (1993) p53 is required for radiation-induced apoptosis in mouse thymocytes Nature 362:847–849

Lu Z, Liu D, Hornia A, Devonish W, Pagano M, Foster DA (1998) Activation of protein kinase C triggers its ubiquitination and degradation. Mol Cell Biol 18:839–845

Lucas M, Solano F, Sanz A (1991) Induction of programmed cell death (apoptosis) in mature lymphocytes. FEBS Lett 1:19–20

Lucas M, Sánchez–Margalet V, Sanz A, Solano F (1994) Protein kinase C activation promotes cell survival in mature lymphocytes prone to apoptosis. Biochem Pharmacol 47:667–672

May WS, Tyler PG, Armstrong DK, Davidson NE (1993) Roll for serine phosphorylation of Bcl-2 in an antiapoptotic signaling pathway triggered by IL-3, EPO and bryostatin. Blood 82:(Suppl 1) 1738

McConkey DJ, Orrenius S, Jondal M (1990) Cellular signaling in programmed cell death (apoptosis). Immunol Today 11:120–121

Meijer L, Borgne A, Mulner O, Chong JP, Blow JJ, Inagaki N, Inagaki M, Delcros JG, Moulinoux JP (1997) Biochemical and cellular effects of roscovitine, a potent and selective inhibitor of the cyclin-dependent kinases cdc2, cdk2 and cdk5. Eur J Biochem 243:527–536

Meikrantz W, Gisselbrecht S, Tam SW, Schlegel R (1994) Activation of cyclin a–dependent protein kinases during apoptosis. Proc Natl Acad Sci USA 91:3754–3758

Migliorati G, Nicoletti I, D'Adamio F, Spreca A, Pagliacci C, Riccardi C (1994) Dexamethasone induces apoptosis in mouse natural killer cells and cytotoxic T lymphocytes. Immunology 81:21–26

Montero M, García-Sancho J, Alvarez J (1993) Transient inhibition by chemotactic peptide of a store-operated Ca^{2+} entry pathway in human neutrophils. J Biol Chem 268:13055–13061

Munchmore SW, Sattler M, Liang H et al. (1996) X-ray and NMR structure of human Bcl-XL, an inhibitor of programmed cell death. Nature 381:225–241

Murata M, Nagai M, Fujita M, Ohmori M, Takahara J (1997) Calphostin C synergistically induces apoptosis with VP,16 in lymphoma cells which express abundant phosphorylated Bcl-2 protein. Cell Mol Life Sci 53:737–743

Murray NR, Fields AP (1997) A typical protein kinase C iota protects human leukemia cells against drug-induced apoptosis. J Biol Chem 272:27521–27524

Newton AC (1997) Regulation of protein kinase C. Curr Opin Cell Biol 9:161–167

Nurse P (1990) Universal control mechanism regulating onset of M-phase. Nature 344:503–509

Obeid LM, Linardic CM, Karolak LA, Hannun YA (1993) Programmed cell death induced by ceramide. Science 259:1769–1771

Oberhammer FA, Hochegger K, Froschl G (1994) Chromatin condensation during apoptosis is accompanied by degradation of lamin A + B without activation of cdc2 kinase. J Cell Biol 126:827–837

O'Brian CA (1998) Protein kinase C-alpha: a novel target for the therapy of androgen-independent prostate cancer. Oncol Rep 5:305–309

Radford IR (1994) Phorbol esters can protect mouse pre–t cell lines from radiation–induced rapid interphase apoptosis. Int J Radiat Biol 65:345–355

Sánchez V, Lucas M, Sanz A, Goberna R (1992) Decreases protein kinase C activity is associated with programmed cell death (apoptosis) in freshly isolated rat hepatocytes. Biosc Report 12:199–206

Sánchez V, Lucas M, Solano F, Goberna R (1993) Sensitivity of insulin–secreting RIN m5F cells to undergoing apoptosis by the protein kinase C inhibitor staurosporine. Exp Cell Res 209:160–163

Shao RG, Shimizu T, Pommier Y (1997a) 7-Hydroxystaurosporine induces apoptosis in human colon carcinoma and leukemia cells independently of p53. Exp Cell Res 234:388–397

Shao RG, Cao CX, Pommier Y (1997b) Activation of PKCalpha downstream from caspases during apoptosis induced by 7-hydroxystaurosporine or the topoisomerase inhibitors, camptothecin and etoposide, in human myeloid leukemia HL60 cells. J Biol Chem 272:31321–31325

Shen SC, Huang TS, Jee SH, Kuo ML (1998) Taxol-induced p34cdc2 kinase activation and apoptosis inhibited by 12-O-tetradecanoylphorbol-13-acetate in human breast MCF-7 carcinoma cells. Cell Growth Differ 9:23–29

Shen L, Glacer IR (1998) Induction of apoptosis in gliobastoma cells by inhibition of protein kinase C and its association with the rapid accumulation of p53 and induction of the insulin-like growth factor-1-binding protein-3. Biochem Pharmacol 55:1711–1719

Shi L, Nishioka WK, Th'ng J, Bradbury EM, Litchfield DW, Greenberg AH (1994) Premature p34^{cdc2} activation required for apoptosis. Science 263:1143–1145

Spiegel S, Foster D, Kolesnick R (1996) Signal transduction through lipid second messengers. Curr Opin Cell Biol 8:159–167

Thornberry NA (1996) The caspase family of cysteine proteases. British Med Bull 53:478–490

Tsushima H, Urata Y, Miyazaki Y, Fuchigami K, Kuriyama K, Kondo T, Tomonaga M (1997) Human erythropoietin receptor increases GATA-2 and Bcl-xL by a protein kinase C-dependent pathway in human erythropoietin-dependent cell line AS-E2. Cell Growth Differ 8:1317–1328

Wang S, Vrana JA, Bartimole TM, Freemerman AJ, Jarvis WD, Kramer LB, Krystal G, Dent P, Grant S (1997) Agents that down-regulate or inhibit protein kinase C circumvent resistance to 1-beta-D-arabinofuranosylcytosine-induced apoptosis in human leukemia cells that overexpress Bcl-2. Mol Pharmacol 52:1000–1009

Wright SC, Zhong J, Larrick JW (1994) Inhibition of apoptosis as a mechanism of tumor promotion FASEB J 8:654–660

Wyllie AH, Kerr JFR, Currie AR (1980). Cell death: the significance of of apoptosis. Int Rev Cytol 68:251–356

Wyllie AH (1997) Apoptosis: an overview. British Med Bull 53:451–465

Yoshida M, Usui T, Tsujimura K, Inagaki M, Beppu T, Horinouchi S (1997) Biochemical differences between staurosporine-induced apoptosis and premature mitosis. Exp Cell Res 232:225–239

CHAPTER 5
How Does Programmed Cell Death Contribute to AIDS Pathogenesis?

M.-L. GOUGEON

A. Introduction

I. The Pathogenesis of HIV Disease

The pathogenesis of human immunodeficiency virus (HIV) infection is complex and multifactorial (Fig. 1). Primary infection with HIV is rapidly followed by dissemination of the virus to the lymphoid organs, in which high virus replication occurs throughout the entire course of infection, even when the patient is clinically asymptomatic (PANTALEO et al. 1993). An intense cellular and humoral immune response is generated, which inhibits viral replication within weeks, but the virus almost invariably escapes from immune control, producing a chronic and persistent infection, and leading to the development of AIDS in the absence of an efficient anti-retroviral therapy (FAUCI 1996). The targets of HIV infection are CD4 expressing cells, such as lymphocytes and monocytes, the first identified receptor for HIV being the CD4 molecule. All strains of HIV infect primary $CD4^+$ T lymphocytes, and many primary isolates (referred to as M-tropic) also replicate well in monocytes, but not in transformed T cell lines. Other isolates that have been passaged in lymphoid cells in vitro infect primary $CD4^+$ T lymphocytes, but not monocytes, and are referred to as T-tropic viruses. The viral determinant of cellular tropism maps to the gp120 subunit of the HIV-1 Env protein and studies to delineate the molecular basis of cellular tropism led to the identification of co-receptors for HIV. The receptor CXCR4 was identified as the co-receptor responsible for the efficient entry of T-tropic strains of HIV-1 into target cells, and the β-chemokine receptor CCR5 was identified as the co-receptor for M-tropic HIV-1. As a corollary, the CXC chemokine SDF-1, the ligand for CXCR4, and the β-chemokines RANTES, MIP-1α and MIP-1β, ligands for CCR5, block infection by T-tropic or M-tropic HIV-1. Other co-receptors have been recently identified which seem to be used at later stages of the disease (FAUCI 1996).

CD4 T lymphocytes are the orchestrators of the immune system. First, through the production of cytokines, they help the effectors of innate immunity, such as natural killer cells (NK), $\gamma\delta$ T lymphocytes or monocytes in the elimination of virus-infected cells. In addition, they are essential to the specific activation and maturation of B lymphocytes into antibody-secreting plasmocytes, and they are required for the differentiation of $CD8^+$ T cells into

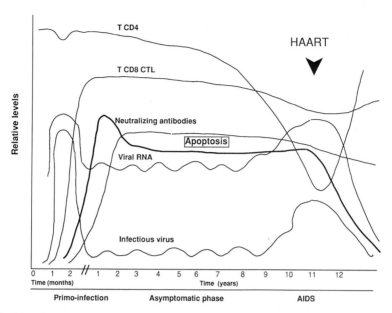

Fig. 1. Kinetics of viral replication and immune response following infection by HIV. Primo-infection with HIV rapidly induces a specific immune response including the activation of CD4 T helper cells, the differentiation of anti-HIV cytotoxic T lymphocytes (*CTL*) and the generation of HIV-specific neutralising antibodies. A progressive decline in CD4 T lymphocytes is observed, concomitant with an increased cell death by apoptosis of patients' lymphocytes. Following HAART (highly active antiretroviral therapy), an efficient control of HIV replication occurs, accompanied by a rapid rise in CD4 T cell number and a drop in the level of apoptosis which reaches normal values

virus-specific cytotoxic T lymphocytes (CTL). Finally, they are a source of chemokines, which are suppressor factors of HIV replication. Therefore, the progressive disappearance of CD4 T lymphocytes leads to the lack of control of HIV replication and to the development of severe immune deficiency responsible for the occurrence of opportunistic infections associated with AIDS.

II. CD4 T Cell Homeostasis in HIV Infection

$CD4^+$ T lymphocyte depletion is the hallmark of HIV infection. CD4 T cell destruction can be mediated directly by HIV replication as a consequence of viral gene expression, or indirectly through priming of uninfected cells to apoptosis when triggered by different agents. In addition to these pathways, a complementary cytopathic effect is probably provided by the immune system, since infected cells may be killed by HIV-specific CTL or antibody-dependent cell-mediated cytotoxicity (ADCC). Despite years of investigation, the dynamic basis for CD4 T cell depletion in HIV infection remains controversial.

Since 1995, a series of studies have increasingly challenged the paradigm that HIV infection induces high CD4 T cell production and turnover as the result of virus-induced T cell destruction. First, it was proposed that a large (10^9) number of CD4 T cells is infected and destroyed every day, and an equal number of CD4 T cells is produced to compensate for the loss (Ho et al. 1995; WEI et al. 1995). The outcome of this process of massive production/destruction of cells is the exhaustion of the T cell regenerative process resulting in the progressive depletion of CD4 T cells. However, this hypothesis is not compatible with the very low frequency of productively HIV-infected T cells in lymphoid tissues (1/100 to 1/1000) (HAASE et al. 1996), or with the estimate of the turnover of CD4 T cells. This was performed by the analysis of T cell telomere length, supposedly a marker for cellular replicative history. The turnover of T cells in the course of HIV infection was found to be considerably increased in CD8 T cells, but much less in CD4 T cells from HIV-infected subjects (WOLTHERS et al. 1996). More recent studies have shown that CD4 T cell turnover is two- to threefold higher in HIV-infected, compared with HIV-negative, subjects (SACHSENBERG et al. 1998; HELLERSTEIN et al. 1999), and that CD4 T cell production in HIV-infected subjects is not significantly different from that in healthy donors (FLEURY et al. 1998; HELLERSTEIN et al. 1999). In fact, virus replication inhibits CD4 T cell production because, following highly active anti-retroviral therapy (HAART), a dramatic increase in CD4 T cell production is observed (HELLERSTEIN et al. 1999). Therefore, the current understanding of CD4 T cell homeostasis in the course of HIV infection is that the progressive depletion of CD4 T lymphocytes is the consequence of both their destruction by several mechanisms dependent on the virus, and the lack of compensation by the production of new CD4 T cells, because of a possible blockade of the CD4 T cell renewal machinery at the level of the bone marrow or of the thymus. In this review, the contribution of programmed cell death (PCD) by apoptosis to the destruction of CD4 T lymphocytes, the mechanisms involved in this process, and the consequences of excessive apoptosis on the effectors of the immune system are discussed.

B. PCD in HIV Infection

I. Influence of HIV-1 Genes on the Induction of Apoptosis

Several HIV-1 gene products can influence directly the survival of the infected cell or of bystander cells. Tat, a viral transcription factor, was found to up-regulate Bcl-2 expression, protecting cells from apoptosis (ZAULI et al. 1995). In contrast, establishment of stable Tat-expressing cell lines or addition of exogenous Tat has been reported to sensitise cells to CD95-, T-cell receptor (TCR)- or CD4-induced apoptosis (LI et al. 1995; WESTENDORP et al. 1995). In these studies, Tat alone was insufficient to induce apoptosis but it appeared to sensitise cells to apoptosis triggered by a second signal, such as CD95 or TCR signalling. The *vpr* gene was also found to induce apoptosis (STEWARD et al.

1997). The vpr protein is required for productive infection of non-dividing cells (HATTORI et al. 1990) and it was recently shown to induce arrest of cells in the G2/M phase of the cell cycle. Following this arrest, *vpr* induces apoptosis in human T cells, peripheral blood lymphocytes and fibroblasts (BARTZ et al. 1996; YAO et al. 1998). Another HIV gene, *vpu*, was analysed for its influence on apoptosis of infected cells and it was found to increase the susceptibility of infected peripheral T cells and Jurkatt T cells to CD95-induced apoptosis (CASELLA et al. 1999). HIV replication in susceptible CD4 T or monocytic cell lines is also controlled by Bcl-2: infection first results in a decrease of Bcl-2, permitting an initial boost of replication, and then the replication is negatively controlled by Bcl-2 to reach a balance characterised by low virus production and a level of Bcl-2 compatible with cell survival (AILLET et al. 1998). Thus Bcl-2 is a critical cellular determinant in the tendency toward an acute or a persistent infection.

Infection of CD4 T cell cultures with HIV is associated with a cytopathic effect of the virus, manifested by ballooning of cells and formation of syncytia, leading to the death by apoptosis of both infected and non-infected cells. Apoptosis is triggered by the viral envelope glycoprotein, gp160, expressed on the surface of infected cells, which binds to accessible CD4 receptors on the surface of neighbouring cells (LAURENT-CRAWFORD et al. 1991; TERAI et al. 1991). During the fusion process, a specific region in the gp120-gp41 complex might become unmasked and thus mediate the onset of apoptosis. Both gp120 and gp41 are required for triggering apoptosis and no other gene besides the envelope is involved (LAURENT-CRAWFORD et al. 1993). Thus, chronically HIV-infected cells can serve as effector cells to induce apoptosis in uninfected target CD4 T cells. During this process, which involves syncytial formation and cell-to-cell spread of HIV infection, the anti-retroviral drug AZT blocks the spread of HIV infection without any apparent effect on apoptosis. On the other hand, cyclosporin A, a powerful suppressor of the immune system, and cycloheximide, which inhibits protein synthesis, do not affect apoptosis. Therefore, by virtue of expression of the gp120-gp41 complex, HIV-producing cells should be considered as potent effector cells for two independent pathological consequences: the first is the cell-to-cell spread of HIV infection, which is inhibited by anti-retroviral drugs; the second is the triggering of apoptosis, which is not affected by AZT. Further studies on the apoptotic pathway involved in gp120-dependent apoptosis of uninfected CD4 T cells showed that it involved caspases, although it was not mediated by the CD95 or TNF-RI molecules (OHNIMUS et al. 1997). These observations have raised the important question in HIV-1 pathogenesis: is virus killing limited to infected T cells in vivo (GOUGEON and MONTAGNIER 1993)?

II. Peripheral T Lymphocytes from HIV-Infected Subjects are Prematurely Primed for Apoptosis

It was reported several years ago that peripheral blood T cells from HIV-infected persons were highly prone to apoptosis induced in vitro (GOUGEON et al. 1991, 1993a; GROUX et al. 1992; MEYAARD et al. 1992). Indeed, while freshly isolated peripheral blood mononuclear cells (PBMCs) from HIV-infected individuals showed a low level of apoptosis (measured by different approaches detecting alterations in membrane permeability, a drop in mitochondrial membrane potential, chromatin condensation or DNA fragmentation) comparable to that of control donors (GOUGEON et al. 1996), their incubation in medium alone induced a rapid spontaneous apoptosis which was detected after a few hours of culture. This premature cell death could affect more than 30% of the lymphocytes from an HIV-infected subject whereas it affected only 2–5% of lymphocytes from control subjects. Moreover, the rate of apoptosis in blood lymphocytes from HIV-infected persons could be significantly increased following stimulation by various stimuli, including ionomycin, mitogens, superantigens or anti-TCR antibodies, whereas these stimuli had a marginal effect on the majority of lymphocytes from control donors (GOUGEON et al. 1991, 1993a; MEYAARD et al. 1992; GROUX et al. 1992).

Although it was first reported that the increased priming for apoptosis in HIV infection exclusively concerned the CD4 subset (GROUX et al. 1992), it became rapidly clear that the CD8 subset is similarly primed for apoptosis (MEYAARD et al. 1992; GOUGEON et al. 1993a; LEWIS et al. 1994). In fact, a phenotypic study of apoptotic cells in a large cohort of HIV-positive patients revealed that not only T cells but all blood mononuclear cells, including B cells, T cells, NK cells, granulocytes and monocytes, had an increased fragility upon short-term culture (GOUGEON et al. 1996). These observations were confirmed at the level of lymph nodes of HIV-infected patients, in which apoptosis was detected not only in CD4 but also in CD8 T cells, B cells and dendritic cells (MURO-CACHO et al. 1995; AMENDOLA et al. 1996), and in tonsillar tissue from HIV-infected donors, which showed increased apoptosis in both CD4 and CD8 T cells compared to uninfected donors (ROSOK et al. 1998).

The central paradox of HIV pathogenesis is that the viral burden, either free or cellular, seems too low to deplete the CD4 population by direct killing. The observation that an important fraction of T cells are prematurely primed for apoptosis in HIV-infected subjects prompted the hypothesis that some indirect mechanisms are responsible for inappropriate cell death and significantly contribute to CD4 T cell depletion (AMEISEN and CAPRON 1991) as well as to CD8 destruction in AIDS (GOUGEON 1995). This hypothesis was later supported by the observation that apoptotic T cells in lymph node sections of HIV-infected children and SIV-infected macaques were dominant in uninfected bystander cells, whereas infected cells were not found to be apoptotic (FINKEL et al. 1995). This has been confirmed by studies highlighting the important number of apoptotic cells in lymph nodes of HIV-infected adults (MURO-CACHO et al. 1995) and revealing the great frequency of T lymphocytes in patients' lymph nodes and blood, which express the tissue transglutaminase (tTG), a Ca^{2+}-dependent enzyme that cross-links intracellular proteins during the apoptotic process and whose expression underlines a pre-apoptotic stage (AMENDOLA et al. 1996).

III. Relationship Between Apoptosis and Immune Activation

Homeostasis is maintained by an extremely complex set of regulatory processes that differ markedly in quiescent and activated cells. For example, during primary viral infection induced by EBV, measles or varicella-zoster virus, T cell lymphocytosis is rapidly detected in the blood, but it is transient as the absolute number of circulating T lymphocytes and the relative proportion of T cell subsets return to normal upon resolution of the disease. It probably occurs via a rapid clearance by apoptosis of the majority of activated T cell blasts in vivo, since the apoptotic cells detected following a short-term culture of patients' lymphocytes express several activation markers (AKBAR et al. 1993; PIGNATA et al. 1998). Apoptosis thus plays a crucial role in the homeostatic control of cell numbers following antigenic stimulation, ensuring the clearance of primed lymphocytes in order to terminate the immune response and to avoid autoimmune reactions (AKBAR and SALMON 1997). Nevertheless, this normal process of elimination of activated cells might be detrimental for the immune system in the case of a chronic infection such as that induced by HIV.

A general state of immune activation is observed in the asymptomatic phase of HIV infection, both in lymphoid tissue and peripheral blood lymphocytes, and persists throughout the entire course of HIV infection. This is reflected by follicular hyperplasia in lymphoid tissue and the expression of activation markers such as HLA-DR, CD45R0 and CD38 in CD4 and CD8 T cells in the lymph-nodes (BOFILL et al. 1995; MURO-CACHO et al. 1995) and in the peripheral blood (LEVACHER et al. 1992; GIORGI et al. 1993). Although HIV replication is dramatically down-regulated under the influence of the specific immune response, HIV is never eliminated, and its persistence associated with the unceasing expression of HIV antigens is probably the primary mechanism for the chronic stimulation of the immune system. In addition, exogenous factors, such as opportunistic pathogens, stimulate the production of proinflammatory cytokines, including TNFα, IL1β and IL-6, which drive cellular activation and viral replication (BLANCHARD et al. 1997).

This unbalanced immune activation might be the primary mechanism responsible for the premature cell death in AIDS. This is suggested by the following observations: (1) apoptotic cells in patients' lymphoid tissues and in blood exhibit an activated phenotype (MURO-CACHO et al. 1995; GOUGEON et al. 1996); (2) there is a statistically significant correlation between the intensity of spontaneous or TCR-triggered apoptosis in both CD4 and CD8 subsets and their in vivo activation state (GOUGEON et al. 1996); (3) recent studies performed in West Africa, comparing patients infected with HIV-1 or HIV-2, showed that the low pathogenicity of HIV-2 infection is associated with a lower level of immune activation and less T cell apoptosis (MICHEL et al. 1999); (4) the lack of chronic immune activation in the non-pathogenic HIV-1 infection in chimpanzees is associated with a very low level of T cell apoptosis (HEENEY et al. 1993; GOUGEON et al. 1993, 1997). At the molecular level, the unbalanced immune activation in HIV-1-infected humans is responsible for the down-regulation of Bcl-2 expression (BOUDET et al. 1996) which is associated with an up-regulation of CD95 and CD95L expression (DEBATIN et al. 1994; KATSIKIS et al. 1995; BOUDET et al. 1996; MITRA et al. 1996; SLOAND et al. 1997) and an alteration in cytokine production (CLERICI and SHEARER 1994), which favours the apoptotic pathway rather than lymphocyte survival.

IV. Relevance of PCD to Disease Progression and AIDS Pathogenesis

A series of observations reported in HIV-infected persons and in simian models of lentiviral infection argue for a correlation between the intensity of T cell apoptosis and the pathogenicity of the infection. First, the proportion of CD4 and CD8 T lymphocytes undergoing apoptosis spontaneously, or after ligation of the TCR or the CD95 receptor, is increasing with disease evolution as evaluated by the in vivo reduction of the CD4 T cell number (BÖHLER et al. 1997; SLOAND et al. 1997; GOUGEON et al. 1999). Second, there is a correlation between the intensity of lymphocyte apoptosis and resistance or susceptibility to AIDS development. Indeed, spontaneous T cell apoptosis is very low in lymphocytes of "long-term non-progressors", a group of persons infected with HIV for at least 8 years but who have maintained normal numbers of CD4 T cells and do not show AIDS-associated symptoms (LIEGLER et al. 1998); and conversely, T cell apoptosis is very high in "rapid progressors", who show a rapid drop in CD4 T cell numbers and develop AIDS within 2 years after HIV primo-infection (M.-L. GOUGEON and H. LECOEUR, unpublished observations). Third, comparative studies in pathogenic models of lentiviral infection, including macaques infected with SIV (GOUGEON et al. 1993; ESTAQUIER et al. 1994), cats infected with FIV (BISHOP et al. 1993; HOLZNAGEL et al. 1998), murine AIDS (COHEN et al. 1993), versus non-pathogenic models, including SIV-infected African green monkeys (ESTAQUIER et al. 1994) or chimpanzees infected with HIV or SIVcpz (HEENEY et al. 1993; GOUGEON et al. 1993, 1997), revealed that increased lymphocyte apoptosis was only observed in pathogenic lentiviral infections. Interestingly, a recent study reported the case of two female chimpanzees, showing a progressive loss of CD4 T cells associated with high viral burdens and increased levels of CD4 T cell apoptosis following inoculation with HIV-1 which was isolated from a chimpanzee infected with the virus for 8 years. Lymph nodes from both animals revealed evidence of immune hyperactivation. By contrast, no apoptosis and no activation was observed in animals without loss of CD4 T cells (DAVIS et al. 1998). These observations provide additional evidence that a correlation exists between immune activation, T cell loss and apoptosis, and that apoptosis can significantly contribute to AIDS pathogenesis. As detailed below, it could be the mechanism responsible for the clearance of activated but healthy T cells, and consequently, could contribute to the impoverishment of the pool of effectors (Th and CTL) and antigen-presenting cells.

C. Molecular Control of HIV-Induced Apoptosis

Regulation of cell survival and death is essential for T cell homeostasis during precursor cell development and termination of an immune response in the periphery. Cell survival may be regulated by default mechanisms in which the expression of anti-apoptotic genes, such as proteins of the Bcl-2 family, is regulated by exogenous survival factors, e.g. cytokines such as IL-2 (BROOME et al. 1995; YANG and KORSMEYER 1996; REED 1997; KROEMER 1997). While the expression of survival genes seems to be critical for further development of precursor cells (positive selection) and T cell survival (VEIS et al. 1993; LINETTE et al. 1994), elimination of T cells in the periphery to down-regulate the immune response may rather involve the active induction of apoptosis through an interaction of death receptors with their respective ligands, including the CD95 system (NAGATA 1997; KRAMMER et al. 1994; DEBATIN 1996).

I. Negative Regulation of Bcl-2 Expression. Consequences on the Anti-Viral Cytotoxic Function

Bcl-2 and its homologous proteins play a key role in the control of cell death of T and B cell lineages during lymphoid development, ensuring their appropriate selection (NUNEZ et al. 1994). In differentiated mature T lymphocytes, regulation of Bcl-2 expression might be crucial for the development and persistence of a memory T cell response following an immune activation (AKBAR et al. 1993; AKBAR and SALMON 1997). In order to determine whether the priming for apoptosis of lymphocytes from HIV-infected donors was associated with a differential expression of Bcl-2, freshly isolated PBMCs from HIV-infected donors at different stages of the disease were analysed by FACS for intracellular Bcl-2 expression (BOUDET et al. 1996). A decreased Bcl-2 expression was consistently detected ex vivo in a fraction of CD8 T lymphocytes from HIV-positive donors, whereas it was never observed in lymphocytes from control donors. Interestingly, the low expression of Bcl-2 molecule in CD8 T lymphocytes primes them for spontaneous apoptosis after a short-term culture, and experiments performed on T lymphocytes from a series of patients showed that a significant correlation exists between the level of Bcl-2 expression and the propensity to undergo apoptosis, either spontaneously or following CD95 ligation (BOUDET et al. 1996). Ex vivo phenotypic characteristics of the low Bcl-2-expressing CD8 T cells suggested that they were cytotoxic, since they were in an activated state and they expressed the TIA-1 granules involved in the cytotoxic function (BOUDET et al. 1996). Interestingly, a parallel study performed by BOFILL et al. (1995) showed that this subset, characterized as $CD8^+$ $CD45R0^+$ $TIA-1^+$ and Bcl-2 low, is highly expressed in lymph nodes of HIV-infected patients.

A strong HIV-specific cytotoxic response is generated rapidly after HIV infection and it persists during the chronic phase of the viral infection. However, this cytotoxic response was reported to be markedly lost with the onset of symptoms (PANTALEO et al. 1990). The molecular mechanisms by which these cytotoxic lymphocytes could be deleted in vivo remain unknown. The lack of survival factors might contribute to the apoptosis of this subset. For example, IL-2 can upregulate in vitro Bcl-2 expression in lymphocytes from acutely EBV-infected patients (AKBAR et al. 1993) and also from chronically HIV-infected patients (GOUGEON et al. 1993; NAORA and GOUGEON 1999). One of the hallmarks of HIV infection is the defective production of IL-2, which is linked to the progressive depletion of CD4 T cells, the major source of IL-2 (CLERICI et al. 1994; LEDRU et al. 1998). The in vivo deficiency in IL-2 production would prevent the up-regulation of Bcl-2 molecules on cytotoxic T lymphocytes, which therefore could not be rescued from apoptosis. Therefore, the loss of anti-viral cytotoxic activity in the course of HIV infection might be related to an abnormal priming for apoptosis of CTL, consequent to both a persistent virus-driven immune stimulation and the gradual loss of survival factors.

II. Upregulation of the CD95 System

The CD95 molecule is a cell surface receptor of the tumour necrosis factor receptor (TNFR) superfamily that includes various molecules involved in immune regulation, such as the TNF receptors I and II, CD27, CD30 and CD40 (TRAUTH et al. 1989; ITOH et al. 1991; OEHM et al. 1992). The CD95 protein structure is characterized by three extracellular cysteine-rich domains (CRDs) found in all family members, a single transmembrane-spanning region and an intracellular part that contains a 70-amino acid region highly homologous to the p55 TNFR. This intracellular "death domain" has been shown to transduce signals for apoptosis through the TNFR and the CD95 molecule (NAGATA 1997; PETER and KRAMMER 1998). The CD95 ligand (CD95L) is a type II transmembrane protein produced by activated T cells and constitutively expressed in a variety of tissues. While the expression of CD95 is likely to be ubiquitous on activated immune cells (WATANABE-FUKUNAGA 1992), that of CD95L is more restricted to activated professional killer cells, such as $CD8^+$ and $CD4^+$ cytotoxic T cells, NK cells and antigen-presenting cells (APC) (SUDA et al. 1995; OSHIMI et al. 1996; BADLEY et al. 1996). A soluble form of CD95L is produced by proteolytic cleavage. A fundamental concept for the importance and the role of the CD95 system in growth control of peripheral T cells has been the demonstration of autocrine and paracrine mechanisms of CD95L-mediated death (DHEIN et al. 1995; ALDERSON et al. 1995; BRUNNER et al. 1995; JU et al. 1995). T cell receptor triggering in activated peripheral T cells may induce apoptosis that involves autocrine suicide or paracrine death mediated via CD95 receptor/ligand interaction. The finding that CD95 and CD95L are mutated in mouse strains suffering from severe autoimmune diseases and lymphoproliferation has greatly facilitated the understanding of the physiological role of the CD95 system in T cell homeostasis (NAGATA and SUDA 1995). Thus, mutations of the CD95 molecule in lpr mice and mutations of the CD95L in gld mice constitute the first genetically defined syndromes of defective apoptosis. Human counterparts of the lpr mutation in mice have been identified (FISHER et al. 1995; RIEUX-LAUCAT et al. 1995) and "deathless" $CD4^+$ and $CD8^+$ single-positive T lymphocytes from these patients fail to undergo apoptosis following stimulation via CD95 or T cell receptor triggering. On the other hand, the exacerbation of CD95-dependent apoptosis might be involved in tissue destruction during viral infections, including AIDS.

The in vivo involvement of the CD95 pathway in T cell apoptosis during HIV infection is supported by a series of observations. An increased expression of CD95 is detected in both CD4 and CD8 T lymphocytes from patients, and at the AIDS stage up to 80–90% of T cells are $CD95^+$ (DEBATIN et al. 1994; KATSIKIS et al. 1995; BOUDET et al. 1996; GOUGEON et al. 1997). This is associated with the existence in patients of cells susceptible to CD95-induced apoptosis, whose proportion increases with disease progression (KATSIKIS et al. 1995; GOUGEON et al. 1997; SLOAND et al. 1997). In addition, serum concentrations of soluble CD95 (MEDRANO et al. 1997) and anti-CD95 auto-antibodies (STRICKER et al. 1998) were found to be predictive markers for progression to AIDS. CD95L is also up-regulated in patients' lymphocytes: CD95L-encoding transcripts (BÄUMLER et al. 1996) and CD95L cell surface molecules (SLOAND et al. 1997) are highly expressed on both CD4 and CD8 T cells from HIV-infected persons, thus becoming possible effectors of apoptosis. In addition, elevated levels of soluble CD95 and CD95L are detected in the plasma of patients, and soluble CD95 concentrations correlate with CD95 expression on apoptotic cells (HOSAKA et al. 1998). Finally, a significant increase in macrophage-associated CD95L is detected in lymphoid tissue from HIV-positive subjects, which is correlated with the degree of tissue apoptosis (DOCKRELL et al. 1998). All these observations suggest that significant dysregulation of both CD95 and CD95L occurs in HIV infection. Experiments performed in HIV-infected chimpanzees, whose resistance to CD4 T cell depletion is associated with the lack of susceptibility of their T lymphocytes to CD95-induced apoptosis, argue for an involvement of the CD95 system in CD4 T cell depletion (GOUGEON et al. 1997).

III. Possible Effectors of CD95-Mediated Apoptosis. Consequences on CD4 T Cell Depletion

Several studies have contributed to the identification of potential effectors of CD95-induced apoptosis in HIV infection. The CD95-based cytotoxic activity could be mediated by both activated CD4 and CD8 T cells and also by HIV-infected APC. The up-regulation on CD4 T cells of CD95L expression through in vitro HIV infection (MITRA et al. 1996) or through the direct effect of viral proteins, such as gp120, Tat or Nef (WESTENDORP et al. 1995; BADLEY et al. 1996), make them possible effectors in killing CD95-expressing cells. This is corroborated by the demonstration that activated CD4 T lymphocytes, expressing CD95L, can kill CD95-expressing CD8 T lymphocytes (PIAZZA et al. 1997). The possible cytotoxic function of macrophages was suggested by studies reporting that CD95L expression was induced on APC either as a consequence of in vitro HIV infection (BADLEY et al. 1996) or following incubation with HIV proteins, gp120 and Tat (WESTENDORP et al. 1995; BADLEY et al. 1996). $CD8^+$ lymphokine-activated killer (LAK) cells were also identified as killers of HIV-infected CD4 T cells in vitro. However, the involvement of the CD95 system in this cytotoxicity was not investigated (WANG et al. 1998). In a recent study, we have asked whether professional CTL, specific to HIV peptides, were potential effectors of the destruction of CD95-expressing activated lymphocytes. Indeed, an anti-Nef HLA class I restricted CTL clone, derived from an HIV-infected subject, was able to mediate both perforin- and Fas-mediated dependent cytotoxic activities on Nef-presenting target cells and on Fas-expressing compliant cells, respectively (GARCIA et al. 1997) (Fig. 2). The biological relevance of this observation in the context of the chronic active HIV infection must be discussed. The high plasmatic viral load (Ho et al. 1995; WEI et al. 1995) associated with active HIV replication in lymphoid organs (PANTALEO et al. 1993) does promote a strong anti-viral CTL response throughout HIV infection (RIVIERE et al. 1989; AUTRAN et al. 1996). Hence, the constant re-stimulation by viral antigens of CTL through the TCR might lead in vivo to the continuous expression of CD95L on these CTL, which then can kill not only HIV-infected cells but also non-infected activated $CD95^+$ cells. Thus, in addition to being protective through the elimination of HIV-infected cells, anti-viral CTL could be deleterious through the destruction of CD95-expressing cells, abundant in HIV-infected patients, because of the persistent stimulation of the immune system.

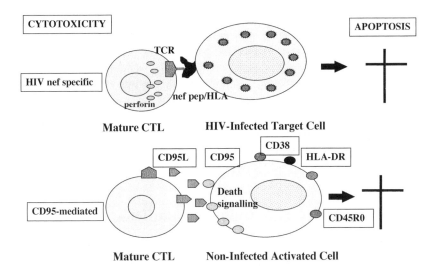

Fig. 2. Potential deleterious effect of HIV-specific CTL. A Nef-specific CTL clone, able to kill target cells which presenting a Nef peptide in the context of an HLA class I molecule, can also kill CD95-positive activated but non-infected T cells. This CD95 pathway is independent of HIV antigen recognition and is not HLA-restricted

IV. Other Cell Death Genes Involved in HIV-Induced Apoptosis

Several examples of CD95-independent apoptosis were reported in HIV infection. In an in vitro system of direct infection with HIV of PBMC or T cell lines, it was observed that the majority of HIV-induced T cell death involves direct loss of infected cells rather than indirect effects on uninfected bystander cells, and this cell death was found to be independent of the CD95 pathway (GANDHI et al. 1998). However, it was also reported that necrosis is the major mechanism involved in the direct killing of CD4 T cells by cytopathic HIV, whereas apoptosis is involved in immune cell-mediated killing (WANG et al. 1998). Thus, it cannot be excluded that in the study by GANDHI et al. (1998), CD4 T cells were mostly dying by necrosis, which would explain the non-involvement of an apoptotic death factor such as CD95. Another example of CD95-independant apoptosis was reported in experiments in which primary uninfected CD4 T cells died of apoptosis when they were in contact with HIV-infected or HIV gp120-expressing cells. Apoptosis was blocked by inhibitors of caspases but not by CD95 or TNF-R1 molecules (OHNIMUS et al. 1997). In fact, several recent studies suggested that, in addition to CD95L, other members of the TNF family are involved in HIV-induced apoptotic cell death. TRAIL (TNF-related-apoptosis-inducing-ligand) was identified as an apoptotic inducing factor in T cells from HIV-infected patients, but not in normal T cells even after prolonged activation in vitro (KATSIKIS et al. 1997; JEREMIAS et al. 1998). Apoptosis in CD8 T cells was reported to involve the TNF/TNF-R system. Indeed, binding of HIV gp120 or SDF-1 (stromal-derived factor 1), the physiologic ligand of the chemokine receptor CXCR4, induces the up-regulation of membrane TNF on macrophages and TNF-RII on peripheral CD8 T cells, leading to apoptosis of CD8 T cells. Apoptosis of CD8 T cells from HIV-infected patients can also be mediated by macrophages through the interaction between membrane TNF and TNF-RII (HERBEIN et al. 1998). The SDF-1 receptor CXCR4, when triggered by HIV gp120, was also found to induce a rapid cell death in normal CD4 T lymphocytes, which was independent of known caspases and lacked oligonucleosomal DNA fragmentation, but showed several features of apoptosis. Apoptosis triggered via CXCR4 was exclusively observed in CD4 but not in CD8 T cells, was independent of CD95, and was inhibited by SDF-1 (BERNDT et al. 1998). The induction of apoptosis through CXCR4 by gp120 or SDF-1 was also reported in human neuronal cells, in the absence of the CD4 molecule (HESSELGESSER et al. 1998).

D. Interrelation of HIV-Induced Apoptosis and Cytokines

I. Dysregulation of Cytokine Synthesis in HIV Infection

Alterations in cytokine production were first reported by CLERICI and SHEARER (1993, 1994) to occur in the course of HIV infection and to be associated with disease progression. The synthesis of two functionally distinct families of cytokines was analysed: type 1 cytokines (IL-2, IFNγ, IL-12, TNFα) mainly involved in cell-mediated immunity and in the destruction of intracellular parasites, and type 2 cytokines (IL-4, IL-5, IL-6, IL-10, IL-13) involved in controlling the activation and differentiation of B cell and of immunity against extracellular pathogens. The pattern of cytokines secreted by PBMC of HIV-infected donors in response to various stimuli, including antigens, mitogens or anti-TCR antibodies, was found altered in HIV-infected patients, since a progressive change in the balance of type 1 cytokines versus type 2 cytokines occurred, and this shift was suggested to contribute to AIDS susceptibility (CLERICI and SHEARER 1993, 1994). However, depending on the methods used for cytokine detection and on the organs analysed, the Th1 to Th2 shift has not been systematically observed. For example, GRAZIOSI et al. (1994) analysed the mRNA expression of Th1 or Th2 cytokines in lymph nodes of HIV-infected persons and the Th1-Th2 dichotomy was not found in these lymphoid organs. The pattern of cytokines produced by an effector population can now be analysed using a single cell analysis method which allows the enumeration of Th1/Th2 subsets derived from peripheral T cells stimulated in short-term cultures and the determination of the number and the phenotype of cells that are potentially capable of producing a given cytokine. Applying this flow cytometry method, it was found that a differential alteration in representation of Th1 subsets, rather than a commitment of T cells to secrete Th2 cytokines, occurs throughout HIV infection. A significant decrease in the number of IL-2 or TNF-α-producing T cells was observed, whereas those producing IFNγ remained preserved (LEDRU et al. 1998). The disappearance of IL-2-producing T cells was correlated with the progressive shrinkage of the naive CD45RA$^+$CD4$^+$ T cell compartment and it was a good indicator of disease progression (LEDRU et al. 1998). With that experimental approach, no increase in the proportion of T cells producing the Th2 cytokines IL-4, IL-5, IL-6, IL-13 was observed in HIV-infected patients compared to control donors, although some HIV-positive patients with hyper-IgE syndrome showed an increased number of IL-13$^+$ T cells. These observations do not exclude the possibility of an increased production of type 2 cytokines such as IL-6 and IL-10 by patients' monocytes, but at the T cell level, HIV infection is rather associated with an alteration of type 1 cytokines, and particularly IL-2, than an imbalance in Th1 and Th2 subsets.

II. The Disappearance of Th1 cells Is Related to Their Priming for Apoptosis

Because the rate of T cell apoptosis is increased early in HIV infection, we have asked whether alteration in the representation of some Th1 subsets was the consequence of a differential susceptibility to activation-induced apoptosis. This was performed by a multiparametric flow cytometry approach, combining at the single-cell level the detection of intracellular cytokines and apoptosis (LECOEUR et al. 1998). Exogeneous cytokines can modulate the susceptibility of lymphocytes to the apoptotic process (AKBAR and SALMON 1997), and we have found that the intrinsic capacity of lymphocytes to produce a given cytokine upon activation can also influence their survival. Indeed, T lymphocytes committed to IFNγ or TNFα production were more sensitive to activation-induced apoptosis than lymphocytes committed to IL-2-production (LEDRU et al. 1998). This gradient of susceptibility to activation-induced apoptosis (IL-2 < IFNγ < TNFα) was detected in both CD4 and CD8 subsets, as well as in control donors and HIV-infected patients. The differential intrinsic apoptosis susceptibility of Th1 effectors was found to be tightly regulated by Bcl-2 expression. In HIV-infected persons, an increased susceptibility to apoptosis was observed in IL-2 producers, which was related to a down-regulation of Bcl-2 expression. The progressive decrease in the proportion of IL-2 synthesising T cells was found correlated with their susceptibility to activation-induced apoptosis and disease progression (LEDRU et al. 1998). These correlations were also observed for TNFα producers. These observations indicate that the exacerbation of PCD in HIV infection probably contributes to the disappearance of Th1 effectors.

III. Regulation of HIV-Induced Apoptosis by Cytokines

Because cytokines can regulate the survival of activated cells, several groups have tested the influence of type 1 versus type 2 cytokines on PCD in T cells of HIV-infected patients. The addition of type 1 cytokine IL-2 was found to block in vitro spontaneous apoptosis (GOUGEON et al. 1993) and activation-induced apoptosis (CLERICI et al. 1994; ESTAQUIER et al. 1996) of T cells from patients. In contrast, the type 2 cytokines, IL-4 and IL-10, either had no effect or enhanced apoptosis (CLERICI et al. 1994; ESTAQUIER et al. 1995). However, activation-induced apoptosis and CD95-mediated apoptosis could be blocked by antibodies against IL-4 and IL-10 and enhanced by anti-IL-12 antibodies (CLERICI et al. 1994; ESTAQUIER et al. 1995). Because IL-15 shares many biological properties with IL-2, we examined the effects of exogenous IL-15 on lymphocytes of HIV-infected individuals (Fig. 3). Although IL-15 failed to inhibit CD3- and Fas-induced lymphocyte apoptosis in vitro, it could act as a potent survival factor in the prevention of spontaneous apoptosis. The greater potency of IL-15 in enhancing lymphocyte survival, as compared with IL-2 when used at an equivalent concentration, was associated with its greater ability to up-regulate Bcl-2 expression. In addition, IL-15 was more potent than IL-2 in stimulating lymphocyte proliferation (NAORA and GOUGEON 1999). These observations indicate that Th1 cytokines, such as IL-2 and IL-15, are able to prevent HIV-dependent apoptosis. IL-2 probably plays a pivotal role in anti-HIV immunity through its involvement in Th and CTL functions, and through its ability to prevent PCD and to promote T cell activation. The requirement of IL-2 for efficient control of HIV infection is suggested by studies performed in HIV-infected chimpanzees, indicating that this non-

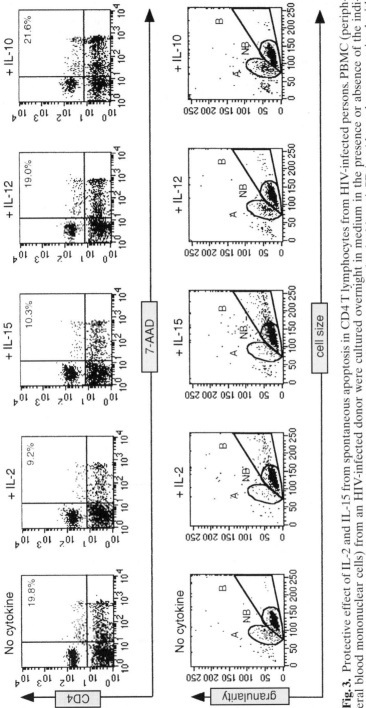

Fig. 3. Protective effect of IL-2 and IL-15 from spontaneous apoptosis in CD4 T lymphocytes from HIV-infected persons. PBMC (peripheral blood mononuclear cells) from an HIV-infected donor were cultured overnight in medium in the presence or absence of the indicated cytokines. Apoptosis was measured by flow cytometry. Lymphocytes were stained with anti-CD4 mAbs and counterstained with 7-AAD, which stains apoptotic cells. Apoptotic cells were also identified according to their FSS/SSC criteria (*A*). These size/granularity criteria were also used to identify blastic (*B*) and non-blastic (*NB*) cells. IL-2 and IL-15 can prevent apoptosis in CD4 T cells, while IL-12 and IL-10 have no effect. The IL-15 protective effect is associated with cell blastogenesis

pathogenic infection was correlated with the maintenance of T cells with a Th1 phenotype, the proportion of IL-2-producing cells being similar in infected and non-infected animals (GOUGEON et al. 1997). Similarly, experiments performed in macaques infected with pathogenic SIV strains or with a Nef-deleted-non-pathogenic SIV strain showed that infection with the latter was associated with the development of a Th1 pattern in lymph nodes, which was predictive of disease outcome (ZOU et al. 1997). The essential role of IL-2 in the control of HIV infection is confirmed by the benefits of in vivo IL-2 infusions into HIV-positive patients, resulting in both a clinical and an immunological improvement, and characterized by an important and stable rise in CD4 T cells (CONNORS et al. 1997).

E. PCD and T Cell Renewal. Influence of HAART

The main role of apoptosis is to maintain the homeostasis through the elimination of activated cells and to limit the clonal expansion of lymphocytes during an immune response (LYNCH et al. 1995). As discussed in this review, a chronic stimulation induces the continuous expression of death factors which could turn lymphocytes, including CD4 T cells (PIAZZA et al. 1996), CTL (GARCIA et al. 1997) or APC (BADLEY et al. 1996), into effectors of apoptosis, leading to the destruction of healthy activated non-infected cells. Thus PCD would significantly contribute to peripheral T cell depletion in AIDS, particularly if the Th cell renewal is impaired. The renewal capacities of the immune system in the context of HIV infection are not known. It was proposed by HEENEY (1995) that the ability of chimpanzees to maintain immunological integrity in the face of persistent HIV infection was associated with the maintenance of the primary and secondary lymphoid environments important for T cell renewal, considering that they are partly destroyed in infected humans. The recent availability of anti-retroviral therapies that reduce viral load to undetectable levels and concomitantly increase CD4 counts will help to determine whether, in the absence of detectable virus in the blood, the immune system can regenerate. In fact, the increase in CD4 T cells is not observed in all patients, and the functional alterations in addition to the skewed TCR repertoire of the CD4 Th subset are only partially corrected under anti-retroviral therapy (KELLEHER et al. 1996; CONNORS et al. 1997). The mechanisms that account for the rise in CD4 T cells in the blood following HAART are currently not completely understood. The initial rise of CD4 T cells would be due to the migration of memory T cells from the lymphoreticular tissues, where they are no longer trapped by the virus, to the blood (PAKKER et al. 1998) and after several weeks of HAART, the sustained rise in CD4 T cells would be the result of their peripheral proliferation, due both to the removal of HIV-induced suppression (HELLERSTEIN et al. 1999) and to the regulation of apoptosis. Indeed, shortly after the initiation of HAART, an important drop in spontaneous, activation-induced and CD95-triggered apoptosis, is observed in both CD4 and CD8 T cells from all treated patients (GOUGEON et al. 1999). This occurs before the decrease in immune activation, and the resistance to CD95-induced apoptosis precedes the down-regulation of CD95 expression. Thus, suppression of the plasmatic viral load is associated with the regulation of apoptosis, which reaches normal values detected in T cells from healthy donors.

F. Concluding Remarks

A provocative question is asked quite often: is apoptosis the cause or the consequence of AIDS pathogenesis? The comparison of HIV infection in chimpanzees, which maintain immunologic integrity in the face of persistent lentiviral infection, and humans, who develop AIDS, provides part of the answer. As summarised in this review, premature apoptosis is the consequence of the chronicity of the lentiviral infection. Continuous production of viral proteins would induce apoptosis either directly, by triggering a cell death signal, or indirectly, by influencing activation of the immune system. In infected chimpanzees, owing to the efficient control of HIV replication, the immune system is not activated and, consequently, inappropriate apoptosis does not occur (GOUGEON et al. 1997). The suggested rapid turn-over of CD4 T cells in HIV-infected persons due to an active regenerative process may contribute significantly to the rate of apoptosis in patients. Owing to an absence of CD4 depletion in chimpanzees, this rapid CD4 cell turnover might not occur in infected chimpanzees (HEENEY 1995). The impaired production of Th1 cytokines, such as IL-2, would prevent cell rescue from apoptosis (GOUGEON et al. 1993; CLERICI et al. 1994; LEDRU et al. 1998). In chimpanzees, no

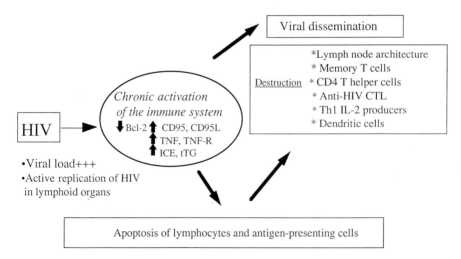

Fig. 4. Influence of apoptosis on the destruction of HIV-specific immune effectors

alteration of the Th1 subset was detected (GOUGEON et al. 1997). Inappropriate signalling by MHC class II APC may contribute to anergy and apoptosis of T cells in infected humans (MEYAARD et al. 1994). While the integrity of the Th MHC class II microenvironment is altered in lymphoid tissues of infected humans, it is preserved in infected chimpanzees (HEENEY 1995). On the other hand, apoptosis can significantly contribute to AIDS pathogenesis. As discussed here, because in vivo apoptosis involves mostly non-infected lymphocytes (FINKEL et al. 1995), it could be the mechanism responsible for the clearance of activated but healthy T cells, such as CD4 Th1 cells, CTL, memory cells, dendritic cells, and consequently could contribute to the impoverishment of the pool of effectors and memory cells, leading to the collapse of the immune system (Fig. 4).

Acknowledgements. I thank all my previous and present collaborators, particularly F. Boudet, E. Ledru, H. Lecoeur, S. Garcia and H. Naora, for their input and discussions. This work was funded by the Agence Nationale de Recherche sur le SIDA (ANRS), the Fondation pour la Recherche Médicale (FRM) (Sidaction), the Centre National de la Recherche Scientifique (CNRS), the Pasteur Institute and the EU (contracts ERB-IC15-CT97–0901 and BMH4-CT97–2055).

References

Aillet F, Masutani H, Elbim C, Raoul H, Chene L, Nugeyre MT, Paya C, Barre-Sinoussi F, Gougerot-Pocidalo MA, Israel N (1998) HIV induces a dual regulation of Bcl-2, resulting in persistent infection of CD4+ T or monocytic lines. J Virol 72:9698–9705

Akbar AN, Salmon M (1997) Cellular environments and apoptosis: tissue microenvironments control activated T-cell death. Immunol Today 18:72–76

Akbar AN, Borthwick N, Salmon M, Gombert W, Bofill M, Shamsadeen N, Pilling D, Pett S, Grundy JE, Janossy G (1993a) The significance of low bcl-2 expression by CD45RO T cells in normal individuals and patients with acute viral infections. The role of apoptosis in T cell memory. J Exp Med 178:427–438

Akbar AN, Salmon M, Savill J, Janossy G (1993b) A possible role for bcl-2 in regulating T-cell memory – a "balancing act" between cell death and survival. Immunol Today 14:526–532

Alderson MR, Tough TW, Davis-Smith T, Braddy S, Falk B, Schooley KA, Goodwin RG, Smith CA, Ramsdell K, Lynch DH (1995) Fas ligand mediates activation induced cell death in human T lymphocytes. J Exp Med 181:71–77

Ameisen JC, Capron A (1991) Cell dysfunction and depletion in AIDS: the program cell death hypothesis. Immunol Today 12:102–105

Amendola A, Gougeon M-L, Poccia F, Bondurand A, Fesus L, Piacentini M (1996) Induction of tissue transglutaminase in HIV pathogenesis. Evidence for a high rate of apoptosis of CD4 T lymphocytes and accessory cells in lymphoid tissues. Proc Natl Acad Sci USA 93:11057–11061

Autran B, Hadida F, Haas G (1996) Evolution and plasticity of CTL responses against HIV. Curr Opin Immunol 8:546–553

Badley AD, McElhinny JA, Leibson PJ, Lynch DH, Alderson MR, Paya CV (1996) Upregulation of Fas ligand expression by human immunodeficiency virus in human macrophages mediates apoptosis of uninfected T lymphocytes. J Virol 70: 199–206

Bartz SR, Rogel ME, Emerman M (1996) HIV-1 cell cycle control: vpr is cytostatic and mediates G2 accumulation by a mechanism which differs from DNA damage checkpoint control. J Virol 70:2324–2331

Bäumler CB, Böhler T, Herr I, Benner A, Krammer PH, Debatin K-M (1996) Activation of the CD95 (APO-1/Fas) system in T cells from human immunodeficiency virus type-1-infected children. Blood 88:1741

Berndt C, Mopps B, Angermuller S, Gierschik P, Krammer PH (1998) CXCR4 and CD4 mediate a rapid CD95-independent cell death in CD4 T cells. Proc Natl Acad Sci USA 95:12556–12561

Bishop SA, Gruffydd-Jones TJ, Harbour DA, Stokes CR (1993) Programmed cell death as a mechanism of cell death in PBMC from cats infected with feline immunodeficiency virus (FIV). Clin Exp Immunol 93:65–71

Blanchard A, Montagnier L, Gougeon M-L (1997) Influence of microbial infections on the progression of HIV disease. Trends Microbiol 5:326–331

Bofill M, Gombert W, Borthwick NJ, Akbar AN, McLaughlin JE et al. (1995) Presence of $CD3^+CD8^+Bcl-2^{low}$ lymphocytes undergoing apoptosis and activated macrophages in lymph nodes of $HIV-1^+$ patients. Am J Pathol 146:1542–1555

Böhler T, Bäumler C, Herr I, Groll A, Kurz T, Debatin K-M (1997) Activation of the CD95 system increases with disease progression in HIV1-infected children and adolescents. Pediatr Inf Dis J 16:754–••

Boudet F, Lecoeur H, Gougeon M-L (1996) Apoptosis associated with ex vivo downregulation of bcl-2 and up-regulation of Fas in potential cytotoxic CD8+ T lymphocytes during HIV infection. J Immunol 156:2282

Broome HE, Dargan CM, Krajewski S, Reed JC (1995) Expression of Bcl-2, Bcl-x and Bax after T cell activation and IL-2 withdrawal. J Immunol 155:2311–2317

Brunner T, Mogil RJ, LaFace D, Yoo NY, Mahboubl A, Echeverri F, Martin SJ, Force WR, Lynch DH, Ware CF, Green DR (1995) Cell-autonomous Fas (CD95)/Fas ligand interaction mediates activation induced apoptosis in T cell hybridomas. Nature 373:441–444

Casella CR, Rapaport EL, Finkel TH (1999) Vpu increases susceptibility of HIV-1 infected cells to Fas killing. J Virol 73:92–100

Clerici M, Shearer GM (1993) A Th1 → Th2 switch is a critical step in the etiology of HIV infection. Immunol Today 14:107–111

Clerici M, Shearer GM (1994) The Th1/Th2 hypothesis of HIV infection : new insights. Immunol Today 15:575–581

Clerici M, Sarin A, Coffman RL, Wynn TA et al. (1994) Type 1/type 2 cytokine modulation of T cells programmed cell death as a model for HIV pathogenesis. Proc Natl Acad Sci USA 91:11811–11817

Cohen DA, Fitzpatrick EA, Barve SS et al. (1993) Activation-dependent apoptosis in $CD4^+$ T cells during murine AIDS. Cell Immunol 151:392

Connors M, Kovacs JA, Krevat S, Banacloche-Gea JC et al. (1997) HIV infection induces changes in CD4+ T cell phenotype and depletions within the CD4+ T cell repertoire that are not immediately restored by antiviral or immune based therapies. Nat Med 5:533

Davis IC, Girard M, Fultz PN (1998) Loss of CD4+ T cells in HIV-1 infected chimpanzees is associated with increased lymphocyte apoptosis. J Virol 72:4623–4632

Debatin K-M (1996) Disturbances of the CD95 (APO-1/Fas) system in disorders of lymphohematopoetic cells. Cell Death Diff 3 2:185

Debatin K-M, Fahrig-Faissner A, Enenkel-Stoodt S, Kreuz W, Benner A, Krammer PH (1994) High expression of APO-1 (CD95) on T lymphocytes from HIV-infected children. Blood 83:3101–3103

Dhein J, Walczak H, Baumler C, Debatin K-M, Krammer PH (1995) Autocrine T-cell suicide mediated by APO-1/(Fas/CD95). Nature 373:438

Dockrell DH, Badley AD, Villacian JS, Heppelmann CJ, Algeciras A, Ziesmer S, Yagita H, Lynch DH, Roche PC, Leibson PJ, Paya CV (1998) The expression of

Fas ligand by macrophages and its upregulation by HIV infection. J Clin Invest 101:2394–2405

Estaquier J, Idziorek T, Zou W, Emilie D, Farber C-M, Bourez JM, Ameisen JC (1995) Th1/Th2 cytokines and T cell death: preventive effect of IL-12 on activation-induced and CD95 (Fas/APO-1)-mediated apoptosis of CD4+ T cells from HIV-infected persons. J Exp Med 182:1759–1765

Estaquier JT, Idziorek F, De Bels F, Barre-Sinoussi F et al. (1994) Programmed cell death and AIDS: significance of T cell apoptosis in pathogenic and nonpathogenic primate lentiviral infections. Proc Natl Acad Sci USA 91:9431

Fauci AS (1996) Host factors and the pathogenesis of HIV-induced disease. Nature 384:529–534

Finkel TH, Tudor-Williams G, Banda NK et al. (1995) Apoptosis occurs predominantly in bystander cells and not in productive cells of HIV- and SIV-infected lymph nodes. Nat Med 1:129

Fisher GH, Rosenberg FJ, Straus SE, Dale JK, Middelton LA, Lin AY, Strober W, Lenardo MJ, Puck JM (1995) Dominant interfering Fas gene mutations impair apoptosis in a human autoimmune lymphoproliferative syndrome. Cell 81:935

Fleury S et al. (1998) Limited CD4+ T cell renewal in early HIV-1 infection : effect of highly active antiviral therapy. Nat Med 4:794–801

Gandhi RT, Chen BK, Straus SE, Dale JK, Lenardo MJ, Baltimore D (1998) HIV-1 directly kills CD4+ T cells by a Fas-independent mechanism. J Exp Med 187:1113–1122

Garcia S, Fevrier M, Dadaglio G, Lecoeur H, Riviere Y, Gougeon M-L (1997) Potential deleterious effect of anti-viral cytotoxic lymphocytes through the CD95 (Fas/APO-1)-mediated pathway during chronic HIV infection. Immunol Lett 57:53–58

Giorgi JV, Liu Z, Hultin LE, Cumberland WG, Hennessey K, Detels R (1993) Elevated level of CD38+CD8+ cells in HIV infection add to the prognostic value of low CD4+ T cell level: results of 6 years follow-up. J Acquir Immune Defic Syndr 6:904–912

Gougeon M-L (1995) Does apoptosis contribute to CD4 T cell depletion in HIV infection? Cell Death Diff 2:1–7

Gougeon M-L, Montagnier L (1993) Apoptosis in AIDS. Science 260:1269–1270

Gougeon M-L, Garcia S, Heeney J, Tschopp R, Lecoeur H, Guetard D, Rame V, Dauguet C, Montagnier L (1993a) Programmed cell death of T lymphocytes in AIDS related HIV and SIV infections. AIDS Res Hum Retrov 9:553–563

Gougeon M-L, Laurent-Crawford AG, Hovanessian AG, Montagnier L (1993b) Direct and indirect mechanisms mediating apoptosis during HIV infection: contribution to in vivo CD4 T cell depletion. Sem Immunol 5:187–194

Gougeon M-L, Lecoeur H, Boudet F, Ledru E, Marzabal S, Boullier S, Roue R, Nagata S, Heeney J (1997) Lack of chronic immune activation in HIV-infected chimpanzees correlates with the resistance of T cells to Fas/Apo-1 (CD95)-induced apoptosis and preservation of a Th1 phenotype. J Immunol 158:2964–2976

Gougeon M-L, Lecoeur H, Dulioust A, Enouf MG, et al (1996) Programmed cell death in peripheral lymphocytes from HIV-infected persons: the increased susceptibility to apoptosis of CD4 and CD8 T cells correlates with lymphocyte activation and with disease progression. J Immunol 156:3509–3519

Gougeon M-L, Lecoeur H, Sasaki Y (1999).The CD95 system in HIV infection. Impact of HAART. Immunol Lett 66:97–104

Gougeon M-L, Olivier R, Garcia S, Guétard D, Dragic T, Dauguet C, Montagnier L (1991) Evidence for an engagement process towards apoptosis in lymphocytes of HIV infected patients. C R Acad Sci 312:529–537

Graziosi C, Pantaleo G, Gantt KR, Fortin JP, Demarest JF, Cohen OJ, Sekaly RP, Fauci AS (1994) Lack of evidence for the dichotomy of TH1 and TH2 predominance in HIV-infected individuals. Science 265:248–251

Groux H, Torpier G, Monté D, Mouton Y, Capron A, Ameisen J-C (1992) Activation-induced death by apoptosis in lymphocytes from human immunodeficiency virus-infected asymptomatic individuals. J Exp Med 175:331–340

Haase AT, et al (1996) Quantitative image analysis of HIV-1 infection in lymphoid tissues. Science 274:985–989

Hattori N, Michaels F, Fargnoli K et al. (1990) The HIV-2 vpr gene is essential for productive infection of human macrophages. Proc Natl Acad Sci USA 87:8080

Heeney J, Jonker J, Koornstra W, Dubbes R, Niphuis H, Di Rienzo A-M, Gougeon M-L, Montagnier L (1993) The resistance of HIV-infected chimpanzees to progression to AIDS correlates with absence of HIV-related dysfunction. J Med Primatol 22:194

Heeney JL (1995) AIDS: a disease of impaired Th-cell renewal? Immunol Today 16:515

Hellerstein M, Hanley MB, Cesar D, Siler S, Papageorgopoulos C, Wieder E, Schmidt D, Hoh R, Neese R, Macallan D, Deeks S, McCune JM (1999) Directly measured kinetics of circulating T lymphocytes in normal and HIV-infected humans. Nat Med 5:83–89

Herbein G, Mahlknecht U, Batliwalla F et al. (1998) Apoptosis of $CD8^+$ T cells is mediated by macrophages through interaction of HIV gp120 with chemokine receptor CXCR4. Nature 395:189–194

Hesselgesser J, Taub D, Baskar P, Greenberg M, Hoxie J, Kolson DL, Horuk R (1998) Neuronal apoptosis induced by HIV-1 gp120 and the chemokine SDF-1α is mediated by the chemokine receptor CXCR4. Curr Biol 8:595–598

Ho DD, Neumann AU, Perelson AS, Chen W, Leonard JM Markowitz M (1995) Rapid turnover of plasma virions and CD4 lymphocytes in HIV-1 infection. Nature 373:123–127

Holznagel E, Hofmann-Lehmann R, Leutenegger CM, Allenspach K, Huettner S et al. (1998) The role of in vitro-induced lymphocyte apoptosis in feline immunodeficiency virus infection: correlation with different markers of disease progression. J Virol 72:9025–9033

Hosaka N, Oyaizu N, Kaplan MH, Yagita H, Pahwa S (1998) Membrane and soluble forms of Fas (CD95) and Fas ligand in PBMC and in plasma from HIV-infected persons. J Inf Dis 178:1030–1039

Itoh N, Yonehara S, Ishii A, Yonehara M, Mizushima S, Sameshima M, Hase A, Seto Y, Nagata S (1991) The polypeptide encoded by the cDNA for human cell surface antigen Fas can mediate apoptosis. Cell 66:233–237

Jeremias I, Herr I, Boehler T, Debatin KM (1998) TRAIL/Apo2-ligand-induced apoptosis in human T cells. Eur J Immunol 28:143–152

Ju ST, Panka DJ, El-Khatib M, Sheer DH, Stanger BZ, Marshak-Rothstein A (1995) Fas (CD95)/FasL interactions required for programmed cell death after T cell activation. Nature 373:444–447

Katsikis PD, Garcia-Ojeda ME, Torres-Roca JF, Tijoe IM, Smith CA, Herzenberg LA, Herzenberg LA (1997) Interleukin-1-beta converting enzyme-like protease involvement in Fas-induced and activation-induced peripheral blood T cell apoptosis in HIV infection-TNF-related apoptosis-inducing ligand can mediate activation-induced T cell death in HIV infection. J Exp Med 186:1365–1372

Katsikis PD, Wunderlich ES, Smith CA, Herzenberg LA, Herzenberg LA (1995) Fas antigen stimulation induces marked apoptosis of T lymphocytes in human immunodeficiency virus-infected individuals. J Exp Med 181:2029–2035

Kelleher AD, Carr A, Zaunders J, Cooper DA (1996) Alterations in the immune response of human immunodeficiency virus (HIV)-infected subjects treated with an HIV-specific protease inhibitor, ritonavir. J Inf Dis 173:321–325

Krammer PH, Dhein J, Walczak H, Behrmann I, Mariani S, Matiba B, Fath M, Daniel PT, Knipping E, Westendorp MO, Stricker K, Bäumler C, Hellbardt S, Germer M, Peter ME, Debatin K-M (1994) The role of APO-1 mediated apoptosis in the immune system. Immunol Rev 142:175–191

Kroemer G (1997) The proto-oncogene Bcl-2 and its role in regulating apoptosis. Nat Med 6:614

Laurent-Crawford AG, Krust B, Muller S, Rivière Y, Rey-Cuillé MA, Béchet J-M, Montagnier L, Hovanessian A (1991) The cytopathic effect of HIV is associated with apoptosis. Virology 185:829–839

Laurent-Crawford AG, Krust B, Riviere Y, Desgranges C, Muller S, Kieni MP, Dauguet C, Hovanessian AG (1993) Membrane expression of HIV envelope glycoproteins triggers apoptosis in CD4 cells. AIDS Res Hum Retrov 9:761

Lecoeur H, Ledru E, Gougeon M-L (1998) A cytofluorometric method for the simultaneous detection of both intracellular and surface antigens on apoptotic peripheral lymphocytes. J Immunol Meth 217:11–26

Ledru E, Lecoeur H, Garcia S, Debord T, Gougeon M-L (1998) Differential susceptibility to activation-induced apoptosis among peripheral Th1 subsets: correlation with Bcl-2 expression and consequences for AIDS pathogenesis. J Immunol 160:3194–3206

Levacher M, Hulstaert F, Tallet S, Ullery S, Pocidalo J-J, Bach BA (1992) The significance of activation markers on CD8 lymphocytes in human immunodeficiency syndrome: staging and prognostic value. Clin Exp Immunol 90:376–382

Lewis DE, Tang DS, Adu-Oppong A, Schober W, Rodgers JR (1994) Anergy and apoptosis in $CD8^+$ T cells from HIV-infected persons. J Immunol 153:412

Li CJ, Friedman DJ, Wang C, Metelev V, Pardee AB (1995) Induction of apoptosis in uninfected lymphocytes by HIV-1 Tat protein. Science 268:429

Liegler TJ, Yonemoto W, Elbeik T, Vittinghoff E, Buchbinder SP, Grene WC (1998) Diminished spontaneous apoptosis in lymphocytes from HIV-infected long term non progressors. J Inf Dis 178:669–679

Linette GP, Grusby MJ, Hedrick SM, Hansen TH, Glimcher LH, Korsmeyer SJ (1994) Bcl-2 is upregulated at the CD4+ CD8+ stage during positive selection and promotes thymocyte differentiation at several control points. Immunity 1:197

Medrano FJ, Leal M, Arienti D, Rey C, Zagliani A, Torres Y, Sanchez-Quijano A, Lissen E, Clerici M (1998) $TNF\beta$ and soluble APO-1/Fas independently predict progression to AIDS in HIV-seropositive patients. AIDS Res Hum Retrov 14: 835–843

Meyaard L, Otto SA, Jonker RR, Mijnster MJ, Keet R, Miedema F (1992) Programmed death of T cells in HIV-1 infection. Science 257:217–219

Meyaard L, Otto SA, Keet IPM, Roos MTL, Miedema F (1994) Programmed death of T cells in HIV infection: no correlation with progression to disease. J Clin Invest 93:982

Michel P, Toure-Balde A, Roussilhon C, Aribot G, Sarthou J-L, Gougeon M-L (1999) Reduced immune activation and T cell apoptosis in HIV-2 compared to HIV-1 infected West African patients. Correlation of T cell apoptosis with seric $\beta 2$ microglobulin and with disease evolution. J Inf Dis. in press

Mitra D, Steiner M, Lynch DH, Staiano-Coico L Laurence J (1996) HIV-1 upregulates Fas ligand expression in CD4+ T cells in vitro and in vivo: association with Fas-mediated apoptosis and modulation by aurintricarboxylic acid. Immunol 87:581

Muro-Cacho CA, Pantaleo G, Fauci A (1995) Analysis of apoptosis in lymph nodes of HIV-infected persons. Intensity of apoptosis correlates with the general state of activation of the lymphoid tissue and not with the stage of disease or viral burden. J Immunol 154:5555

Nagata S, Suda (1995) Fas and Fas ligand: lpr and gld mutations. Immunol Today 16:39

Nagata S (1997) Apoptosis by death factor. Cell 88:355

Naora H, Gougeon M-L (1999) Influence of interleukin-15 on apoptosis of CD4 and CD8 lymphocytes of HIV-infected individuals: benefits of enhanced lymphocyte survival counter-balanced by increased lymphocyte activation and poor inhibition of CD3- and Fas- induced apoptosis. Cell Death and Diff. in press

Nunez G, Merino R, Grillot D, Gonzalez-Garcia M (1994) Bcl-2 and Bcl-x: regulatory switches for lymphoid death and survival. Immunol Today 15:582

Oehm A, Behrmann I, Falk W et al. (1992) Purification and molecular cloning of the APO-1 cell surface antigen, a member of the tumor necrosis factor/nerve growth factor receptor superfamily. Sequence identity with the Fas antigen. J Biol Chem 267:10709

Ohnimus H, Heinkelein M, Jassoy C (1997) Apoptotic cell death upon contact of $CD4^+$ T lymphocytes with HIV glycoprotein-expressing cells is mediated by caspases but bypasses CD95 (Fas/Apo-1) and TNF receptor 1. J Immunol 159:546–5252

Oshimi Y, Oda S, Honda Y, Nagata S, Miyazaki S (1996) Involvement of Fas ligand and Fas-mediated pathway in the cytotoxicity of human natural killer cells. J Immunol 157:2909

Pakker NG, Notermans DW, Boer RJ de et al. (1998) Biphasic kinetics of peripheral blood T cells after triple combination therapy in HIV-1 infection: a composite redistribution and proliferation. Nat Med 4:208–214

Pantaleo G, De Maria A, Koenig S, Butini L, Moss B, Baseler M, Lane HC, Fauci AS (1990) CD8+ T lymphocytes of patients with AIDS maintain normal broad cytolytic function despite the loss of human immunodeficiency virus-specific cytotoxicity. Proc Natl Acad Sci USA 87:4818

Pantaleo G, Graziosi C, Demarest JF, Butini L, Montroni M, Fox CH, Orenstein JM, Kotler DP, Fauci AS (1993) HIV infection is active and progressive in lymphoid tissue during the clinically latent stage of disease. Nature 362:355

Peter ME, Krammer PH (1998) Mechanisms of CD95 (APO-1/Fas)-mediated apoptosis. Curr Opin Immunol 10:545–551

Piazza C, Montani MS, Gilardini S, Moretti S, Cundari E, Piccolella E (1997) CD4+ T cells kill CD8+ T cells via Fas/Fas ligand-mediated apoptosis. J Immunol 158:1503

Pignata C, Fiore M, De Filippo S, Cavalcanti M, Gaetaniello L, Scotese I (1998). Apoptosis as a mechanism of peripheral blood mononuclear cell death after measles and varicella-zoster virus infections in children. Ped Res 43:77–83

Reed JC (1997) Double identity for proteins of the Bcl-2 family. Nature 387:773

Rieux-Laucat F, Le Deist F, Hivroz C, Roberts IA, Debatin KM, Fischer A Villartay JP de (1995) Mutations in Fas associated with human lymphoproliferative syndrome and autoimmunity Science 268:1347

Riviere Y, Tanneau-Salvadori F, Regnault A, Lopez O, Sansonetti P, Guy B, Kieny MP, Fournel JJ, Montagnier L (1989) HIV-specific cytotoxic responses of seropositive individuals: distinct types of effector cells mediate killing of targets expressing gag and env proteins. J Virol 63:2270

Stent G, Bjerknes R, Voltersvik P, Olofsson J, Asjo B (1998) Correlates of apoptosis of CD4+ and CD8+ T cells in tonsillar tissue in HIV type 1 infection. AIDS Res Hum Retrov 14:1635–1643

Sachsenberg N, et al (1998) Turnover of CD4 and CD8 T lymphocytes in HIV-1 infection as measured by Ki67 antigen. J Exp Med 187:1295–1303

Sloand EM, Young NS, Kumar P, Weichold FF, Sato T, Maciejewski JP (1997) Role of Fas ligand and receptor in the mechanism of T-cell depletion in AIDS: effect on $CD4^+$ lymphocyte depletion and HIV replication. Blood 89:135

Stewart SA, Poon B, Jowett JBM, Chen ISY (1997) HIV-1 Vpr induces apoptosis following cell cycle arrest. J Virol 71:5579–5583

Suda T, Okazaki T, Naito Y, Yokota Y, Arai N, Ozaki S, Nakao K, Nagata S (1995) Expression of the Fas ligand in cells of T cell lineage. J Immunol 154:3806

Terai C, Kornbluth, RS, Pauza CD, Richman DD, Carson DA (1991) Apoptosis as a mechanism of cell death in cultured T lymphoblasts acutely infected with HIV-1. J Clin Invest 87:1710–1715

Trauth BC, Klas C, Peters AMJ, Matzku S, Müller P, Falk W, Debatin K-M, Krammer PH (1989) Monoclonal antibody-mediated tumor regression by induction of apoptosis. Science 245:301

Veis DJ, Sorenson CM, Shutter JR, Korsmeyer SJ (1993) Bcl-2 deficient mice demonstrate fulminant lymphoid apoptosis, polycystic kidneys, and hypopigmented hair. Cell 75:229

Wang LQ, Klimpel GR, Planas JM, Li HB, Cloyd MW (1998) Apoptotic killing of CD4$^+$ T lymphocytes in HIV-1 infected PHA-stimulated PBL cultures is mediated by CD8$^+$ LAK cells. Virology 241:169–180

Watanabe-Fukunaga R, Brannan CI, Itoh N, Yonehara S, Copeland NG, Jenkins NA, Nagata S (1992) The cDNA structure, expression, and chromosomal assignment of the mouse Fas antigen. J Immunol 148:1274

Wei X, Ghosh SK, Taylor ME, Johnson VA, Emini EA, Deutsch P, Lifson JD, Bonhoeffer S, Nowak MA, Hahn BH, Saag MS, Shaw GM (1995) Viral dynamics in human immunodeficiency virus type 1 infection. Nature 373:117–320

Westendorp MO, Frank R, Oschsenbauer C, Stricker K, Dhein J, Walczak H, Debatin K-M, Krammer PH (1995) Sensitization of T cells to CD95-mediated apoptosis by HIV-1 Tat and gp120. Nature 375:497

Wolthers KC, Bea G, Wisman A et al. (1996) T cell telomere length in HIV-1 infection: no evidence for increased CD4+ T cell turnover. Science 274:1543–1545

Yang E, Korsmeyer SJ (1996) Molecular thanatopsis: a discourse on the Bcl-2 family and cell death. Blood 88:386

Yao XJ, Mouland AJ, Subbramanian RA, Forget J, Rougeau N, Bergeron D, Cohen EA (1998) Vpr stimulates viral expression and induces cell killing in HIV-1 infected dividing Jurkatt T cells. J Viro 72:4686–4693

Zauli G, Gibellini D, Caputo A, Bassini A, Negrini M, Monne M, Mazzoni M, Capitani S (1995) The HIV-1 tat protein upregulates bcl-2 gene expression in Jurkat T-cell lines and primary peripheral blood mononuclear cells. Blood 86:3823

Zou W, Lackner AA, Simon M, Durand-Gasselin I et al. (1997) Early cytokine gene expression in lymph nodes of macaques infected with SIV is predictive of disease outcome and vaccine efficacy. J Virol 71:1227

CHAPTER 6
Apoptotic Cell Phagocytosis

J. Savill and C. Bebb

A. Introduction

Given the enormous interest currently shown in apoptosis, it may seem remarkable that this form of cell death was overlooked for so long. However, apoptosis is histologically inconspicuous because of the speed and efficiency by which cells undergoing this programmed form of cell death are recognised, ingested and degraded by nearby phagocytes. Where the load of apoptotic cells is small the clearance job can be done by neighbouring cells of the same type acting as "semi-professional" phagocytes, but even where many cells are undergoing apoptosis these can be efficiently cleared by the professional phagocytes of the body, cells of the macrophage line. Indeed, recent studies have emphasised that the clearance of apoptotic cells may not only be "silent", preventing inflammation due to leakage from dying cells of injurious contents, but by re-programming phagocytes can also be actively anti-inflammatory. Nevertheless, it appears that there may also be circumstances by which particular phagocyte types can incite potentially deleterious immune responses by presentation to T lymphocytes of antigens borne by ingested apoptotic cells. This chapter will therefore attempt to dissect this apparent paradox through a detailed discussion of the cellular and molecular mechanisms mediating phagocytic clearance of cells being eliminated by cell death.

B. Tissue Consequences of Cell Death
I. Necrosis and Incitement of Inflammatory Injury

There is a general consensus that cell death by necrosis is not a safe means of cell clearance (Wyllie et al. 1980). Where cells have been exposed to "murderous" stimuli such as extremes of temperature, severe hypoxia or high doses of ionising radiation, fields of stricken cells lose the ability to regulate membrane permeability, swell and then disintegrate. Presumably because local defence mechanisms such as antiproteases and neutral interstitial pH are overwhelmed or perturbed by sudden local release of noxious contents from necrotic cells, further tissue injury and inflammatory responses are triggered. Primary necrosis is clearly a messy and dangerous form of cell death.

Extensive study of cultured cells undergoing apoptosis also emphasises that cellular disintegration is an undesirable event in cell death. Although there is persuasive evidence that tissue transglutaminase irreversibly cross-links membrane proteins during apoptosis in some cell types (FESUS et al. 1987, 1989), suggesting that the contents of apoptotic cells may be safely sealed inside an insoluble keratin-like cocoon (PIREDDA et al. 1997), most cell types undergoing apoptosis in culture in the absence of phagocytes remain intact for only a few hours before undergoing "secondary necrosis", swelling and then leaking their contents (REN and SAVILL 1998). Indeed, in co-culture experiments in which macrophage phagocytosis of neutrophils undergoing apoptosis was non-specifically blocked with colchicine, non-ingested apoptotic neutrophils undergoing secondary necrosis released large quantities of potent degrative enzymes (such as elastase) which eluded endocytic clearance by macrophages (KAR et al. 1993).

Furthermore, secondary necrosis of apoptotic cells may be doubly deleterious, not only causing direct local tissue injury through leakage of noxious contents but also stimulating macrophages to release cytokines capable of amplifying inflammatory responses. For example, cultured human monocyte-derived macrophages ingesting apoptotic eosinophils did not release pro-inflammatory mediators, but large-scale release of such molecules occurred when macrophages were "fed" cell debris from eosinophils which had undergone secondary necrosis in culture (STERN et al. 1996). To conclude, it would appear that cellular necrosis, whether primary in previously healthy cells or secondary to apoptosis, threatens tissue injury by a number of mechanisms.

II. Silent and Anti-Inflammatory Cell Clearance by Apoptosis

A histological hallmark of cell death by apoptosis in living tissues is the complete absence of inflammation. This does not merely reflect a small "load" of dying cells in comparison with necrosis, because clearance of huge numbers of cells undergoing apoptosis can also occur without inciting inflammation, as exemplified by lymphoid organs in which "tingible body" macrophages are apparently stuffed with lymphoid cells undergoing apoptosis after failing selection (WYLLIE et al. 1980; SURH and SPRENT 1994)). Indeed, kinetic calculations indicate that a very high proportion of would-be lymphocytes meet this fate, over 90% in the thymus.

Therefore, on the basis of histological observations, it would appear that phagocytes taking up apoptotic cells are unlikely to make a pro-inflammatory response. We tested this hypothesis in vitro, employing various phagocyte types and various "target" cell types undergoing apoptosis. Our findings were similar whether we employed monocyte-derived macrophages as model "professional" phagocytes or glomerular mesangial cells to represent "semi-professional" phagocytes; uptake of apoptotic cells did not incite increased release of a range of pro-inflammatory mediators including eicosanoids, granule enzymes or chemokines (MEAGHER et al. 1992; STERN et al. 1996; HUGHES et al. 1997). A recent report (KUROSAKA et al. 1998) asserting that there is release of IL-8 and IL-1 from a macrophage cell line after uptake of apoptotic cells probably reflects the use of dying cells in which a high proportion of ingested targets exhibited signs of necrosis. Indeed, in systems with minimal evidence of necrosis in the apoptotic "meal" we frequently observed a small but statistically significant suppression of mediator release compared with unstimulated phagocytes (MEAGHER et al. 1992). Nevertheless, we obtained definitive evidence that apoptotic cells were not a "poisoned meal". First, macrophages ingesting apoptotic cells were still able to mount vigorous pro-inflammatory responses to a subsequent meal of particles opsonised with immunoglobulin and complement. Second, when apoptotic neutrophils were deliberately opsonised and conditions altered so that phagocytosis was exclusively via macrophage Fc receptors, large-scale release of pro-inflammatory mediators was observed (MEAGHER et al. 1992). It was clear that suppression of background mediator release was not due to some non-specific toxic effect of ingested apoptotic cells.

More recently, FADOK et al. (1998a) have extended these observations to emphasise that clearance of apoptotic cells by activated macrophages may be positively "anti-inflammatory" rather than merely silent. Both unstimulated and LPS-stimulated macrophages released the immunosuppressive cytokine transforming growth factor $\beta 1$ (TGF-$\beta 1$) after uptake of apoptotic cells, but the amount released was considerably greater when deliberately activated macrophages were used. Furthermore, stimulated release of the potent pro-inflammatory cytokine tumour necrosis factor-α (TNFα) was dramatically downregulated when activated macrophages took up apoptotic cells but not control particles. Inhibitor studies suggested that this specific down-regulation of pro-inflammatory responses occurred by a complex paracrine/autocrine signalling loop involving macrophage release of TGF-$\beta 1$, prostaglandin E2 and platelet activating factor.

Indeed, Ferguson's group have recently presented data which argue that cells undergoing Fas-mediated apoptosis may actively synthesise the anti-inflammatory cytokine interleukin-10 (IL-10), which can potently downregulate pro-inflammatory responses from stimulated macrophages. The group had shown that when antigen-coupled cells were introduced into the immunologically privileged eye, systemic tolerance to the antigen ensued if the tagged cells underwent Fas-mediated apoptosis (GRIFFITH et al. 1996) but if the administered apoptotic cells were from IL-10 "knockout" mice (but not controls) then tolerisation failed (GAO et al. 1998). Furthermore, lymphocytes undergoing Fas-mediated apoptosis expressed IL-10. Whether these observations are generally applicable to a wide range of cell types undergoing apoptosis remains to be seen, but the data raise the intriguing concept that phagocyte clearance of apoptotic cells may result in the delivery to phagocytes (as if via a Trojan horse) of "packets" of IL-10. Further developments are awaited with interest.

Finally, however, the discovery that engagement of apoptosis may result in expression of IL-10 by the doomed cell could explain apparent differences between studies of Fadok et al. (1998a) and Voll et al. (1997). Whereas both studies agreed that uptake of apoptotic cells by stimulated macrophages inhibited TNFα release from the phagocytes, only in the study by Voll and colleagues was there additional evidence of IL-10 release. Rather than reflecting secretion of this immunosuppressive cytokine from macrophages, the IL-10 detected in the system of Voll et al. (1997) may have been released from apoptotic cells losing viability in an assay system where dying targets and phagocytes were co-cultured for protracted periods.

Nevertheless, despite nuances which may reflect experimental differences, a considerable body of in vitro evidence now emphasises that, in addition to being "silent", phagocytic clearance of apoptotic cells has the potential for active suppression of inflammatory and perhaps immune responses. Moreover, at least in the specialised system of the eye, there is in vivo evidence to support this concept. However, as we shall see, administration of apoptotic cells by an alternative route may have very different consequences.

III. Provocation of Immune Responses During Clearance of Apoptotic Cells

In addition to threatening direct tissue injury (Fig. 1), the contents of apoptotic cells have potential to incite (auto)immune responses. First, cleavage of internucleosomal DNA by apoptotic endonucleases yields oligonucleosomes within dying cells (Wyllie 1980). Unscheduled release of oligonucleosomes from non-ingested cells is potentially deleterious, stimulating lymphocyte proliferation (Bell et al. 1991), triggering pro-inflammatory responses from phagocytes (Emlen et al. 1992), and acting as a nidus for deposition of antibody and complement in organs such as the glomerulus, where cationic histones in nucleosomes bind the anionic filtration structures (Koutouzov et al. 1996). Second, a series of beautiful studies by Rosen's group (Casciola-Rosen et al. 1994, 1995, 1996) have emphasised that the processes of apoptosis may not only redistribute nuclear autoantigens such as Ro and La into sites vulnerable to leakage, such as cell surface blebs, but also have potential to generate neoantigens by enzymatic cleavage of proteins.

Furthermore, by contrast with studies where apoptotic cells were administered into the immunologically privileged eye (Griffith et al. 1996), intravenous administration to normal mice of 10^7 syngeneic thymocytes, irradiated so as to trigger apoptosis, stimulated transient appearance in blood of nuclear autoantibodies, anticardiolipin and anti-single stranded DNA autoantibodies (Mevorach et al. 1998a). Far lower levels of such autoantibodies were observed in animals receiving either viable non-apoptotic splenocytes or lysates from viable thymocytes as controls. Intriguingly, although animals receiving apoptotic thymocytes did not develop proteinuria, a high proportion exhibited IgG deposits in glomeruli several months later. Although this obser-

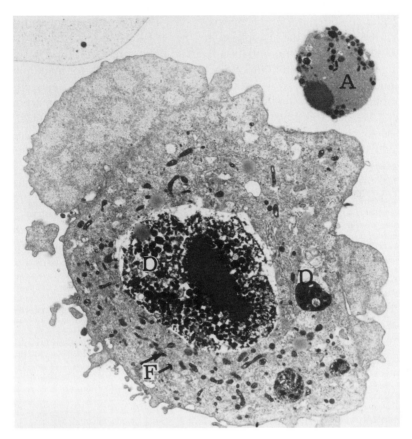

Fig. 1. Retention of cellular contents during apoptosis leading to phagocytic clearance. Electron micrograph (×13,000) of cultured glomerular mesangial cell (identifiable microfilaments are at F) which has ingested two apoptotic granulocytes (D); the example on the *left* is recently ingested and exhibits a full complement of granules; that on the *right* is at an advanced stage of degradation. A non-ingested apoptotic leucocyte fragment is seen at A

vation emphasises the pathogenic potential of autoantibodies induced by intravenous administration of apoptotic cells, no consistent histological abnormality was observed in the kidney. Furthermore, neither the fate of administered apoptotic cells nor the mechanisms eliciting autoantibody production are yet apparent.

How could clearance of apoptotic cells trigger immune responses? It now seems likely that an answer to this question may relate to the important discovery that apoptotic cells can be ingested by dendritic cells derived in culture from bone marrow (RUBARTELLI et al. 1997). Such phagocytes are well-established models of myeloid-derived dendritic cells, which are specialised for presentation of antigen to T lymphocytes by virtue of high expression of

MHC and co-stimulatory molecules (AUSTYN 1998). Furthermore, translocation of CD83 from the cytoplasm to the cell surface has recently been identified as marking differentiation from an immature dendritic cell capable of antigen capture by retention of macrophage-like capacity for macropinocytosis and phagocytosis to a mature phenotype which excels at antigen presentation, but exhibits diminished capacity to ingest fluid phase or particulate matter (ALBERT et al. 1998b). Immature dendritic cells exhibited capacity for large-scale phagocytosis of apoptotic cells and were able to present antigen derived from apoptotic cells to naive T cells in a co-culture system (ALBERT et al. 1998a). This was demonstrated by exploiting the propensity of influenza-infected monocytes to undergo apoptosis and thereby serve as an antigen-laden "meal" for maturing dendritic cells. The latter were then able to present 'flu peptides via MHC Class I and thereby induce CD8-positive cytotoxic lymphocytes (CTLs) specific for influenza. Importantly, CTLs were not induced when necrotic flu-infected monocytes were fed to dendritic cells, nor if uptake of apoptotic monocytes was blocked by cytochalasin D, emphasising dependence upon phagocytosis of apoptotic cells as a substrate for antigen presentation. Furthermore, subsequent reports have documented MHC Class II-mediated presentation of apoptotic cell-derived antigens to CD4 positive T lymphocytes by phagocytic dendritic cells (INABA et al. 1998).

Could phagocytes other than dendritic cells present antigen derived from apoptotic cells? One would predict that this is unlikely given the typically "silent" nature of cell clearance by apoptosis. Certainly, despite antigen-presenting capacity and expression of MHC Class I and Class II, it is reassuring that some observers find that macrophages are unable to present apoptotic cell-derived antigen to naive T cells (ALBERT et al. 1998a). Furthermore, although there is a report that "macrophages" can present antigen via this route to primed T cells, the phagocytes concerned were obtained by prolonged culture in GMCSF, which can promote differentiation into immature myeloid dendritic cells (BELLONE et al. 1997). Further work is obviously needed to pursue these exciting leads. Nevertheless, perhaps unexpectedly, there is now persuasive evidence to suggest that apoptotic cells can be cleared in such a way that (auto)immune responses ensue. Clearly, therefore, there is a need to reconcile these data with those indicating that the clearance of apoptotic cells may suppress inflammatory and immune responses.

IV. Resolving the Clearance Paradox for Apoptotic Cells

On first principles it would seem that immunogenic clearance of apoptotic cells ought to be an unusual consequence of this mode of cell death which, in turn, should be kept under tight control. Such regulation might be as simple as ensuring that dendritic cells are usually "hedged around" by efficient phagocytes which do not elicit immune responses to antigens derived from apoptotic cells. For example, ALBERT et al. (1998a) included monocyte-derived macrophages in co-culture of dendritic cells and 'flu-infected apoptotic monocytes, and presentation of antigen was inhibited. In addition to ingesting apoptotic cells efficiently before these can be taken up by neighbouring dendritic cells, release of immunosuppressive cytokines (such as TGF-β1) from macrophages that ingest apoptotic cells might serve either to downregulate antigen presentation by immature dendritic cells or to inhibit their maturation to full antigen-presenting capacity even if they are successful in taking up apoptotic cells.

Control of dendritic cell responses by load of apoptotic cells is a potentially important regulatory mechanism suggested by the fascinating in vitro studies of ROVERE et al. (1998). Employing an experimental system in which immature dendritic cells taking up apoptotic cells were assessed for capacity to present antigen via either MHC Class I or Class II, they found that presentation was increasingly efficient the greater the ratio of apoptotic cells to phagocytes, being greatest when this was 5 to 1. Furthermore, this correlated with increasing release from phagocytes of the pro-inflammatory cytokines TNFα and IL-Iβ and associated maturation of dendritic cells evidenced by increased expression of CD86 etc. Interestingly, although both "early" and "late" apoptotic cells were able to sustain antigen presentation, frankly necrotic cells did not do so, as in other reports (ALBERT et al. 1998a,b). The group suggests that in situations where large numbers of cells undergo apoptosis, perhaps overwhelming anti-inflammatory phagocyte defences, the scene is set for uptake by immature dendritic cells and autocrine/paracrine promotion of presentation of antigens from ingested cells via TNFα/IL-1β-driven maturation of phagocytic dendritic cells. These intriguing ideas need formal testing in vivo, but they suggest that presentation of antigens by dendritic cells may only occur when very large numbers of cells undergo apoptosis relatively synchronously. This might ultimately be desirable, for example in a viral infection where beneficial cytotoxic responses might be promoted.

The rate of IL-10 release in co-cultures of apoptotic cells and myeloid cells might also impinge on the apparent paradox of anti-inflammatory *vs* immunogenic cell clearance. Whether released from the dying cells (GAO et al. 1998) or the phagocytes (VOLL et al. 1997), ROVERE et al. (1998) detected small quantities of extracellular IL-10 when apoptotic cells were co-cultured with immature dendritic cells at a ratio of 5:1, and yet maturation of dendritic cells ensued (perhaps because of much larger release of TNFα) despite the reported capacity of this cytokine to inhibit such maturation (AUSTYN 1998). However, release of IL-10 may explain the appearance of autoantibodies in mice receiving intravenous apoptotic cells (MEVORACH et al. 1998a) since IL-10 augments B cell activation and is produced at high concentration in patients with autoimmune conditions such as systemic lupus erythematosus (SLE) (LLORENTE et al. 1995). Thus an apparently immunosuppressive and anti-inflammatory cytokine could contribute to B cell-mediated immunity.

Finally, as we shall see later in this chapter, different recognition mechanisms could prove to be the molecular substrate for flexible or apparently contradictory responses made by phagocytes ingesting apoptotic cells. Nevertheless, much work remains before we can be sure of the immune and inflammatory consequences of the phagocytic clearance of apoptotic cells in health and disease.

C. Molecular Mechanisms by Which Phagocytes Recognise Cells Undergoing Apoptosis

Over the last few years, in vitro studies have demonstrated that many cell surface molecules may mediate interaction between cells dying by apoptosis and phagocytes (Fig. 2). Because this large body of data has been extensively reviewed elsewhere (SAVILL 1997; FADOK et al. 1998b) this chapter will focus in detail only on the newest data and points of controversy or growth in the field.

I. "Eat Me" Signals Displayed by Apoptotic Cells

Early scanning electron microscopy studies of cells undergoing apoptosis revealed dramatic structural changes including loss of microvilli and the development of "pits" or invaginations (MORRIS et al. 1984). Such structural alterations obviously suggest biochemical changes likely to mark intact dying cells for removal by phagocytes. Although new protein synthesis appears to be a requirement for apoptosis in some systems, this is not generally the case (WYLLIE et al. 1980), implying that synthesis and expression at the cell surface of new "eat me" proteins is unlikely to mediate recognition by phagocytes. By contrast, accumulating data point to modification or rearrangement of existing plasma membrane components as the most likely way(s) to reveal "eat me" signals. These will now be considered in diminishing order of their degree of characterisation.

1. Exposure of Phosphatidylserine

Early studies of the phagocytosis by human monocyte-derived macrophages (Mϕ) of apoptotic neutrophils indicated that apoptotic cells exposed sites which were recognised by phagocytes (SAVILL et al. 1989). Thus recognition was inhibited by cationic aminosugars, which pre-incubation studies and varying the pH of the interaction with Mϕ showed to be acting by masking of anionic sites on the apoptotic cell. These sites were resistant to broad spectrum proteases but were not characterised further. However, indirect evidence supports the candidacy of anionic phospholipids or sulfated lipids (on the grounds of involvement of bridging thrombospondin or phagocyte scavenger receptors/CD14, as described below) or possibly oxidised lipids/proteins (implicated by involvement of phagocyte CD36 and CD68, see below). Nevertheless, the anionic phospholipid phosphatidylserine is well established as an "eat me" signal (FADOK et al. 1998b).

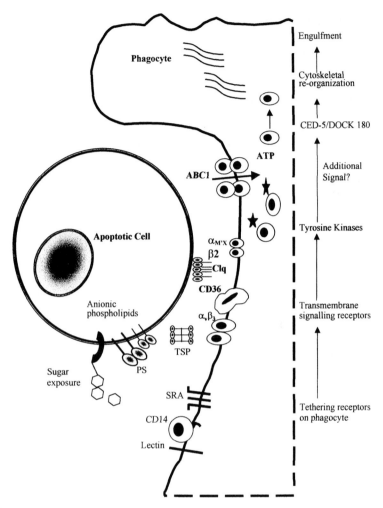

Fig. 2. Summary of molecules implicated in phagocytosis of apoptotic cells. Apoptotic cells display phosphatidylserine (PS), altered sugar side chains and, possibly, anionic phospholipids which may bind "bridging" thrombospondin (TSP). Some phagocyte receptors, such as GPI-linked CD14 and the class A scavenger receptor (SRA) may lack capacity to signal to the cytoskeleton and could function as "tethering" receptors; it is proposed that yet-to-be characterised lectins and phosphatidylserine receptors (not shown) may have similar properties. Bridging TSP binds the $\alpha_v\beta_3$ integrin and CD36, which can activate tyrosine kinases. Binding of Clq to apoptotic cells may bridge apoptotic cells to discrete Clq receptors on phagocytes (not shown) or, by activating the complement cascade to generate opsonic fragments such as iC3b (not shown), ligate either the $\alpha_m\beta_2$ integrin (CR3) or $\alpha_x\beta_2$ integrin (CR4) which can also signal phagocytosis. ABC-1 and ATP-binding cassette molecule may form a pore which also signals to the cytoskeleton; a possible intermediate is the CED-5/DOCK-180 molecule implicated in cytoskeletal reorganisation. (Adapted from SAVILL 1998)

Phosphatidylserine (PS) is normally confined to the inner leaflet of the plasma membrane bilayer of viable cells but is preferentially exposed on the surface of apoptotic cells as evidenced by procoagulant activity in the PS-dependent Russell viper venom assay (FADOK et al. 1992a). While care is needed to ensure that dying cells have not developed the leaky membranes of early necrosis (a typical criterion is exclusion of propidium iodide), the capacity of annexin V protein to bind to cell surface PS has been extensively exploited as an easy means of detecting PS exposure in cells undergoing apoptosis in culture (KOOPMAN et al. 1994). Study of various cell types emphasises that PS exposure may be an early feature of apoptosis in some (MARTIN et al. 1995a) but not all (HOMBERG et al. 1995) cell types, suggesting that recognition of PS by phagocytes (see below) in vivo could lead to removal of dying cells before these have the chance to display later morphological changes such as chromatin condensation.

The mechanisms responsible for PS exposure during apoptosis appear to involve downregulation of an ATP-dependent aminophospholipid translocase, which normally directs PS to the inner membrane leaflet, and upregulation of calcium-dependent flippases which promote PS exposure (VERHOVEN et al. 1995). In addition to calcium fluxes (BRATTON et al. 1997), caspases (MARTIN et al. 1995b), and apoptosis inducing factor released from mitochondria (ZAMZANI et al. 1996; SUSIN et al. 1999) have been implicated in driving PS exposure, and cleavage of submembrane cytoskeletal elements such as fodrin may contribute to the preferential binding of annexin V to plasma membrane "blebs" (CASCIOLA-ROSEN et al. 1996). However, work with mitochondrial poisons such as antimycin has shown that there may be a discrete "membrane subprogram" in apoptosis which enables such reagents to inhibit exposure of annexin V binding sites and related recognition by phagocytes, whilst allowing activation of caspase 3 and chromatin condensation (ZHUANG et al. 1998). It therefore seems likely that much work remains to be done before we can understand how PS is exposed by apoptotic cells, and we will need to remember that exposure of other anionic moieties may also be as important.

2. Sites Which Bind "Bridging" Proteins

In vivo, recognition of apoptotic cells is likely to take place in microenvironments which may communicate with interstitial fluid or plasma. Consequently, protein systems which have evolved to bind foreign or altered cell surfaces are candidates for "tagging" apoptotic cells for removal. Thrombospondin 1 (TSP1), a trimeric glycoprotein secreted by many cell types and known to mediate the binding of activated platelets to monocytes (SILVERSTEIN et al. 1992), binds apoptotic cells and "bridges" them to phagocytes (SAVILL et al. 1992). Although the involvement of TSP1 in clearance of dying cells in vivo has yet to be demonstrated, it is intriguing that TSP1–/– "knockout" animals develop persistent inflammatory responses which could relate to impaired phagocytic clearance of apoptotic leucocytes (LAWLER et al. 1998). However, although sulfated lipids (or "sulfatides") are known to bind TSP1 and are therefore attractive candidate TSP1-binding sites, there is very limited understanding as to how TSP1 may bind selectively to apoptotic cells in such a way that phagocytosis is promoted through phagocyte receptors for TSP1 such as the $\alpha_v\beta_3$ integrin and CD36 (see Sect. C.II.1, Thrombospondin Receptors: $\alpha_v\beta_3$ and CD36).

While there is growing evidence that PS could be recognised directly by phagocyte receptors (see below), some data also suggest that PS could bind plasma proteins which then "bridge" PS to phagocyte receptors, including the plasma protein β_2GPI (PRICE et al. 1996; MANFREDI et al. 1998b) and opsonic complement fragments such as C3bi (MEVORACH et al. 1998b). Initial studies suggesting that PS on apoptotic cells may fix complement used target cell systems likely to contain a significant proportion of cells undergoing secondary necrosis (TAKIZAWA et al. 1996), but MEVORACH et al. (1998b) were careful to use apoptotic populations with very high viability as assessed by exclusion of vital dyes, implicating PS in activation of complement because annexin V could partially inhibit C3bi binding to apoptotic cells. The potential involvement of complement components in "bridging" may include mechanisms other than PS-mediated complement activation. KORB and AHEARN (1997) discovered that the first component of the classical pathway, C1q, can specifically and directly bind to cell surface blebs on apoptotic keratinocytes. While this might serve to activate complement via the classical pathway, there is also evidence of C1q receptors to which direct bridging might occur. Very intriguingly, C1q–/– "knockout" mice (see below) exhibited increased numbers of apoptotic cells in apparently normal glomeruli consistent with, but not definitively diagnostic of, impaired clearance of apoptotic cells in vivo (BOTTO et al. 1998). Lastly, activation of the coagulation cascade by PS might also result in coating of apoptotic cells by "sticky" proteins (CASCIOLA-ROSEN et al. 1996).

3. Carbohydrate Changes on Apoptotic Cells

However, even earlier than data pointing to a role for bridging TSP1 or exposed PS, Wyllie's group initiated studies which now support the possibility that apoptotic cells display changes in cell surface carbohydrates likely to promote clearance by phagocyte lectins. Initially, data obtained by using the tricky technique of cell micro-electrophoresis were consistent with loss of sialic acid from cells undergoing apoptosis leading to the unmasking of sugar residues such as N-acetylglucosamine and N-acetylgalactosamine, which could be recognised by phagocyte lectins (MORRIS et al. 1984). Furthermore, "apoptotic envelopes", detergent-resistant intracellular structures formed during apoptosis by cross-linking of membrane proteins catalysed by tissue transglutaminase (FESUS et al. 1989) selectively bound fluorescent lectins (DINI et al. 1992). However, of more relevance to recognition of apoptotic cells by phagocytes were subsequent data demonstrating lectin binding to the surface of intact apoptotic cells (DINI et al. 1995) and inhibition of phagocytosis by simple sugars (DINI et al. 1995; DUVALL et al. 1985). To date, the molecules bearing sugar rich chains in which residues are exposed remain obscure, but a candidate is described in the next section.

4. Intercellular Adhesion Molecule (ICAM)-3

ICAM-3 is a transmembrane protein encoded by a member of the immunoglobulin (Ig) supergene family encoding five Ig-like extracellular domains. ICAM-3 is normally expressed by leucocytes and mediates well-characterised interactions with β_2 integrins including $\alpha_L\beta_2$ (LFA-1) and the novel leucointegrin $\alpha_d\beta_2$. Gregory's group screened a large panel of leucocyte antibodies for inhibition of the binding of apoptotic B lymphocytes to human Mϕ and found that two ICAM-3 mAbs, 3A9 and BU68, inhibited recognition (FLORA and GREGORY 1995; GREGORY et al. 1998). These mAbs were shown by epitope mapping to recognise similar epitopes on domain 1 of ICAM-3, and various approaches confirmed that the Mϕ counter-receptor for ICAM-3 on apoptotic cells was not a β_2 integrin. ICAM-3 expression is generally limited to leucocytes, so it was possible to obtain definitive evidence of a role in recognition of apoptotic cells by transfecting 293T kidney cells (usually ICAM-3 negative) with cDNA encoding ICAM-3. When such cells were triggered into apoptosis their recognition by Mϕ could be inhibited by the 3A9 and BU68 ICAM-3 mAbs. Intriguingly, rather than being upregulated on apoptotic cells there seems to be a general reduction in ICAM-3 expression as leucocytes undergo apoptosis and display PS, suggesting that there is preferential exposure of the "recognition directing" epitope(s) defined by 3A9 or BU68 and distinct from those binding β_2 integrins. Since ICAM-3 is heavily glycosylated it seems appropriate to pursue the possibility that ICAM-3 is a molecule displaying altered carbohydrate "eat-me" signals.

II. Phagocyte Receptors for Apoptotic Cells

Recent studies in vitro have identified a number of receptors expressed by phagocytes which are candidate mediators of the uptake of apoptotic cells (SAVILL 1997, 1998). Initially, the data suggested that apoptotic cells of different lineages may "look the same" to phagocytes, the differences in recognition mechanisms being attributable to the phagocyte, so that any one phagocyte type predominantly used a single recognition mechanism (SAVILL et al. 1993; FADOK et al. 1992b,c). However, it has been suggested that such selectivity is illusory, merely reflecting different facets of a single mechanism (PRADHAN et al. 1997). Nevertheless, careful studies suggest that a particular phagocyte type may deploy a range of receptor types in different combinations to achieve an apparent specificity of recognition (FADOK et al. 1998c). Furthermore, there are indications that the lineage of the apoptotic target may influence the recognition mechanism employed (HART et al. 1997; FADOK et al. 1998c). What seems to be emerging is a rich molecular substrate upon which phagocytes may be able to draw in order to achieve a wide and flexible range of responses to the uptake of apoptotic cells.

Nevertheless, an important limitation upon our understanding is the lack of in vivo data supporting a role for particular molecules in the clearance of apoptotic cells from mammalian tissues. Important insights are already arising from the characterisation of genes controlling clearance of cells undergoing developmental cell death in the nematode *Caenorrhabditis elegans* (ELLIS et al. 1991; and see below), but it should be noted that this organism does not possess professional phagocytes, so major mechanisms for clearance of apoptotic cells may have evolved in higher animals. Furthermore, establishing that there is a defect in clearance of apoptotic cells may be more difficult than we imagine. Thus, in the only mammalian model in which there is currently evidence of impaired clearance of apoptotic cells, the Clq–/– "knockout" mouse (BOTTO et al. 1998), there is evidence that deliberate induction of inflammation in order to identify possible defects in leucocyte clearance results in initial recruitment of more leucocytes than in wild type animals, cells which are programmed to die by apoptosis. This means that direct comparison of counts of apoptotic cells between Clq–/– and wild type animals may be misleading, so that methods will need to be developed to track the kinetics of leucocyte infiltration of tissues and the routes and rates of removal of leucocytes from the inflamed site. Much work remains to be done before we can be sure that particular phagocytic mechanisms are indeed involved in the clearance of apoptotic cells in vivo.

1. Thrombospondin Receptors: $\alpha_v\beta_3$ and CD 36

In addition to evidence that TSP1 may bind and "bridge" apoptotic cells to phagocytes, there is strong evidence (SAVILL et al. 1990, 1992) that, in human monocyte-derived Mϕ recognition of apoptotic neutrophils, TSP1 must be co-ordinately bound by the $\alpha_v\beta_3$ "vitronectin receptor" integrin (in a manner dependent on the arg-gly-asp- [RGD] tripeptide in TSP1) and by the transmembrane phagocyte monomer CD36 (which is RGD-independent). Indeed, transfection of CD36 into $\alpha_v\beta_3$ + ve CD36-ve human Bowes melanoma cells can reconstitute the $\alpha_v\beta_3$/TSP1/CD36 recognition mechanism as defined by inhibitory peptides and mAbs (REN et al. 1995a). Furthermore, homologue substitution mutagenesis of CD36 indicates that, in order to confer capacity for phagocytosis of apoptotic cells, CD36 needs to bear a domain involved in low affinity binding of TSP1 (PUENTE-NAVAZO et al. 1996). The proposed $\alpha_v\beta_3$/TSP1/CD36 recognition mechanism appears to be deployed by human monocyte-derived Mϕ and by murine bone marrow-derived Mϕ (although no blocking CD36 antibody is yet available in mice) in recognition of apoptotic cells of a number of different lineages (FADOK et al. 1992b; REN et al. 1999).

However, CD36 is not essential for $\alpha_v\beta_3$ /TSP1-mediated recognition. Glomerular mesangial cells, "semi-professional" phagocytes capable of taking up apoptotic cells more slowly and to a lower degree than Mϕ, employ a CD36-

independent but $\alpha_v\beta_3$/TSP1-mediated recognition mechanism in uptake of apoptotic neutrophils (HUGHES et al. 1997). It is therefore attractive to propose that CD36 might be some form of "amplifying element" for $\alpha_v\beta_3$-mediated recognition akin to CD47, the integrin associated protein which amplifies adhesive and phagocytic functions of this and other integrins (LINDBERG et al. 1993), but which is not involved in the uptake of apoptotic cells as evidenced by the use of blocking antibody and macrophages from CD47–/– mice (REN et al. 1999). The idea of CD36 as an amplifying element gains credence from more than a functional analogy with CD47; there is strong evidence that CD36 can associate with cytoplasmic tyrosine kinases of the *src* family (SILVERSTEIN et al. 1992). However, recent studies (REN et al. 1999) of the large-scale phagocytosis by macrophages of defined populations of "late" apoptotic neutrophils, (HEBERT et al. 1996) indicate that this is mediated by a CD36-independent mechanism involving $\alpha_v\beta_3$ and TSP1; clearly putative CD36-directed amplification is not required in this model.

An alternative role for CD36 in recognition of apoptotic cells could be to provide "suppressive" signals to the phagocyte which promote silent or anti-inflammatory clearance of apoptotic cells, since CD36 has been implicated in signalling the suppressive effects of TSP1 upon angiogenesis (SILVERSTEIN et al. 1992) and antibody ligation of CD36 reduces TNFα secretion from LPS-stimulated macrophages (VOLL et al. 1997). Indeed, this is in keeping with recent data indicating that CD36 may participate not only in RGD-dependent recognition of apoptotic cells via the $\alpha_v\beta_3$/TSP1/CD36 mechanism (SAVILL et al. 1992) but also in PS-dependent recognition (FADOK et al. 1998c), both of which have been implicated in "anti-inflammatory" clearance of apoptotic cells. However, "silent" clearance can occur without CD36, as exemplified by CD36-independent uptake of apoptotic cells by mesangial cells (HUGHES et al. 1997). Furthermore, recent data indicate that CD36 participates in the pro-immunogenic uptake of apoptotic cells by dendritic cells (ALBERT et al. 1998b), which appears coupled to proinflammatory cytokine release (ROVERE et al. 1998). However, this difference could reflect coupling of CD36 with $\alpha_v\beta_5$ rather than $\alpha_v\beta_3$ (ALBERT et al. 1998b). Indeed, this theme of potential alternative partners for CD36 and $\alpha_v\beta_3$ gains support from the apparent dissociation of these two receptors in PS dependent recognition (which involves CD36 but not $\alpha_v\beta_3$) (FADOK et al. 1998c), and the capacity of $\alpha v \beta_3$ to co-operate with lectin-dependent recognition mechanisms in fibroblasts (HALL et al. 1994) and, possibly, also macrophages (FADOK et al. 1998c).

Clearly it will be of considerable interest to make a detailed study of phagocyte clearance of apoptotic cells in mice deleted for CD36 and α_v. Unfortunately, the latter exhibit neonatal lethality (BADER et al. 1998), so alternative gene deletion strategies may be necessary. Nevertheless, useful information may come from genetic abnormalities/manipulation in non-mammalian species, such as *Drosophila*, since this fly has professional phagocytes and a CD36 homologue demonstrated to function in vitro in phagocytosis of apoptotic cells and therefore dubbed "croquemort" (the "eater of death") (FRANC et al. 1996).

2. Scavenger Receptors

In addition to being a thrombospondin receptor, CD36 can also function as a so-called "class B" scavenger receptor (see below), being able to mediate endocytosis of oxidised low density lipoprotein (oxLDL) and free fatty acids (ENDEMANN et al. 1993). Indeed, in vitro studies of CD36 homologues such as SRB-1 emphasise dual functions in endocytosis of altered lipoproteins and uptake of apoptotic cells (FUKASAWA et al. 1996). Furthermore, a body of data points to a role for receptors that recognise oxidised cells and lipoproteins, one of which appears to be macrosialin or CD68 (SAMBRANO et al. 1994; SAMBRANO and STEINBERG 1995). Moreover, there is persuasive evidence that the classical ~220 kD scavenger receptors (designated "class A" and exhibiting different specificity for polyanions to that of class B scavenger receptors) also mediate recognition of apoptotic cells by thymic and peritoneal macrophages. The data (PLATT et al. 1996) are particularly compelling; quite apart from specific inhibition of macrophage uptake of apoptotic cells by polyanion ligands of the scavenger receptor (SRA) and the anti-murine SRA mAb 2F8, macrophages from SRA–/– "knockout" mice exhibit ~50% less uptake of apoptotic cells than wild-type macrophages, and transfection of COS cells with SRA confers capacity for binding and uptake of apoptotic cells. However, although a candidate phosphatidylserine receptor, the specificity of inhibition by polyanions suggests that the SRA may provide additional phagocytic capacity.

3. CD14

The list of phagocyte receptors involved in uptake of apoptotic cells and having specificity for charged lipids has recently been extended by elegant studies from the Gregory laboratory, which implicate the myeloid lineage receptor for bacterial polysaccharide (LPS). In immunological terminology this is CD14, and its involvement in uptake of apoptotic cells is particularly interesting on two counts. First, this receptor is generally linked to highly efficient activation of myeloid phagocytes rather than the "silent" lack of response to CD14-mediated uptake of apoptotic cells made by human monocyte-derived Mϕ from donors that do not employ the $\alpha_v\beta_3$/TSP1/CD36 recognition mechanism. Second, CD14 is not a transmembrane molecule – it is a GPI-linked receptor which might therefore represent a highly mobile "tethering" device available to Mϕ for initial binding of apoptotic cells and ferrying to phagocytic receptors. Indeed, a "tethering" role for CD14 is suggested by the way its role in recognition of apoptotic cells was discovered. Employing an assay of Mϕ interaction with apoptotic lymphocytes in which a major component was tethering rather than phagocytosis, FLORA and GREGORY (1994) found that this was specifically blocked by mAb 61D3. Expression cloning of the 61D3 antigen revealed it to be CD14 and another CD14 mAb sharing an epitope close to the LPS-binding site also inhibited binding of apoptotic cells by Mϕ. Furthermore, expression of CD14 cDNA in COS cells specifically conferred capacity for interaction with apoptotic B cells (DEVITT et al. 1998). However, although CD14 may be involved in phagocytosis of apoptotic targets of non-lymphoid origin, some data point to a relative preference for lymphocytes (FADOK et al. 1998c).

4. Phosphatidylserine Receptors (PSRs)

There is very strong evidence for the existence of stereospecific PSRs which mediate recognition of apoptotic cells by particular phagocyte types: murine thioglycollate-elicited peritoneal macrophages (FADOK et al. 1992b); human THP-1 monocytic cells induced with phorbol ester (FADOK et al. 1992b); vascular smooth muscle cells (BENNETT et al. 1995): murine bone marrow-derived macrophages stimulated with β glucan particles (FADOK et al. 1992c); and, most recently, human monocyte-derived macrophages stimulated with β glucan (FADOK et al. 1998c). Study of the latter, in which CD36 was blocked by mAbs or oxidised LDL, indicates that, although some data suggest that CD36 itself could act as a PSR (RYEOM et al. 1994), CD36 probably acts as a permissive partner to another molecule which is a "professional" PSR. The Fadok group are working hard to characterise PSR(s) and their findings are awaited with interest. While CD14 and class A and class B scavenger receptors are all candidates, their polyanion specificity suggests that they may not be the quarry hunted.

5. Complement Receptors

As described above, there are both in vitro and in vivo data suggesting that the first component of the classical pathway of complement activation C1q may act as a "bridging" molecule in phagocyte recognition of apoptotic cells (KORB and AHEARN 1997; BOTTO et al. 1998). A direct interaction with phagocyte C1q receptors is an important candidate for the "residual eat" resistant to RGD peptide, PS liposomes and sugars described in ostensibly "serum-free" studies of human monocyte-derived Mϕ recognition of apoptotic neutrophils which had been cultured in heat-inactivated complement-depleted fetal calf serum (FADOK et al. 1998c).

However, under serum-replete conditions in which complement components are available, a recent report indicates up to fourfold greater recognition of apoptotic cells than under serum-free conditions (MEVORACH et al. 1998b), although the assay employed seems to have included a large "tethering" component. In addition to demonstrating deposition of the opsonic complement fragment C3bi on apoptotic cells, antibody blockade experiments indicated that both β_2 integrin complement receptors ($\alpha_M\beta_2$ or CR3/Mac1 and $\alpha_X\beta_2$ or CR4/p150,95) mediated Mϕ interaction with apoptotic cells. Intriguingly, these findings may not be inconsistent with "silent" clearance since previous studies have indicated that complement-mediated phagocytosis need not activate phagocytes (WRIGHT and SILVERSTEIN 1983; YAMAMOTO and JOHNSTON 1984). Nevertheless, further work will be needed to reconcile these studies with reports that verified blockage of CR3 and CR4 had no effect on uptake of apoptotic cells (SAVILL et al. 1992), and that such phagocytosis proceeded apparently normally in macrophages from an individual with congenital β_2 deficiency (DAVIES et al. 1991). While serum was not deliberately added to these systems, both cell types were cultured in serum beforehand. Clearly, studies in β_2 –/– knockout mice may help resolve the debate, as may studies of inflammatory responses in β_2-deficient Leucocyte Adhesion Deficiency type 1 (LAD-1) patients. In such individuals β_2-independent migration of leucocytes into the lung can occur (ANDERSON and SPRINGER 1987), so if CR3 and CR4 are indeed important in clearance of apoptotic cells, one would expect to see an excess of apoptosis in the inflammatory infiltrate.

6. Murine ABC1 and C. Elegans CED-7 Proteins

Murine macrophage ABC-1 is a member of the ATP-binding cassette superfamily of membrane transporters, which includes the multi drug resistance P glycoprotein expressed by cancer cells. In development, ABC-1 +ve macrophages cluster at sites of cellular apoptosis, and mAbs against the ATP-binding cytoplasmic domain of ABC-1 introduced into elicited peritoneal Mϕ inhibit uptake of apoptotic cells (LUCIANI and CHIMINI 1996). How ABC-1 interacts with other receptors for dying cells in unclear, although one can speculate that it acts as a conductance which provides a "second signal" to assist phagocytically competent receptors such as integrins to engage the cytoskeleton. ABC-1 has an intriguing relationship to a homologous protein CED-7 in the nematode *C. elegans*. Mutations in the *ced-7* gene result in diminished clearance by neighbours of cells undergoing apoptosis-like developmental cell death. However, in the nematode the evidence implies that wild type *ced-7* expression occurs on both the phagocyte and target (WU and HORVITZ 1998b), reminding us that the aminophospholipid translocase inactivated in apoptotic cells is also a member of the ATP-binding cassette family.

7. Intraphagocyte Signalling; CED-5 and CED-6

Given the well-defined genetic abnormalities which affect phagocyte clearance of dying cells in *C. elegans*, there has been intense interest in cloning the seven genes believed to be responsible, which segregate into two potentially redundant groups (ELLIS et al. 1991). The first to be published was CED-5, which proved to be homologous with Myoblast City and DOCK-180, two adaptor proteins bearing the SH2 domain "passport" for interaction with cytoplasmic tyrosine kinases and apparently involved in mediating downstream cytoskeletal reorganisation following kinase activation (WU and HORVITZ 1998a; RUSHTON et al. 1995). CED-6 also proved to be an intracellular signalling molecule bearing a phosphotyrosine-binding (PTB) domain (LIU and HENGART-

NER 1998). While the role proteins play in phagocytosis remains speculative, these findings are a timely reminder that successful phagocytosis of apoptotic cells not only requires "eat me" signals on the dying cells and phagocyte receptors for these, but also involves dramatic cytoskeletal changes to enable the phagocyte to "swallow" dying cells.

III. Why So Many Recognition Mechanisms?

At the moment it would be conventional to invoke redundancy in explaining the growing number of phagocyte receptors and "eat me" signals being uncovered; one could argue that, because clearance of apoptotic cells is so essential to health, a range of mechanisms have evolved to ensure safe clearance of apoptotic cells. Nevertheless, the data hint at other "explanations". Some receptors, such as the GPI-linked CD14 in macrophages, may be specialised for "tethering" apoptotic cells before "handing" these on to phagocytic receptors such as integrins. CD14 may also be an example of a receptor which is relatively specialised for clearance of a particular cell type, in this case lymphocytes (FADOK 1998c), although most other receptors seem not to be so choosy (FADOK et al. 1992b). Another possibility is that cells dying by apoptosis express a sequential series of "eat me" signals as they progress toward eventual secondary necrosis. Thus there is evidence of a very early caspase-dependent PS-independent "tethering" signal (KNEPPER-NICOLAI et al. 1998), early exposure of PS recognisable to PS receptors (see section Exposure of Phosphatidylserine above) and, as a last resort, $\alpha_v\beta_3$/TSP1-mediated recognition of late apoptotic cells (REN et al. 1999). We have already referred to the possibility that particular recognition mechanisms deliver different signals into phagocytes; compare $\alpha_v\beta_3$/CD36 in macrophages (MEAGHER et al. 1992; STERN et al. 1996) with $\alpha_v\beta_5$/CD36 in dendritic cells (ALBERT et al. 1998b). But we must also consider that certain microenvironments demand particular recognition mechanisms – perhaps C1q is especially important in the high pressure/high flow glomerulus. Lastly, recognition mechanisms could act as "back-ups", being recruited into action when phagocytes are stimulated to increase efficiency of clearance; an example is the capacity of CD44 ligation to increase rapidly and specifically the uptake of apoptotic neutrophils by recruitment of a yet-to-be characterised recognition mechanism (HART et al. 1997). Ultimately it seems that we will only be able to understand the significance of the in vitro data when various combinations of receptors and "eat me" signals have been thoroughly characterised and targeted in animal models. Studies of C1q–/– and SRA–/– mice (BOTTO et al. 1998; PLATT et al. 1990) point in the direction that this work will go.

D. Perturbations of Clearance in Disease

There are now tantalising indications of specific factors which could contribute to disease states by perturbing safe clearance of apoptotic cells. Until recently, workers in the field have speculated that relatively non-specific factors might

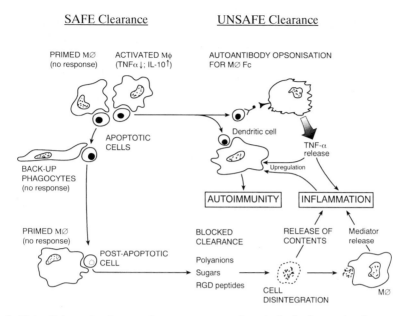

Fig. 3. Potential mechanisms and consequences of perturbed phagocyte clearance of cells undergoing apoptosis. Apoptotic cells are safely cleared by primed macrophages (Mφ) and "back-up" semi-professional phagocytes such as glomerular mesangial cells without eliciting a pro-inflammatory response; furthermore primed Mφ can also take up intact "post-apoptotic" or "late apoptotic" cells without making such responses. Indeed, uptake of apoptotic cells by activated Mφ inhibits Mφ release of pro-inflammatory TNFα and may trigger release of anti-inflammatory IL-10. Unsafe clearance of apoptotic cells leading to autoimmunity and inflammation could occur by a number of mechanisms. For example, antiphospholipid autoantibodies binding apoptotic cells can opsonise apoptotic cells for Mφ Fc receptors and thereby trigger TNFα release, and such antibodies could also promote uptake of apoptotic cells by dendritic cells which may then present self (neo)antigens and fuel autoimmunity. Inflammation could also be exacerbated should factors such as polyanions, sugar moieties or RGD peptides block phagocyte receptors such as scavenger receptors, lectins and integrins, since this would lead to secondary necrosis and disintegration of non-ingested apoptotic cells with release of proinflammatory contents and/or indirect incitement of inflammation due to release of phlogistic mediators from Mφ ingesting cell debris. (Reproduced from REN and SAVILL 1998, with permission)

inhibit safe clearance, such as microenvironmental changes in pH (SAVILL et al. 1992) or accumulation of extracellular matrix protein fragments (SAVILL et al. 1990). These potential inhibitory mechanisms could still be important. However, there now follows a description of specific factors, although their potential significance in disease remains uncertain. A summary of "safe" and "unsafe" clearance of apoptotic cells is shown in Fig. 3.

I. C1q Deficiency

Patients with C1q deficiency are at high risk of developing the multisystem autoimmune disorder systemic lupus erythematosus (SLE) in which circulating oligonucleosomes strongly hint at failed clearance of apoptotic cells. As alluded to above, in keeping with abnormalities in humans with SLE, a proportion of Clq–/– mice develop severe crescentic glomerulonephritis. Given the capacity of Clq to bind apoptotic cells (KORB and AHEARN 1997), and the excess of apoptotic cells in apparently non-inflamed glomeruli of knockouts vs wild types (BOTTO et al. 1998), it is tempting to ascribe tissue injury and autoimmunity to failure of safe clearance, so that potentially injurious or antigenic apoptotic cell contents leak and/or are presented to T-lymphocytes by dendritic cells gaining access to apoptotic cells which slip past incompetent phagocytes.

II. Antiphospholipid Autoantibodies

Around 40% of patients with SLE develop autoantibodies to phospholipids (aPL), the specificity of which includes epitopes involving phosphatidylserine (PS) (HUGHES and KAMASHATA 1994). An important study from Levine's group has demonstrated that aPL specifically bind to the surface of apoptotic cells by a PS-dependent mechanism involving the abundant serum protein β_2 glycoprotein-I (β_2GPI) (PRICE et al. 1996). As one might expect, other workers discovered that aPL could opsonise apoptotic cells for macrophage Fc receptors so that uptake of dying cells triggered apparently undesirable release of pro-inflammatory TNFα from the phagocytes (MANFREDI et al. 1998a). Equally alarming is the capacity of aPL to promote uptake of apoptotic cells by dendritic cells (MANFREDI et al. 1998b). However, aPL are clearly neither a "fast track" to multisystem autoimmune disease nor sufficient for the development of autoimmunity as demonstrated by patients with primary antiphospholipid antibody syndrome. Such individuals do not display conventional features of immune disease, but their tendency to thrombosis could still represent an undesirable consequence of a PL binding to apoptotic cells. Thus, should minor endothelial injury lead to exposure of PS by apoptotic endothelial cells, deposition of aPL and fixation of complement could amplify vascular injury and propagate thrombosis.

E. Promotion of Safe Clearance

In view of the growing evidence that clearance of apoptotic cells may be perturbed in autoimmune and inflammatory disease states, attention has turned to the possible therapeutic applications of strategies aimed at promotion of safe clearance. However, it should be noted that there may be "spare clearance capacity" in some situations, an example being experimental eosinophilic airway inflammation (TSUYUKI et al. 1995). Administration to the airways of an aerosolised ligand for Fas on eosinophils resulted in a wave of eosinophil apoptosis which appeared to be safely cleared by the existing complement of macrophages.

I. Glucocorticoids

Until recently, despite evidence of multiple anti-inflammatory effects and capacity to direct apoptosis in eosinophils and lymphoid cells, the influence of glucocorticoids on clearance of apoptotic cells was unknown. Nevertheless, with the caveat of spare clearance capacity, clinical observations in asthma hinted that glucocorticoids might co-ordinately delete infiltrating eosinophils and promote their safe clearance (WOOLLEY et al. 1996). We found in vitro that glucocorticoids were able to increase, by around fourfold, the capacity of various types of phagocyte (including professional macrophages and semi-professional glomerular mesangial cells) to ingest apoptotic leucocytes of both myeloid and lymphoid lineage (LIU et al. 1999). This effect of glucocorticoids was specific for apoptotic cells in that uptake of opsonised particles was not promoted and required the phagocyte glucocorticoid receptor. Furthermore, of particular importance was the observation that glucocorticoid enhancement was not bought at the cost of a pro-inflammatory response in that increased macrophage and mesangial cell uptake of apoptotic cells did not result in release of chemokines such as IL-8. The mechanisms mediating this potentially beneficial effect of glucocorticoids upon clearance of apoptotic cells require clarification since they might represent a new therapeutic target in inflammatory disease.

II. Other Factors

By contrast with glucocorticoids it seems unlikely that there is clinical efficacy in the capacity of granulocyte/macrophage colony stimulating factor (GMCSF) and other pro-inflammatory cytokines to increase macrophage uptake of apoptotic leucocytes (REN and SAVILL 1995b). Nevertheless, this observation suggests that increased clearance capacity may be programmed into the inflammatory response. However, proinflammatory cytokines and glucocorticoids take a few hours to begin to increase clearance capacity (REN and SAVILL 1995b; LIU et al. 1999). It will therefore be of great interest to dissect mechanisms mediating the much more rapid potentiation of macrophage ingestion of apoptotic cells which follows ligation of CD44, particularly since this effect apparently makes recruitment of a novel recognition mechanism with selectivity for apoptotic granulocytes (HART et al. 1997). Lastly, given that transfection of cDNAs for CD36, CD14 and SRA confers increased capacity for phagocytosis upon "amateur"/semi professional phagocytes (REN et al. 1998; DEVITT et al. 1998; PLATT et al. 1996), it may not be outlandish to explore "pro-phagocytic gene therapy" approaches.

F. Conclusions and Future Prospects

Since the first descriptions of apoptosis, the potential importance of safe phagocytic clearance of dying cells has been evident. However, until relatively recently there has been little interest in the mechanisms involved. As described in this contribution, this situation is changing as investigators realise that the fate of dying cells may be pivotal in regulating inflammatory and immune processes.

Clearly we need to understand much more about phagocyte response to – and handling of – ingested apoptotic cells, and the potential for presentation of antigen requires careful dissection. The mechanisms mediating uptake of apoptotic cells may be central to governing phagocyte responses, but the dissection now needs a "frame shift" from the culture dish to in vivo models. However, there are now exciting prospects that these lines of enquiry may yield new insights into the pathogenesis of inflammatory and immune disease. Furthermore new therapeutic approaches seem likely to arise.

Acknowledgements. CB was supported by a Wellcome Trust Research Training Fellowship for Medical Graduates, and JS has received extensive support from The Wellcome Trust, Medical Research Council and National Kidney Research Fund. Mrs Dorothy May typed the manuscript.

References

Albert ML, Sauter B, Bhardwaj N (1998a) Dendritic cells acquire antigen from apoptotic cells and induce class 1-restricted CTLs. Nature 392:86–89

Albert ML, Pearce FA, Francisco LM, Sauter B, Roy P, Silverstein RL, Bhardwaj N (1998b) Immature dendritic cells phagocytose apoptotic cells via $\alpha_v\beta_5$ and CD36, and cross-present antigens to cytotoxic T lymphocytes. J Exp Med 188:1359–1368

Anderson DC, Springer TC (1987) Leucocyte adhesion deficiency: an inherited defect in the Mac-1, LFA-1 and p150,95 glycoproteins. Annual Review of Medicine 38:175–198

Austyn JM (1998) Dendritic cells. Current Opinion in Immunology 5:3–15

Bader BL, Rayburn H, Crowley D, Hynes RO (1998) Extensive vasculogenesis, angiogenesis and organogenesis precede lethality in mice lacking all αv integrins. Cell 93:1159–1170

Bell DA, Morrison B (1991) The spontaneous apoptotic death of normal human lymphocytes in vitro: the release of and immunoproliferative response to nucleosomes in vitro. Clin Immunol Immunopath 60:13–26

Bellone M, Iezzi G, Rovere P, Galati G, Ronchetti A, Protti MP, Davoust J, Rugarli C, Manfredi AA (1997) Processing of engulfed apoptotic bodies yields T cell epitopes. J Immunol 159:5391–5399

Bennett MR, Gibson DF, Schwartz SM, Tait JF (1995) Binding and phagocytosis of apoptotic vascular smooth muscle cells is mediated in part by exposure of phosphatidylserine. Circ Res 77:1136–1142

Botto M, Agnola CD, Bygrave AE, Thompson EM, Cook HT, Petry F, Loos M, Pandolfi PP, Walport MJ (1998) Homozygous C1q deficiency causes glomerulonephritis associated with multiple apoptotic bodies. Nature Genetics 19:56–59

Bratton DL Fadok VA, Richter DA, Kailey JM, Guthrie LA, Henson PM (1997) Appearance of phosphatidylserine on apoptotic cells requires calcium-mediated nonspecific flip-flop and is enhanced by loss of the aminophospholipid translocase. J Biol Chem 272:26159–26165

Casciola-Rosen LA, Anhalt G, Rosen A (1994) Autoantigens targeted in systemic lupus erythematosus are clustered in two populations of surface structures on apoptotic keratinocytes. J Exp Med 179:1317–1330

Casciola-Rosen LA, Annhalt GJ, Rosen A (1995) DNA-dependent protein kinase is one of a subset of autoantigens specifically cleaved early during apoptosis. J Exp Med 182:1625–1634

Casciola-Rosen L, Rosen A, Petri M, Schlissel M (1996) Surface blebs on apoptotic cells are sites of enhanced procoagulant activity: implications for coagulation events and antigenic spread in systemic lupus erythematosus. Proc Natl Acad Sci USA 93:1624–1629

Davies KA, Toothill VJ, Savill J, Hotchin N, Peters AM, Pearson JD, Haslett C, Burke M, Law SK, Mercer NF, Webster ADB (1991) A 19-year-old man with leucocyte adhesion deficiency. In vitro and in vivo studies of leucocyte function. Clin Exp Immunol 84:223–231

Devitt A, Moffatt OD, Raykundalia C, Capra JD, Simmons DL, Gregory CD (1998) Human CD14 mediates recognition and phagocytosis of apoptotic cells. Nature 392:505–509

Dini L, Autori F, Lentini A, Olivierio S, Piacentini M (1992) The clearance of apoptotic cells in the liver is mediated by the asialoglycoprotein receptor. FEBS Lett 296: 174–178

Dini L, Lentini A, Diez GD, Rocha M, Falasca L, Serafino L, Vidal-Vanachocha F (1995) Phagocytosis of apoptotic bodies by liver endothelial cells. J Cell Sci 108:967–973

Duvall E, Wyllie AH, Morris RG (1985) Macrophage recognition of cells undergoing programmed cell death. Immunology 56:351–358

Emlen W, Holers M, Arend WP, Kotzin B (1992) Regulation of nuclear antigen expression on the cell surface of human monocytes. J Immunol 148:3042–3048

Endemann G, Stanton LW, Madden KS, Bryant CM, White RT, Protter AA (1993) CD36 is a receptor for oxidised low density lipoprotein. J Biol Chem 268:11,811–11,816

Fadok VA, Bratton DL, Konowal A, Freed PW, Westcott JY, Henson PM (1998a) Macrophages that have ingested apoptotic cells in vitro inhibit proinflammatory cytokine production through autocrine/paracrine mechanisms involving TGF-β, PGE2, and PAF. J Clin Invest 101:890–898

Fadok VA, Bratton DL, Frasch SC, Warner ML, Henson PM (1998b) The role of phosphatidylserine in recognition of apoptotic cells by phagocytes. Cell Death and Differentiation 5:557–563

Fadok VA, Voelker DR, Campbell PA, Cohen JJ, Bratton DL, Henson PM (1992a) Exposure of phosphatidylserine on the surface of apoptotic lymphocytes triggers specific recognition and removal by macrophages. J Immunol 148:2207–2216

Fadok VA, Savill JS, Haslett C, Bratton DL, Doherty DE, Campbell PA, Henson PM (1992b) Different populations of macrophages use either the vitronectin receptor or the phosphatidylserine receptor to recognise and remove apoptotic cells. J Immunol 149:4029–4035

Fadok VA, Laszlo DJ, Noble PE, Weinstein L, Riches DWH, Henson PM (1992c) Particle digestibility is required for induction of the phosphatidylserine recognition mechanism used by murine macrophages to phagocytose apoptotic cells. J Immunol 151:4274–4285

Fadok VA, Warner ML, Bratton DL, Henson PM (1998c) CD36 is required for phagocytosis of apoptotic cells by human macrophages that use either a phosphatidylserine receptor or the vitronectin receptor ($\alpha_v\beta_3$). J Immunol 161:6250–6257

Fesus L, Thomazy V, Falus A (1987) Induction and activation of tissue transglutaminase during programmed cell death. FEBS Lett 224:104–108

Fesus L, Thomazy V, Autuouri F, Ceru MP, Tarcsa E, Piacentini M (1989) Apoptotic hepatocytes become insoluble in detergents and chaotropic agents as a result of transglutaminase action. FEBS Lett 245:150–154

Flora PK, Gregory CD (1995) Recognition pathways in the interaction of macrophages with apoptotic B cells. In: Schlossman SF, Boumsell L, Gilks W et al. (eds) Leuko-

cyte typing: white cell differentiating antigens. Oxford University Press, Oxford, p 1675
Flora PK, Gregory CD (1994) Recognition of apoptotic cells by human macrophages: inhibition by a monocyte/macrophage-specific monoclonal antibody. Eur J Immunol 24:2625–2632
Franc NC, Dimarq J-L, Lagueux M, Hoffmann J, Ezekowitz RAB (1996) Croquemort, a novel *Drosophila* hemocyte/macrophage receptor that recognises apoptotic cells. Immunity 4:431–443
Fukasawa M, Adachi H, Hirota K, Tsujimoto M, Arai H, Inoue K (1996) SRB1, a class B scavenger receptor, recognises both negatively charged liposomes and apoptotic cells. Exp Cell Res 222:246–250
Gao Y, Herndon JM, Zhang H, Griffith TS, Ferguson TA (1998) Antiinflammatory effects of CD95 ligand (FasL)-induced apoptosis. J Exp Med 188:887–896
Gregory CD, Devitt A, Moffatt O (1998) Roles of ICAM-3 and CD14 in the recognition and phagocytosis of apoptotic cells by macrophages. Biochem Soc Trans 26:644–649
Griffith TS, Yu X, Herndon JM, Green DR, Ferguson TA (1996) CD95-induced apoptosis of lymphocytes in an immune privileged site induces immunological tolerance. Immunity 5:7–16
Hall S, Savill J, Henson P, Haslett C (1994) Apoptotic neutrophils are phagocytosed by fibroblasts with participation of the fibroblast vitronectin receptor and involvement of a mannose/fucose-specific lectin. J Immunol 153:3218–3227
Hart SP, Dougherty GJ, Haslett C, Dransfield I (1997) CD44 regulates phagocytosis of apoptotic neutrophil granulocytes, but not apoptotic lymphocytes by human macrophages. J Immunol 159:919–925
Hébert M-J, Takano T, Holthöfer H, Brady HR (1996) Sequential morphological events during apoptosis of human neutrophils. J Immunol 157:3105–3115
Homberg CE, de Haas M, von dem Borne AEGK, Verhoeven AJ, Reutelingsperger CPM, Roos D (1995) Human neutrophils lose their surface FcγRIII and acquire annexin V binding sites during apoptosis in vitro. Blood 85:432–540
Hughes J, Liu Y, Van Damme J, Savill J (1997) Human glomerular mesangial cell phagocytosis of apoptotic neutrophils: mediation by a novel CD36-independent vitronectin receptor/thrombospondin recognition mechanism that is uncoupled from chemokine secretion. J Immunol 158:4389–4397
Hughes GRV, Khamashta MA (1994) The antiphospholipid syndrome. J Roy Coll Phys Lond 28:301–304
Inaba K, Turley S, Yamaide F, Iyoda T, Mahnke K, Inaba M, Pack M, Subklewe M, Sauter B, Sheff D, Albert M, Bhardwaj N, Mellman I, Steinman RM (1998) Efficient presentation of phagocytosed cellular fragments on the major histocompatability complex class II products of dendritic cells. J Exp Med 188:2163–2173
Kar S, Ren Y, Savill JS, Haslett C (1993) Inhibition of macrophage phagocytosis in vitro of aged neutrophils increases release of neutrophil contents. Clin Sci 85:27p (Abstract)
Knepper-Nicolai B, Savill J, Brown SB (1998) Constitutive apoptosis in human neutrophils requires synergy between calpains and the proteasome downstream of caspases. J Biol Chem 273:30530–30536
Koopman G, Reutelingsperger CPM, Kuijten GAM, Keehen RMJ, Pals ST, van Oers MHJ (1994) Annexin V for flow cytometric detection of phaosphatidylserine expression on B cells undergoing apoptosis. Blood 84:1415–1420
Korb LC, Ahearn JM (1997) C1q binds directly and specifically to surface blebs of apoptotic human keratinocytes. J Immunol 158:4525–4528
Koutouzov S, Cabrespones A, Amoura Z, Chabre H, Lotton C, Bach J-F (1996) Binding of nucleosomes to a cell surface receptor: redistribution and endocytosis in the presence of lupus antibodies. Eur J Immunol 26:472–486
Kurosaka K, Watanabe N, Kobayashi Y (1998) Production of proinflammatory cytokines by phorbol myristate acetate-treated THP-1 cells and monocyte-derived

macrophages after phagocytosis of apoptotic CTLL-2 cells. J Immunol 161:6245–6249
Lawler J, Sunday M, Thibert V, Duquette M, George EL, Rayburn H, Hynes RO (1998) Thrombospondin-1 is required for normal murine pulmonary homeostasis and its absence causes pneumonia. J Clin Invest 101:982–992
Lindberg FP, Gresham HD, Schwarz E, Brown EJ (1993) Molecular cloning of integrin-associated protein: an immunoglobin family member with multiple membrane-spanning domains implicated in $\alpha_v\beta_3$-dependent ligand binding. J Cell Biol 123:485–496
Liu Y, Cousin JM, Hughes J, Van Damme J, Seckl JR, Haslett C, Dransfield I, Savill J, Rossi AG (1999) Glucocorticoids promote nonphlogistic phagocytosis of apoptotic leukocytes. J Immunol 162:3639–3646
Liu QA, Hengartner MO (1998) Candidate adaptor protein CED-6 promotes the engulfment of apoptotic cells in C. elegans. Cell 93:961–972
Llorente L, Zou W, Levy Y, Richaud-Patin Y, Wijdenes J, Alcocer-Varela J, Morel-Fourrier B, Brouet J-C, ALarcon-Segovia D, Galanaud P, Emilie D (1995) Role of interleukin 10 in the B lymphocyte hyperactivity and autoantibody production of human systemic lupus erythematosus. J Exp Med 181:839–844
Luciani MF, Chimini G (1996) The ATP binding cassette transporter ABC1, is required for engulfment of corpses generated by apoptotic cell death. EMBO J 15:226–235
Manfredi AA, Rovere P, Heltai S, Galati G, Nebbie G, Tincani A, Balestrieri G, Sabbadina MG (1998b) Apoptotic cell clearance in systemic lupus erythematosus. II Role of β_2-glycoprotein I. Arthritis Rheum 41:215–223
Manfredi AA, Rovere P, Galati G, Heltai S, Bozzolo E, Soldini L, Davoust J, Balestrieri G, Tincani A, Sabbadini MG (1998a) Apoptotic cell clearance in systemic lupus erythematosus. I. Opsonization by antiphospholipid antibodies. Arthritis Rheum 41:205–214
Martin SJ, Reutelingsperger CPM, McGahon AJ, Rader JA, van Schie RC, LaFace DM, Green DR (1995a) Early redistribution of plasma membrane phosphatidylserine is a general feature of apoptosis regardless of the initiating stimulus: inhibition by overexpression of Bcl-2 and Abl. J Exp Med 182:1545–1556
Martin SJ, O'Brien GA, Nishioka WK, McGahon AJ, Saido T, Green DR (1995b) Proteolysis of Fodrin (non-erythroid spectrin) during apoptosis. J Biol Chem 270: 6425–6428
Meagher LC, Savill JS, Baker A, Haslett C (1992) Phagocytosis of apoptotic neutrophils does not induce macrophage release of Thromboxane B2. J Leuk Biol 52:269–273
Mevorach D, Zhou JL, Song X, Elkon KB (1998a) Systemic exposure to irradiated apoptotic cells induces autoantibody production. J Exp Med 188:387–392
Mevorach D, Mascarenhas JO, Gershov D, Elkon KB (1998b) Complement-dependent clearance of apoptotic cells by human macrophages. J Exp Med 188:2313–2320
Morris RG, Hargreaves AD, Duvall E, Wyllie AH (1984) Hormone induced cell death. 2. Surface changes in thymocytes undergoing apoptosis. Am J Pathol 115:426–436
Piredda L, Amendola A, Colizzi V, Davies PJA, Farrace MG, Fraziano M, Gentile V, Uray I, Piacentini M, Fesus L (1997) Lack of "tissue" transglutaminase protein cross-linking leads to leakage of macromolecules from dying cells: relationship to development of autoimmunity in MRLlpr/lpr mice. Cell Death and Differentiation 4:463–472
Platt N, Suzuki H, Kurihara Y, Kodama T, Gordon S (1996) Role for the class A macrophage scavenger receptor in the phagocytosis of apoptotic thymocytes in vitro. Proc Natl Acad Sci USA 93:12456–12460
Pradhan D, Krahling S, Williamson P, Schiegel RA (1997) Multiple systems for recognition of apoptotic lymphocytes by macrophages. Mol Biol Cell 8:767–778
Price BE, Rauch J, Shia MA, Walsh MT, Lieberthal W, Gillegan HM, O'Laughlin T, Koh JS, Levine JS (1996) Anti-phospholipid autoantibodies bind to apoptotic, but

not viable, thymocytes in a β_2-glycoprotein 1-dependent manner. J Immunol 157: 2201–2208

Puente Navazo MD, Daviet L, Savill J, Ren Y, Leung LLK, McGregor JL (1996) Identification of a domain (155–183) on CD 36 implicated in the phagocytosis of apoptotic neutrophils. J Biol Chem 271:15381–15385

Ren Y, Savill J (1998) Apoptosis: the importance of being eaten. Cell Death and Differentiation 5:563–568

Ren Y, Silverstein RL, Allen J, Savill J (1995a) CD36 gene transfer confers capacity for phagocytosis of cells undergoing apoptosis. J Exp Med 181:1857–1862

Ren Y, Savill J (1995b) Pro-inflammatory cytokines potentiate thrombospondin-mediated phagocytosis of neutrophils undergoing apoptosis. J Immunol 154:2366–2374

Ren Y, Lindberg FP, Rosenkranz AR, Kar S, Chen Y, Haslett C, Mayadas TN, Savill J (1999) Specific recognition and non-phlogistic clearance of late apoptotic neutrophils by macrophages (submitted)

Rovere P, Vallinoto C, Bondanza A, Crosti MC, Rescigno M, Ricciardi-Castagnoli P, Rugarli C, Manfredi AA (1998) Bystander apoptosis triggers dendritic cell maturation and antigen-presenting function. J Immunol 161:4467–4471

Rubartelli A, Foggi A, Zocchi MK (1997) The selective engulfment of apoptotic bodies by dendritic cells is mediated by the $\alpha_v\beta_3$ integrin and requires intracellular and extracellular calcium. Eur J Immunol 27:1893–1900

Rushton E, Drysdale R, Abmayr SM, Michelson AM, Bate M (1995) Mutations in a novel gene, *myoblast city*, provide evidence in support of the founder cell hypothesis for Drosophila muscle development. Development 121:1979–1988

Ryeom SW, Sparrow JW, Silverstein RL (1994) CD36 is expressed on retinal pigment epithelium and mediates phagocytosis of photoreceptor outer segments. Clinical Research 42:113a

Sambrano GR, Parthasarathy S, Steinberg D (1994) Recognition of oxidatively damaged erythrocytes by a macrophage receptor with specificity for oxidised low density lipoprotein. Proc Natl Acad Sci USA 91:3265–3269

Sambrano GR, Steinberg D (1995) Recognition of oxidatively damaged and apoptotic cells by an oxidized low density lipoprotein receptor on mouse peritoneal macrophages: role of membrane phosphatidylserine. Proc Natl Acad Sci USA 92: 1396–1400

Savill J (1997) Recognition and phagocytosis of cells undergoing apoptosis. British Medical Bulletin 53:491–508

Savill JS, Henson PM, Haslett C (1989) Phagocytosis of aged human neutrophils by macrophages is mediated by a novel 'charge sensitive' recognition mechanism. J Clin Invest 84:1518–1527

Savill JS, Hogg N, Ren Y, Haslett C (1992) Thrombospondin cooperates with CD36 and the vitronectin receptor in macrophage recognition of neutrophils undergoing apoptosis. J Clin Invest 90:1513–1522

Savill J (1998) Apoptosis: phagocytic docking without shocking. Nature 392:442–443

Savill J, Fadok VA, Henson PM, Haslett C (1993) Phagocyte recognition of cells undergoing apoptosis. Immunol Today 14:131–136

Savill JS, Dransfield I, Hogg N, Haslett C (1990) Vitronectin receptor mediated phagocytosis of cells undergoing apoptosis. Nature 343:170–173

Silverstein RL, Baird M, Lo SK, Yesner LM (1992) Sense and antisense cDNA transfection of CD36 (glycoprotein IV) in melanoma cells. Role of CD36 as a thrombospondin receptor. J Biol Chem 267:16607–16612

Stern M, Savill J, Haslett C (1996) Human monocyte-derived macrophage phagocytosis of senescent eosinophils undergoing apoptosis: mediation by $\alpha_v\beta_3$/CD36/ thrombospondin recognition mechanism and lack of phlogistic response. Am J Pathol 149:911–921

Surh CD, Sprent JJ (1994) T-cell apoptosis detected in situ during positive and negative selection in the thymus. Nature 372:100–103

Susin SA, Lorenzo HK, Zamzami N, Marzo I, Snow BE, Brothers GM, Mangion J, Jacotot E, Costantini P, Loeffler M, Larochette N, Goodlett DR, Aebersold R, Siderovski DP, Penninger JM, Kroemer G (1999) Molecular characterization of mitochondrial apoptosis-inducing factor. Nature 397:441–446

Takizawa F, Tsuji S, Nagasawa S (1996) Enhancement of macrophage phagocytosis upon iC3b deposition on apoptotic cells. FEBS Lett 397:269–272

Tsuyuki S, Bertrand C, Erard F, Trifilieff A, Tsuyuki J, Wesp M, Anderson GP, Coyle AJ (1995) Activation of the Fas receptor on lung eosinophils leads to apoptosis and the resolution of eosinophilic inflammation of the airways. J Clin Invest 96:2924–2931

Verhoven B, Schlegel RA, Williamson P (1995) Mechanisms of phosphatidylserine exposure, a phagocyte recognition signal, on apoptotic T lymphocytes. J Exp Med 182:1597–1601

Voll RE, Hermann M, Roth EA, Stach C, Kalden JR (1997) Immunosuppressive effects of apoptotic cells. Nature 390:350–351

Woolley KL, Gibson PG Carty K, Wilson AJ, Twaddell SH, Woolley MJ (1996) Eosinophil apoptosis and the resolution of airway inflammation in asthma. Am J Respir Crit Care Med 154:237–243

Wright SD, Silverstein SC (1983) Receptors for C3b and C3bi promote phagocytosis but not the release of toxic oxygen from human phagocytes. J Exp Med 158:2016–2023

Wu YC, Horvitz HR (1998a) *C. elegans* phagocytosis and cell migration protein CED-5 is similar to human DOCK180. Nature 392:501–504

Wu YC, Horvitz HR (1998b) The *C. elegans* cell corpse engulfment gene *ced-7* encodes a protein similar to ABC transporters. Cell 93:951–960

Wyllie AH, Kerr JFR, Currie AR (1980) Cell death: the significance of apoptosis. Int Rev Cytol 68:251–306

Wyllie AH (1980) Glucocorticoid-induced thymocyte apoptosis is associated with endogenous endonuclease activation. Nature 284:555–556

Yamamoto K, Johnston RB (1984) Dissociation of phagocytosis from stimulation of the oxidative metabolic burst in macrophages. J Exp Med 159:405–416

Zamzani N, Susin SA, Marchetti P, Hirsch T, Gomez-Monterrey I, Castedo M, Kroemer G (1996) Mitochondrial control of nuclear apoptosis. J Exp Med 183:1533–1544

Zhuang J, Ren Y, Snowden RT, Zhu H, Gogvadze V, Savill JS, Cohen GM (1998) Dissociation of phagocyte recognition of cells undergoing apoptosis from other features of the apoptotic program. J Biol Chem 273:15628–15632

CHAPTER 7
T Cell Apoptosis and Its Role in Peripheral Tolerance

R. CAMERON and L. ZHANG

A. Introduction

Apoptosis of T lymphocytes has a central role in developmental, physiologic and pathologic processes including deletion of T cell clones expressing self-antigens in the thymus, elimination of T cells which are infected with viruses, and homeostasis of T cell populations that have expanded following high dose antigen exposures. In this chapter, we will analyze the mechanisms of apoptosis of peripheral T lymphocytes, discuss the role of T cell apoptosis in the induction of transplantation tolerance, and suggest the possibility of modulation by drugs.

B. Phenotypically Different Types of Apoptosis of T Lymphocytes

I. Activation Induced Cell Death

This type of apoptosis of T lymphocytes is a multi-step process involving activation, clonal expansion of T cells, and cell death. Models to study this phenomenon have been developed both in vitro and in vivo. Activation involves a powerful immune stimulus such as bacterial super-antigen or the male HY antigen. Clonal expansion involves a marked clonal proliferation of antigen specific T cells. By 48–96h, there is clonal deletion which involves antigen specific T cells, especially the $CD4^+$ and $CD8^+$ T cells. Activation induced cell death or AICD is mediated by CD95 molecules. Following clonal deletion, there is a period of unresponsiveness or tolerance to specific antigens which lasts for 4–6 weeks (WEBB et al. 1990, 1994; MACDONALD et al. 1991; ZHANG et al. 1992; MCCORMACK et al. 1993; MIETHKE et al. 1994; RHODE et al. 1996; WACK et al. 1997).

II. Veto Cell Phenomenon

Veto cells were identified as a subpopulation of T cells. This is a one-step process which involves the binding of cytotoxic lymphocyte precursor cells to the veto cell and direct apoptosis of the cytotoxic lymphocyte. The binding interaction of the veto cell and the cytotoxic lymphocyte is antigen specific and also MHC restricted. The cytotoxic lymphocytes are only sensitive to the veto cell action at 24–48h in culture and not after that time. The veto cell phenomenon has been studied primarily as an in vitro process (MILLER and DERRY 1979; MARASKA et al. 1984; MILLER 1986; KIZIROGLU and MILLER 1991; SAMBHARA and MILLER 1991).

III. Programmed Cell Death

Programmed cell death is best exemplified in thymus in which immature thymocytes are deleted (COHEN 1991). This phenomenon has been studied both in vivo and in vitro. In some in vivo models, it has been noted that cell death involving $CD4^+CD8^+$ thymocytes, and B-cells in germinal centers is by pyknosis and does not involve fragmentation of DNA directly, but only after the dead cells are phagocytosed by local macrophages (ITOH et al., Chap. 15, this volume). In addition, when caspase I and caspase III knockout mice were studied which show caspase deficiencies, there was a special type of cell death involving thymocytes which did not require the cooperation of caspases as well (KUIDA et al. 1995, 1996). In addition, this type of cell death appears to be mediated by E2/CD99 molecules and not by CD95, and requires up to 18h to complete (BERNARD et al. 1997).

IV. Activation Induced Cell Death of Human Peripheral T Cells

AICD involving human peripheral T cells is very similar in nature to AICD involving mouse T lymphocytes in that there is activation by exposure to antigen and clonal proliferation of antigen specific T cells followed by apoptosis. This process has been studied in vivo. In the human, apoptosis of peripheral T cells is CD2 mediated and involves CD58, CD59, and CD48 ligands (WESSELBORG et al. 1993; MOLLEREAU et al. 1996; LI-WEBER et al. 1998).

C. Molecules Involved in T Cell Apoptosis

I. TNF Receptor Family

TNF receptor family molecules broadly consist of two groups of receptor molecules, namely those such as CD40 and CD27 that either induce B-cell activation or enhance T-cell proliferation, respectively (KOOPMAN, Chap. 17, this volume), and those molecules that carry "death domains" and induce apoptosis (NAGATA 1997). Activation induced cell death of T lymphocytes has been shown to be mediated by the tumor necrosis factor or TNF family of receptors, most notably Fas or Apo-1 which has been named as CD95. This type of cell death can be neutralized by anti-CD95 antibodies (HARGREAVES et al. 1997). During the course of in vivo studies using mice which are Fas defective, namely the mtr/lpr mutant mice, mature $CD4^+$ T lymphocytes were resistant to activation induced cell death, i.e., dependent on Fas for apoptosis. Therefore, in this model, the Fas gene was shown to be essential for activation induced cell death in peripheral T lymphocytes (ROUVIER et al. 1993; SINGER and ABBAS 1994; NAGATA and GOLSTEIN 1995). The active component of the Fas ligand was shown to be a type II membrane protein which is predominantly expressed in activated T cells (TANAKA et al. 1998). It was also shown that activation through the T cell receptor or TCR of peripheral T cells induced Fas ligand expression and simultaneously induced resistance to Fas ligand in naïve T cells. The activated T cells may use the Fas ligand to kill their targets such as virus infected cells. This mechanism also ensured that bystander T cells are not activated in an antigen nonspecific manner (SUDA et al. 1996). Fas and TNF R1 mediated apoptosis occur in the presence of inhibitors of either RNA or protein synthesis and even enucleated cells undergo apoptosis upon Fas activation, suggesting that all components necessary for apoptotic signal transduction are present de novo and that Fas activation simply triggers this machinery (Chap. 15, this volume). Apoptosis occurs in various cells and various tissues and Fas is found abundantly in cells in the thymus, liver, heart, and kidney. Fas ligand is predominantly expressed only in activated T lymphocytes and natural killer cells (NAGATA 1997). Mature T cells from lpr or gld mice do not die after activation and activated cells accumulate in the lymph nodes and spleens of these mice. When T cell hybridomas are activated in the presence of a Fas neutralizing molecule, they do not die. These results indicate that Fas is involved in activation induced cell death of T lymphocytes and is part of the down-regulation of the immune reaction (SINGER and ABBAS 1994; NAGATA and GOLSTEIN 1995). Con A activated mature mouse T lymphocytes showed a specific resistance to CD95 or Fas induced apoptosis during the S phase of their cell cycle (DAO et al. 1997, 1998).

II. Bcl-2 Family

Members of the Bcl-2 gene family encode proteins that function either to promote or to inhibit apoptosis (ADAMS and CORY 1998). Anti-apoptotic members such as Bcl-2 and Bcl-x_L prevent programmed cell death in response to a wide variety of stimuli. Conversely, pro-apoptotic proteins, exemplified by Bax and Bak, can accelerate death and in some instances are sufficient to cause apoptosis independent of additional signals. Bcl-2 related proteins are localized to the outer mitochondrial, outer nuclear, and endoplasmic reticular membranes (CHAO and KORSMEYER 1997). The ability of Bcl-2 to prevent apoptosis was clearly shown in experiments with knockout mice which show apoptosis of thymocytes and spleen cells (VEIS et al. 1993). Up-regulation of the Bcl-2 gene product as in cytokine deprived activated T cells leads to apoptosis (AKBAR et al. 1996). Bcl-2 was shown to block cell-mediated cytotoxicity by allospecific cytotoxic lymphocytes when apoptosis was induced by degranulation as in the action of perforin and granzymes, but not with apoptosis induced by cytotoxic lymphocytes by means of the Fas pathway (CHIU et al. 1995).

III. Caspases

Caspases are a family of cysteine proteases that cleave their target proteins at aspartic acid residues in a defined cascade sequence. Caspase-3 and caspase-8 are involved in cytotoxic T cell induced apoptosis, both of which are mediated by granzyme B (DARMAN et al. 1995; ENARI et al. 1996; BOLDIN et al. 1996; MUZIO et al. 1996; MEDEMA et al. 1997; AMARANTE-MENDES et al. 1998). Caspase-8 can also induce apoptosis in response to the anti-cancer drugs betubinic acid and etoposide in the absence of CD95 receptor-ligand interaction, i.e., CD95-independent (PETER and KRAMMER 1998). Activated caspases cleave a multitude of cellular substrates and finally allow caspase-activated DNase to enter the nucleus to cut DNA between the nucleosomes (PETER and KRAMMER 1998; THORNBERRY and LAZEBNIK 1998).

D. Regulators of T Cell Apoptosis

I. Cytokines (IL-2, IL-4, Interferon gamma, etc.)

Cytokines such as IL-2 can increase or up-regulate Bcl-2 expression and prevent apoptosis in activated T cells. Using human IL-2 deprived activated T cells, it was possible to show that other cytokines such as IL-4, IL-7, and IL-15 could also prevent apoptosis of activated T cells in the absence of IL-2 (AKBAR et al. 1996). In contrast, sensitivity to the priming step for activation induced cell death was dependent on the cytokine interleukin-2 but not on cytokines IL-4, IL-7, or IL-15 (WANG et al. 1996). Furthermore, it was shown, using transgenic mice which have a deficiency in the ability to use IL-2, that their T cells were resistant to Fas-mediated activation induced cell death and that this defect could only be corrected by similar cytokines like IL-15 (VAN PARIJS et al. 1997a,b). The kinetics of IL-2 production are as follows: messenger RNA is detectable within 3–5h and cytokine protein is also seen at this early time, cytokine mRNA is rapidly down-regulated shortly after it reaches a peak level at 6–12h, and the amount of cytokine produced is at least ten times that seen in naïve cells with the same receptor (SWAIN et al. 1996). TCR stimulation of T lymphocytes that are activated in cycline in the presence of IL-2 leads to programmed cell death. This effect was shown to be due mostly to the ability of IL-2 to increase expression of mRNAs which encode ligands and receptors that mediate apoptosis (ZHENG et al. 1998).

II. Co-Stimulatory Molecules (B7, CD28, CTLA-4, etc.)

C28/B7 ligation provides co-stimulatory signals important for the development of T cell responses and CD28 is a principal co-stimulatory receptor for T cell activation. CD28 co-stimulation markedly enhances the production of lymphokines, especially of IL-2. In addition, CD28 sustains the late proliferative response of naïve T cell populations and enhances their long-term survival (SPERLING et al. 1996; TAI et al. 1997). CD28 deficient T cells were shown to be enhanced in their long term survival by cultures with IL-4 (STACK et al. 1998). Circulating T cells which express B7, a novel cell surface glycoprotein, were found to be independent of co-stimulation by using anti CD28 antibodies (SOARES et al. 1997). Further studies showed that in fact cells expressing high levels of CD28 were entirely resistant to apoptosis by the CD95 pathway (MCLEOD et al. 1998). CD28 co-stimulation was also shown to promote T cell survival by enhancing the expression of Bcl-x_L (BOISE et al. 1995a,b; RADVANYI et al. 1996).

SIGAL et al. (1998) showed using monoclonal antibodies to B7-1 and B7-2 co-stimulatory molecules in MHC class II-deficient mice lacking most CD4$^+$ T cells compared to wild-type mice that the generation of viral Ag-specific CD8$^+$ CTLs was Th cell independent and dependent on B7-co-stimulation for activation. In contrast to co-stimulatory actions of B7 or CD28 molecules, CTLA-4 acts as a negative regulator of T cells by binding to the TCR complex and inhibiting tyrosine, phosphoregulation after T cell activation (LEE et al. 1998; ALEGRE et al. 1998).

III. Effect of Viral Infection

GOUGEON (Chap. 5, this volume) has shown that death of CD4$^+$ T lymphocytes in HIV infection can occur either directly by viral replication or indirectly through priming of uninfected T cells to apoptosis both in vitro and also observed in lymph node tissue of HIV-infected donors. The rate of apoptosis in non-infected blood lymphocytes from HIV-infected persons could be increased in response to drugs such as ionomycin, superantigens, or mitogens. Th1 effector cells were found to be more sensitive to activation-induced apoptosis than Th2 cells, and this was controlled by down-regulation of Bcl-2 expression.

E. Mechanisms Involved in Peripheral Tolerance

I. Clonal Deletion

1. Bacterial Superantigen-Induced AICD

WEBB et al. (1990) showed that exposure of mature (peripheral) T cells in vivo to a powerful immune stimulus, namely Mls^a antigen, led to marked clonal expansion of Vβ6$^+$ T cells, followed by their deletion, and specific tolerance that persisted for at least six weeks. Similar results were found by MACDONALD et al. (1991) using SEB superantigen exposures to mice in vivo which led to marked clonal expansion of CD4$^+$ and CD8$^+$ T cells in lymphoid tissue at 2–4 days and then clonal deletion of Vβ8$^+$ T cells, and tolerance that lasted at least 30 days.

WEBB et al. (1994) showed further that the elimination of mature T cells in vivo was correlated with strong high avidity T cell-APC interactions.

McCormack et al. (1993) showed that chronic exposures to SEA superantigen caused virtual complete deletion of Ag-reactive T cells, even if doses were as low as 1 mg of SEA. At chronic low dose exposures deletion of T cells occurred but clonal expansion by proliferation did not occur.

Miethkle et al. (1994) exposed the clonally diverse T cells of normal mice to graded doses of SEB of 0.001–10 mg, and showed that V$\beta 8^+$ T cells became anergic within 6–16 h, and had three dose related patterns.

Anergy induced by low concentrations of SEB (0.001–0.1 mg) was transient and overcome by clonal growth. At higher concentrations of SEB (0.1–10 mg) the anergy induced was long-lasting and resistant to the effects of cell cycle progression. At very high dose exposures to SEB of 1–10 mg, most anergic V$\beta 8^+$ T cells down-regulated their TCR with loss of CD2, 4, and 8, and a subset, V$\beta 8$ low $CD3^+$ cells, underwent apoptosis within 1 h.

2. Alloantigen-Induced AICD

Studies by Zhang et al. (1992, 1995) followed the fate of mature Ag-specific T cells in vivo using female transgenic mice that contain a large population of male H-Y Ag-specific T cells. The number of Ag-reactive $CD8^+$ transgenic T cells in the periphery began to decrease by two days of in vivo exposure to male Ag and remained low for at least six weeks. Non-deleted Ag-reactive $CD8^+$ cells were fully responsive to repeat stimulation by male Ag in vitro. Their findings present evidence of the importance of the nature of the antigen-presenting cells (APCs) in determining the outcome, e.g., "stimulatory" APCs can initiate an active immune response whereas "functionally deleting" APCs act as veto cells to delete clonally Ag-reactive T cells.

Zhang et al. (1996a,b) showed that peripheral tolerance could be induced by means of clonal deletion with activation-induced apoptosis of antigen specific T cells in a transgenic mouse model. In this model, anti-major histocompatibility complex Class I Ld^+ T cell or TCR transgenic cells were adoptively transferred into severe combined immunodeficient mice which express the Ld^+ antigen on all nucleated cells and the fate of transferred antigen specific T cells could be followed in vivo. Apoptotic antigen specific T cells could be identified in vivo using a technique developed by Zhang et al. (1995b) which combined labeling with the cell surface marker to an apoptotic marker, namely in situ NICK translation assay. It was found that after encountering antigen in vivo, the number of antigen specific T cells increased 10–15-fold followed by a decline in number to a value that was still above the starting value. The expansion of antigen specific T cells could be prevented by blocking CD28 co-stimulatory molecules on the T cells prior to the antigen stimulation. Using the double label technique for marking apoptotic specific T cells, it was found that the antigen specific T cells disappear from the periphery and died by activation induced apoptosis. Not all of the antigen specific T cells were killed by apoptosis and those that survived showed down-regulation of both their TCR and CD8 on their cell surface and were fully unresponsive when cultured with L^{d+} cells, even in the presence of exogenous interleukin-2 and

IL-4. These cells, however, were still susceptible to apoptosis when transferred into a secondary host to provide a new source of antigen and antigen-presenting cells. These studies indicated that peripheral T cell tolerance could be induced by multiple mechanisms in which activation induced antigen specific T cell apoptosis played a major role. Further studies by ZHANG et al. (1996b) showed that a possible mechanism for the survival of antigen specific T cells may be their expression of a high level of Th2 type of cytokines. In addition, these residual antigen specific T cells were able to suppress proliferation of other antigen specific T cells, suggesting that they in fact prolong tolerance in vivo.

3. Clonal Anergy

T-cell anergy is proposed by SCHWARTZ (1996) to occur in specific situations and to be defined by specific molecular mechanisms. The anergic state is induced by a TCR occupancy event that stimulates the production of several inhibitors, one that blocks $p21^{ras}$ activation and another (Nil-2a) that blocks cytokine transcription. These inhibitors prevent transcription of IL-2 and other cytokines, and they block proliferative pathways when the cell is reactivated. The induction of these inhibitors is normally antagonized by co-stimulation involving signaling through receptors such as CD28, and proliferation involved by signaling through the IL-2 receptor. Human T-cell clones respond to high concentration of peptides with down-modulation of the TCR and CD28 receptors and "calcium-blocked" anergy with inhibition of the calcium and calcineurin signaling. Co-stimulation (with B7 linked to CD28 receptor) can block anergy induction even 2h after TCR occupancy. Unresponsiveness of T-cell clones induced by anergic pathways is not just a slow form of cell death since anergic cells can be recovered and activated by exposure to exogenous IL-2. Experiments by GROUX et al. (1996) have shown that IL-10 can promote the induction of anergy either by blocking co-stimulatory signals or inducing inhibitors of $p21^{ras}$ or Nil-2a. The critical biological question is what role do the anergic cells play in an immune response or in tolerance induction. Human T-cells in an anergic state fail to produce IL-2 but IL-4 and IFN-γ production are similar in responsive or unresponsive T-cells. Recent experiments by VAN PARIJS et al. (1997b) have clearly identified two processes that regulate the induction of clonal anergy in vivo. T-cell tolerance was induced in recipients of adoptively transferred T-cells, from T-cell receptor transgenic mice. The combination of IL-12, exogenous administration, and antibodies to CTLA-4 converted this tolerant state to an activated and immunogenic one. CTLA-4 engagement promotes antigen-specific T-cell proliferation, whereas IL-12 stimulates Th_0 conversion to Th_1 effector cells.

II. Suppression, Regulatory (Suppressor) T Cells

Peripheral tolerance can also be induced by active suppression by means of regulatory or suppressor T cells. In experiments by MILLER et al. (1992) the low dose oral exposure to myelin basic protein as antigen was effective in inducing oral tolerance and in suppressing experimental autoimmune encephalomyelitis. They found that the T cells generated by oral tolerance mediated suppression both in vitro and in vivo by means of the release of the cytokine transforming growth factor beta. TGFβ has been demonstrated to be secreted by a variety of cells including macrophages, natural killer cells, B cells, and both CD4$^+$ and CD8$^+$ T cells. Further studies by WEINER et al. (1993, 1994) have shown that active suppression is mediated by regulatory T cells, including Th$_2$ cells which secrete IL-4 and IL-10 and Th$_3$ cells which secrete TGFβ. In more recent studies it was found that interleukin-4 cytokine could prevent regulatory T cells from apoptosis (ZHANG et al. 1999).

During investigations of the responses of regulatory T cells after oral administration of low doses of myelin basic proteins, it was found that the regulatory T cells induced by oral antigens would secrete antigen non-specific cytokines after being triggered by the fed antigen which would then suppress inflammation in the local environment (WEINER 1997). This bystander suppression has also been found in other experimental models of autoimmune disease including experimental autoimmune encephalomyelitis, arthritis, and diabetes. This process, whereby anti-inflammatory cytokines could be targeted to an organ and in so doing suppress inflammation in a local environment, has been proposed as a treatment of a variety of organ specific inflammatory conditions of either autoimmune or other type such as psoriasis, in which immune manipulation could induce Th$_2$ or Th$_3$ type of regulatory cells to suppress the inflammatory responses in these diseases (WEINER 1997).

III. Immune Deviation (Th1 to Th2 Switching)

MOSMANN et al. (1986) characterized two distinct T helper cell clonal populations, each with unique cytokine patterns, and each with differing sensitivity to apoptosis, e.g., Th1 > Th2 cells. Th1 cells produce IL-2, IFN-γ and GM-CSF whereas Th2 cells produce IL-4, 5, 6, and 10 (MOSMANN and COFFMAN 1989). A strong Th1 response results in enhancement of several cytotoxic mechanisms including macrophage activation, phagocytosis, and delayed type hypersensitivity reactions (MOSMANN and COFFMAN 1989). A predominant Th2 response leads to high antibody levels, especially IgE, and proliferations of mast cells and eosinophils.

SWAIN et al. (1996) have shown that in vivo patterns of cytokines can also be highly polarized as with Th1 and Th2 cells but individual T cells can still produce a broad range of cytokines. Within 1 to 2 days of antigen stimulation they found a 10-fold increase in IL-2 production and from 100 to 1000 times increases in other cytokines, e.g., Th1 effector cells made 4000–6000 units/ml of IFN-γ in vitro compared to 100–800 units/ml by naïve T cells. This cytokine production is rapidly down-regulated within hours. They postulate that the selection of the cytokine pattern happens early on, during the primary response.

WONG et al. (1993) were able to show that the co-stimulator B7 was effective in stimulating cytokine production of Th1 cells but not Th2 cells. RAMSDELL et al. (1994) found Fas and Fas-L on Th1 cells and observed AICD of Th1 in culture, whereas Th2 cells did not express appreciable amounts of Fas, Fas-L and did not show AICD.

LINTON et al. (1996) showed that aging mice had a shift in cytokine production and functional patterns of T cells with age. They found a shift towards the memory cell phenotype (CD44) but with hyporesponsiveness and low proliferative capacity to antigen, and reduced IL-4, IL-2, or IFN-γ cytokine response to antigen stimulation.

F. Role of T Cell Apoptosis in Oral Tolerance and Autoimmunity

MILLER et al. (1992) gave 1 mg of myelin basic protein (MBP) of guinea pigs orally to rats every 2–3 days for 5 doses. Rat splenic T cells removed at 7–14 days later were shown to suppress the development of experimental autoimmune encephalomyelitis (EAE) in vivo and to suppress proliferative responses to MBP in vitro (tolerance). Anti-sera against TGF-β could abrogate these protective effects, suggesting that oral tolerance induction was dependent on TGF-β secretion by splenic T cells.

In a double blind clinical trial reports by TRENTHAM et al. (1993), 28 patients with active rheumatoid arthritis (RA) of about 10 years duration were given "chick" type II collagen (100 mg) for 3 months daily orally, and compared to 31 RA patients on placebo. Of the patients receiving oral collagen, most showed improvements in joint tenderness and joint swelling and four patients had complete remissions whereas no such effects were seen in RA patients on placebo. These data demonstrated the clinical efficacy of an oral toleration approach for treatment of the autoimmune disease rheumatoid arthritis.

HANCOCK et al. (1993) showed that oral exposure to alloantigen prevented accelerated allograft rejection by selective intragraft Th2 cell activation in LEW rats. When these LEW rats received (LEW X BN) F_1 hearts as transplants, then rejection occurs in 6–8 days. If they receive in addition BN skin grafts 7 days before, then rejection is accelerated to 1–2 days post heart transplant. Oral exposure to BN splenic T cells between skin and heart grafts prevents early rejection, and was shown to suppress Th1 cells function as measured by increased IL-2 and IFN-γ production.

WEINER et al. (1994, 1997a,b) discusses immune mechanisms of oral tolerance, and current usage in treatment of autoimmune diseases by oral administration of autoantigens. He found that at low dose oral exposures to antigens, tolerance was by means of induction of "regulatory" TGF-β secreting cells and Th2 cells producing IL-4 and IL-10. At high doses of antigens orally, tolerance is by anergy and clonal deletion of Th1 cells, and cells which secrete TGF-β (Th3 cells) were resistant to deletion (CHEN et al. 1995a).

CHEN et al. (1995b) went further to show that oral tolerance induced by MBP could be modified in mice depleted of $CD8^+$ T cells in vivo with anti-CD8 monoclonal antibodies but without significant changes in active suppression of oral tolerance suggesting a dominant role of $CD4^+$ cells in oral tolerance. In a similar model in rats, KELLY and WHITACRE (1996) showed that oral tolerance to MBP could be reversed by exposures to IL-4 and IL-5 cytokines.

SODO et al. (1997) found the intestinal bacterial flora was essential to the health and competence of Th2 cells and their susceptibility to oral tolerance induction in mice.

WEINER (1997b) describes "bystander suppression" in association with regulatory cells (Th2 or Th3) induced by oral antigen in which anti-inflammatory cytokines act on organs distant to the organ-specific site of the autoantigen.

STROBEL and MOWAT (1998) describe the details of immune response to dietary antigens and oral tolerance. Antigen-specific suppression induced by oral tolerance can be induced by 24h of a single feed, and with DTH responses can last up to 17 months. APCs must be fully competent for the induction of oral tolerance. Dose and frequency of antigen exposures is also critical to the outcome.

G. Role of T Cell Apoptosis in Transplantation Tolerance

I. Mechanisms of Transplantation Tolerance

Transplants of organs or skin across a complete MHC mismatch are rejected unless the recipient is immunosuppressed. Passenger leukocytes within the graft are the main stimulators of this rejection. A local increase in the cytokines IL-2 and interferon γ occurs in the rejection of a transplant whereas a reduction in their expression is associated with graft tolerance (BISHOP et al. 1997). The balance of graft rejection vs graft tolerance seems to be maintained by the conditions which would favor an immune response involving Th1 cytokines such as interleukin-2 and interferon γ or the Th2 cytokines (FIELD et al. 1997). A specific cytokine such as IL-12 appears to promote Th1 responses and at the same time inhibit Th2 differentiation whereas cytokines such as IL-4 have a central role in the development of Th2 responses (PICCOTTI et al. 1997). FIELD et al. (1997) have developed a hypothetical model of how regulatory $CD4^+$ cells maintain tolerance. Memory Th2 T cells of the $CD4^+$ type regulate the ability of APCs to direct maturation of naïve $CD4^+$ cells and effector $CD8^+$ cytotoxic lymphocytes by altering the activation state of the APCs. Th2 T cells secrete anti-inflammatory cytokines IL-4, IL-10, TGFβ which interfere with expression of co-stimulatory molecules such as B7 and block the APC production of IL-12. These deactivated APCs fail to trigger naïve $CD4^+$ cells to differentiate into Th1 cells and also promote Th2 differentiation. $CD8^+$ cytotoxic lymphocytes fail to develop in the absence of the proper helper T cell or co-stimulatory function.

Liver transplants across major barriers of MHC break these general rules of rejection vs tolerance and Th1 and Th2 switching because they are often not rejected even in the absence of immunosuppression. In addition, liver passenger leukocytes seem to be required for this spontaneous form of graft acceptance which is accompanied by rapid immune activation shortly after liver transplant (BISHOP et al. 1996). BISHOP et al. (1996) further showed that spontaneous acceptance of liver grafts seems to be due to rapid migration of large numbers of donor cells to recipient lymphoid tissues followed by rapid immune activation in the lymphoid tissues giving rise to tolerance of the graft. This was felt to be akin to the high dose tolerance associated with exposure to Class I antigen in the soluble form which can prevent rejection by neutralizing graft specific antibodies or by inhibiting graft reactive cytotoxic T cells (BISHOP et al. 1997). There are four lines of evidence to support this theory of high dose associated graft tolerance: (a) liver tolerance associated with greater cytokine production than liver rejection; (b) reduction of the immunostimulatory cells of the graft (to the passenger leukocytes) causes rejection of livers that are otherwise tolerated; (c) treatment of tolerant strain combinations with hydrosteroids at the time of transplantation reduces tolerance; and (d) increasing the amount of kidney and heart tissue and donor leukocytes leads to acceptance in these organs similar to that in the liver. GORCZYNSKI et al. (1997) made use of the concept of Th1 cytokines as playing a critical role in the induction of graft rejection and developed a model using gamma delta TCR^+ hybridoma cells in which the infusion of anticytokines antibodies were used to decrease graft prolongation. When both anti-IL-10 and anti-TGFβ antibodies were used together, graft prolongation was abolished and allograft rejection developed. Similar results were found in an MHC incompatible renal allograft model in mice (GORCZYNSKI et al. 1997).

The concept of immunologic tolerance arose from bone marrow transplantation in neonatal or irradiated mice in which the predominant mechanism is clonal deletion of donor specific T cells by donor hematopoietic cells in the recipient thymus (QIN et al. 1989, 1993). A short term treatment with nonlytic CD4 and CD8 monoclonal antibodies can induce tolerance to tissue allografts or reversal of spontaneous autoimmunity (QIN et al. 1989, 1993). It was recently shown by BEMELMAN et al. (1998) that a large dose of donor bone marrow produces significant deletion of antigen reactive T cells whereas a much lower dose of bone marrow produces tolerance to the graft with little evidence of clonal deletion. It is this low dose tolerance which can be transferred by $CD4^+$ T cells and passed on to naïve T cells as if infectious, and can act to suppress rejection of third party antigens when linked on F1 grafts. SYKES et al. (1997) developed a method that allowed bone marrow engrafting without toxic or myelosuppressive host conditioning. B6 mice received depleting anti-CD4 and anti-CD8 monoclonal antibodies, local thymic irradiation, and a high dose of major histocompatibility mismatched bone marrow cells spread over four days. This treatment was not associated with significant myelosuppression, toxicity or graft vs host disease. This was the first demonstration that high levels of allogeneic hematopoietic repopulation and central deletional tolerance could be achieved with a conditioning regimen that excludes myelosuppressive treatment.

II. Potential of Immunosuppressive Drugs to Modulate T Cell Apoptosis and Induce Transplantation Tolerance

A number of drugs have been developed to date which suppress the immune response to an allograft and each of these drugs has been shown to function by interfering with a number of specific graft rejection mechanisms: (a) inhibition of activation-induced cell death or apoptosis by prevention of the up-regulation of Fas ligand and interaction with Fas as shown by 9-*cis*-retinoic acid or glucocorticoids (YANG et al. 1995); (b) toxicity to specific cytotoxic T lymphocyte populations in renal allograft recipients by the experimental immunotoxin FN18-CRM9 (NEVILLE et al. 1996; FECHNER et al. 1997); (c) inhibition of IL-2 expression by cyclosporine and daclizumab (SIGAL and DUMONT 1992; ZHENG et al. 1998; (d) immune deviation with a shift from Th1 cytokine pattern to Th2 cytokine pattern by rapomycin, CTLA4 immunoglobulin, anti-CD4 antibody, and cyclosporine (cited in KABELITZ 1998); and (e) induction of activation-induced cell death in activated T cells by anti-CD3 antibody OKT3 or FK506 (SIGAL and DUMONT 1992; KABELITZ 1998).

H. Apoptosis and Immune Privilege

Immune privilege involves sites such as the eye, brain, and reproductive organs where immune responses either do not proceed, or proceed in a manner different from other areas. This process is related not only to physical barriers such as the blood drained vascular barrier but also active processes such as apoptosis of lymphoid cells (GRIFFITH et al. 1995; FERGUSION and GRIFFITH 1997). GRIFFITH et al. (1995) showed that the CD95 and CD95 ligand normally expressed on activated T cells was also constitutively expressed in cells of the eye and testes. It was found further that the apoptotic cells could be recognized phagocytosed and removed from these sites without the induction of inflammatory or immune reactions. WILDNER and THURAU (1995) found in experimental autoimmune uveoretinitis that, once inflammation had been initiated in the retina, orally induced bystander suppression was not effective in suppressing inflammation in the eye. A prominent feature of immune privilege is T cell unresponsiveness which can be due to clonal deletion, clonal anergy, immune deviation, or T cell suppression (NIEDERKORN 1990; GRIFFITH et al. 1995; WILDNER and THURAU 1995; FERGUSON and GRIFFITH 1997). In addition to T cell unresponsiveness, B cell regulation as well as mechanisms of innate immunity involving natural killer cells, macrophages, and complement are also important for the maintenance of immune privilege (FERGUSON and GRIFFITH 1997).

I. Conclusions

Activation induced cell death or AICD of T cells in the periphery is of central importance to homeostasis of the immune system. An effective response to foreign invaders, especially powerful antigenic stimuli such as bacterial superantigens, is an extensive T cell proliferation with tremendous expansion of antigen-specific T cell clones and efficient immune-mediated clearance AICD of the majority of these Ag-specific T cells then follow to return the numbers of T cells in the periphery back towards normal.

A major objective of the study of T cell apoptosis is the practical application of knowledge to the prevention of graft rejection and the lasting induction of transplantation tolerance. Existing immunosuppressive drugs do have specific effects on immune processes but in general are very broad in their actions and also inhibit protective functions of the immune system which allows opportunistic infections to appear. In addition, toxicity to immune reactive cells can lead in some instances to lymphoproliferative disorders and lymphoma. The ideal immunosuppressive agent would be one that targets the specific part of the adaptive immune response responsible for causing the tissue injury. One approach which favors the switch from graft rejection to graft tolerance has been the manipulation of the cytokine environment from a Th_1 pattern expressing IL-2 to the Th_2 pattern expressing IL-4 and IL-10 cytokines.

Another major shortcoming of existing immunosuppressive therapy is that the effects of these drugs are only transient and require daily drug therapy for the lifetime of the graft. Optimal therapy would be to attempt to tolerize, delete, or anergize specific donor reactive T cells early in the transplantation process and thus avoid the need for chronic drug therapy.

Acknowledgement. Thanks to Marie Maguire for her excellent work and formatting of this manuscript.

References

Adams JM, Cory S (1998) The Bcl-2 protein family: arbiters of cell survival. Science 281:1322–1326

Akbar AN, Borthwich NJ, Wickremasinghe RG, Panayiotidis P, Pilling D, Bofill M, Krajewski S, Reed JC, Salmon M (1996) Interleukin-2 receptor common γ-chain signaling cytokines regulate T cell apoptosis in response to growth factor withdrawal: selective induction of anti-apoptotic (Bcl-2, Bcl-x_L) but not pro-apoptotic (Bax, Bcl-x_S) gene expression. Eur J Immuno 26:294–299

Alegre M-L, Shiels H, Thompson CB, Gajewski TF (1998) Expression and function of CTLA-4 in Th1 and Th2 cells. J Immunol 161:3347–3356

Amarante-Mendes GP, Kim CN, Liu L, Huang Y, Perkins CL, Green DR, Bhalla K (1998) Bcr-Abl exerts its antiapoptotic stimuli through blockade of mitochondrial release of cytochrome C and activation of caspase-3. Blood 91:1700–1705

Bemelman F, Honey K, Adams E, Cobbold S, Waldmann H (1998) Bone marrow transplantation induces either clonal deletion or infectious tolerance depending on the dose. J Immunol 160:2645–2648

Bernard G, Breittmayer J-P, de Matteis M, Tranpout P, Hofman P, Senik A, Bernard A (1997) Apoptosis of immature thymocytes mediated by E2/CD99. J Immunol 158:2543–2550

Beverly B, Kong S-M, Lenardo MJ, Schwartz RH (1992) Reversal of in vitro T cell clonal anergy by IL-2 stimulation. Int Immunol 4:661–671

Bishop GA, Sun J, DeCruz DJ et al. (1996) Tolerance to rat liver allografts. III. Donor cell migration and tolerance-associated cytokine production in peripheral lymphoid tissues. J Immunol 156:4925

Bishop GA, Sun J, Sheil AGR, McCaughan GW (1997) High-dose/activation-associated tolerance. Transplantation 64:1377–1382

Boise LH, Minn AJ, Noel PJ, June CH, Accavitti MA, Lindsten T, Thompson CB (1995) CD28 co-stimulation can promote T cell survival by enhancing the expression of Bcl-x_L. Immunity 3:87–98

Boldin MP, Goncharov TM, Goltsev YV, Wallach D (1996) Involvement of MACH, a novel MORT1/FADD-interacting protease, in Fas/APO-1-and TNF receptor-induced cell death. Cell 85:803–815

Chao DT, Korsmeyer SJ (1997) Bcl-x_L-regulated apoptosis in T cell development. Int Immunol 9:1375–1384

Chen Y, Inobe J-I, Marks R, Gonnella P, Kuchroo VK, Weiner HL (1995) Peripheral deletion of antigen-reactive T cells in oral tolerance. Nature 376:177–180

Chen Y, Inobe J-I, Weiner HL (1995) Induction of oral tolerance to myelin basic protein in CD8-depleted mice: both CD4+ and CD8+ cells mediate active suppression. J Immunol 155:910–916

Chen Y, Kuchroo VK, Inobe J-I, Hafler DA, Weiner HL (1994) Regulatory T cell clones induced by oral tolerance: suppression of autoimmune encephalomyelitis. Science 265:1237–1240

Chiu VK, Walsh CM, Liu C-C, Reed JC, Clark WR (1995) Bcl-2 blocks degranulation but not Fas-based cell-mediated cytotoxicity. J Immunol 154:2023–2032

Cobbold SP, Adams E, Marshall SE, Davies JD, Waldmann H (1995) Mechanisms of peripheral tolerance and suppression induced by monoclonal antibodies to CD4 and CD8. Immunol Rev 148:1–29

Cohen JJ (1991) Programmed cell death. Adv Immunol 50:55–85

Crispe IN (1994) Fatal interactions: Fas-induced apoptosis of mature T cells. Immunity 1:347–349

Darmon AJ, Nicholson DW, Bleackley RC (1995) Activation of the apoptotic protease CPP32 by cytotoxic T-cell-derived granzyme B. Nature 377:446–448

Dao T, Huleatt JW, Hingorami R, Crispe IN (1997) Specific resistance of T cells to CD95-induced apoptosis during S phase of the cell cycle. J Immunol 159:4261–4267

Dao T, Mehal WZ, Crispe IN (1998) IL-18 augments perforin-dependent cytotoxicity of liver NK-T cells. J Immuno 161:2217–2222

DeSilva DA, Urdahl KB, Jenkins MK (1991) Clonal anergy is induced in vitro by T cell receptor occupancy in the absence of proliferation. J Immunol 147:3261–3267

Enari M, Jug H, Nagata S (1995) Involvement of an ICE-like protease in Fas-mediated apoptosis. Nature 375:78–81

Fechner JH, Vargo DJ, Geissler EK, Graeb C, Wang J et al. (1997) Split tolerance induced by immunotoxin in a rhesus kidney allograft model. Transplantation 63:1339–1345

Ferguson TA, Griffith TS (1997) A vision of cell death: insights into immune privilege. Immunol Rev 156:167–184

Field EH, Gao Q, Chen N, Rouse TM (1997) Balancing the immune system for tolerance. Transplantation 64:1–7

Gorczynski AM, Chen Z, Zeng H, Fu XM (1997) Specificity for in vivo graft prolongation in $\gamma\delta$T cell receptor hybridomas derived from mice given portal vein donor-specific pre-immunization and skin allografts. J Immunol 159:3698–3706

Griffith TS, Brunner T, Fletcher SM, Green DR, Ferguson TA (1995) Fas ligand-induced apoptosis as a mechanism of immune privilege. Science 270:1189–1192

Hancock WW, Sayegh MH, Kwok CA, Weiner HL, Carpenter CB (1993) Oral, but not intravenous alloantigen prevents accelerated allograft rejection by selective intragraft TH2 cell activation. Transplantation 55:1112–1118

Hargreaves RG, Borthwick NJ, Montani MSG, Piccolella E, Carmichael P et al. (1997) Dissociation of T cell anergy from apoptosis by blockade of Fas/Apo-1 (CD95) signaling. J Immunol 158:3099–3107

Hurtado JC, Kim Y-J, Kwon BS (1997) Signals through 4-1 BB are co-stimulatory to previously activated splenic T cells and inhibit activation-induced cell death. J Immunol 158:2600–2609

Jenkins MK, Schwartz RH (1987) Antigen presentation by chemically modified splenocytes induces antigen-specific T cell unresponsiveness in vitro and in vivo. J Exp Med 165:302–319

Jenkins MK, Pardoll DM, Mizuguchi J, Chused TM, Schwartz RH (1987) Molecular events in the induction of a nonresponsive state in interleukin 2-producing helper T-lymphocytes clones. Proc Natl Acad Sci 84:5409–5413

Kabelitz D (1998) Apoptosis, graft rejection and transplantation tolerance. Transplantation 65:869–875

Kiziroglu F, Miller RG (1991) In vivo functional clonal deletion of recipient CD4+ T helper precursor cells that can recognize class II MHC on injected donor lymphoid cells. J Immunol 146:1104–1112

Lee K-M, Chuang E, Griffin M et al. (1998) Molecular basis of T cell inactivation by CTLA-4. Science 282:2263–2266

Li-Weber M, Laur O, Hekele A, Coy J, Walezak H, Krammer PH (1998) A regulatory element in the CD95 (APO-1/Fas) ligand promoter is essential for responsiveness to TCR-mediated activation. Eur J Immunol 28:2373–2383

Linton P-J, Haynes L, Klinman NR, Swain SL (1996) Antigen-independent changes in naïve CD4 T cells with aging. J Exp Med 184:1891–1900

MacDonald HR, Baschieri S, Lees RK (1991) Clonal expansion precedes anergy and death of V beta 8^+ peripheral T cells responding to staphylococcal enterotoxin B in vivo. Eur J Immunol 21:1963–1966

MacDonald TT (1982) Immunosuppression caused by antigen feeding. Eur J Immunol 12:767–713

McCormack JE, Callahan JE, Kappler J, Marrack PC (1993) Profound deletion of mature T cells in vivo by chronic exposure to superantigen. J Immunol 150:3785–3792

McLeod JD, Walker LSK, Patel YI, Boulougouris G, Sansom DM (1998) Activation of human T cells with superantigen (staphylococcal enterotoxin B) and CD28 confers resistance to apoptosis in CD95. J Immunol 160:2072–2079

Medema JP, Toes REM, Scaffidi C, Zheng TS, Flavell RA et al. (1997) Cleavage of FLICE (caspase 8) by granzyme B during cytotoxic T lymphocyte-induced apoptosis. Eur J Immunol 27:3492–3498

Miethke T, Wahl C, Gaus H, Hug K, Wagner H (1994) Exogenous superantigens acutely trigger distinct levels of peripheral T cell tolerance/immunosuppression: dose-response relationship. Eur J Immunol 24:1892–1902

Miethke T, Wahl C, Hug K, Wagner H (1993) Acquired resistance to superantigen-induced T cell shock. J Immunol 150:3776–3784

Miller A, Lider O, Weiner HL (1991) Antigen-driven bystander suppression following oral administration of antigens. J Exp Med 174:791–798

Miller A, Lider O, Roberts AB, Sporn MB, Weiner HL (1992) Suppressor T cells generated by oral tolerization to myelin basic protein suppress both in vitro and in vivo immune responses by the release of transforming growth factor beta after antigen-specific triggering. Proc Natl Acad Sci 89:421–425

Miller RG, Derry H (1979) A cell population in nu/nu spleen can prevent generation of cytotoxic lymphocytes by normal spleen cells against self antigens of the nu/nu spleen. J Immunol 122:1502–1509

Miller SD, Hanson DG (1979) Inhibition of specific immune responses by feeding protein antigens. J Immunol 123:2344–2350

Mollereau B, Dechart M, Deas O, Rieux-Lancat F, Hirsch F et al. (1996) CD2-induced apoptosis in activated human peripheral T cells. J Immunol 156:3184–3190

Mosmann TR, Coffmann RL (1989) Th_1 and Th_2 cells: different patterns of lymphokine secretion lead to different functional properties. Ann Rev Immunol 7:145–173

Mosmann TR, Cherwinski H, Bond MW, Giedlin MA, Coffmann RL (1986) Two types of murine helper T clone. J Immunol 136:2348–2357

Mowat AM, Strobel S, Drummond HE, Ferguson A (1982) Immunological responses to fed proteins in mice. Immunology 45:105–113

Mueller DL, Jenkins MK, Schwartz RH (1989) Clonal expansion versus functional clonal inactivation. Ann Rev Immunol 7:445–480

Nagata S (1997) Apoptosis by death factor. Cell 88:355–365

Nagata S, Golstein P (1995) The Fas death factor. Science 267:1449–1455

Neville DM Jr, Scharff J, Hultz, Rigaut K, Shiloach J et al. (1996) A new reagent for the induction of T-cell depletion, anti-CD3-CRM9. J Immunotherapy 19:85–92

Neiderkorn JY (1990) Immune privilege and immune regulation in the eye. Adv Immunol 48:191–226

Peter ME, Krammer PH (1998) Mechanisms of CD95 (APO-1-/Fas)-mediated apoptosis. Curr Op Immunol 10:545–551

Piccotti JR, Chan SY, Vanbuskirk AM, Eichwald EJ, Bishop DK (1997) Are Th2 helper T lymphocytes beneficial, deleterious, or irrelevant in promoting allograft survival? Transplantation 63:619–624

Qin S, Cobbold S, Benjamin R, Waldmann H (1989) Induction of classical transplantation tolerance in the adult. J Exp Med 169:779–794

Qin S, Cobbold S, Poju H, Elliott J, Kioussis D, Davies J, Waldmann H (1993) "Infectious" transplantation tolerance. Science 259:974–976

Radvanyi LG, Shi Y, Vazier H, Sharma A, Dhale R, Mills GB, Miller RG (1996) CD28 co-stimulation inhibits TCR-induced apoptosis during a primary T cell response. J Immunol 156:1788–1798

Rammensee H-G, Kroschewski R, Frangoulis B (1989) Clonal anergy induced in mature Vβ6+ T lymphocytes on immunizing Mls-1^b mice with Mls-1^a expressing cells. Nature 339:541–544

Rellahan BL, Jones LA, Kruisbeck AM, Fry AM, Matis LA (1990) In vivo induction of anergy in peripheral Vβ8+ T cells by staphylococcal enterotoxin B. J Exp Med 172:1091–1100

Rhode PR, Burkhardt M, Jiao J-a, Siddiqui AH, Huang GP, Wong HC (1996) Single-chain MHC Class II molecules induce T cell activation and apoptosis. J Immunol 157:4885–4891

Rocha B, Von Boehmer H (1991) Peripheral selection of the T cell repertoire. Science 251:1225–1228

Rouvier E, Luciani M-F, Golstein P (1993) Fas involvement in Ca^{2+}-independent T cell-mediated cytotoxicity. J Exp Med 177:195–200

Sambhara SR, Miller RG (1991) Programmed cell death of T cells signaled by the T cell receptor and the $\alpha 3$ domain of class I MHC. Science 252:1424–1427

Sigal LJ, Reiser H, Roch KL (1998) The role of B7–1 and B7–2 co-stimulation for the generation of CTL responses in vivo. J Immunol 161:2740–2745

Sigal NH, Dumont FJ (1992) Cyclosporin A, FK-506, and rapamycin: pharmacologic probes of lymphocyte signal transduction. Ann Rev Immunol 10:519–560

Singer GG, Abbas AK (1994) The Fas antigen is involved in peripheral but not thymic deletion of T lymphocytes in T cell receptor transgenic mice. Immunity 1:365–371

Soares LAB, Rivas A, Tsavaler L, Engelman EG (1997) Ligation of the V7 molecule on T cells blocks anergy induction through a CD28-independent mechanism. J Immunol 159:1115–1124
Speiser DE, Sebzda E, Ohteki T, Bachmann MF, Pfeffer K, Mak TW, Ohashi PS (1996) Tumor necrosis factor receptor p55 mediates deletion of peripheral cytotoxic T lymphocytes in vivo. Eur J Immunol 26:3055–3060
Sperling AI, Auger JA, Ehst BD, Bluestone JA (1996) C28/B7 interactions deliver a unique signal to naïve T cells that regulates cell survival but not early proliferation. J Immunol 157:3909–3917
Stack RM, Thompson CB, Fitch FW (1998) IL-4 enhances long-term survival of CD28-deficient T cells. J Immunol 160:2255–2262
Streilein JW, Niederkorn JY, Shadduck JA (1980) Systemic immune unresponsiveness induced in adult mice by anterior chamber presentation of minor histocompatibility antigens. J Exp Med 152:1121–1125
Strobel S, Mowat AM (1998) Immune responses to dietary antigens: oral tolerance. Immunol Today 19:173–181
Suda T, Tanaka M, Miwa K, Nagata S (1996) Apoptosis of mouse naïve T cells induced by recombinant soluble Fas ligand and activation-induced resistance to Fas ligand. J Immunol 157:3918–3924
Sudo N, Sawamura S, Tanaka K, Aiba Y, Kubo C, Koga Y (1997) The requirement of intestinal bacterial flora for the development of an IgE production system fully susceptible to oral tolerance induction. J Immunol 159:1739–1745
Swain SL, Croft M, Dubey C, Haynes L, Rogers P, Zhang X, Bradley LM (1996) From naïve to memory T cells. Immunol Rev 150:143–167
Sykes M, Szot GL, Swenson KA, Pearson DA (1997) Induction of high levels of allogeneic hematopoietic reconstitution and donor-specific tolerance without myelosuppressive conditioning. Nature Med 3:783–789
Tai X-G, Toyooka K, Yashiro Y, Abe R, Park C-S, Hamaoka T et al. (1997) CD9-mediated co-stimulation of TCR-triggered naïve T cells leads to activation followed by apoptosis. J Immunol 159:3799–3807
Tamura T, Ishihara M, Lamphier MS, Tanaka N, Oishi I, Aizawa S et al. (1995) An IRF-1-dependent pathway of DNA damage-induced apoptosis in mitogen-activated T lymphocytes. Nature 376:596–599
Tanaka M, Itai T, Adachi M, Nagata S (1998) Downregulation of Fas ligand by shedding. Nature Med 4:31–36
Thornberry NA, Lazebnik Y (1998) Caspases: enemies within. Science 281:1312–1316
Trentham DE, Dynesius-Trentham RA, Oraw EJ, Combitchi D, Lorenzo C, Sewell KL, Hafler DA, Weiner HL (1993) Effects of oral administration of type II collagen on rheumatoid arthritis. Science 261:1727–1730
Van Parijs L, Abbas AK (1998) Homeostasis and self-tolerance in the immune system: turning lymphocytes off. Science 280: 243–248
Van Parijs L, Bruckians A, Ibraginow A, Alt FW, Willerford DM, Abbas AK (1997) Functional responses and apoptosis of CD25 (IL-2Rα)-deficient T cells expressing a transgenic antigen receptor. J Immunol 158:3738–3745
Wack A, Corbella P, Harker N, Crispe IN, Kioussis D (1997) Multiple sites of post-activation CD8+ T cell disposal. Eur J Immunol 27:577–583
Wang R, Rogers AM, Rush BJ, Russell JH (1996) Induction of sensitivity to activation-induced death in primary CD4+ cells: a role for interleukin-2 in the negative regulation of responses by mature CD4+ T cells. Eur J Immunol 26:2263–2270
Webb SR, Hutchison J, Hayden K, Sprent J (1994) Expansion/deletion of mature T cells exposed to endogenous superantigens in vivo. J Immunol 152:586–597
Webb S, Morris C, Sprent J (1990) Extrathymic tolerance of mature T cells: clonal elimination as a consequence of immunity. Cell 63:1249–1256
Weiner HL (1997) Oral tolerance for the treatment of autoimmune diseases. Ann Rev Med 48:341–351

Weiner HL, Friedman A, Miller A, Khoury SJ et al. (1994) Oral tolerance. Ann Rev Immunol 12:809–837

Weiner HL, Mackin GA, Matsui M, Oraw EJ, Khoury SJ et al. (1993) Double-blind pilot trial of oral tolerization with myelin antigens in multiple sclerosis. Science 259:1321–1324

Wesselborg S, Janssen O, Kabilitz D (1993) Induction of activation-driven death (apoptosis) in activated but not resting peripheral blood T cells. J Immunol 150:4338–4345

Wildner G, Thurau SR (1995) Orally induced bystander suppression in experimental autoimmune uveoretinitis occurs only in the periphery. Eur J Immunol 25:1292–1297

Yang Y, Mercep M, Ware CF, Ashwell JD (1995) Fas and activation-induced Fas ligand mediate apoptosis of T cell hybridomas: inhibitors of Fas ligand expression by retinoic acid and glucocorticoids. J Exp Med 181:1673–1682

Zhang L (1996) The fate of adoptively transferred antigen-specific T cells in vivo. Eur J Immunol 26:2208–2214

Zhang L, Fung-Leung W, Miller RG (1995) Down-regulation of CD8 on mature antigen-reactive T cells as a mechanism of peripheral tolerance. J Immunol 155:3464–3471

Zhang L, Martin DR, Fung-Leung W-P, Teh H-S, Miller RG (1992) Peripheral deletion of mature CD8+ antigen-specific T cells after in vivo exposure to male antigen. J Immunol 148:3740–3745

Zhang L, Miller RG, Zhang J (1996) Characterization of apoptosis-resistant antigen-specific T cells in vivo. J Exp Med 183:2065–2073

Zhang L, Wang C, Radvanyi LG, Miller RG (1995) Early detection of apoptosis in defined lymphocyte populations in vivo. J Immunol Methods 181:17–27

Zheng L, Trageser CL, Willerford DM, Lenardo MJ (1998) T cell growth cytokines cause the superinduction of molecules mediating antigen-induced T lymphocyte death. J Immunol 160:763–769

Zheng TS, Schlosser SF, Dao T, Hingorani R, Crispe IN et al. (1998) Caspase-3 controls both cytoplasmic and nuclear events associated with Fas-mediated apoptosis in vivo. Proc Natl Acad Sci USA 95:13618–13623

CHAPTER 8
Apoptosis of Nerve Cells

A.-M. WOODGATE and M. DRAGUNOW

A. Introduction

One of the major challenges facing neuroscientists is to understand the molecular basis of nerve cell death in the brain and spinal cord. This information will provide the basis for a rationale drug design strategy to treat acute (e.g., stroke, traumatic brain injury, status epilepticus, perinatal asphyxia) and chronic (e.g., Alzheimer's disease, Parkinson's disease, Huntington's disease, amyotrophic lateral sclerosis) neurodegenerative disorders. For many years it was thought that nerve cells die in these diseases by a passive necrotic lysis-type mechanism. More recently, data from a number of sources including human brain material, and in vitro and in vivo models, have suggested that degenerative nerve cell death might be caused by an active apoptotic mechanism (reviewed in DRAGUNOW et al. 1998).

One of the first indications that neurons die via an active process came from in vitro studies which showed that RNA and protein synthesis inhibitors prevent the death of sympathetic neurons following deprivation of nerve growth factor (NGF, MARTIN et al. 1988). This result suggested that NGF promotes neuronal survival by suppressing an endogenous death program. BATISTATOU and GREENE (1991) then demonstrated that death of NGF-deprived sympathetic neurons and PC12 cells was associated with DNA cleavage and could be prevented with an endonuclease inhibitor. These studies were followed by a wealth of in vitro data which showed that neurons undergo apoptosis in response to a variety of pathological insults (reviewed in DRAGUNOW and PRESTON 1995). Subsequently, evidence from in vivo studies has emerged which supports a role for apoptosis in status epilepticus, hypoxia-ischemia, and a number of neurodegenerative disorders (reviewed in DRAGUNOW and PRESTON 1995; DRAGUNOW et al. 1998).

Apoptosis and necrosis can be distinguished on a morphological basis (for review see BAR 1996; KERR et al. 1972; LEIST and NICOTERA 1997; WEBB et al. 1997; WYLLIE et al. 1980). Necrosis is characterized by cellular swelling, rapid loss of internal homeostasis, damage to organelles and, finally, cell lysis. The release of cytoplasmic components from the damaged cell provokes an inflammatory response which harms nearby, otherwise healthy tissue. In contrast, during apoptotic death, the cytoplasm and nucleus of the dying cell condense with preservation of organellar structure. Other hallmarks include the compaction of chromatin against the nuclear membrane, nuclear breakdown, plasma membrane blebbing and the eventual "budding off" of membrane-bound fragments known as apoptotic bodies. The apoptotic bodies are rapidly phagocytosed by macrophages or parenchymal cells before they lose membrane integrity, enabling the cell death process to take place without inflammation or damage to surrounding tissue.

The morphological characteristics of apoptosis are frequently accompanied by activation of calcium-dependent endonucleases which cleave the genome into equal-size fragments (COHEN and DUKE 1984; WYLLIE 1980). While apoptosis is classically associated with fragmentation of DNA into 180–200 base pair multimers, recent studies suggest that breakdown into larger fragments (50kb) takes place before internucleosomal cleavage (BROWN et al. 1993; WALKINSHAW and WATERS 1994; MACMANUS et al. 1997). The regularly degraded DNA fragments from apoptotic cells can be visualized as a characteristic DNA ladder following agarose gel electrophoresis. In contrast, in necrotic cells the random degradation of DNA by lysosomes produces a smear on an agarose gel. Alternatively, DNA fragmentation can be detected using the TUNEL stain (TdT-mediated dUTP biotin nick end labeling), although it should be noted that this method, in some cases, labels necrotic as well as apoptotic cells (NISHIYAMA et al. 1996; THOMAS et al. 1995).

Although evidence of DNA fragmentation has been observed in some neurodegenerative diseases, whether this process is an essential component of the apoptotic program remains controversial. Several studies have shown that inhibition of endonuclease activity using aurintricarboxylic acid can attenuate apoptosis, suggesting that oligonucleosomal DNA fragmentation is critically involved in the cell death process (BATISTATOU and GREENE 1991; WALKINSHAW and WATERS 1994). In contrast, SCHULZ et al. (1998) found that DNA fragmentation takes place in trophic factor-deprived rat cerebellar granule cells but is not absolutely required for apoptotic death. Furthermore, the morphological characteristics of apoptosis in the absence of internucleosomal DNA cleavage has been reported in a mouse embryonal cell line following serum deprivation (COLLINS et al. 1992; TOMEI et al. 1993); in PC12 cells exposed to etoposide (SAURA et al. 1997), nerve growth factor withdrawal and serum deprivation (MESNER et al. 1992); and in cultured rat hippocampal neurons after glucocorticoid treatment (MASTERS et al. 1989). The issue is further confounded by the finding that oligonucleosomal cleavage can take place following necrotic insults (TOMINAGA et al. 1993). Thus, these observations suggest that DNA fragmentation should not be used as the sole determinant of apoptosis, but rather as an adjunct to support morphological observations.

Apoptosis is frequently attributed to the expression of so-called "cell death" genes which subsequently cause the cells to self-destruct and, for this reason, is often equated with "programmed cell death" (PCD). In fact these terms should not be used interchangeably as apoptosis does not necessarily require de novo protein synthesis (MESNER et al. 1992; SCHWARTZ and OSBORNE 1993; WEIL et al. 1996). Furthermore, within the category of programmed cell death, not all dying cells exhibit apoptotic morphology (SCHWARTZ and OSBORNE 1993). Thus, for the purposes of this review, the term apoptosis refers to a morphological description of cell death, whereas PCD indicates the involvement of de novo protein synthesis in the cell death process.

I. Programmed Cell Death

The term "programmed cell death" was first coined to describe the death of cells within a developmental context in response to the appearance or loss of an external signal, but more recently has been expanded to include all types of cell death which require activation of a genetic program. Although the genes which mediate programmed cell death have not been clearly defined in vertebrates, this process has been well characterized in the nematode *C. elegans* (ELLIS and HORVITZ 1986; HEDGECOCK et al. 1983). In the central nervous system, programmed cell death takes place extensively during development where a surplus of post-mitotic neurons compete for a limited supply of target-derived neurotrophic factors. Those neurons which acquire sufficient amounts will survive, while the unrequired nerve cells will die via an apoptotic mechanism. Support for the involvement of gene expression in this process is derived from studies in primary neuronal cultures, which show that trophic factor withdrawal-induced apoptosis is attenuated by inhibitors of transcription and translation (D'MELLO et al. 1993; MARTIN et al. 1988). As genetically "programmed" apoptosis plays a key role in CNS development, it has been suggested that dysregulated activation of these pathways underlies the pathogenesis of degenerative neuronal loss. Indeed in support of this notion, mounting evidence in vivo implicates a role for apoptosis in a variety of neuropathological conditions.

B. Apoptosis in the Brain

I. Alzheimer's Disease

Alzheimer's disease (AD) is a neurodegenerative disorder characterized clinically by personality changes, memory loss and deterioration of cognitive function. The classic neuropathological symptoms include the presence of senile plaques and neurofibrillary tangles, as well as neuronal loss in regions of the brain associated with memory. Senile plaques represent extracellular deposits containing β-amyloid protein, while the neurofibrillary tangles consist of paired helical filaments composed of the hyperphosphorylated microtubule-associated protein tau.

Studies using in situ DNA end-labeling (TUNEL) have observed an increase in TUNEL-positive cells in post-mortem human AD hippocampus compared to that of non-AD controls (ANDERSON et al. 1996; DRAGUNOW et al. 1995; LASSMANN et al. 1995; LI et al. 1997; SMALE et al. 1995; SU et al. 1994). While this technique can, in some cases, detect necrosis, the morphology of at least a portion of the positively stained cells is consistent with an apoptotic mechanism. Double labeling studies have demonstrated that the TUNEL-positive cells are a mixture of neuronal and glial cells, although some controversy surrounds which cell type predominates (LASSMANN et al. 1995; LI et al. 1997; SMALE et al. 1995; SU et al. 1994). Several attempts have been made to correlate TUNEL-positive cells with the pathological features of AD (senile plaques and neurofibrillary tangles). While immunohistochemical staining has shown that the majority of TUNEL-positive nuclei are not located within amyloid deposits or in tangle-bearing neurons (BANCHER et al. 1997; DRAGUNOW et al. 1998), BANCHER et al. (1997) reported a significant increase in DNA fragmentation in tangle-bearing neurons compared to non-tangle-bearing neurons, and in cells located within amyloid plaques compared to those in unaffected tissue. A recent study has shown that TUNEL-positive neurons are co-localized with nitrotyrosine (SU et al. 1997), suggesting that peroxynitrite-induced apoptosis may be involved in Alzheimer's disease. Other apoptotic markers, such as clusterin, which are expressed in senile plaques may be involved in the production of neurotoxic amyloid peptides (LAMBERT et al. 1998). Indeed, amyloid peptides may play a more general role in neuronal apoptosis since GALLI et al. (1998) found that these peptides were secreted by neurons during apoptosis.

II. Parkinson's Disease

Parkinson's disease (PD) is a neurodegenerative disorder characterized by tremor, rigidity and akinesia. These symptoms are generally attributed to loss of the dopaminergic neurons in the substantia nigra pars compacta. In contrast to AD, TUNEL staining on human post-mortem PD tissue has yielded a mixture of positive and negative results (DRAGUNOW et al. 1995; KOSEL et al. 1997; MOCHIZUKI et al. 1996; TOMPKINS et al., 1997). While several studies have failed to observe TUNEL-positive nuclei in post-mortem PD substantia nigra (DRAGUNOW et al. 1995; KOSEL et al. 1997), others have reported a small increase in DNA fragmentation in the PD cases compared to controls (MOCHIZUKI et al. 1996; TOMPKINS et al. 1997). In support of a role for apoptosis in PD, a recent ultrastructural study observed cellular shrinkage and chromatin condensation in a small percentage of nigral neurons from post-mortem PD tissue, but no apoptotic-like changes in neurons from control brains (ANGLADE et al. 1997). The discrepancies between these results may reflect different stages of the disease process.

The pathological features of PD can be reproduced in vivo using the mitochondrial complex I inhibitors, 6OHDA and MPTP. However, studies based on these models have failed to clarify the cell death mechanism underlying PD. The absence of apoptotic morphology in nigral neurons has been observed after both 6OHDA lesion (JEON et al. 1995) and MPTP treatment (JACKSON-LEWIS et al. 1995). In agreement with these results, TUNEL staining and agarose gel electrophoresis failed to detect evidence of apoptosis in the substantia nigra following transection of the medial forebrain bundle (VENERO et al. 1997). However in contrast, TATTON and KISH (1997) observed apoptotic cells in mice following chronic exposure to low dose MPTP. Only a small number of apoptotic nuclei per section were observed at each time point, probably a reflection of the short life span of apoptotic nuclei in vivo where they are rapidly engulfed by macrophages and phagocytosed (TATTON and KISH 1997).

III. Cerebral Ischemia

Cerebral ischemia results from a blockage in the flow of blood to the brain. When a specific brain region is affected, the insult is classified as focal. In contrast, in global ischemia the blood supply to the entire brain is obstructed. While ischemic nerve cell death is conventionally considered necrosis, recent evidence suggests that it has an apoptotic component. Indeed, a multitude of studies have reported TUNEL-positive cells in vulnerable neuronal populations in models of both global and focal ischemia (BEILHARZ et al. 1995; CHARRIAUT-MARLANGUE et al. 1996; KIHARA et al. 1994; LI et al. 1995a; LINNIK et al. 1995; MACMANUS et al. 1993; MACMANUS et al. 1994; NITATORI et al. 1995; SCHMIDT-KASTNER et al. 1997; SEI et al. 1994). While not all studies have examined the morphology of the positively stained cells, at least in some cases, chromatin condensation, nuclear segmentation and apoptotic bodies have been reported (BEILHARZ et al. 1995; CHARRIAUT-MARLANGUE et al. 1996; LI et al. 1995a; LI et al. 1995b; NITATORI et al. 1995). Further support for an apoptotic mechanism is derived from DNA fragmentation analysis which demonstrates the presence of oligonucleosomal-sized fragments in some model systems (LI et al. 1995a; LINNIK et al. 1995; BEILHARZ et al. 1995; MACMANUS et al. 1993; SEI et al. 1994), as well as the appearance of other markers such as clusterin and annexin V (WALTON et al. 1996, 1997). Evidence for apoptotic nerve cell death has also been reported in human brain after hypoxia (LOVE et al. 1998).

The apoptotic component of ischemic nerve cell loss is thought to account primarily for the delayed death which occurs some hours after the initial insult. In support of this notion, signs of apoptosis have been reported at the penumbra of a focal ischemic insult and thus may contribute to the development of the infarct (LI et al. 1995a). Furthermore, TUNEL-positive cells have been reported after moderate ischemia which activates a delayed cell death mechanism (BEILHARZ et al. 1995; KIHARA et al. 1994). In contrast, necrotic death is observed following more severe insults which trigger rapid nerve cell loss (BEILHARZ et al. 1995).

IV. Status Epilepticus

Status epilepticus (SE) is characterized by prolonged or frequently repeated seizure activity. Studies based on several models of SE have found evidence of DNA fragmentation, as detected by TUNEL, and electrophoresis in selectively vulnerable populations (DRAGUNOW and PRESTON 1995; FILIPKOWSKI et al. 1994; POLLARD et al. 1994).

V. Huntington's Disease

Huntington's disease (HD) is an autosomal dominant neurodegenerative disorder characterized by progressive loss of specific neuronal groups in the basal ganglia. Studies on human HD tissue have reported evidence of DNA fragmentation in the striatum as shown by TUNEL staining and a correlation between TUNEL staining and the length of the polyglutamine repeat (BUTTERWORTH et al. in press). However, the majority of TUNEL-positive cells are non-neuronal and do not exhibit apoptotic morphology (DRAGUNOW et al. 1995; PORTERA-CAILLIAU et al. 1995; THOMAS et al. 1995). Thus, neuronal apoptosis has been observed after exposure to 3-nitropropionic acid and quinolinic acid, two compounds which reproduce HD-like cell loss in vivo (DURE et al. 1995; SATO et al. 1997). Some in vitro studies also support a role for apoptosis in HD (BEHRENS et al. 1995, 1996), although others implicate an excitotoxic mechanism (FINK et al. 1996; ZEEVALK et al. 1995).

VI. Other Brain Disorders

Apoptotic cells as identified by TUNEL (and the presence of p53, DE LA MONTE et al. 1998) have been observed in ALS, a disease characterized by progressive degeneration of motoneurons (YOSHIYAMA et al. 1994). This form of cell death may also be involved in Creuzfeld-Jacob disease (LUCAS et al. 1997), HIV encephalitis (ADLE-BIASSETTE et al. 1995; GELBARD et al. 1995) and measles virus infection of the CNS (MCQUAID et al. 1997). In addition, cannabis has been recently shown to induce neuronal apoptosis (CHAN et al. 1998).

Because of the wealth of evidence implicating a role for apoptosis in nerve cell death, research at present is directed towards unraveling the cell death pathways underlying this process in neurons. As mechanistic issues are difficult to address in vivo, a number of cell culture models of developmental and degenerative neuronal death have been established.

C. Models of Neuronal Apoptosis

I. Developmental Nerve Cell Death

Developmental nerve cell death is reproduced in cell culture by removal of survival factors from the culture medium. The most extensively characterized paradigms are based on NGF withdrawal from cultured sympathetic neurons (DECKWERTH and JOHNSON 1993; DESHMUKH and JOHNSON JR 1997; EDWARDS et al. 1991; EDWARDS and TOLKOVSKY 1994), and potassium/serum withdrawal from cerebellar granule cells (D'MELLO et al. 1993; GALLI et al. 1995; MILLER and JOHNSON 1996). Studies utilizing these model systems have shown that the morphological characteristics of apoptosis are first evident 8–18h after trophic factor withdrawal. At the 48h timepoint, the majority of the neuronal population have died through an apoptotic mechanism. The cell death process is accompanied by oligonucleosomal DNA fragmentation and de novo protein synthesis.

The molecular basis underlying NGF withdrawal-induced apoptosis has also been extensively investigated using a rat pheochromocytoma cell line, PC12 (MESNER et al. 1992, 1995; PITTMAN et al. 1993). Although they are not neuronal per se, PC12 cells differentiate like sympathetic nerve cells in response to NGF. Following NGF withdrawal, PC12 cells undergo genetically programmed apoptotic cell death with similar characteristics to sympathetic neurons.

II. Degenerative Nerve Cell Death

While the role of apoptosis in developmental nerve cell death is well established, it is not clear whether this process underlies the pathogenesis of neurodegeneration. However, numerous studies have shown that toxins implicated in neurodegenerative diseases can trigger apoptotic death in cell culture. For instance, β-amyloid, the major component of senile plaques in AD, induces apoptosis in primary hippocampal and cortical cultures but has no effect on GABAergic neurons, which are largely preserved in AD (ANDERSON et al. 1995; COTMAN and ANDERSON 1995; ESTUS et al. 1997; FORLONI et al. 1993; LOO et al. 1993). It has been suggested that β-amyloid exerts its neurotoxic effect via tau phosphorylation (LE et al. 1997). The typical morphological and biochemical features of apoptosis are also observed in vitro following treatment with 6OHDA and MPTP, toxins which reproduce PD-like cell loss in vivo (DIPASQUALE et al. 1991; HARTLEY et al. 1994; MOCHIZUKI et al. 1994; SHEEHAN et al. 1997; WALKINSHAW and WATERS 1994). In addition, cell culture studies have found that apoptotic nerve cell death can be induced by glutamate, hydrogen peroxide and heavy metals (ANKARCRONA et al. 1995; DESOLE et al. 1996; WHITTEMORE et al. 1994), compounds implicated in a variety of neurodegenerative processes. The time course and magnitude of apoptosis in these paradigms varies depending on the cell type and the nature of the neurotoxic stimuli. However, the cell death process generally evolves over several days and is characterized by delayed membrane lysis, cellular shrinkage, compaction of chromatin and DNA laddering.

As studies based on in vitro models have shown that apoptosis can be attenuated by inhibition of transcription and translation, the cell death process, in many cases, appears to be dependent on the synthesis of new proteins. Although the precise signal transduction pathways underlying apoptosis remain unclear, a number of candidate "cell death" genes have been identified. Within the area of developmental and degenerative nerve cell death, most studies implicate the involvement of four families, namely the inducible transcription factors, the caspases, the bcl-related genes, and the cell cycle regulators.

D. Biochemical Apoptosis Pathways in Neurons

I. The Inducible Transcription Factors

Inducible transcription factors (ITFs) play an important role in the transduction of extracellular signals into long-term changes in neuronal phenotype (for reviews see ANGEL and KARIN 1991; HUGHES and DRAGUNOW 1995; MORGAN and CURRAN 1991; RAHMSDORF 1996). Following stimulation of cell surface receptors, the activation of cytoplasmic second messenger systems triggers an early wave of ITF transcription. Once translated, these proteins re-enter the nucleus and regulate transcription by binding to specific sequences in the DNA of late response genes. In this way, ITF expression is rapid, transient and not dependent on de novo protein synthesis.

The most extensively characterized ITFs are the Fos (c-Fos, Fra-1 and Fra-2) and Jun (c-Jun, JunD and JunB) proteins, which bind to the AP-1 consensus sequence in the promoter region of target genes. As members of the leucine zipper superfamily, the transcriptional activity of Fos and Jun is dependent on the formation of homo- (Jun only) or heterodimeric complexes, deemed AP-1 complexes. The composition of the AP-1 complex determines its DNA binding affinity and, thus, its transactivational potency. The most transcriptionally active complex is composed of a c-Fos/c-Jun heterodimer, while complexes of lower transcriptional efficacy are formed from dimerization of c-Fos with either JunB or JunD, or Jun homodimers. The interaction between the AP-1 complex and its consensus sequence is modulated by several proteins such as IP-1 and CREB, which competitively antagonize AP-1 binding. Once bound to the AP-1 site, the activity of Fos and Jun proteins is further regulated by post-translational modifications which alter transcriptional ability. In addition, ITFs can dimerize with other transcription factor families such as the CREB/ATF family, MyoD and Rel/NFkB proteins. The resulting complexes have increased affinity for promoter sites other than AP-1, thereby expanding the array of potential target genes. Thus, the specificity of the ITF response is determined by a multitude of factors, including the stoichiometry of ITFs induced, the presence of other potential dimer partners and the available target genes.

II. The Role of the ITFs in Apoptosis

1. During CNS Development

c-Fos and c-Jun immunoreactivity in the developing rat brain temporally and spatially correlates with the distribution of cells destined to undergo apoptosis, implicating a key role for these proteins in CNS development (FERRER et al. 1996; GONZALEZ-MARTIN et al. 1992; SMEYNE et al. 1992, 1993). However the finding that neuronal apoptosis occurs normally in c-Jun, c-Fos and c-Fos/c-Jun null mice (most c-Jun knockout mice die in mid-gestation) suggests that induction of these transcription factors is not an absolute requirement of developmental nerve cell death (ROFFLER-TARLOV et al. 1996). It is possible that the functions of c-Fos and c-Jun in knockout animals are assumed by other inducible transcription factors.

2. Degenerative Nerve Cell Death

A wealth of correlative evidence implicates the involvement of ITFs in various neuropathological conditions. For instance, studies on human postmortem Alzheimer's disease tissue revealed a co-localization between c-Jun immunoreactivity with TUNEL-positive cells (ANDERSON et al. 1994), paired helical filaments (ANDERSON et al. 1996), and β-amyloid plaques (FERRER et al. 1996), suggesting that this protein is centrally involved in the disease pathology. However, while other studies support an increase in c-Jun immunoreactivity in AD brains (MACGIBBON et al. 1997), the expression of the remaining ITF family members in AD remains controversial (MACGIBBON et al. 1997).

Numerous studies have reported induction of Jun and Fos family members following both global and focal ischemic insults (DRAGUNOW et al. 1993, 1994; GASS et al. 1992; GUBITS et al. 1993; HSU et al. 1993; KIESSLING et al. 1993; NEUMANN-HAEFELIN et al. 1994; WESSEL et al. 1991). While there is considerable variation in the ITF family members induced, c-Fos and c-Jun induction are observed in the majority of model systems. Whether ITF expression is associated with nerve cell death or survival in cerebral ischemia remains unclear. One study found that potassium channel openers concurrently reduce nerve cell death and ITF expression, providing correlative evidence that these two phenomenon are related (HEURTEAUX et al. 1993). DRAGUNOW et al. (1994) reported that severe hypoxia-ischemia, which caused mainly necrosis, did not induce ITF proteins, whereas a moderate insult which lead to apoptotic death produced extensive ITF protein expression in the selectively vulnerable areas, implicating a direct role for ITFs in apoptotic cell death processes (DRAGUNOW et al. 1994). However, in contrast, other studies have observed ITF expression in less vulnerable or resistant neuronal populations (FERRER et al. 1997; GASS and HERDEGEN 1995; KIESSLING et al. 1993). In addition to cerebral ischemia, induction of c-Fos and c-Jun has been reported in dying hippocampal neurons in two models of status epilepticus, suggesting that ITF expression is involved in seizure-related nerve cell death (DRAGUNOW and PRESTON 1995; DRAGUNOW et al. 1993).

It has been suggested that the temporal pattern of ITF expression may be an important determinant of cellular fate (DRAGUNOW et al. 1994; DRAGUNOW and PRESTON 1995; KAMME et al. 1995). Indeed, several studies have observed a generalized transient wave of ITF expression in resistant neuronal populations occurring rapidly after an ischemic insult or SE, followed by a delayed, prolonged expression (24–72 h) of, predominantly, c-Jun restricted to the neurons which subsequently undergo apoptosis.

Thus, while ITFs are expressed in response to various neuropathological stimuli, the inconsistencies between studies prevents the establishment of a clear relationship between ITF induction and nerve cell death. Furthermore, as an array of ITFs are expressed in many paradigms, exactly which family members mediate the cell death process is unclear. These mechanistic issues have recently been addressed using well characterized in vitro models of neuronal apoptosis.

3. Evidence of a Role for c-Jun and c-Fos in Apoptotic Nerve Cell Death

The first study to propose a role for c-Jun in nerve cell death was based upon the observation that c-Jun was selectively induced in medial septal neurons after axotomy (DRAGUNOW 1992). Subsequently, this protein was implicated in delayed nerve cell death after status epilepticus and ischemia (DRAGUNOW et al. 1993). In support of these in vivo studies, mounting in vitro evidence has demonstrated a central role for c-Jun and, to a lesser extent, c-Fos in the apoptotic nerve cell death process. For instance, ESTUS et al. (1994), found that injection of neutralizing antibodies specific for c-Jun protects rat sympathetic neurons against NGF withdrawal-induced apoptosis, whereas neutralization of JunD and JunB proteins was ineffective (ESTUS et al. 1994). As these authors also showed that Fos antibodies reduced NGF withdrawal-induced apoptosis, it is tempting to speculate that the combination of c-Jun and Fos mediate the cell death process. Along a similar vein, SCHLINGENSIEPEN et al. (1994) demonstrated that inhibition of c-Jun expression using *c-jun* antisense oligonucleotides markedly increased survival of cultured hippocampal neurons, whereas suppression of JunB expression reduced survival (SCHLINGENSIEPEN et al. 1993, 1994). In addition, overexpression of a c-Jun dominant negative mutant attenuates apoptosis triggered by NGF withdrawal in sympathetic neurons (EILERS et al. 1998; HAM et al. 1995) and PC12 cells (XIA et al. 1995), potassium/serum deprivation in cerebellar granule cultures (WATSON et al. 1998) and dopamine exposure in striatal nerve cells (LUO et al. 1998). Further support of a role for c-Jun and c-Fos in apoptosis is derived from transfection studies which show that overexpression of these proteins is sufficient to activate the cell death machinery in several neuronal and non-neuronal cell types (BOSSY-WETZEL et al. 1997; HAM et al. 1995; PRESTON et al. 1996).

4. How Might c-Jun Mediate Neuronal Apoptosis?

a. Upstream Mediators

Attempts to unravel the pathways which influence cellular death and survival have focused primarily on c-Jun and its upstream mediators. Indeed, a growing body of evidence suggests that the MEKK1/SEK1/JNK pathway, which mediates c-Jun activation, is centrally involved in apoptotic nerve cell death. Recent studies have shown that JNK inhibition blocks motoneuron apoptosis (MARONEY et al. 1998). Furthermore, JNK3 knockout mice are resistant towards kainic acid-induced apoptosis, a response which is associated with dephosphorylation of c-Jun (YANG et al. 1997). XIA et al. (1995) reported that expression of a constitutively active MEKK1 mutant markedly increased the number of apoptotic PC12 cells in the presence of NGF (XIA et al. 1995). The observation that the cell death process was blocked by a c-Jun dominant negative mutant implicates c-Jun as a downstream mediator of MEKK1-induced apoptosis. Along a similar vein, EILERS et al. (1998) found that MEKK1 increased apoptotic death and expression of c-Jun and phosphorylated c-Jun in sympathetic neurons via a SEK1-dependent mechanism (EILERS et al. 1998). Furthermore, another study reported that inhibition of SEK1 expression attenuates dopamine-induced JNK activation and apoptosis in striatal nerve cell cultures (LUO et al. 1998). Thus, these results suggest that sequential activation of MEKK, SEK1 and JNK lead to c-Jun activation and, subsequently, apoptosis. Interestingly, it has recently been suggested that JNK3 can regulate neuronal apoptosis by phosphorylating MADD (ZHANG et al. 1998).

The p38 kinase pathway, which increases *c-jun* transcription via phosphorylation of activating transcription factor-2 (ATF-2), has also been implicated in nerve cell death. Studies with mutant forms of MKK3, a selective activator of the p38 signaling cascade have established that this pathway is involved in NGF withdrawal-induced apoptosis in PC12 cells (XIA et al. 1995). However, in contrast to these results, EILERS et al. (1998) found that this system was not activated in NGF-deprived sympathetic neurons.

b. Downstream Mediators

While these studies have shed some light on the upstream mediators of c-Jun expression in the apoptotic cell death cascades, the downstream targets of this protein have not been clearly established. As c-Jun expression is frequently accompanied by cleavage of ICE-like proteases in models of nerve cell death (ELDADAH et al. 1997; SCHULZ et al. 1996; STEFANIS et al. 1996), it has been suggested that caspase activation is a downstream mediator of the c-Jun/JNK pathway. In support of this notion, several studies have observed that inhibition of caspase activity prevents neuronal apoptosis but has no effect on c-Jun expression or JNK activation (DESHMUKH et al. 1996; PARK et al. 1996; STEFANIS et al. 1996). While a direct relationship between caspase activity and the c-Jun/JNK pathway has not yet been established in neuronal cultures, SEIMIYA et al. (1997) reported that JNK1 antisense prevents caspase activation and apoptosis in U937 cells. In addition, using a conditionally active c-Jun allele dependent on the presence of β-estradiol, BOSSY-WETZEL et al. (1997) demonstrated that c-Jun-mediated apoptosis in NIH 3T3 cells involved cleavage of ICE-like proteases (BOSSY-WETZEL et al. 1997). It has been suggested that c-Jun regulates caspases by increasing their gene expression (WALTON et al. submitted). In cerebellar granule neurons, during low potassium-induced apoptosis, the following apoptotic pathway has been proposed: induction and phosphorylation of c-Jun, activation of Bax, activation of caspase activity, DNA fragmentation and death (MILLER et al. 1997). Another sequence of events has been suggested by TANABE et al. (1998): c-Jun activation, de novo RNA synthesis, mitochondrial permeability transition, activation of caspase 3, nuclear shrinkage and death.

As c-Jun has been implicated in cell cycle progression, it has been suggested that induction of this protein may trigger apoptosis by initiating an abortive attempt in post-mitotic neurons to re-enter the cell cycle. Indeed, induction of c-Jun in paradigms of neuronal apoptosis is frequently accompanied by increases in cell cycle regulatory proteins (FREEMAN et al. 1994; GAO and ZALENKA 1995; KRANENBURG et al. 1996). FREEMAN et al. (1994) demonstrated that the increase in c-Jun preceded induction of cyclin D1 in sympathetic neurons undergoing NGF withdrawal-induced apoptosis, indicating that the attempt to re-enter the cell cycle is downstream of c-Jun activation. Consistent with this hypothesis, PARK et al. (1996) found that inhibition of cell cycle progression prevents NGF withdrawal-induced death in PC12 cells, but has no effect on JNK activity (PARK et al. 1996, 1997). Although it is possible that activation of c-Jun and cyclin D1 occur via separate pathways, HERBER et al. (1994) demonstrated the *cyclin D1* gene is a potential target of c-Jun, as its promoter region contains potential AP1 sites which are activated by c-Jun overexpression. Another downstream target of c-Jun during apoptosis may be the amyloid precursor protein 751 (WALTON et al. in press). Interestingly, a recent study has implicated AP-1 activation in cell necrosis (XU et al. 1997), further complicating and obscuring the distinction between apoptosis and necrosis.

III. The Caspase Family

The caspases (also known as ICE-related proteases) first gained attention as mammalian homologues of *ced-3*, the pro-apoptotic gene found in nematodes (for review see NICHOLSON and THORNBERRY 1997; SCHWARTZ and MILLIGAN 1996). Like *ced-3*, caspases are synthesized as dormant pro-enzymes which, following proteolytic activation, cleave specific proteins at aspartate residues. Currently this family comprises ten members, which can be broadly divided into three subgroups based on structural similarities: (1) the *ced*-3-like subfamily, including CPP32 (also as caspase 3, apopain and Yama), Mch2, Mch3 (also known as CMH-1 and ICE-LAP-3), and Mch4; (2) the ICE-like subfamily, including ICE, Tx (also known as ICE rel II and ICH-2), and ICE rel III; and (3) the MEDD-2 family members, including ICH-1, Nedd2 (murine) and Mch6 (also known as ICE-LAP6). Of these proteins, the most extensively characterized family members are CPP32, which cleaves poly(ADP-ribose) polymerase, and ICE, which cleaves and activates pIL-1β to generate active IL-1β. Numerous studies using a wide variety of cell types have reported that caspases are selectively cleaved during apoptosis and that inhibition of this process using peptide-based molecules or viral proteins, *crmA* and p35, attenuates cell death (DRAGUNOW et al. 1998). These findings have lead researchers to suggest that the caspases are essential components of a proteolytic cascade which is triggered in response to an apoptotic stimuli.

1. Evidence of a Role for Caspases in Apoptotic Nerve Cell Death

a. Developmental Nerve Cell Death

In vitro studies provided the first evidence of caspase involvement in developmental nerve cell death. Using chick ganglion nerve cells, GAGLIARDINI et al. (1994) demonstrated that caspase inhibition by *crmA* suppressed nerve growth factor-withdrawal apoptosis (GAGLIARDINI et al. 1994). Similar findings were reported by MILLIGAN et al. (1995), who found that peptide inhibitors of ICE blocked programmed cell death of trophic factor-deprived motoneurons both in vitro and in vivo (MILLIGAN et al. 1995). Subsequently, studies on knockout mice further clarified the role of the caspase family in developmental nerve cell death. Mice deficient in CPP32 contained supernumerary neurons and exhibited disorganized brain structure, although other structures such as thymus were normal (KUIDA et al. 1996), whereas mice deficient in ICE showed no developmental abnormalities in any major organs including the brain, but had a deficit in the inflammatory response and IL-1β secretion (LI et al. 1995). Thus, these results implicate CPP32 as a key mediator of neuronal apoptosis during development, while ICE activity is, in contrast, not absolutely required. In support of this notion, NI et al. (1997) recently reported the presence of a caspase with high homology to CPP32 in neuron-rich regions of the developing and adult rat brain, although expression was profoundly down-regulated in the adult CNS (NI et al. 1997).

b. Degenerative Nerve Cell Death

Although the role of caspases in degenerative nerve cell death in vivo has not been extensively investigated, a growing body of evidence suggests that these enzymes are activated following ischemic brain injury. Several studies have reported up-regulation of CPP32 and Nedd2 mRNAs following ischemia and seizures (ASAHI et al. 1997; GILLARDON et al. 1997; NAMURA et al. 1998). Furthermore, inhibitors selective for CPP32 and ICE can decrease infarct size and oligonucleosomal DNA fragmentation (ENDRES et al. 1998; HARA et al. 1997). In addition, a number of studies have observed a reduction in ischemic brain injury in ICE-deficient mice compared to wild type (HARA et al. 1997; SCHIELKE et al. 1998). However, it is possible that the beneficial effects of ICE inhibition in these paradigms is related to a decrease in the ICE-mediated inflammatory response rather than to inhibition of apoptotic death. In support of this hypothesis, BHAT et al. (1996) observed selective localization of ICE in microglial cells, not neurons, following global forebrain ischemia. In addition to cerebral ischemia, activation of CPP32 has been observed during neuronal apoptosis induced by traumatic brain injury (YAKOVLEV et al. 1997). The administration of a CPP32-selective inhibitor in this model markedly reduced the nerve cell death and improved neurological function.

Although the expression of caspases and their cleavage products in neurodegenerative diseases is not well characterized, TETER et al. (1996) did observe PARP cleavage associated with apoptotic hippocampal neurons and neurofibrillary tangles in the AD brain. Furthermore, mutant forms of the presenilins, associated with early onset AD, are cleaved at alternative sites by caspases. As this alternative cleavage increases production of β-amyloid (1–42), it may contribute to the pathogenesis of AD (KIM et al. 1997). It has also been suggested that caspases are involved in Huntington's disease, as the Huntington protein is a substrate for CPP32, with the extent of cleavage dependent on the length of the polyglutamine repeat (ROSEN 1996).

2. Which Caspases Mediate Apoptotic Nerve Cell Death?

In vitro studies have attempted to further clarify the role of the caspases in developmental and degenerative nerve cell death. To date, caspase activation has been associated with neuronal apoptosis due to a wide variety of stimuli including NGF withdrawal and 6-hydroxydopamine treatment in PC12 cells (HAVIV et al. 1997; OCHU et al. 1998; STEFANIS et al. 1996; STEFANIS et al. 1997; TROY et al. 1997), NGF withdrawal in sympathetic neurons (DESHMUKH et al. 1996; STEFANIS et al. 1997; TROY et al. 1996), β-amyloid exposure in hippocampal neurons (JORDAN et al. 1997), potassium/serum deprivation and MPTP exposure in cerebellar granule cells (ARMSTRONG et al. 1997; D'MELLO et al. 1998; DU et al. 1997; ELDADAH et al. 1997) and staurosporine exposure in neuroblastoma cells (POSMANTUR et al. 1997). Furthermore, increased caspase activity is not observed in neuronal cultures following necrotic stimuli, indicating that this phenomenon is a specific biochemical marker of apoptosis (ARMSTRONG et al. 1997; DU et al. 1997; OCHU et al. 1998). Notably, not all apoptotic cell death pathways involve caspase activation, as MILLER et al. (1996) found that caspase inhibition only marginally reduced trophic factor withdrawal-induced apoptosis in cerebellar granule cells.

As the majority of caspases are not well characterized, exactly which family members mediate apoptotic nerve cell death remains unclear. However, in line with the studies on knockout mice, considerable in vitro evidence implicates a central role for CPP32. For instance, cleavage of CPP32 and its substrate PARP, as well as a specific increase in CPP32 enzyme activity, has been reported in many paradigms (ARMSTRONG et al. 1997; ELDADAH et al. 1997; KEANE et al. 1997; NI et al. 1997; POSMANTUR et al. 1997; STEFANIS et al. 1996). Furthermore, inhibition of CPP32 using z-DEVD-fmk attenuates apoptosis due to trophic factor withdrawal and neurotoxin exposure (DU et al. 1997; ELDADAH et al. 1997; STEFANIS et al. 1996). Interestingly, however, D'MELLO et al. (1998) observed that DEVD-fmk decreased apoptosis in trophic factor-deprived cerebellar granule cells but had no effect on CPP32 or PARP cleavage, indicating that the effects of this compound may, in some cases, be mediated by other members of the caspase family.

In contrast to CPP32, ICE activity does not, for the most part, dramatically change following apoptotic stimulation, and inhibition of this enzyme is relatively ineffective against apoptotic nerve cell death (D'MELLO et al. 1998; DU et al. 1997; ELDADAH et al. 1997; HAVIV et al. 1997; POSMANTUR et al. 1997; STEFANIS et al. 1996; TROY et al. 1996). However, a recent study reported that ICE activity increased during β-amyloid-induced apoptosis in hippocampal neurons (JORDAN et al. 1997), and that a selective ICE inhibitor, Ac-YVAD-CMK, prevented cell death in this paradigm. Similarly, Ac-YVAD-CMK attenuates apoptosis due to superoxide dismutase (SOD1) downregulation in PC12 cells (TROY et al. 1996). In addition, FRIEDLANDER et al. (1997) found that neurons cultured from ICE deficient mice were resistant towards trophic factor withdrawal-induced apoptosis. Nedd2 has also been implicated in some types of apoptotic nerve cell death (STEFANIS et al. 1997; TROY et al. 1997). Using Nedd2 antisense oligonucleotides, TROY et al. (1997) found that inhibition of this protease blocked trophic factor withdrawal-induced apoptosis in sympathetic neurons and PC12 cells, but had no effect on SOD1 downregulation. Thus, while overwhelming evidence implicates the involvement of CPP32 in apoptosis, the observation that ICE and Nedd2 are key mediators of the cell death program in some paradigms supports the existence of parallel caspase pathways, which are selectively activated in response to specific stimuli.

3. Regulation of Apoptosis by the Caspases

Although the regulation of caspase activity is not well characterized, several studies support a role for the Bcl-2-related proteins. For instance, overexpression of Bcl-2 prevents apoptosis and Nedd2 cleavage in the GT1–7 neuronal cell line (SRINIVASAM et al. 1996). Similarly, in trophic factor-deprived PC12 cells, Bcl-2 blocks apoptotic death and the increase in CPP32 activity. In addition, MILLER et al. (1997) found that potassium deprivation of Bax-deficient cerebellar granule cells failed to increase caspase activity or trigger apoptotic death. However, while these observations implicate the Bcl-2 family as upstream regulators of caspase activation, the pro-apoptotic genes, Bax and Bak, can induce apoptosis in the presence of caspase inhibitors, indicating that the effects of these proteins can be mediated via caspase-independent pathways (MCCARTHY et al. 1997; XIANG et al. 1996).

Interestingly, recent evidence suggests Bcl-2 family members can also act as caspase substrates. Several studies have shown that CPP32 cleaves Bcl-2, and its truncated version, Bcl-X_L, during apoptosis, thereby converting these anti-apoptotic genes into potent cell death effectors (CHENG et al. 1997; CLEM et al. 1998). As the cleaved products further activate downstream caspases, it has been suggested that Bcl-2/Bcl-X_L cleavage by CPP32 establishes a positive feedback cycle which ensures the inevitability of cell death.

IV. The Bcl-2 Family

The Bcl-2-related proteins (for reviews see KROEMER 1997; MERRY and KORSMEYER 1997; REED 1994) are an expanding family of apoptosis regulatory genes which act as either death agonists (Bax, Bak, Bcl-X_S, Bad, Bid, Bik and Hrk) or antagonists (Bcl-2, Bcl-X_L, Bcl-w, Bfl-1, Brag-1, Mcl-1 and A1). The anti-apoptotic properties of the founding member, Bcl-2, were first recognized by VAUX et al. (1988), who showed that Bcl-2 overexpression prolonged survival of immature B cells in the absence of interleukin-3 (IL-3). Subsequent studies, which demonstrated that Bcl-2 decreased apoptosis of sympathetic and sensory neurons deprived of NGF and brain-derived neurotrophic factor (BDNF), confirmed its ability to act as a death suppressor gene (ALLSOPP et al. 1993; GARCIA et al. 1992). Since these initial discoveries, Bcl-2 has been shown to block or markedly reduce cell death induced by a wide variety of stimuli, giving rise to the hypothesis that this protein inhibits the final common pathway leading to apoptosis (REED 1994). However, the failure of Bcl-2 to protect cells in some paradigms supports the existence of Bcl-2-independent, as well as dependent, cell death mechanisms.

Apart from Bcl-2 itself, the most extensively characterized Bcl-2-related genes are Bax and Bcl-X. Alternative splicing of Bcl-X gives rise to three transcripts, Bcl-X_S, Bcl-X_L, and Bcl-$X\beta$, which have opposing effects on cellular fate. Overexpression of Bcl-X_L and Bcl-$X\beta$ attenuates trophic factor withdrawal-induced apoptosis, whereas, in contrast, Bcl-X_S renders cells more susceptible to a death stimulus. Like Bcl-X_S, Bax facilitates apoptosis when over-produced, and can antagonize the protective effect of Bcl-2.

1. Regulation of Bcl-2-Related Genes

The effect of Bcl-2-related proteins on cellular fate is determined, at least in part, by the relative abundance of pro-apoptotic and anti-apoptotic family members. An excess of cell death antagonists promotes survival, whereas an excess of death effectors renders cells more vulnerable to apoptosis. This lifedeath rheostat is mediated via competitive dimerization between selective pairs of agonists and antagonists (e.g. Bax/Bax, Bcl-2/Bax, Bcl-2/Bcl-2). While it is not clear which dimer combinations determine whether a cell survives or dies, the most extensively characterized interaction involves Bax and Bcl-2/Bcl-X_L. The increased susceptibility towards apoptosis in response to Bax overexpression is generally attributed to formation of Bax/Bax homodimers, while Bcl-2 is thought to suppress cell death through competitive inhibition of Bax homodimerization. In support of this model, several studies demonstrated that mutations in Bcl-2 and Bcl-X_L, which interfere with their ability to bind Bax, also block their anti-apoptotic function (CHENG et al. 1996; YININ et al. 1994). However, as other mutations in Bcl-X_L, which affect its interaction with Bax or Bak, do not abolish its anti-apoptotic activity, these cell death suppressor molecules can clearly function independently (CHENG et al. 1996). Consistent with this hypothesis, KNUDSON and KORSMEYER (1997) demonstrated, using transgenic mice, that the formation of Bcl-2/Bax heterodimers is not required for either Bcl-2 repression or Bax promotion of apoptosis. In addition to the Bcl-2-related genes, Bcl-2 and Bcl-X_L can dimerize with other proteins such as Raf-1, Ced-4 and calcineurin. However, the functional significance of these interactions is unclear.

As the levels of Bcl-2 and Bax do not change in cells undergoing apoptosis, exactly what regulates the ratio of Bcl-2-related genes is unclear. However, recent evidence suggests that the activity of Bcl-2-related proteins is regulated post-translationally. Indeed, it has been shown that the anti-apoptotic properties of Bcl-2 are neutralized by phosphorylation at serine/threonine residues (HALDAR et al. 1996). Similarly, serine/threonine phosphorylation of Bad blocks its death effector activity by inhibiting dimerization with Bcl-X_L (ZHA et al. 1996). Taken together, these studies suggest that the Bcl-2-related proteins are part of a complex system which is regulated at a variety of levels.

2. Evidence of a Role for Bcl-2-Related Genes in the Nervous System

a. Developmental Nerve Cell Death

Immunocytochemical studies have characterized the pattern of Bcl-2, Bcl-X and Bax expression during central nervous system development. Bcl-2 is expressed at high levels in the developing brain and is downregulated after birth (FERRER et al. 1994; MARTINOU et al. 1994; MERRY et al. 1994), whereas, in contrast, Bcl-X expression increases postnatally, reaching peak levels in the adult brain (GONZALEZ-GARCIA et al. 1995). High levels of *Bax* mRNA are observed in the sympathetic cervical ganglion and motor neurons at a time when these neuronal populations are susceptible to apoptosis (DECLWERTH et al. 1996). In addition, Bax expression is apparent in the developing trigeminal motor nucleus and cerebellum. In the adult, widespread expression of Bax is observed in most neuronal populations, and it has been suggested that this protein may contribute to the vulnerability of post-mitotic neurons to a variety of insults (KRAJEWSKI et al. 1994). Studies with transgenic animals have shed some light on the functional roles of the Bcl-2-related genes during development. Despite the widespread expression of Bcl-2 during CNS development, *bcl-2*-deficient mice show no overt abnormalities in developmental nerve cell death in the prenatal stage (VEIS et al. 1993). However progressive degeneration of sympathetic, sensory and motoneurons is observed postnatally, indicating that Bcl-2 is critical for the maintenance of certain neuronal populations (MICHAELIDIS et al. 1996). In contrast to Bcl-2, *bcl-X*-deficient mice die at approximately embryonic day-13 and show widespread nerve cell loss. The cell death occurs mainly in differentiating neurons, which have not yet made synaptic connections, indicating that Bcl-X is absolutely required for neuronal survival during differentiation and maturation (MOTOYAMA et al. 1995). In accordance with its expression during neurogenesis, developmental sympathetic and motor neuronal death is reduced in *Bax*-deficient mice, implicating

Bax as a critical mediator of trophic factor withdrawal-induced apoptosis in these populations (DECKWERTH et al. 1996).

b. Degenerative Nerve Cell Death

In line with their opposing roles in apoptosis, a number of studies have observed differential regulation of Bax/Bcl-X_S and Bcl-2/Bcl-X_L after both global and focal ischemia. While there is some variation between models, up-regulation of Bax is generally observed in vulnerable neurons which subsequently undergo apoptosis (CHEN et al. 1996; GILLARDON et al. 1996; HARA et al. 1996; ISENMANN et al. 1998; KRAJEWSKI et al. 1995; MACGIBBON et al. 1997; MATSUSHITA et al. 1998). In contrast, Bcl-2 expression is restricted to neuronal populations which survive the insult (CHEN et al. 1995; ISENMANN et al. 1998; MATSUSHITA et al. 1998). Interestingly, CHEN et al. (1997) reported an increase in *bcl-2* and *bcl-x_l* mRNA in both surviving and dying neurons following global ischemia, but had previously found that their proteins were expressed only in neurons destined to survive (CHEN et al. 1995). Based on these observations, it has been suggested that the failure to translate *bcl-2* and *bcl-x_l* mRNA contributes to the initiation of the apoptotic cell death program in vulnerable neuronal populations (CHEN et al. 1997). In addition to ischemia, an increase in Bax expression and a decrease in Bcl-2 has been reported following kainic acid treatment in mice (GILLARDON et al. 1995), and during β-amyloid-induced apoptosis in human neurons (PARADIS et al. 1996). Up-regulation of Bax has also been associated with 6-hydroxydopamine toxicity in PC12 cells (BLUM et al. 1997) and MPTP treatment in mice (HASSOUNA et al. 1996). Taken together, these results suggest that alterations in the Bcl-2/Bax ratio play a critical role in determining whether post-mitotic neurons survive or die. This notion, however, does not appear to extend to the developing brain as no change in levels of Bcl-2, Bax, Bcl-X_S or Bcl-X_L were observed in 8-day-old rats following hypoxia-ischemia (FERRER et al. 1997). Furthermore, a recent study has shown Bax is necessary for apoptosis induced by low potassium, but not for NMDA receptor-mediated excitotoxicity of cerebellar granule cells (MILLER et al. 1997). While a decrease in Bcl-2 protein in dying neurons has been observed in animal models of nerve cell death, studies using human post-mortem PD and AD tissue have observed elevated Bcl-2 expression in affected neuronal populations (MARSHALL et al. 1997; MIGHELI et al. 1994; MOGI et al. 1996; O'BARR et al. 1996; SATOU et al. 1995; SU et al. 1996). Since Bcl-2 appears to prevent nerve cell death, it has been hypothesized that this increase reflects a compensatory mechanism instigated in response to cellular degeneration. In support of this notion, SU et al. (1996) co-localized Bcl-2 expression with neurons exhibiting DNA fragmentation in post-mortem AD brains. These authors also demonstrated that Bcl-2 expression was down-regulated in tangle-bearing neurons, implicating the loss of Bcl-2 protein in the formation of neurofibrillary tangles and subsequent nerve cell death.

The expression of Bax has also been investigated in post-mortem AD brains (MACGIBBON et al. 1997; SU et al. 1997). Although Bax is expressed at relatively high levels in neurologically normal brains, an increase in Bax immunoreactivity is apparent in neurons and microglia of AD hippocampi. Furthermore, co-localization studies demonstrated that Bax immunoreactivity was associated with senile plaques, tau-positive tangles, and TUNEL-positive neurons, lending strong support to a role for this protein in the disease pathogenesis. However, as the activity of Bax (and Bcl-2) is strongly influenced by the presence of other Bcl-2-related genes, further research is required to elucidate the role of this system in degenerative nerve cell death.

As Bcl-2 overexpression prevents nerve cell death induced by a variety of toxic stimuli in vitro (LAWRENCE et al. 1996; MYERS et al. 1995; OFFEN et al. 1997; ZHONG et al. 1993), several studies have examined whether its protective effect extends to the in vivo situation. Indeed, a reduction in infarct size was observed in transgenic mice overexpressing Bcl-2 compared to wild type (MARTINOU et al. 1994). Furthermore, LAWRENCE et al. (1997) and LINNIK et al. (1995) found that delivery of a herpes simplex virus (HSV) expressing Bcl-2 markedly reduced nerve cell loss following a focal ischemic insult, implicating a potential role for this protein in the treatment of stroke.

3. How Does Bcl-2 Exert Its Neuroprotective Effects?

A multitude of theories have been formulated to account for the anti-apoptotic effects of Bcl-2, including free radical scavenging, ion flux regulation and caspase inhibition. Most recently, it has been suggested that Bcl-2 prevents the early mitochondrial changes associated with apoptosis, in particular the mitochondrial permeability transition (KROEMER 1997; REED 1997). The mitochondrial permeability transition, which occurs almost universally during apoptotic death, involves the opening of a large channel in the inner mitochondrial membrane. This alteration has a variety of consequences which may contribute to induction of apoptosis, including free radical generation, the release of stored Ca^{2+} and mitochondrial proteins into the cytosol, and subsequently caspase activation. Overexpression studies have shown that Bcl-2 can inhibit the mitochondrial permeability transition, whereas it is induced by Bax (SUSIN et al. 1996; XIANG et al. 1996). Whether these proteins directly control pore opening or influence it indirectly by regulating other mitochondrial functions remains at present unclear.

V. Cell Cycle Regulators

It been suggested that cellular susceptibility to apoptosis results from dysregulated expression of conflicting or inappropriate growth and survival signals. This hypothesis is based on the observation that cell cycle regulatory proteins such as p53 and the cyclin family are often induced in paradigms of apoptotic death. The p53 tumor suppressor gene encodes a nuclear phosphoprotein which functions as an important regulator of cellular proliferation and apoptosis (for reviews see BELLAMY 1997; HUGHES et al. 1997; WEBB et al. 1997). Often dubbed the "guardian of the genome", p53 accumulates via increased translation in cells in response to genotoxic damage and then enables DNA repair through inhibition of cell cycle progression at the late G_1 phase. If DNA damage is severe and irreversible, p53 induces the cell to undergo apoptosis, thereby preventing replication of a damaged genome. The anti-proliferative and pro-apoptotic actions of p53 are thought to be mediated via transcriptional regulation of a specific set of target genes including Bax, Gadd45 and WAF/p21, which contain a p53 consensus sequence. p53 can also interact directly with cellular proteins and is, itself, the target of several viral proteins.

Numerous studies have implicated p53 in the cellular apoptotic response to genotoxic damage (BELLAMY 1997). While p53-mediated apoptosis in this scenario is necessary to prevent tumorogenesis, it is possible that inappropriate accumulation of wild type p53 may induce undesirable cell death. Thus, the expression of this protein in paradigms of developmental and degenerative nerve cell death has recently been examined.

1. Evidence of a Role for p53 in Neuronal Apoptosis

Evidence derived from several sources suggests that p53 does not play a major part in developmental nerve cell death. Firstly, normal CNS development is observed in mice deficient in p53, although these animals are susceptible to spontaneous tumors (DONEHOWER et al. 1992). In addition, NGF withdrawal from sympathetic and sensory neurons cultured from p53 null embryos induces apoptosis in the usual fashion (DAVIES and RISENTHAL 1994). Furthermore, WOOD and YOULE (1995) observed that the cerebellar granule cells undergo normal developmental cell death in p53 null mice, but were not, unlike wild type, susceptible to γ-irradiation-induced cell death, implicating the existence of p53-dependent and independent pathways in the CNS.

While it does not appear to be involved in developmental nerve cell death, increased expression of p53 has been reported following a variety of neurotoxic insults, including cerebral ischemia (CHOPP et al. 1992; LI et al. 1997), photochemical brain injury (MANEV et al. 1994), kainic acid-induced seizures (SAKHI et al. 1994, 1996), excitotoxic lesions (HUGHES et al. 1996) and adrenalectomy (SCHREIBER et al. 1994). Although p53 immunoreactivity is frequently co-localized with cells containing fragmented DNA, it is not clear whether induction of this protein is a cause or result of DNA damage. However, as p53 knockout mice are resistant towards neuronal injury in many of the aforementioned paradigms, it is likely that this gene plays a role in the apoptotic cell death program which subsequently leads to DNA breakdown (CRUMRINE et al. 1994; MORRISON et al. 1996; SAKHI et al. 1996; TRIMMER et al. 1996). In support of this notion, in vitro studies have provided direct evidence of a role for p53 in nerve cell death processes. XIANG et al. (1996) showed that both kainic acid and glutamate treatment triggered massive death in hippocampal and cortical neurons containing at least one p53 allele but had no effect on p53 (−/−) cultures. However, re-introduction of p53 to the p53-deficient cultures was sufficient to promote nerve cell death, even in the absence of a toxin. Notably, it has been suggested that the effect of p53 is cell type-dependent, as overexpression of this protein does not induce apoptosis in sympathetic nerve cells (SADOUL et al. 1996).

While the role of p53 in neurodegenerative diseases has not been extensively researched, several studies have observed increased p53 levels in post-mortem AD tissue compared to that of controls (DE LA MONTE et al. 1997; KITAMURA et al. 1997). DE LA MONTE et al. (1997) found that p53 was associated with senile plaques and some, but not all, tau-positive neurites, while another study reported that p53 immunoreactivity was present in glial cells. In addition, an increase in p53 has been observed in PC12 cells following exposure to 6-hydroxydopamine, a neurotoxin implicated in PD (BLUM et al. 1997).

2. How Does p53 Mediate Neuronal Apoptosis?

Although the downstream mediators of p53 are not well characterized, it has been suggested that this protein triggers apoptosis by altering the Bcl-2/Bax ratio. Indeed, several studies have found that p53 can increase expression of Bax and decrease Bcl-2 levels (MIYASHITA et al. 1994; MIYASHITA and REED 1995). Induction of Bax has been reported following p53 expression in apoptotic neurons after quinolinic acid treatment (HUGHES et al. 1997), and p53-mediated apoptosis of hippocampal neurons is blocked in Bax deficient neurons (XIANG et al. 1998). In addition, expression of both p53 and Bax has been reported following cerebral ischemia (LI et al. 1997) and after 6-hydroxydopamine treatment in PC12 cells (BLUM et al. 1997). However, it should be noted that p53-mediated apoptosis can also occur in the absence of Bax induction (ALLDAY et al. 1995; CANMAN et al. 1995). Furthermore, Bax-deficient mice show a normal p53-dependent apoptotic response to ionizing radiation (KNUDSON et al. 1995). Thus, these findings suggest that regulation of the Bcl-2/Bax family may be involved in some, but not all, forms of p53-mediated apoptosis.

3. Cyclins and Cyclin-Dependent Kinases

The cyclins regulate progression through the cell cycle by stimulating activity of the cyclin-dependent kinases. In vitro studies lend strong support to a role for this family in developmental nerve cell death. Induction of cyclins, in particular cyclin D1 and cyclin-dependent kinases, has been observed following trophic factor withdrawal-induced apoptosis in PC12 cells, sympathetic neurons (FREEMAN et al. 1994; GAO and ZALENKA 1995) and N1E-115-derived nerve cells (KRANENBURG et al. 1996). Furthermore, agents which inhibit cell cycle progression at the G_1/S phase promote neuronal survival in these paradigms (FARINELLI and GREENE 1996; KRANENBURG et al. 1996; PARK et al. 1997;

Rydel and Greene 1988). In addition, Park et al. (1997) demonstrated that expression of dominant negative forms of cyclin-dependent kinases protects against NGF withdrawal-induced apoptosis in sympathetic neurons, suggesting that CDKs play an essential role in the cell death process. Interestingly, while cyclin D1 is a key mediator of trophic factor withdrawal-induced apoptosis in sympathetic neurons and N1E-115 nerve cells (Freeman et al. 1994; Kranenburg et al. 1996), levels of this protein do not change following potassium/serum deprivation of cerebellar granule cells. Rather, a decrease in cyclin A mRNA and protein is observed in this paradigm, indicating that the cyclin proteins are differentially regulated during apoptotic death.

Increased expression of various cyclins and cyclin-dependent kinases has been observed in vivo following quinolinic acid lesions (Henchcliffe and Burke 1997) and cerebral ischemia (Kuroiwa et al. 1998; Li et al. 1997). However, it is not yet clear whether these proteins are associated with nerve cell death or survival. While Kuroiwa et al. (1998) observed that cyclin D1 was preferentially expressed in dying cells, others have reported that this protein was localized to morphologically intact neurons (Li et al. 1997). Furthermore, Weissner et al. (1996) found increased levels of cyclin D1 in microglia, implicating a role for this protein in microglial proliferation, rather than nerve cell death processes. Several studies have also observed increased cyclin and cyclin-dependent kinase expression in AD brains compared to controls. While there is some variation in the cyclin family members induced, an increase in cyclin B is the most consistent change (Busser et al. 1998; Nagy et al. 1997; Vincent et al. 1997). Thus, although cyclins are critical for proliferation in premitotic cells, their expression in the aforementioned paradigms gives rise to the possibility that apoptosis results from an abortive attempt to activate the cell cycle in terminally differentiated neurons.

E. Conclusion

Apoptosis is clearly an important mechanism of cell death in the nervous system, both during brain development and in neurodegenerative diseases. Understanding the biochemical pathways responsible for nerve cell apoptosis will provide novel targets for drug development to treat brain diseases.

Acknowledgements. This work was supported by grants from the Health Research Council of New Zealand, the New Zealand Neurological Foundation, the Auckland Medical Research Foundation, the Auckland University Research Committee and Lotteries Health to MD.

References

Adle-Biassette HYL, Colombel M, Poron F, Natchev S, Keohane C, Gray F (1995) Neuronal Apoptosis in HIV infection in adults. Neuropathol Appl Neurobiol 21:218–227

Allanson J, Bond AB, Morton J, Hunt SP (1994) Transfection of proto-oncogene c-Jun into a mouse neuroblastoma cell line potentiates differentiation. Gene Ther 1 (Suppl)

Allday MJ, Inman GJ, Crawford DH, Farrell PJ (1995) DNA damage in human B cells can induce apoptosis, proceeding from G_1/S when p53 is transactivation competent and G_2/M when it is transactivation defective. EMBO J 14:4994–5005

Allsopp TE, Wyatt S, Paterson HF, Davies AM (1993) The proto-oncogene bcl-2 can selectively rescue neurotrophic-dependent neurons from apoptosis. Cell 73:295–307

Anderson AJ, Cummings BJ, Cotman CW (1994) Increased immunoreactivity for Jun- and Fos-related proteins in Alzheimer's disease: Association with pathology. Exp Neurol 125:286–295

Anderson AJ, Pike CJ, Cotman CW (1995) Differential induction of immediate early gene proteins in cultured neurons by b-amyloid (Ab): Association of c-jun with Ab-induced apoptosis. J Neurochem 65:1487–1498

Anderson AJ, Su JH, Cotman, CW (1996) DNA damage and apoptosis in Alzheimer's disease: Colocalisation with c-jun immunoreactivity, relationship to brain area, and effect of post mortem delay. J Neurosci 16:1710–1719

Angel P, Karin M (1991) The role of Jun, Fos and the AP1 complex in cell-proliferation and transformation. Biochim Biophys Acta 1072:129–157

Anglade P, Vyas S, Javoyagid F, Herrero MT, Michel PP, Marquez J, Mouattprigent A, Ruberg M, Hirsch EC, Agid Y (1997) Apoptosis and autophagy in nigral neurons of patients with Parkinson's disease. Histol Histopathol 12:25–3

Ankarcrona M, Dypbukt JM, Bonfoco E, Zhivotovsky B, Orrenius S, Lipton SA, Nicotera P (1995) Glutamate-induced neuronal death – A succession of necrosis or apoptosis depending on mitochondrial function. Neuron 15:961–973

Armstrong RC, Aja TJ, Hoang KD, Gaur S, Bai X, Alnemri ES, Litwack G, Karenewsky DS, Fritz LC, Tomaselli KJ (1997) Activation of CED3/ICE-related proteases CPP32 in cerebellar granule neurons undergoing apoptosis but not necrosis. J Neurosci 17:553–562

Asahi M, Hoshimaru M, Uemura Y, Tokime T, Kojima M, Ohtsuka T, Matsuura N, Aoki T, Shibahara K, Kikuchi H (1997) Expression of interleukin-1b converting enzyme gene family and bcl-2 gene family in the rat brain following permanent occlusion of the middle cerebral artery. J Cereb Blood Flow Metab 17:11–18

Bancher C, Lassman H, Breitschopf H, Jellinger KA (1997) Mechanisms of cell death in Alzheimer's disease. J Neural Transm 50:141–152

Bar PR (1996) Apoptosis – The cell's silent exit. Life Sci 59:369–378

Batistatou A, Greene LA (1991) Aurintricarboxylic acid rescues PC12 cells and sympathetic neurons from cell death caused by nerve growth factor deprivation: Correlation with suppression of endonuclease activity. J Cell Biol 115:461–471

Behrens MI, Koh JY, Muller MC, Choi DW (1996) NADPH diaphorase-containing striatal or cortical neurons are resistant to apoptosis. Neurobiol Dis 3:72–75

Behrens MI, Koh J, Canzoniero LMT, Sensi S, Csernansky CA, Choi DW (1995) 3-Nitropropionic acid induces apoptosis in cultured striatal and cortical neurons. Neuroreport 6:545–548

Beilharz EJ, Williams CE, Dragunow M, Sirimanne ES, Gluckman PD (1995) Mechanisms of delayed cell death following hypoxic-ischemic injury in the immature rat: evidence for apoptosis during selective neuronal loss. Mol Brain Res 29:1–14

Bellamy COC. (1997) p53 and apoptosis. Br Med Bull 53:522–538

Bhat RV, DiRocco R, Marcy VR, Flood DG, Zhu Y, Dobrzanski P, Siman R, Scott R, Contreras PC, Miller M (1996) Increased expression of IL-1b converting enzyme in hippocampus after ischemia: selective localisation in microglia. J Neurosci 16: 4146–4154

Blum D, Wu Y, Nissou M-F, Arnaud S, Benabid A-L, Verna J-M (1997) p53 and bax activation in 6-hydroxydopamine-induced apoptosis in PC12 cells. Brain Res 751:139–142

Bossy-Wetzel E, Bakiri L, Yaniv M (1997) Induction of apoptosis by the transcription factor c-Jun. EMBO J 16:1695–1709

Brecht S, Martin-Villalba M, Zimmerman M, Herdegen T (1995) Transection of fimbria-fornix induces lasting expression of c-jun in axotomised septal neurons immunonegative for choline acetyltransferase and nitric oxide synthase. Exp Neurol 134:112–125

Brown DG, Sun XM, Cohen GM (1993) Dexamethasone-induced apoptosis involves cleavage of DNA to large fragments prior to internucleosomal fragmentation. J Biol Chem 268:3037–3039

Busser J, Geldmacher DS, Herrup K (1998) Ectopic cell cycle proteins predict the sites of neuronal cell death in Alzheimer's disease brain. J Neurosci 18:2801–2807

Butterworth NJ, Williams L, Bullock JY, Love DR, Faull RLM, Dragunow M (1998) Trinucleotide (CAG) repeat length is positively correlated with the degree of DNA fragmentation in Huntington's disease striatum. Neuroscience (In Press)

Canman CE, Gilmer TM, Coutts SB, Kastan MB (1995) Growth factor modulation of p53-mediated growth arrest versus apoptosis. Genes Devel 9:600–611

Chan G C-K, Hinds TR, Impey S, Storm DR (1998) Hippocampal neurotoxicity of delta^9-tetrahydrocannabinol. J Neurosci 18(14):5322–5332

Charriaut-Marlangue C, Margaill I, Represa A, Popovici T, Plotkine M, Ben-Ari Y (1996) Apoptosis and necrosis after reversible focal ischemia: An in situ DNA fragmentation analysis. J Cereb Blood Flow Metab 16:186–194

Chen J, Graham SH, Chan PH, Lan J, Zhou RL, Simon RP (1995) Bcl-2 is expressed in neurons that survive focal ischemia in the rat. Neuroreport 6:394–398

Chen J, Graham SH, Nakayama M, Zhu RL, Jin K, Stetler RA, Simon RP (1997) Apoptosis repressor genes Bcl-2 and Bcl-x_l are expressed in the rat brain following global ischemia. J Cereb Blood Flow Metab 17:2–10

Chen J, Zhu RL, Nakayama M, Kawaguchi K, Jin K, Stsler RA, Simon RP, Graham SH (1996) Expression of apoptosis-effector gene Bax, is up-regulated in vulnerable hippocampal neurons following global ischemia. J Neurochem 67:64–71

Cheng EH-Y, Kirsch DG, Clem RJ, Ravi R, Kastan MB, Bedi A, Ueno K, Hardwick JM (1997) Conversion of Bcl-2 to a Bax-like death effector by caspases. Science 278:1966–1968

Cheng EH-Y, Levine B, Boise LH, Hardwick JM (1996) Bax-independent inhibition of apoptosis by Bcl-x_l. Nature 379:554–556

Chopp M, Li Y, Zhang ZG, Freytag SO (1992) p53 expression in brain after middle cerebral artery occlusion in the rat. Biochem Biophys Res Comm 182:1201–1207

Clem RJ, Cheng EH, Karp CL, Kirsch DG, Ueno K, Takahashi A, Kastan MB, Griffin DE, Earnshaw WC, Veliunoa MA, Hardwick JM (1998) Modulation of cell death by Bcl-XL through caspase interaction. Proc Natl Acad Sci USA 95:554–559

Cohen JJ, Duke RC (1984) Glucocorticoid activation of a calcium-dependent endonuclease in thymocyte nuclei leads to cell death. J Immunol 132:38–42

Collins RJ, Harmon BV, Gobe GC, Kerr JFR (1992) Internucleosomal DNA cleavage should not be the sole criterion for identifying apoptosis. Int J Radiat Biol 61:451–453

Colotta F, Polentarutti N, Sironi M, Mantovani A (1992) Expression and involvement of c-fos and c-jun proto-oncogenes in programmed cell death induced by growth factor deprivation in lymphoid cell lines. J Biol Chem 267:18278–18283

Cotman CW, Anderson AJ (1995) A potential role for apoptosis in neurodegeneration and Alzheimer's disease. Mol Neurobiol 10:19–45

Crumrine R, Thomas AL, Morgan PF (1994) Attenuation of p53 expression protects against focal ischemic brain damage in transgenic mice. J Cereb Blood Flow Metab 14:887–891

D'Mello SR, Galli C, Ciotti T, Calissano P (1993) Induction of apoptosis in cerebellar granule neurons by low potassium: inhibition of death by insulin-like growth factor I and cAMP. Proc Natl Acad Sci USA 90:10989–10993

D'Mello S, Aglieco F, Roberts MR, Borodezt K, Haycock JW (1998) A DEVD-inhibited caspase other than CPP32 is involved in the commitment of cerebellar granule neurons to apoptosis induced by K+ deprivation. J Neurochem 70:1809–1818

Davies AM, Rosenthal A (1994) Neurons from mouse embryos with a null mutation in the tumour suppressor gene p53 undergo normal cell death in the absence of neurotrophins. Neurosci Lett 182:112–114

de Groot RP, Kruijer W (1991) Up-regulation of Jun/AP1 during differentiation of N1E-115 neuroblastoma cells. Cell Growth Differ 2:631–636

de Groot RP, Kruyr FAE, van der Saag PT, Kruijer W (1990) Ectopic expression of c-jun leads to differentiation of P19 embryonal carcinoma cells. EMBO J 9:1831–1837

de Groot RP, Schoorlemmer JTvGS, Kruijer W (1990) Differential expression of jun and fos genes during differentiation of mouse P19 embryonal carcinoma cells. Nucleic Acids Res 18:3195–3202

Monte SM de la, Sohn YK, Wands JR (1997) Correlates of p53- and Fas (CD95)-mediated apoptosis in Alzheimer's disease. J Neurol Sci 152:73–83

Monte SM de la, Sohn YK, Ganju N, Wands JR (1998) p53- and CD95-associated apoptosis in neurodegenerative diseases. Lab Invest 78:401–411

Deckwerth TL, Johnson EM (1993) Temporal analysis of events associated with programmed cell death (apoptosis) of sympathetic neurons deprived of nerve growth factor (NGF) J Cell Biol 123:1207–1222

Deckwerth TL, Elliot JL, Snider WD, Korsmeyer SJ (1996) BAX is required for neuronal death after trophic factor deprivation and during development. Neuron 17:401–411

Deshmukh M, Johnson EM Jr (1997) Programmed cell death in neurons: Focus on the pathway of nerve growth factor deprivation-induced death of sympathetic neurons. Mol Pharmacol 51:897–906

Deshmukh M, Vasilakos J, Lampe PA, Shivers BD, Johnson EM Jr. (1996) Genetic and metabolic status of NGF-deprived sympathetic neurons saved by an inhibitor of the ICE family proteases. J Cell Biol 135:1341–1454

Desole MS, Sciola L, Delogu MR, Sirana S, Migheli R (1996) Manganese and 1-methyl-4(2'-ethylphenyl)-1,2,3,6-tetrahydropyridine induce apoptosis in PC12 cells. Neurosci Lett 209:193–196

Dipasquale B, Marini AM, Youie RJ (1991) Apoptosis and DNA degradation induced by 1-methyl-4-phenylpyridinium in neurons. Biochem Biophys Res Comm 181:1442–1448

Donehower LA, Harvey M, Slagle BL, McArthur MJ, Montgomery J, Butel JS, Bradley A (1992) Mice deficient for p53 are developmentally normal but susceptible to spontaneous tumours. Nature 356:215–221

Dragunow M, Preston K (1995) The role of inducible transcription factors in apoptotic nerve cell death. Brain Res Rev 21:1–28

Dragunow M, Beilharz E, Sirimanne E, Lawlor P, Williams C, Bravo R, Gluckman P (1994) Immediate-early gene protein expression in neurons undergoing delayed death, but not necrosis, following hypoxic-ischemic injury to the young rat brain. Brain Res Mol Brain Res. 25:19–33

Dragunow M, Faull RLM, Lawlor P, Beilharz E, Singelton K, Walker EB, Mee E (1995) In situ evidence for DNA fragmentation in Huntington's disease striatum and Alzheimer's disease temporal lobes. Neuroreport 6:1053–1057

Dragunow M, MacGibbon GA, Lawlor PA, Butterworth N, Connor B, Henderson C, Walton M, Woodgate A, Hughes P, Faull RLM (1998) Apoptosis, Neurotrophic Factors and Neurodegeneration. Rev Neurosci 8:223–265

Dragunow M, Young D, Hughes P, MacGibbon G, Lawlor P, Singleton K, Sirimanne E, Beilharz E, Gluckman P (1993) Is c-jun involved in nerve cell death following status epilepticus and hypoxic-ischemic brain injury? Mol Brain Res 18:347–352

Du Y, Dodel RC, Bales KR, Jemmerson R, Hamilton-Byrd E, Paul S (1997) Involvement of a caspase-3-like protease in 1-methyl-4-phenylpyridinium-mediated apoptosis of cultured cerebellar granule neurons. J Neurochem 69:1382–1388

Dungy LJ, Siddiqi TA, Bravo R, Kiessling M (1993) C-jun and junB oncogene expression during placental development. Am J Obstet Gynecol 165:1853–1856

Dure LS, Weiss S, Standaert DG, Rudolf G, Testa CM, Young AB (1995) DNA fragmentation and immediate early gene expression in rat striatum following quinolinic acid lesion. Exp Neurol 133:207–214

Edwards SN, Tolkovsky AM (1994) Characterisation of apoptosis in cultured rat sympathetic neurons after nerve growth factor withdrawal. J Cell Biol 124:537–546

Edwards SN, Buckmaster AE, Tolkovsky AM (1991) The death programme in cultured sympathetic neurons can be suppressed at the post-translational level by nerve growth factor, cyclic AMP, and depolarisation. J Neurochem 57:2140–2143

Eilers A, Whitfield J, Babij C, Rubin LL, Ham J (1998) Role of the jun kinase pathway in the regulation of c-jun expression and apoptosis in sympathetic neurons. J Neurosci 18:1713–1724

Eldadah BA, Yakoviev AG, Faden AI (1997) The role of CED-3-related cysteine proteases in apoptosis of cerebellar granule cells. J Neurosci 17:6105–6113

Ellis HM, Horvitz HR (1986) Genetic control of programmed cell death in the nematode Caenorhabditis elegans. Cell 44:817–829

Endres M, Namura S, Shimizu-Sasmata M, Waeber C, Zhang L, Gomez-Isla T, Hyman BT, Moskowitz MA (1998) Attenuation of delayed neuronal death after mild focal ischemia in mice by inhibition of the caspase family. J Cereb Blood Flow Metab 18:238–247

Estus S, Tucker HM, van Rooyen C, Wright S, Brigham EF, Wogulis M, Rydel RE (1997) Aggregated Amyloid-b protein induces cortical neuronal apoptosis and concomitant "apoptotic" pattern of gene induction. J Neurosci 17:7736–7745

Estus S, Zaks WJ, Freeman RS, Gruda M, Bravo R, Johnson EM Jr (1994) Altered gene expression in neurons during programmed cell death: Identification of c-jun as necessary for neuronal apoptosis. J Cell Biol 127:1717–1727

Farinelli SE, Greene LA (1996) Cell cycle blockers mimosine, ciclopirox and deferoxamine prevent the death of PC12 cells and post mitotic sympathetic neurons after withdrawal of trophic support. J Neurosci 16:1150–1162

Ferrer I, Ballabriga J, Pozas E (1997) Transient forebrain ischemia in the adult gerbil is associated with a complex c-Jun response. Neuroreport 8:2483–2487

Ferrer I, Olive M, Ribera J, Planas AM (1996) Naturally occurring (programmed) and radiation-induced apoptosis are associated with selective c-jun expression in the developing rat brain. Eur J Neurosci 8:1286–1298

Ferrer I, Pozas E, Lopez E, Ballabriga J (1997) Bcl-2, Bax, Bcl-x expression following hypoxia-ischemia in the infant rat brain. Acta Neuropathol 94:583–589

Ferrer I, Segui J, Planas AM (1996) Amyloid deposition is associated with c-jun expression in Alzheimer's disease and amyloid angiopathy. Neuropathol Appl Neurobiol 22:521–526

Ferrer I, Tortosa A, Condom E, Blanco R (1994) Increased expression of bcl-2 immunoreactivity in the developing cerebral cortex of the rat. Neurosci Lett 179: 13–16

Filipkowski RK, Hetman M, Kamiska B, Kacmarek L (1994) DNA fragmentation in the rat brain after intraperitoneal administration of kainate. Neuroreport 5:1538–1540

Fink SL, Ho DY, Sapolsky RM (1996) Energy and glutamate dependency of 3-nitropropionic acid neurotoxicity in culture. Exp Neurol 138:298–304

Forloni G, Chiesa R, Smiroldo S, Verga L, Salmona M, Tagliavini F, Angeretti N (1993) Apoptosis mediated neurotoxicity induced by chronic application of b amyloid fragment 25–35. Neuroreport 4:523–526

Freeman RS, Estus S, Johnson EM (1994) Analysis of cell cycle-related gene expression in post-mitotic neurons: selective induction of cyclin D1 during programmed cell death. Neuron 12:343–355

Friedlander RM, Gagliardini V, Hara H, Fink KB, Li W, MacDonald G, Fishman MC, Greenberg AH, Moskowitz MA, Yuan J (1997) Expression of a dominant negative mutant of interleukin-1 beta converting enzyme in transgenic mice prevents neuronal cell death induced by trophic factor withdrawal and ischemic brain injury. J Exp Med 185:933–940

Gagliardini V, Fernandez PA, Lee RKK, Drexler HCA, Rotello RJ, Fishman MC, Yuan J (1994) Prevention of vertebrate neuronal death by the crmA gene. Science 263:826–828

Galli C, Meucci O, Scorziella A, Werge T, Calissano P, Schettini G (1995) Apoptosis in cerebellar granule cells is blocked by high KCl, forskolin and IGF-1: the involvement of intracellular calcium and DNA synthesis. J Neurochem 15:1172–1179

Galli C, Piccini A, Ciotti MT, Castellani L, Calissano P, Zaccheo D, Tabaton M (1998) Increased amyloidogenic secretion in cerebellar granule cells undergoing apoptosis. Proc Natl Acad Sci USA 95:1247–1252

Gao CY, Zalenka PS (1995) Induction of cyclin B and H1 kinase activity in apoptotic PC12 cells. Exp Cell Res 219:612–618

Garcia I, Martinou Y, Tsuijimoto Y, Martinou J-C (1992) Prevention of programmed cell death of sympathetic neurons by the bcl-2 proto-oncogene. Science 258:302–304

Gass P, Herdegen T (1995) Neuronal expression of AP-1 proteins in excitotoxic-neurodegenerative disorders and following nerve fibre lesions. Prog Neurobiol 47:257–290

Gass P, Spranger M, Herdegen T, Bravo R, Kock P, Hacke W, Kiessling M (1992) Induction of fos and jun proteins after focal ischemia in the rat: differential effect of N-methyl-D-aspartate receptor antagonist MK801. Acta Neuropathol 84:545–553

Gelbard HA, James HJ, Sharer LR, Perry SW, Saito Y, Kazee AM, Blumberg BM, Epstein LG (1995) Apoptotic neurons in brains from pediatric patients with HIV-1 encephalitis and progressive encephalopathy. Neuropathol Appl Neurobiol 21:208–217

Gillardon F, Bottiger B, Scmitz B, Zimmermann M, Hossmann K-A (1997) Activation of CPP-32 protease in hippocampal neurons following ischemia and epilepsy. Mol Brain Res 50:16–22

Gillardon F, Lenz C, Waschke KF, Krajewski S, Reed JC, Zimmermann M, Kuschinsky W (1996) Altered expression of Bcl-2, Bcl-X, Bax, and c-Fos colocalises with DNA fragmentation and ischemic cell damage following middle cerebral artery occlusion in rats. Mol Brain Res 40:254–260

Gillardon F, Wickert H, Zimmermann M (1995) Up-regulation of bax and down-regulation of bcl-2 is associated with kainate-induced apoptosis in mouse brain. Neurosci Lett 192:85–88

Gonzalez-Garcia M, Garcia I, Ding L, O'Shea S, Boise LH (1995) bcl-x is expressed in embryonic and postnatal neural tissues and functions to prevent neuronal cell death. Proc Natl Acad Sci USA 92:4304–4308

Gonzalez-Martin C, de Diego I, Crespo D, Fairen A (1992) Transient c-fos expression accompanies naturally occurring cell death in the developing interhemispheric cortex of the rat. Devel Brain Res 68:83–95

Gubits RM, Burke RE, Casey-MacIntosh G, Bandele A, Munell F (1993) Immediate early gene induction after neonatal hypoxia-ischemia. Mol Brain Res 18:228–238

Haldar S, Chintapalli J, Croce C. M (1996) Taxol induces bcl-2 phosphorylation and death of prostate cancer cells. Cancer Res 56:1253–1255

Ham J, Babij C, Whitfield J, Pfarr CM, Lallemand D, Yaniv M, Rubin LL (1995) A c-jun dominant negative mutant protects sympathetic neurons against programmed cell death. Neuron 14:927–939

Hara A, Iwai T, Niwa M, Uematsu T, Yoshimi N, Tanaka T, Mori H (1996) Immunohistochemical detection of Bax and Bcl-2 proteins in gerbil hippocampus following transient forebrain ischemia. Brain Res 711:249–253

Hara H, Fink K, Endres M, Friedlander RM, Gagliardini V, Yuan J, Moskowitz MA (1997) Attenuation of transient focal cerebral ischemic injury in transgenic mice expressing mutant ICE inhibitory protein. J Cereb Blood Flow Metab 17:370–375

Hara H, Friedlander RM, Gagliardini V, Ayata C, Fink K, Huang Z, Shimizu-Sasamata M, Yuan J, Moskowitz MA (1997) Inhibition of interleukin-1beta converting enzyme family proteases reduced ischemic and excitotoxic neuronal damage. Proc Natl Acad Sci USA 94:2007–2012

Hartley A, Stone JM, Heron C, Cooper JM, Shapira AHV (1994) Complex I inhibitors induce dose-dependent apoptosis in PC12 cells: Relevance to Parkinson's disease. J Neurochem 63:1987–1990

Hassouna I, Wickert H, Zimmerman M, Gillardon F (1996) Increase in bax expression in substantia nigra following 1-methyl-4-phenyl-1,2,3,6-tetrahydropyridine (MPTP) treatment of mice. Neurosci Lett 204:85–88

Haviv R, Lindenboim L, Li H, Yuan J, Stein R (1997) Need for caspases in apoptosis of trophic factor-deprived PC12 cells. J Neurosci Res 50:69–80

Hedgecock EM, Sulston JE, Thomson JN (1983) Mutations affecting programmed cell deaths in the nematode Caenorhabditis elegans. Science 220:1277–1279

Henchcliffe C, Burke RE (1997) Increased expression of cyclin-dependent kinases 5 in induced apoptotic neuron death in rat substantia nigra. Neurosci Lett 230:41–44

Hengartner MO, Ellis RE, Horvitz HR (1992) Caenorhabditis elegans gene ced-9 protects cells from programmed cell death. Nature 356:494–499

Herber B, Truss M, Beato M, Muller R (1994) Inducible regulatory elements in the human cyclin D1 promoter. Oncogene 9(7):2105–2107

Herdegen T, Brecht S, Kummer W, Mayer B, Leah J, Bravo R, Zimmerman M (1993) Persisting expression of JUN and KROX transcription factors and increase of nitric oxide synthase immunoreactivity in rat central neurons following axotomy. J Neurosci 13:4130–4146

Heurteaux C, Bertaina V, Widmann C, Lazdunski M (1993) K+ channel openers prevent global ischemia-induced expression of c-fos, c-jun, heat shock protein and b-amyloid precursor protein genes and neuronal death in the hippocampus. Proc Natl Acad Sci USA 90:9431–9435

Hsu CY, An G, Liu JS, Xue JJ, He YY, Lin TN (1993) Expression of immediate early gene and growth factor mRNAs in a focal cerebral ischemia model in the rat. Stroke 24 (Suppl I), I78–I81

Hughes PE, Alexi T, Schreiber SS (1997) A role for the tumour suppressor gene p53 in regulating neuronal apoptosis. Neuroreport 8:5–12

Hughes PE, Alexi T, Yoshida T, Schreiber SS, Knusel B (1996) Excitotoxic lesion of rat brain with quinolinic acid induced expression of p53 messenger RNA and protein and p53-inducible genes bax and gadd-45 in brain areas showing DNA fragmentation. Neuroscience 74:1143–1160

Hughes P, Dragunow M (1995) Induction of immediate-early genes and the control of neurotransmitter-regulated gene expression within the nervous system. Pharmacol Rev 47:133–178

Ignatowicz E, Vezzini AM, Rizzi M, D'Incalci M (1991) Nerve cell death induced in vivo by kainic acid and quinolinc acid does not involve apoptosis. Neuroreport 2:651–654

Isenmann S, Stoll G, Schroeter M, Krajewski S, Reed JC, Bahr M (1998) Differential regulation of Bax, Bcl-2, and Bcl-X proteins in focal cortical ischemia in the rat. Brain Pathol 8:49–62

Jackson-Lewis V, Jakowec M, Burke RE, Przedborski S (1995) Time course and morphology of dopaminerigc neuronal death caused by the neurotoxin 1-methyl-4-phenyl-1,2,3,6-tetrahydropyridine. Neurodegeneration 4:257–269

Jenkins R, Hunt SP (1991) Long term increase in the levels of c-jun mRNA and c-Jun protein immunoreactivity in motor and sensory neurons following axon damage. Neurosci Lett 129:107–110

Jenkins R, Tetzlaff W, Hunt SP (1993) Differential expression of immediate early genes in rubrospinal neurons following axotomy in the rat. Eur J Neurosci 5:203–209

Jeon BS, Jackson-Lewis V, Burke RE (1995) 6-Hydroxydopamine lesion of the rat substantia nigra: time course and morphology of cell death. Neurodegeneration 4:131–137

Jordan J, Galindo MF, Miller RF (1997) Role of calpain and interleukin-1b converting enzyme-like proteases in the b-amyloid-induced death of rat hippocampal neurons in culture. J Neurochem 68:1612–1621

Kamme F, Campbell K, Wieloch T (1995) Biphasic expression of the Fos and Jun families of transcription factors following transient forebrain ischemia in the rat. Effect of hypothermia. Eur J Neurosci 7:2007–2016

Keane RW, Srinivasan A, Foster LM, Testa M-P, Ord T, Nonner D, Wang H-G, Reed JC, Bredesen DE, Kayalar C (1997) Activation of CPP32 during apoptosis of neurons and astrocytes. J Neurosci Res 48:168–180

Kerr JFR, Wyllie AH, Currie AR (1972) Apoptosis: a basic biological phenomenon with wide-ranging implications in tissue kinetics. Br J Cancer 26:239–257

Kiessling M, Stumm G, Xie Y, Herdegen T, Aguzzi A, Bravo R, Gass P (1993) Differential transcription and translation of immediate early genes in the gerbil hippocampus after transient global ischemia. J Cereb Blood Flow Metab 13:914–924

Kihara S, Shiraishi T, Nakagawa S, Toda K, Tabuchi K (1994) Visualisation of DNA double strand breaks in the gerbil CA1 following transient ischemia. Neurosci Lett 175:133–136

Kim T-W, Pettingell WH, Jung Y-K, Kovacs DM, Tanzi RE (1997) Alternative cleavage of Alzheimer's-associated presenelins during apoptosis by a caspase-3 family protease. Science 277:373–376

Kitamura Y, Shimohama S, Kamoshima W, Matsuoka Y, Nomura Y, Taniguchi T (1997) Changes in p53 in the brains of patients with Alzheimer's disease. Biochem Biophys Res Comm 232:418–421

Knudson CM, Korsmeyer SJ (1997) Bcl-2 and bax function independently to regulate cell death. Nat Genet 16:358–363

Knudson MC, Tung KSK, Tourtellotte WG, Brown GAJ, Korsmeyer SJ (1995) Bax-deficient mice with lymphoid hyperplasia and male germ cell death. Science 1995:96–98

Kosel S, Egensperger R, Voneitzen U, Mehraein P, Graeber MB (1997) On the question of apoptosis in the Parkinsonian substantia nigra. Acta Neuropathologica 93:105–108

Krajewski S, Krajewska M, Shabaik A, Miyashita T, Wang HG, Reed JC (1994) Immunohistochemical determination of in vivo distribution of Bax, a dominant inhibitor of Bcl-2. Am J Pathol 145:1323–1336

Krajewski S, Mai JK, Krajewska M, Mossakowski MJ, Reed JC (1995) Upregulation of Bax protein levels in neurons following cerebral ischemia. J Neurosci 15:6364–6376

Kranenburg O, van der Eb AJ, Zantema A (1996) Cyclin D1 is an essential mediator of apoptotic neuronal cell death. EMBO J 15:46–54

Kroemer G (1997) The proto-oncogene Bcl-2 and its role in regulating apoptosis. Nat Med 3:614–620

Kuida K, Zheng TS, Na SQ, Kuan CY, Yang D, Karasuyama H, Rakic P, Flavell RA (1996) Decreased apoptosis in the brain and premature lethality in CPP32-deficient mice. Nature 384:368–372

Kuroiwa S, Katai N, Shibuki H, Kurokawa T, Umihira J, Nikaido T, Kametani K, Yoshimura N (1998) Expression of cell cycle-related genes in dying cells in retinal ischemic injury. Invest Opthalmol Vis Sci 39:610–617

Lambert MP, Barlow AK, Chromy BA, Edwards C, Freed R, Liosatos M, Morgan TE, Rozovsky I, Trommer B, Viola KL, Wals P, Zhang C, Finch CE, KraffT GA, Klein WL (1998) Diffusable, nonfibrillar ligands derived from $A\beta_{1-42}$ are potent central nervous system neurotoxins. Proc Natl Acad Sci USA 95:6448–6453

Lassmann H, Bancher C, Breitschopf H, Wegiel J, Bobinski M, Jellinger K, Wisniewski HM (1995) Cell death in Alzheimer's disease evaluated by DNA fragmentation in situ. Acta Neuropathol 89:35–41

Lawrence MS, Ho DY, Sun GH, Steinberg GK, Sapolsky RM (1996) Overexpression of Bcl-2 with herpes simplex virus vectors protects CNS neurons against neurological insults in vitro and in vivo. J Neurosci 16:486–496

Lawrence MS, McLaughlin JR, Sun GH, Ho DY, McIntosh L, Kunis DM, Sapolsky RM, Steinberg GK (1997) Herpes simplex viral vectors expressing Bcl-2 are neuroprotective when delivered after a stroke. J Cereb Blood Flow Metab 17:740–744

Le W, Xie WJ, Kong R, Appel SH (1997) β-Amyloid-induced neurotoxicity of a hybrid septal cell line associated with increased Tau phosphorylation and expression of β-Amyloid precursor protein. J Neurochem 69:978–985

Leah JD, Herdegen T, Murashov A, Dragunow M, Bravo R (1993) Expression of immediate-early gene proteins following axotomy and inhibition of axonal transport in the rat CNS. Neuroscience 57:53–66

Leist M, Nicotera P (1997) The shape of cell death. Biochem Biophys Res Comm 236:1–9

Li P, Allen H, Banerjee S, Franklin S, Herzog L, Johnston C, McDowell J, Paskind M, Rodman L, Salfeld J, Towne E, Tracey D, Wardwell S, Wei F-Y, Wong W, Kamen R, Seshadri T (1995) Mice deficient in IL-1b-converting enzyme are defective in production of mature IL-1b and are resistant to endotoxic shock. Cell 80:401–411

Li WP, Chan WY, Lai HWL, Yew DT (1997) Terminal dUTP nick end labeling (TUNEL) positive cells in the different regions of the brain in normal aging and Alzheimer's patients. J Mol Neurosci 8:75–82

Li Y, Chopp M, Powers C (1997) Granule cell apoptosis and protein expression in hippocampal dentate gyrus after forebrain ischemia in the rat. J Neurol Sci 150:93–102

Li Y, Chopp M, Jiang N, Zaloga C (1995b) In situ detection of DNA fragmentation after focal cerebral ischemia in mice. Mol Brain Res 28:164–168

Li Y, Chopp M, Jiang N, Yao F, Zaloga C (1995a) Temporal profile of in situ DNA fragmentation after transient middle cerebral artery occlusion in the rat. J Cereb Blood Flow Metab 15:389–397

Li Y, Chopp M, Powers C, Jiang N (1997) Apoptosis and protein expression after focal cerebral ischemia in rat. Brain Res 765:301–312

Li Y, Chopp M, Powers C, Jiang N (1997) Immunoreactivity of cyclin D1/cdk4 in neurons and oligodendrocytes after focal ischemia in rat. J Cereb Blood Flow Metab 17:846–856

Linnik MD, Miller JA, Sprinkle-Cavallo J, Mason PJ, Thompson FY, Montgomery LR, Schroeder KK (1995) Apoptotic DNA fragmentation in the rat cerebral cortex induced by permanent middle cerebral artery occlusion. Mol Brain Res 32:116–124

Linnik MD, Zahos P, Geschwind MD, Federoff HJ (1995) Expression of bcl-2 from a defective simplex virus-1 vector limits neuronal death in focal cerebral ischemia. Stroke 26:1670–1674

Loo DT, Copani AG, Pike CJ, Whittemore ER, Walencewicz AJ, Cotman CW (1993) Apoptosis is induced by beta-amyloid in cultured central nervous system neurons. Proc Natl Acad Sci USA 90:7951–7955

Love S, Barber R, Wilcock GK (1998) Apoptosis and expression of DNA repair proteins in ischaemic brain injury in man. Neuroreport 9:955–959

Lucas M, Izquierdo G, Munoz C, Solano F (1997) Internucleosomal breakdown of the DNA of brain cortex in human spongiform encephalopathy. Neurochem Int 31:241–244

Luo Y, Umegaki H, Wang X, Abe R, Roth GS (1998) Dopamine induces apoptosis through an oxidation-involved SAPK/JNK activation pathway. J Biol Chem 273 6:3756–3764

MacGibbon GA, Lawlor PA, Sirimanne ES, Walton MR, Connor B, Young D, Williams C, Gluckman P, Faull RLM, Hughes P, Dragunow M (1997) Bax expression in mammalian neurons undergoing apoptosis and in Alzheimer's disease hippocampus. Brain Res 750:223

MacManus JP, Hill IE, Huang Z-G, Rasquinha I, Xue D, Buchan AM (1994) DNA damage consistant with apoptosis in transient focal ischemic neocortex. Neuroreport 5:493–496

MacManus JP, Rasquinha I, Tuor U, Preston E (1997) Detection of higher-order 50- and 10-kbp DNA fragments before apoptotic interneucleosomal cleavage after transient cerebral ischemia. J Cereb Blood Flow Metab 17:376–387

Manev H, Khariamov A, Armstrong DM (1994) Photochemical brain injury in rats triggers DNA fragmentation p53 and HSP72. Neuroreport 5:2661–2664

Maroney AC, Glicksman MA, Basma AN, Walton KM, Knight Jr E, Murphy CA, Bartlett BA, Finn JP, Angeles T, Matsuda Y, Neff NT, Dionne CA (1998) Motoneuron apoptosis is blocked by CEP-1347 (KT 7515), a novel inhibitor of the JNK signalling pathway. J Neurosci 18(1):104–111

Marshall KA, Daniel SE, Cairns N, Jenner P, Halliwell B (1997) Upregulation of the anti-apoptotic protein Bcl-2 may be an early event in neurodegeneration: studies on Parkinson's and incidental Lewy body disease. Biochem Biophys Res Comm 240:84–87

Martin DP, Schmidt RE, Distefano PS, Lowry OH, Carter JG, Johnson EM (1988) Inhibitors of protein and RNA synthesis prevent neuronal cell death caused by NGF deprivation. J Cell Biol 106:829–844

Martinou J-C, Dubois-Dauphin M, Staple JK, Rodriguez I, Frankowski H, Missotten M, Albertini P, Talabot D, Catsicas S, Pietra C, Huarte J (1994) Overexpression of BCL-2 in transgenic mice protects neurons from naturally occuring cell death and experimental ischemia. Neuron 13:1017–1030

Martinou J-C, Frankowski H, Missotten M, Martinou I, Potier L, Dubois-Dauphin M (1994) Bcl-2 and neuronal selection during development of the nervous system. J Physiol 88:209–211

Masters JN, Finch CE, Sapolsky RM (1989) Glucocorticoid endangerment of hippocampal neurons does not involve deoxyribonucleic acid cleavage. Endocrinol 124:3083–3088

Matsushita K, Matsuyama T, Matsumoto M, Yanaghara T, Sugita M (1998) Alternations of Bcl-2 family proteins precede cytoskeletal proteolysis in the penumbra, but not infarct centres, following focal cerebral ischemia in mice. Neuroscience 83:439–448

McCarthy NJ, Whyte MKB, Gilbert CS, Evan GI (1997) Inhibition of Ced-3/ICE-related proteases does not prevent cell death induced by oncogenes, DNA damage of the Bcl-2 homologue, Bak. J Cell Biol 136:215–227

McQuaid S, McMahon J, Herron B, Cosby SL (1997) Apoptosis in measles virus-infected human central nervous system tissues. Neuropathol Appl Neurobiol 23:218–224

Merry DE, Korsmeyer SJ (1997) Bcl-2 gene family in the nervous system. Annual Rev Neurosci 1997:245–267

Merry DE, Veis DJ, Hickey WF, Korsmeyer SJ (1994) Bcl-2 protein expression is widespread in the developing central nervous system and retained in the adult PNS. Development 120:301–311

Mesner PW, Winters TR, Green SH (1992) Nerve growth factor withdrawal-induced cell death in neuronal PC12 cells resembles that in sympathetic neurons. J Cell Biol 119:1669–1680

Mesner P, Epting CL, Hegarty JL, Green SH (1995) A timetable of events during programmed cell death induced by trophic factor withdrawal from neuronal PC12 cells. J Neurosci 15:7357–7366

Michaelidis TM, Sendtner M, Cooper JD, Airaksinen MS, Holtmann B, Meyer M, Thoenen H (1996) Inactivation of bcl-2 results in progressive degeneration of motoneurons, sympathetic and sensory neurons during early postnatal development. Neuron 17:75–89

Migheli A, Cavalla P, Piva R, Giordana MT, Schiffer D (1994) Bcl-2 protein expression in the aged brain and neurodegenerative diseases. Neuroreport 5:1906–1908

Miller TM, Johnson EM (1996) Metabolic and genetic analyses of apoptosis in potassium/serum-deprived rat cerebellar granule cells. J Neurosci 16:7487–7495

Miller TM, Moulder DL, Knudson CM, Creedon DJ, Deshmukh M, Korsmeyer SJ, Johnson Jr EM (1997) Bax deletion further orders the cell death pathway in cerebellar granule cells and suggests a caspase-independent pathway to cell death. J Cell Biol 139:205–217

Milligan CE, Prevette D, Yaginuma H, Homma S, Cardwell C, Fritz LC, Tomaselli KJ, Oppenheim RW, Schwartz TZ (1995) Peptide inhibitors of the ICE family arrest programmed cell death of motoneurons in vivo and in vitro. Neuron 15:385–393

Miyashita T, Reed JC (1995) Tumour suppressor p53 is a direct transcriptional activator of the human bax gene. Cell 80:293–299

Miyashita T, Harigai M, Hanada M, Reed JC (1994) Identification of a p53-dependent negative response element in the bcl-2 gene. Cancer Res 54:3131–3135

Miyashita T, Krajewski S, Krajewska M, Wang HG, Lin HK, Liebermann DA, Hoffman B, Reed JC (1994) Tumour suppressor p53 is a regulator of bcl-2 and bax expression in vivo and in vitro. Oncogene 9:1799–1805

Mochizuki H, Goto K, Mori H, Mizuno Y (1996) Histochemical detection of apoptosis in Parkinson's disease. J Neurol Sci 137:120–123

Mochizuki H, Nakamura N, Nishi K, Mizuno Y (1994) Apoptosis is induced by 1-methyl-4-phenylpyridinium ion (MPP+) in ventral mesencephalic-striatal co-culture in rat. Neurosci Lett 170:191–194

Mogi M, Harada M, Kondo T, Mizuno Y, Narabayashi H, Riederer P, Nagatsu T (1996) Bcl-2 protein is increased in the brain from parkinsonian patients. Neurosci Lett 215:137–139

Morgan JI, Curran T (1991) Stimulus-transcription coupling in the nervous system: involvement of inducible proto-oncogenes fos and jun. Ann Rev Neurosci 14:421–451

Morrison RS, Wenzel HJ, Kinoshita Y, Robbins CA, Donehower LA, Schwartzkroin PA (1996) Loss of p53 tumour suppressor gene protects neurons from kainic acid-induced cell death, J Neurosci 16(4):1337–1345

Motoyama N, Wang F, Roth KA, Sawa H, Nakayama K-I, Nakayama K, Negishi I, Senju S, Zhang Q Fujii S, Loh D (1995) Massive cell death of immature haematopoietic cells and neurons in Bcl-x-deficient mice. Science 267:1506–1510

Myers KM, Fiskum G, Yuanbin L, Simmens SJ, Bredesen DE, Murphy AN (1995) Bcl-2 protects neural cells from cyanide/aglycemia-induced lipid oxidation, mitochondrial injury and loss of viability. J Neurochem 65:2432–2440

Nagy Z, Esiri MM, Cato AM, Smith AD (1997) Cell cycle markers in the hippocampus in Alzheimer's disease. Acta Neuropathol 94:6–15

Namura S, Zhu J, Fink K, Endres M, Srinivasan A, Tomaselli KJ, Yuan J, Moskowitz MA (1998) Activation and cleavage of caspase-3 in apoptosis induced by experimental cerebral ischemia. J Neurosci 18(10):3659–3668

Neumann-Haefelin T, Weissner C, Vogel P, Back T, Hossmann K-A (1994) Differential expression of the immediate early genes c-fos, c-jun, junB and NGFI-B in the rat

brain following transient forebrain ischemia. J Cereb Blood Flow Metab 14:206–216
Ni,B, Wu X, Du Y, Su Y, Hamilton-Byrd E, Rockey PK, Rosteck P, Poirer GG, Paul SM (1997) Cloning and expression of a rat brain interleukin-1b-converting enzyme (ICE)-related protease (IRP) and its possible role in apoptosis of cultured cerebellar granule neurons. J Neurosci 17:1561–1569
Nicholson DW, Thornberry NA (1997) Caspases: Killer proteases. Trends Biochem Sci 22:299–306
Nishiyama K, Kwak S, Takekoshi S, Watanabe K, Kanazawa I (1996) In situ nick endlabelling detects necrosis of hippocampal pyramidal cells induced by kainic acid. Neurosci Lett 212:139–142
Nitatori T, Sato N, Waguri S (1995) Delayed neuronal death in the CA1 pyramidal cell layer of the gerbil hippocampus following transient ischemia is apoptosis. J Neurosci 12:1001–1011
O'Barr S, Schultz J, Rogers J (1996) Expression of the proto-incogene bcl-2 in Alzheimer's disease brain. Neurobiology of Aging 17:131–136
Ochu EE, Rothwell NJ, Waters CM (1998) Caspases mediate 6-hydroxydopamine-induced apoptosis but not necrosis in PC12 cells. J Neurochem 70:2637–2640
Offen D, Ziv I, Panet H, Wasserman L, Stein R, Melamed E, Barzilai A (1997) Dopamine-induced apoptosis is inhibited in PC12 cells expressing Bcl-2. Cell Mol Neurobiol 17:289–304
Paradis E, Douillard H, Koutroumanis M, Goodyer C, LeBlanc A (1996) Amyloid b peptide of Alzheimer's disease downregulates Bcl-2 and upregulates Bax expression in human neurons. J Neurosci 16:7533–7539
Park DS, Levine B, Ferrari G, Greene L (1997) Cyclin dependent kinase inhibitors and dominant negative cyclin dependent kinase 4 and 6 promote survival of NGF-deprived sympathetic neurons. J Neurosci 17:8975–8983
Park DS, Stefanis L, Yan CYI, Farinelli SE, Greene LA (1996) Ordering the cell death pathway. J Biochem 271:21898–21905
Pfarr CM, Mechta F, Spyrou G, Lallemand D, Carillo S, Yaniv M (1994) Mouse JunD negatively regulates fibroblast growth and antagonises transformation by ras. Cell 76:747–760
Pittman RN, Wang S, DiBenedetto AJ, Mills JC (1993) A system for characterising cellular and molecular events in programmed neuronal cell death. J Neurosci 13:3669–3680
Pollard H, Cantagrel S, Charriaut-Marlangue C, Moreau J, Ben-Ari Y (1994) Apoptosis associated DNA fragmentation in epileptic brain damage. Neuroreport 5:1053–1055
Portera-Cailliau C, Hedreen JC, Price DL, Koliatsos VE (1995) Evidence for apoptotic cell death in Huntington disease and excitotoxic animal models. J Neurosci 15:3775–3787
Posmantur R, McGinnis K, Nadimpalli R, Gilbertsen RB, Wang KKW (1997) Characterisation of CPP32-like protease activity following apoptotic challenge in SH-SY5Y neuroblastoma cells. J Neurochem 68:2328–2337
Preston GA, Lyon TT, Yin Y, Lang JE, Soloman G, Annab L, Srinivasan DG, Alcorta DA, Barrett JC (1996) Induction of apoptosis by c-Fos protein. Mol Cell Biol 16:211–218
Rahmsdorf HJ (1996) Jun: Transcription factor and oncoprotein. J Mol Med 74:725–747
Reed JC (1994) Bcl-2 and the regulation of programmed cell death. J Cell Biol 124:1–6
Reed JC (1997) Double identity for proteins of the Bcl-2 family. Nature 387:773–776
Roffler-Tarlov S, Gibson Brown JJ, Tarlov E, Stolarov J, Chapman DL, Alexiou M, Papaioannou VE (1996) Programmed cell death in the absence of c-Fos and c-Jun. Development 122:1–9

Rosen A (1996) Huntingtin: a new marker along the road to cell death. Nat Genet 13:380–382

Rydel RE, Greene LA (1988) cAMP analogs promote survival and neurite outgrowth in cultures of rat sympathetic and sensory neurons independently of nerve growth factor. Proc Natl Acad Sci USA 85:1257–1261

Sadoul R, Quiquerez AL, Matinou I, Fernandz PA, Martinou JC (1996) p53 protein in sympathetic neurons: cytoplasmic location and no apparent function in apoptosis. J Neurosci Res 43:594–601

Sakhi S, Bruce A, Sun N, Tocco G, Baudry M, Schreiber SS (1994) p53 induction is associated with neuronal damage in the central nervous system. Proc Natl Acad Sci USA 91:7525–7529

Sakhi S, Gilmore W, Tran ND, Schrieber SS (1996) p53-deficient mice are protected against adrenalectomy-induced apoptosis. Neuroreport 8(1):233–235

Sakhi S, Sun N, Wing LL, Mehta P, Schreiber SS (1996) Nuclear accumulation of p53 protein following kainic acid-induced seizures. Neuroreport 7:493–496

Sato S, Gobbel GT, Honkaniemi J, Li Y, Kondo T, Murakami K, Sato M, Copin JC, Chan PH (1997) Apoptosis in the striatum of rats following intraperitoneal injection of 3-nitropropionic acid. Brain Res 745:343–347

Satou T, Cummings BJ, Cotman CW (1995) Immunoreactivity for Bcl-2 protein within neurons in the Alzheimer's disease brain increases with disease severity. Brain Res 697:35–43

Saura J, MacGibbon G, Dragunow M (1997) Etoposide-induced PC12 cell death: apoptotic morphology without oligonucleosomal DNA fragmentation or dependency on de novo protein synthesis. Mol Brain Res 48:382–388

Schielke GP, Yang GY, Shivers BD, Betz AL (1998) Reduced ischemic brain injury in interleukin-1 beta converting enzyme deficient mice. J Cereb Blood Flow Metab 18:180–185

Schlingensiepen K-H, Schlingensiepen R, Kunst M, Klinger I, Gerdes W, Seifert W, Brysch W (1993) Opposite function of jun-B and c-jun in growth regulation and neuronal differentiation. Devel Genet 14:305–312

Schlingensiepen K-H, Wollnik F, Kunst M, Schlingensiepen R, Herdegen T, Brysch W (1994) The role of jun transcription factor expression and phosphorylation in neuronal differentiation, neuronal cell death and plastic adaptions in vivo. Cell Mol Neurobiol 14:487–505

Schmidt-Kastner R, Fliss H, Hakim AM (1997) Subtle neuronal death in striatum after short forebrain ischemia in rats detected by in situ end-labelling for DNA damage. Stroke 28:163–170

Schreiber SS, Sakhi S, Dugich-Djordevic MM, Nicholas NR (1994) Tumour suppressor p53 induction and DNA damage in hippocampal granule cells after adrenalectomy. Exp Neurol 130:368–376

Schulz JB, Beinroth S, Weller M, Wullner U, Klockgether T (1998) Endonucleolytic DNA fragmentation is not required for apoptosis of cultured rat cerebellar granule neurons. Neurosci Lett 245:9–12

Schulz JB, Weller M, Klockgether T (1996) Potassium deprivation-induced apoptosis of cerebellar granule neurons: A sequential requirement for new mRNA and protein synthesis, ICE-like protease activity, and reactive oxygen species. J Neurosci 16:4696–4706

Schwartz LM, Milligan CE (1996) Cold thoughts of death: the role of ICE proteases in neuronal cell death. Trends Neurosci 19:555–561

Schwartz LM, Osborne BA (1993) Programmed cell death, apoptosis and killer genes. Immunol Today 14:582–590

Sei Y, Von Lubitz DKJE, Basile AS, Borner MM, Lin RC-S, Skolnick P, Fossom LH (1994) Internucleosomal DNA fragmentation in gerbil hippocampus following forebrain ischemia. Neurosci Lett 171:179–182

Seimiya H, Mashima T, Toho M, Tsuruo T (1997) c-Jun NH_2-terminal kinase-mediated activation of interleukin-1b converting enzyme/CED-3-like protease during anti-cancer drug-induced apoptosis. J Biol Chem 272:4631–4636

Sheehan JP, Palmer PE, Helm GA, Tuttle JB (1997) MPP$^+$-induced apoptotic cell death in SH-SY5Y neuroblastoma cells – an electron microscope study. J Neurosci Res 48:226–237
Smale GNRN, Brady DR, Finch CE, Horton WE (1995) Evidence for apoptotic cell death in Alzheimer's disease. Exp Neurol 133:225–230
Smeyne RJ, Morgan T, Morgan RJ (1992) Temporal and spatial expression of a fos-lacZ transgene in the developing nervous system. Mol Brain Res 16:158–162
Smeyne RJ, Vendrell M, Hayward M, Baker SJ, Miao GG, Schilling K, Robertson LM, Curran T, Morgan JI (1993) Continuous c-fos expression precedes programmed cell death in vivo. Science 363:166–169
Srinivasam A, Foster LM, Testa M-P, Ord T, Keane RW, Bredesen DE, Kayalar C (1996) Bcl-2 expression in neural cells blocks activation of ICE/CED-3 family proteases during apoptosis. J Neurosci 16:5654–5660
Stefanis L, Park DS, Yan CYI, Farinelli SE, Troy CM, Shelanski ML, Greene LA (1996) Induction of CPP32-like activity in PC12 cells by withdrawal of trophic support. J Biol Chem 271:30663–30671
Stefanis L, Troy CM, Qi H, Greene LA (1997) Inhibitors of trypsin-like serine proteases inhibit processing of caspase Nedd-2 and protect PC12 cells and sympathetic neurons from death evoked by withdrawal of trophic support. J Neurochem 69:1425–1437
Su JH, Anderson AJ, Cummings BJ, Cotman CW (1994) Immunohistochemical evidence for apoptosis in Alzheimer's disease. Neuroreport 5:2529–2533
Su JH, Deng G, Cotman CW (1997) Bax protein expression is increased in Alzheimer's brain: Correlations with DNA damage, Bcl-2 expression and brain pathology. J Neuropath Exp Neurol 56:86–93
Su JH, Deng G, Cotman CW (1997) Neuronal DNA damage precedes tangle formation and is associated with up-regulation of nitrotyrosine in Alzheimer's disease brain. Brain Res 774:193–199
Su J, Satou T, Anderson AJ, Cotman CW (1996) Up-regulation of Bcl-2 is associated with neuronal DNA damage in Alzheimer's disease. Neuroreport 7:437–440
Susin S, Zamzami N, Castedo M, Hirsch T, Marchetti P, Macho A, Daugas E, Geukens M, Kroemer G (1996) Bcl-2 inhibits the mitochondrial release of an apoptogenic protease. J Exp Neurol 184:1331–1342
Tanabe H, Eguchi Y, Shimizu S, Martinou J-C, Tsujimoto Y (1998) Death-signalling cascade in mouse cerebellar granule neurons. Eur J Neurosci 10:1403–1411
Tatton NA, Kish SJ (1997) In situ detection of apoptotic nuclei in the substantia nigra compacta of 1-methyl-4-phenyl-1,2,3,6-tetrahydropyridine-treated mice using terminal deoxynucleotidyl transferase labelling and acridine orange staining. Neuroscience 77:1037–1048
Teter B, Yang F, Wasterlain C, Smulson M, Frautschy SA, Cole GM (1996) Cleavage site-specific cell death switch antibody labels Alzheimer's lesions. Society for Neuroscience [Abstr] 22:258
Thomas LB, Gates DJ, Richfield EK, O'Brien TF, Schweitzer JB, Steindler DA (1995) DNA end labelling (TUNEL) in Huntington's disease and other neuropathological conditions. Exp Neurol 133:265–272
Tomei D, Shapiro J, Cope F (1993) Apoptosis in C3H/101/2 mouse embryonic cells: evidence for internucleosomal DNA modification in the absence of double-strand cleavage. Proc Natl Acad Sci USA 90:853–857
Tominaga T, Kure S, Narisawa K, Yoshimoto T (1993) Endonuclease activation following focal ischemic injury in the brain. Brain Res 608:21–26
Tompkins MM, Basgall EJ, Zamrini E, Hill WD (1997) Apoptotic-like changes in lewybody-associated disorders and normal aging in substantia nigral neurons. Am J Pathol 150:119–131
Trimmer PA, Smith TS, Jung AB, Bennett JP (1996) Dopamine neurons from transgenic mice with a knockout of the p53 gene resist MPTP neurotoxicity. Neurodegeneration 5(3):233–239

Troy CM, Stefanis L, Greene LA, Shelanski ML (1997) Nedd2 is required for apoptosis after trophic factor withdrawal but not superoxide dismutase(SOD1) down-regulation in sympathetic neurons and PC12 cells. J Neurosci 17:1911–1918

Troy CM, Stefanis L, Prochiantz A, Greene LA, Shelanski ML (1996) The contrasting roles of ICE family proteases and interleukin-1b in apoptosis induced by trophic factor withdrawal and copper/zinc dismutase down-regulation. Proc Natl Acad Sci USA 93:5635–5640

Vaux DL, Weissman IL, Kim SK (1992) Prevention of programmed cell death in Caenorhabditis elegans by human bcl-2. Science 258:1955–1957

Vaux D, Cory S, Adams J (1988) Bcl-2 gene promotes haematopoietic cell survival and cooperates with c-myc to immortalise pre-B cells. Nature 335:440–442

Veis DJ, Sorenson CM, Shutter JR, Korsmeyer SJ (1993) Bcl-2-deficient mice demonstrate fulminant lymphoid apoptosis, polycystic kidneys, and hypopigmented hair. Cell 75:229–240

Venero JL, Revuelta M, Cano J, Machado A (1997) Timecourse changes in the dopaminergic nigrostriatal system following transection of the medial forebrain bundle – detection of oxidatively modified proteins in substantia nigra. J Neurochem 68:2458–2468

Vincent I, Jicha G, Rosado M, Dickson DW (1997) Abherrant expression of mitotic cdc2/cyclin B1 kinase in degenerating neurons of Alzheimer's disease brain. J Neurosci 17:3588–3598

Walkinshaw G, Waters CM (1994) Neurotoxin-induced cell death in neuronal PC12 cells is mediated by induction of apoptosis. Neuroscience 63:975–987

Walton M, Sirimanne E, Reutelingsperger C, Williams C, Gluckman P, Dragunow M (1997) Annexin V labels apoptotic neurons following hypoxia-ischemia. Neuroreport 8:3871–3875

Walton M, Young D, Sirimanne E, Dodd J, Christie D, Williams C, Gluckman P, Dragunow M (1996) Induction of clusterin in the immature brain following a hypoxic-ischemic injury. Mol Brain Res 39:137–152.

Watson A, Eilers A, Lallemand D, Kyriakis J, Rubin LL, Ham J (1998) Phosphorylation of c-Jun is necessary for apoptosis induced by survival signal withdrawal in cerebellar granule neurons. J Neurosci 18:751–762

Webb SJ, Harrison DJ, Wyllie AH (1997) Apoptosis: An overview of the process and its relevance in disease. Adv Pharmacol 41:1–34

Weil M, Jacobson MD, Coles HSR, Davies TJ, Gardner RL, Raff KD, Raff MC (1996) Constitutive expression of the machinery for programmed cell death. J Cell Biol 133:1053–1059

Wessel TC, Joh TH, Volpe BT (1991) In situ hybridization analysis of c-fos and c-jun expression in the rat brain following transient forebrain ischemia. Brain Res 567:231–240

Whittemore ER, Loo DT, Cotman CW (1994) Exposure to hydrogen peroxide induces cell death via apoptosis in cultured rat cortical neurons. Neuroreport 5:1485–1488

Wilkinson DG, Bhatt S, Ryseck RP, Bravo R (1989) Tissue-specific expression of c-jun and junB during organogenesis in the mouse. Development 106:465–471

Wood KA, Youle RJ (1995) The role of free radicals and p53 in neuron apoptosis in vivo. J Neurosci 15:5851–5857

Wyllie AH (1980) Glucocorticoid-induced thymocyte apoptosis is associated with endogenous endonuclease activation. Nature 284:555–556

Wyllie AH, Kerr JFR, Currie AR (1980) Cell death: the significance of apoptosis. Int Rev Cytol 68:251–306

Xia Z, Dickens M, Raingeaud J, Davus RJ, Greenberg ME (1995) Opposing effects of ERK and JNK-p38 MAP kinases on apoptosis. Science 270:1326–1331

Xiang H, Hochman DW, Saya H, Fujiwara T, Schwartzkroin PA, Morrison RS (1996) Evidence for p53-mediated modulation of neuronal viability. J Neurosci 16:6753–6765

Xiang H, Kinoshita Y, Knudson CM, Korsmeyer SJ, Schwartzkroin PA, Morrison RS (1998) Bax involvement in p53-mediated neuronal cell death. J Neurosci 18(4):1363–1373

Xiang J, Chao DT, Korsmeyer SJ (1996) Bax-induced cell death may not require interleukin-1b converting enzyme-like proteases. Proc Natl Acad Sci USA 93:14559–14563

Xu Y, Bradham C, Brenner DA, Czaja MJ (1997) Hydrogen peroxide-induced liver cell necrosis is dependent on AP-1 activation. Am J Physiol 27 (Gastrointestinal Liver Physiology 36), G795–G803

Yakovlev AG, Knoblach SM, Fan L, Fox GB, Goodnight R, Faden AL (1997) Activation of CPP32-like caspases contributes to neuronal apoptosis and neurological dysfunction after traumatic brain injury. J Neurosci 17:7415–7424

Yang DD, Kuan C-Y, Whitmarsh AJ, Rincon M, Zheng TS, Davis RJ, Rakic P, Flavell RA (1997) Absence of excitotoxicity-induced apoptosis in the hippocampus mice lacking the Jnk3 gene. Nature 389:865–870

Yin X-M, Oltvai ZN, Korsmeyer SJ (1994) BH1 and BH2 domains of Bcl-2 are required for inhibition of apoptosis and heterodimerisation with Bax. Nature 369:321–318

Yoshiyama Y, Yamada T, Asanuma K, Asahi T (1994) Apoptosis related antigen, Le^Y and nick-end labeling are positive in spinal motor neurons in amyotrophic lateral sclerosis. Acta Neuropathol 88:207–211

Yuan J, Horvitz HR (1990) The Caenorhabditis elegans genes ced-3 and ced-4 act cell-autonomously to cause programmed cell death. Devel Biol 138:33–41

Yuan J, Shaham S, Ledoux S, Ellis HM, Horvitz HR (1993) The C. elegans cell death gene ced-3 encodes a protein similar to a mammalian interleukin-1 beta-converting enzyme. Cell 75:641–652

Zeevalk GD, Derr-Yellin E, Nicklas WJ (1995) NMDA receptor involvement in toxicity to dopamine neurons in vitro caused by the succinate dehydrogenase inhibitor 3-nitropropionic acid. J Neurochem 64:455–458

Zha JP, Harada H, Yang E, Jockel J, Korsmeyer SJ (1996) Serine phosphorylation of death agonist BAD in response to survival factor results in binding to 14-3-3 not BGL-X(L) Cell 87:619–628

Zhang Y, Zhou L, Miller CA (1998) A splicing variant of a death domain protein that is regulated by a mitogen-activated kinase is a substrate for c-Jun N-terminal kinase in the human central nervous system. Proc Natl Acad Sci USA 95:2586–2591

Zhong LT, Sarafian T, Kane DJ, Charles AC, Mah SP, Edwards RH, Bredesen DE (1993) Bcl-2 inhibits death of central neural cells induced by multiple agents. Proc Natl Acad Sci USA 90:4533–4537

CHAPTER 9
Use of p53 as Cancer Cell Target for Gene Therapy

C. THIEDE, T.D. KIM, and A. NEUBAUER

A. Introduction

Cancer is the second most frequent cause of death in developed countries, with rising prevalence. There are, basically, three different types of cancer therapies, and these have not changed over the last 40 years: surgery, irradiation, and chemotherapy. While surgery can frequently cure cancer in the early stages, no treatment provides a certain cure for most advanced human cancers, except for rare forms such as testicular cancer or lymphomas. Most patients suffering from cancer will thus die from tumor progression. There was hope that a thorough knowledge of the genetics of human cancer would eventually lead to better cure rates. However, despite considerable success in the understanding of the mechanisms leading to human cancer, therapeutic interventions based on this knowledge for specific approaches are limited to selected cancer types only. One example clearly is all-*trans*-retinoic acid therapy of acute promyelocytic leukemias carrying the translocation t(15;17) (RAELSON et al. 1996). Another paradigm is cure of *H. pylori* infection in early gastric lymphomas of the mucosa associated lymphoid tissue (WOTHERSPOON et al. 1993; BAYERDORFFER et al. 1995). However, in most tumors, chemotherapy is still a result of empiric data, which are based on large clinical studies, rather than in vitro test results.

Cancer is caused by genetic alterations affecting oncogenes and tumor suppressor genes. In adults, most of these genetic aberrations are not a result of inherited germ line mutations, but rather acquired during the life span of single cells and thus represent somatic mutations. During tumor progression, most human cancers additionally acquire defects in the cellular response towards chemotherapeutic substances, i.e., detection of DNA damage and apoptosis. Since chemotherapeutic agents induce cell death via induction of apoptosis, and regular induction of apoptosis is inhibited in cancer cells, understanding the biology of apoptosis may eventually lead to better chemotherapeutic protocols for treatment of human cancer. Thus as a first step, one may try to correct the defects in apoptosis acquired in cancer cells during tumor progression via genomic instability. Restoring components of the normal response pathway in a cancer cell would represent an attractive goal in treating cancer by correction of genetic defects.

One gene of particular interest is the tumor suppressor gene p53, which acts in several ways to protect normal cells from genotoxic hits. The following will focus on the role of gene therapy using the tumor suppressor gene p53 as a promising first step in this context.

B. Genetic Changes in Tumor Development

The last twenty years have seen important progress in the understanding of the cellular mechanisms important for tumor formation. According to these data, the process of tumor development in adults is a stepwise process characterized by the accumulation of multiple changes in oncogenes as well as tumor suppressor genes. KINZLER and VOGELSTEIN (1997) recently suggested that the genes important for oncogenesis may be divided into *caretakers* and *gatekeepers*. Whereas *caretakers* are mainly responsible for cellular DNA repair after respective injury, *gatekeeper* proteins may be much more important for tumorigenesis. Mutations in *caretaker* genes result in the accelerated accumulation in other, presumably critical genes, eventually leading to genetic instability. Examples of this group are the are the mismatch repair genes (hMLH1, hMSH2, hPMS1, hPMS2). Germ line mutations of these genes are found in patients suffering from the hereditary non-polyposis colorectal cancer syndrome (HNPCC) (PAPADOPOULOS and LINDBLOM 1997), which accounts for up to 10% of all colorectal cancer cases. In contrast, inactivation of *gatekeeper* proteins, i.e., genes that control keys steps in the growth control of a cell, is the first and most important step in the initiation of aberrant growth. An example for a typical *gatekeeper* protein is the adenomatosous polyposis coli (APC) gene on chromosome 5q. Alterations of this gene seem to be the first and most important lesion in colorectal cancer. Another important example is the p53 tumor suppressor gene. p53 plays important roles in different cellular pathways, all dealing with cellular reaction towards stress. Due to its decisive role in this context, this protein was termed guardian of the genome (LANE 1992) and became the "molecule of the year" in 1993 (KOSHLAND 1993).

C. The p53 Tumor Suppressor Gene

I. p53: From Structure to Function

The tumor suppressor gene p53 codes for a 53kD nuclear phosphoprotein. The genomic sequence is organized in 11 exons and 10 introns and covers 20kb on the short arm of chromosome 17 band 13. The 393 amino acids of the protein can be subdivided into three major functional domains: (I) the amino-terminal transcriptional activation domain (AA 20–42); (II) the central region (AA 102–292), which is essential for sequence specific DNA binding and carries more than 90% of all missense point mutations; and (III) a multi-functional c-terminal domain (AA 300–393), carrying motifs important for homodimerization and nuclear transport (Ko and PRIVES 1996). In principle, p53 can induce two major pathways: (i) cell cycle arrest; and (ii) induction of apoptosis.

In addition to the SV40 large T antigen, which was found to bind to p53 at the first description 20 years ago (LANE and CRAWFORD 1979; LINZER and LEVINE 1979) numerous cellular and viral proteins have been shown to interact with specific parts of the p53-protein. The MDM-2 protein, the antagonist of p53 on the cellular level, binds to the NH_2-terminal part of TP53 and induces rapid degradation of the protein (HAUPT et al. 1997; KUBBUTAT et al. 1997; NIELSEN and MANEVAL 1998). This part of the p53 protein is important for the binding of several transcription factors as well as p300 and CREB binding protein (CBP), two proteins which have been shown to enhance p53 mediated transcription (SCOLNICK et al. 1997). Other viral proteins include the oncogenic E6 and E7 proteins of the Human Papilloma virus (HPV) strains 16 and 18, which also bind the amino terminal part of p53 and induce degradation. In addition, several proteins of the adenoviral early region (E1B 55 kD, E4orf6) have been shown to interact with p53, thereby inactivating the protein and allowing viral replication (YEW and BERK 1992; STEEGENGA et al. 1998) (for details see below).

As illustrated in Fig. 1, a number of stimuli can activate TP53 directly or indirectly. DNA damage, hypoxia, activation of cellular oncogenes, infection with several oncogenic viruses (HPV, HBV), oxidative stress, and depletion of cellular ribonucleotide pools have been shown to induce nuclear accumulation of p53 in normal cells (GIACCIA and KASTAN 1998). DNA damage, induced either by ionizing radiation, UV-light, or certain chemotherapeutic agents, is one of the best characterized mechanisms for p53 induction. Upon DNA damage, p53 accumulates rapidly through a posttranscriptional mechanism. p53 monomers homotetramerize to form the functional complex. This tetrameric complex is then translocated to the nucleus. Here the p53 binds to its recognition sequence and induces the transcription of downstream target genes which can then arrest the cell cycle at the critical transition step between the G1 and the S-phase. Table 1 gives a list of genes which are transcriptionally activated by wild-type (wt) p53. This list clearly illustrates that, besides mdm-2, which is part of a feedback loop to control its own function, most of the genes listed are either important in the process of cell cycle control, response to DNA damage, or are important in the process of apoptosis.

One of the most important effector proteins for cell cycle arrest induced by p53 is the $p21^{WAF1/CIP1}$ gene. This gene belongs to the group of cyclin dependent kinase inhibitor (CDK-I) proteins. Besides p21, this family of CDK inhibitors includes $p15^{INK4B/MTS2}$, $p16^{INK4A/MTS1}$, $p27^{KIP1}$, and $p57^{KIP2}$ (KING and CIDLOWSKI 1998). Inactivation of these important regulators of the cell cycle is a common mechanism in cancer development. Upon induction, p21 binds to one of the cyclin/CDK complexes, which can no longer phosphorylate one of the members of the retinoblastoma (Rb) protein family of tumor suppres-

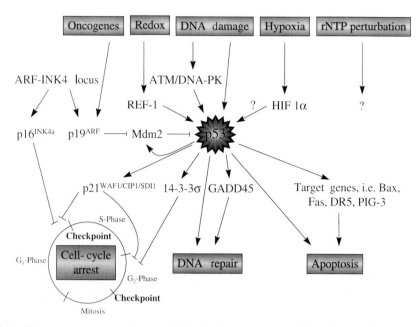

Fig. 1. Upstream events and major biological outcomes of p53 activation. For details see text

Table 1. Genes transcriptionally induced by wt-p53

Gene target	Function	Reference
mdm-2	Regulation of p53 function	BARAK et al. (1993)
p21 (WAF1, CIP1)	Cell cycle control	EL-DEIRY et al. (1994); XIONG et al. (1993)
GADD45	DNA-repair	KASTAN et al. (1992)
Bax	Apoptosis	MIYASHITA and REED (1995)
IGF-BP3	Apoptosis	BUCKBINDER et al. (1995)
KILLER/DR5	Apoptosis	WU et al. (1997)
FAS/Apo-1	Apoptosis	OWEN-SCHAUB et al. (1995)
PIG-3	Apoptosis	POLYAK et al. (1997)

sors (KING and CIDLOWSKI 1998). This phosphorylation is necessary to hinder Rb from the binding of the E2F-transcription factor, which would normally induce the transcription of genes which are necessary to enter the S-phase. Loss of $p21^{CIP1/WAF1}$ function results in loss of p53 mediated G1-arrest. Furthermore, it has recently been shown that p21 is also important to sustain the G2-arrest (BUNZ et al. 1998).

Recent data indicate that phosphorylation of serine residues (S15 and S37) in the c-terminal part of the protein is a critical event for activation of p53. This modification alleviates binding by mdm2 (SHIEH et al. 1997), thus rescuing p53 from degradation. Both the DNA-dependent protein kinase (WOO et al. 1998) and the Ataxia Teleangiectasia Mutated (ATM) (SILICIANO et al. 1997; CANMAN et al. 1998) protein are responsible for this posttranslational modification. Only very recently has it been shown that another key molecule controlling this step is the p19^{ARF} protein (POMERANTZ et al. 1998).

II. p53 and Induction of Apoptosis

Cancer, in many ways, is the result of a lack of equilibrium between cellular proliferation and senescence, or apoptosis. In recent years it has been recognized that lesions in genes playing crucial roles in the apoptotic process are frequently observed in cancer cells (EVAN and LITTLEWOOD 1998).

Apoptosis or programmed cell death is one of the key mechanisms in maintaining the cellular homeostasis in multicellular organisms. It is an energy dependent process, which is characterized by several specific morphological and molecular changes, i.e., chromatin condensation, blebbing of the cell membrane, vesicularization of the cell contents, and finally internucleosomal fragmentation of the DNA. The process of apoptosis is important in embryonal development, the maturation of the immune system, defense against viral infections, and the prevention of tumor formation. For instance, apoptosis plays a crucial role in the normal development of the lymphatic tissue; lymphoid cells not fitting certain requirements (specific signals induced by antigens, bystander signals) must undergo death in order to protect the organism from cells which may be potentially harmful, i.e., autoreactive.

Apoptosis requires the stepwise activation of proteins. Considerable progress has been made in the identification of proteins that regulate the apoptotic pathway at each level. Important proteins can be assigned to two major groups. The Bcl-2 family of proteins and the proteins belonging to the TNF-receptor family. The Bcl-2 family proteins can be subdivided into *pro-* (i.e., Bax, Bcl-XS, Bad) and *anti-*apoptotic (Bcl-2, Bcl-X$_L$, Mcl-1) molecules; for recent review see CHAO and KORSMEYER (1998). The decision as to whether a cell will survive or enter the process of apoptosis is based on the quantity of either pro- and anti-apoptotic members of this family, which form homo or heterodimers via conserved domains. Bax appears to be a key player in the p53 mediated response towards chemotherapy, which will be discussed below.

Another key molecule in this context is the Fas-receptor (APO-1, CD95) and its ligand (FasL). Fas belongs to the TNF-receptor protein family. Upon engagement of FAS by binding of the FAS-ligand molecule, Fas induces the downstream apoptotic program through activation of several proteins via binding to an 80 amino acid part of the protein, called the death domain, including FADD and Flice1/caspase-8 (ZHANG et al. 1998). This binding activates a branching cascade of other caspase proteins, i.e., caspases 3, 6, 7, finally inducing the characteristic changes seen in apoptosis; for recent review see KIDD (1998).

As stated above, apoptosis is the key mechanism in chemotherapy. Many chemotherapeutic substances induce direct changes of the DNA, thus inducing gatekeepers like p53, which subsequently start the cellular program for apoptosis. The p53 tumor suppressor gene is involved in induction of apoptosis in many ways. Wild type p53 induces the expression of the FAS/APO 1 protein (Owen-Schaub et al. 1995; Sheard et al. 1997) thereby sensitizing cells for autocrine or paracrine interaction with the FasL molecule and subsequent elimination. In addition, p53 upregulates the expression of the Bax protein, which in turn forms heterodimers with and blocks the antiapoptotic Bcl-2 protein, thereby inducing apoptosis (Chao and Korsmeyer 1998). Bax-deficient cells show a reduction of about 50% in the rate of p53 induced apoptosis (McCurrach et al. 1997). Thus Bax is considered to be a tumor suppressor gene (Yin et al. 1997). In support of this, inactivating mutations of the Bax gene have been demonstrated in several human tumors (hematological malignancies) (Meijerink et al. 1998), colon (Rampino et al. 1997), and may explain the frequently observed resistance towards chemotherapy in these tumors. In addition, p53 upregulates the expression of the Bax protein (Miyashita and Reed 1995). As another pathway, myc induced apoptosis seems to be dependent on p53 function and appears to be at least partly mediated via the p19^{ARF}-interface (Hermeking and Eick 1994), although there are reports indicating that myc can also induce apoptosis via p53 independent mechanisms (Sakamuro et al. 1995).

Besides this direct interference with apoptosis, wt-p53 also downregulates the expression of the multi drug resistance-1 (mdr) gene (Chin et al. 1992; de-Kant et al. 1996). The mdr-1 gene codes for the multiple drug transporter glycoportein (P-GP). Overexpression of this protein prevents the intracellular accumulation of many chemotherapeutic drugs and is responsible for the multi drug resistance phenotype observed in many cancers.

III. Alterations of p53 in Human Cancers

Alterations of the p53 gene are the most frequent genetic change in human cancer (Hussain and Harris 1998). It is assumed, that about 50% of all human malignancies contain inactivating mutations of the p53 gene (Hollstein et al. 1991). p53 mutations are common genetic alterations in tumors of the lung, the colorectum, the pancreas, the breast the stomach, and the prostate. Inactivation most frequently occurs via missense point mutations in the central part (AA 102–292) of the protein. These mutations frequently involve amino residues in evolutionary highly conserved domains, which are important for DNA-binding, e.g., AA 175, 248, and 273. Loss of the remaining wt-p53 allele (loss of heterozygosity, LOH) seems to occur subsequently. In most cancers, inactivation of p53 is a late event in the multistep process of tumor development and, in the majority of the studies published to date, the presence of p53 mutations has been associated with a significantly worse prognosis (Kirsch and Kastan 1998). However, the importance of p53 mutations in the patients' outcome has to be seen in the tissue context. Conflicting results on the role of p53 mutations, for example in breast and colorectal cancer, may be sometimes caused by methodological problems. However, there are also reports indicating that the presence of a p53 inactivation can be associated with increased sensitivity towards chemotherapy in certain malignancies (RUSSELL et al. 1995; HAWKINS et al. 1996; TADA et al. 1998), indicating that the complex role of p53 in this context is still not fully understood.

IV. p53 Homologues

The p53 tumor suppressor gene was thought for a long time to be a unique protein. This, however, would have been quite unusual, the p53 pathway being central for cell survival. And, indeed, recent data indicate that p53 has several homologues capable of substituting its function, at least partially. P73 is located on chromosome 1p36, a region frequently deleted in neuroblastoma and other tumors (KAGHAD et al. 1997). This protein was shown to be able to induce the expression of p53 induced genes, e.g., $p21^{CIP1}$ and apoptosis (JOST et al. 1997), and therefore closely resembles the function of p53. The group of p53 homologous genes was recently extended to P40 (TRINK et al. 1998), P51 A and B (OSADA et al. 1998), and P63 (YANG et al. 1998). Future research must clarify, whether there is any tissue restriction or functional specificity, which may distinguish these proteins from p53.

D. p53 and Gene Therapy

I. Introduction

As stated above, tumor development, especially solid tumors, is a process of accumulation of genetic defects. These changes also frequently involve genes which are important in the cellular processes of detection of DNA damage. Since many drugs currently used in cancer chemotherapy exert their beneficial effect via DNA damage, either directly or via interfering with the cellular nucleotide pool, cancer chemotherapy essentially relies on the presence of functional systems for the detection of DNA damage. Abrogation of these key control steps in the malignant cells may thus explain resistance to chemotherapy which is frequently observed, especially in late stages of disease.

Given these considerations, p53 is a very attractive target for gene replacement therapy. First, this protein is inactivated in about 50% of all human malignancies, making a potential therapy applicable for a wide variety of human cancer. Second, the essential function of TP53 in controlling such important tumor suppressive mechanisms like DNA repair, cell cycle control, and apoptosis qualifies p53 and all downstream effectors (e.g., $p21^{Waf1}$, BAX) as ideal targets for intervention in tumor cells. Transfection of wt-p53 into tumor cell lines has shown to reverse many of the changes, thus making the cell responsive to anticancer therapy.

II. Gene Therapy: General Remarks

The idea to restore a genetic defect has inspired genetic analysis for a long time. Gene transfer in general can be defined as the introduction of a DNA fragment (encoding either a foreign gene from another species or an endogenous gene, which is either defective or missing completely) into a cell. The expression of this gene would then either complement a missing normal protein or serve to facilitate treatment with selective drugs, e.g., use of herpes simplex thymidin kinase (TK) for treatment with gancyclovir. Gene therapy of genetic defects caused by a single genetic alteration (e.g., hemophilia or cystic fibrosis) can be achieved by expressing the missing gene. However, in diseases involving multiple genetic changes, like cancer, the achievement of similar effects may be more challenging.

III. Rationale for p53-Targeting in Gene Therapy

In principle, several approaches may be used for p53 based gene therapy. The most straightforward technique is the transfection of wt-p53 into tumor cells in order to restore the function of the wild type protein. The majority of currently performed trials are focusing on this issue. Similarly, the function of p53 can be restored using vectors expressing downstream effector proteins like $p21^{WAF1, CIP1}$ (MENG et al. 1998). Expression of effector proteins may overcome a primary resistance of the tumor cells to cancer therapy, since the ultimate goal of p53-gene therapy must be the restoration of normal apoptotic response towards DNA-damage. Thus high expression of molecules necessary for this response is intended. A problem that may arise when p53 is restored in cancer cells, is the dominant negative effect of mutant p53 (MILNER and MEDCALF 1991), which potentially may block the effects of wt-p53 expressed in the tumor cells. However, in vitro data indicate that the expression of wt-p53 in the tumor cells after transfection can overcome this dominant negative effect (GJERSET et al. 1995). Whether the level of expression plays a causal role is not known. An elegant approach to overcome this potential problem and to achieve a really selective targeting of cells carrying mutant p53 was recently published (BISCHOFF et al. 1996). This group used a human adenovirus lacking a functional E1B 55K-protein (see below).

IV. Trials Reconstituting Wild-Type p53

Although several potential targets have been identified in the p53 pathway, the most attractive approach for p53 based gene therapy is to reconstitute wt-p53 expression. Major problems in this context are to target specifically the tumor tissue, to transduce successfully the tumor cells with the p53 gene, and to achieve sufficient and lasting expression. All these factors are influenced by the type of delivery system used. These can be divided into viral and non-viral systems. Most of the currently performed clinical trials aiming to reconstitute p53 expression in the tumor cells use viral approaches.

1. Adenoviruses

Recombinant adenoviral constructs are the most frequently used method for this purpose. For use in gene therapy, the viruses are rendered replication defective by deletion of the E1 A and B regions. The use of this DNA-virus has several advantages for cancer gene therapy; for a recent review see VERMA and SOMIA (1997). Adenoviruses can infect dividing and non-dividing cells and are usually not integrating into the host genome, thus avoiding the risk of insertional mutagenesis. In addition, adenoviruses can infect a wide variety of tissues. One of the most important advantages of adenoviruses compared to retroviruses is the possibility of producing high concentrations of infectious particles, which is important to achieve sufficient concentration at the tissue of interest. A major drawback in the use of adenoviral vectors is the fact that about 80% of humans react against adenoviral proteins. Although the profound immune reaction induced upon infection, consisting of specific humoral response and cytotoxic T-cells, may increase the primary antitumor effect in the patient, it may also limit the possibility of repeated use of this therapy. Efforts are currently being made to define further those parts of the viral genome which are important for eliciting the host reactivity, in order to design recombinant vectors which are less immunogenic.

2. Retroviruses

Retroviral vectors are the second transfection method frequently applied for p53 gene therapy. These vectors are used in many gene therapy protocols for non-malignant diseases currently performed; however they are less frequently taken for cancer gene therapy (ANDERSON 1998). Retrovirus vectors can infect only dividing cells, which has the advantage of conferring some degree of selectivity, since predominantly cancer cells are infected. Concerns are focused on the problem of insertional mutagenesis which may be problematic in genetically highly unstable cancer cells. Furthermore, expression of the transgene is frequently limited to only a few days, because cellular factors inhibit transcription of the inserted viral gene, presumably by DNA-hypermethylation.

3. Non-Viral Gene Delivery Systems

Non-viral methods for the transmission of genes have several potential advantages, most important being the ease of manufacturing and the safety issues, which will make them the preferred delivery system in the future (ANDERSON 1998). However, up to now they have been rarely used for gene therapy targeting p53 due to the limitations in transfection efficacy and expression. LESOON et al. (1995) described the use of a liposome-p53 complex for systemic therapy of nude mice challenged with breast carcinoma cells and reported significant reduction of the tumor size. Subsequent analysis revealed that transduction efficacy in this setting is only 5%; however, it appears that a bystander mechanism which significantly reduces tumor vascularization may be involved (Xu et al. 1997). On the basis of recently published results, trials are attempting to use this observation and to increase inhibition of tumor vascularization by cotransfection with a fragment of thrombospondin I (Xu et al. 1998), showing synergistic effects in tumor growth inhibition.

4. In Vitro Data and Preclinical Trials

The applicability of these kind of constructs has been shown in several in vitro and animal models (BADIE et al. 1995; CIRIELLI et al. 1995; HARRIS et al. 1996; MUJOO et al. 1996; WILLS et al. 1994; YANG et al. 1995; FUJIWARA et al. 1994b; CARBONE and MINNA 1994); for an excellent recent review on this issue the reader is referred to NIELSEN and MANEVAL (1998). Transfection of wt-p53 into cell lines carrying p53 mutants induces growth arrest and apoptosis in a dose dependent manner in the majority of cell lines tested, including several epithelial malignancies (cervical-, head and neck-, bladder-, and skin cancer), adenocarcinomas (breast-, prostate-, and colorectal cancer), and lymphomas and leukemias. In contrast, no growth inhibitory effect was observed in the majority of normal tissues transfected with wt-p53 in hematopoietic stem cells (SETH et al. 1997; SCARDIGLI et al. 1997), fibroblasts (CLAYMAN et al. 1995), and mammary epithelium (KATAYOSE et al. 1995).

In mouse xenograft models, a marked reduction of tumor growth and the induction of apoptosis was observed compared to mock transfected animals (CLAYMAN et al. 1995; EASTHAM et al. 1995; FUJIWARA et al. 1994a; NIELSEN et al. 1997; ZHANG et al. 1994).

Recently, the efficiency of p53 gene transfer in combination with conventional chemotherapy was tested. NGUYEN et al. (1996) were able to demonstrate, that combination therapy with cisplatin could enhance the antitumor effect of an Ad-p53 construct. According to their data, the time of administration of cisplatin was essential for this effect, with no enhancement when given in parallel or after administration of Ad-p53, but with a pronounced effect when used 2–3 days prior to gene transfer. These results confirm published in vitro data on enhanced chemosensitivity when cells are preincubated with cisplatin prior to p53 transfer (FUJIWARA et al. 1994a). This approach might be useful to treat cancers that are frequently resistant towards chemotherapy, as recently demonstrated in the mouse model (OGAWA et al. 1997).

Another currently followed direction is the combination of genes for anticancer gene therapy. SANDIG et al. (1997) recently published results on the combination of the cell cycle inhibitory protein $p16^{INK4/CDKN2}$ and wt-p53 in an adenoviral vector. Using this construct, they were able to demonstrate enhanced apoptosis and reduced tumor growth in a nude mice model. Thus, combining several genes in one vector might be an attractive tool to increase the effect of the therapy.

5. Clinical Trials

Based on these data, several clinical protocols using different settings to target tumor cells have been activated. Table 2 shows a summary of currently performed trials, which have been presented at the 1998 Meeting of the American Society of Clinical Oncology (ASCO).

ROTH et al. (1996) were the first group reporting in vivo results of a phase I study using a retroviral wt-p53 construct. To achieve a wide expression of p53 in different tissues, this group used a β-actin promoter to control the expression of the wt-p53 cDNA. Nine patients suffering from advanced non-small cell lung cancer (NSCLC) were treated in a phase I study by either bronchoscopic or percutaneous injections of the vector. The data indicate that this setting is feasible. No side effects attributable to the vector were seen, however, due to the way of administration complications were observed in the patients (pneumothorax, intubation, and mechanical ventilation). In this group of patients with progressed stages of disease, one patient had objective clinical evidence of tumor regression, whereas another three patients showed stable disease and one patient progressed under therapy. Although six of seven tumors showed evidence of increased apoptosis after therapy, arguing for p53 mediated effects, no p53 expression was detected by reverse transcription polymerase chain reaction (RT-PCR), and the levels of transfection as documented by in situ hybridization were quite low (1%–3%). Thus, this group also switched to adenoviral vectors for phase II studies to achieve higher transfection rates of tumor tissues.

CLAYMAN et al. (1998) published data on 33 patients with advanced recurrent head and neck squamous cell carcinoma (HNSCC) who were treated by local injection with a recombinant Ad-p53 vector. The construct consisted of a replication defective adenovirus type 5 with a wt-p53 expression cassette replacing the E1 region. Delivery and in vivo expression of the wt-p53 construct was documented in the tumor. No adverse side effects were observed and in several patients a clinical response was seen in these end stage tumors, with one patient showing a complete remission. Data on a phase 1 trial in patients with advanced NSCLC were also published by SCHULER et al. (1998). This group treated 15 patients with a commercial preparation of a recombinant adenoviral vector containing wt-p53 under the control of a CMV promoter (rAd/p53, SCH 58500, Schering-Plough, Kenilworth NJ) by either bronchoscopic intratumoral injection or CT guided percutaneous intratumoral injection. Increasing doses of vector were administered (10^7–10^{10} plaque forming units (PFU)). Expression of vector specific p53 cDNA was detected in six out of eight patients treated with doses of 10^9 and 10^{10} PFU, although only in one out of six patients below this dose level, illustrating a clear dose relationship between the efficacy of the gene transfer and the concentration

Table 2. Currently performed trials using p53 in different malignancies based on the ASCO-meeting 1998

Institute/Company	Vector[a]	Tumor-Type[b]	Administration[b]	patients (N)[b]	Clinical Phase[b]	Citation[c]
Schering-Plough	(1) Repl. deficient adenovirus (2) wt-p53 (3) CMV-Promoter (4) SCH 58500	Head and neck cancer Colorectal cancer/ hepatic metastases	Local Locoregional	16 16	Phase I Phase I	1479 1661
ONYX Pharmaceuticals	(1) Mutant Adenovirus group C (del 55KD E1B) (4) ONYX-015	Pancreatic cancer Head and neck cancer Hep. metastasis of GI cancer	Local Local Local	16 21 16	Phase I Phase II Phase I	815 1509 814
Introgen-Therapeutics/ Rhône Poulonc Rorer	(1) repl. deficient Adenovirus (2) wt-p53 (4) INGN 201	NSCLC*	local	52	Phase I/II	1660/1659

[a] (1), virus; (2), insert; (3), promoter; (4), company name of product.
[b] Details of currently performed clinical trials using the vectors described in Vector column.
[c] Numbers refer to respective abstract numbers as given in the Program and Proceedings of the 1998 Annual Meeting of the American Society of Clinical Oncology.
*Non small cell lung cancer.

of vector used. Side effects with mild flu-like symptoms, fever, dyspnea, hypertension, and tachycardia were also more pronounced in the group of patients treated with these high titers. Due to the phase I design of this study, a clinical response was documented only in one patient showing a stable disease for 6 months.

Currently, this vector is being evaluated in phase I and II studies in patients with non-small cell lung cancer (NSCLC), bladder cancer, ovarian cancer, and pancreatic cancer. Introgen therapeutics/Rhône Poulonc Rorer also recently announced the onset of several trials based on wt-p53 transfer mediated by a recombination deficient adenovirus (Table 2).

V. The ONYX-015 Virus

As discussed above, a general problem of any somatic gene therapeutic approach is to reach as many cancer cells as possible. Another major concern is that, by introducing a tumor suppressor gene, the transfected cancer cells will have a growth disadvantage compared to their untransfected counterparts. Since cancer development is generally considered to be a process of microevolution, when selecting cancer cells for their ability to overcome growth control those cells without transfection will outgrow the transfected cells and thus will limit the efficacy of the whole therapy. McCormick and his group used an approach, which turns the advantage of the tumor cell, having a loss of p53-function, into a disadvantage in the process of cellular microevolution.

The theoretical background of this setting is schematically depicted in Fig. 2. Adenovirus infection of a normal cell induces p53, which in turn inhibits replication of viral genes. The adenovirus E1B 55K binds to and inactivates wt-p53, thereby allowing viral replication. BISCHOFF et al. (1996) constructed a mutant group C adenovirus lacking a functional E1B 55K protein. As a consequence, this virus (ONYX-015) should not be able to replicate in cells carrying wt-p53. In contrast, tumor cells carrying mutant p53-protein would not be able to inhibit viral replication and would subsequently undergo viral lysis. A panel of tumor cell lines was tested for sensitivity to cytopathic effects induced by infection with ONYX-015 as compared to transfection of the wt-adenovirus as positive control. The ONYX-015 virus induced cytolysis comparable to the wt-vector in all p53-deficient cell lines, including tumor cells of the brain, breast, cervix, colon, larynx, liver, lung, ovary, and pancreas. As an unexpected finding, however, seven out of ten tumor cell lines without evidence of p53 mutations were also sensitive to the ONYX-virus (HEISE et al. 1997). Some recent reports (ROTHMANN et al. 1998; HALL et al. 1998) doubted the selectivity of ONYX-replication, since these groups showed that a functional p53 is necessary for efficient cell lysis, illustrating that not all mechanisms involved in this process have been fully understood yet. In contrast to these in vitro data, preliminary clinical data phase I and phase II trials using this vector in head and neck cancer squamous cell carcinoma (HNSCC) show

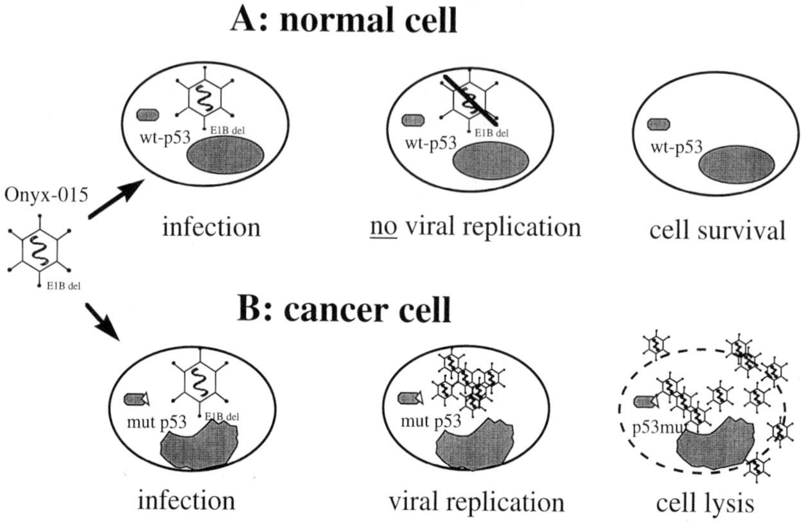

A: normal cell

B: cancer cell

Fig. 2. Illustration of ONYX-015 infection in cells with wt- and mutant p53. The ONYX-015 virus is a group C adenovirus carrying a deletion of the E1B 55 kD protein (E1B del). This renders normal cells (A) insensitive for ONYX-015 infection, since this protein is needed to inactivate wt-p53. However, cancer cells (B) with inactivating mutations of p53 (mut p53) are unable to block viral replication and are thus susceptible to the cytopathic effects induced by the virus

encouraging results (KIRN et al. 1998). Further trials have to look for the applicability of this approach in the clinical situation.

VI. Future Directions

All p53 gene therapy approaches aiming to restore p53 function are limited because they can only be used for locoregional therapy. This limitation is due to the fact that the viral vectors expressing wt-p53 used so far do not have any specificity for the tumor tissue; extremely high amounts would be needed to transfect circulating cells or to use systemic therapy. Efforts to target specific cell types have centered on attempts to engineer the natural viral envelope proteins of retroviruses to confirm tissue specificity (ANDERSON 1998); however the results obtained so far are not sufficient, since engineering of the capsid proteins frequently alters the efficient uptake of the virus.

Currently great efforts are made to develop novel vector constructs for efficient delivery and expression of genes. Adeno-associated viruses (AAV) are promising because they are non-pathogenic, have a small size, and can also target non-dividing cells. Furthermore, they integrate into the host genome and have been shown to induce long lasting gene expression. In contrast to retroviruses, AAV show preferential integration on the short arm of chromosome 19, and thus may hold promise for a "safe" integrating virus (ANDERSON

1998). Constructs based on recombinant AAV are not immunogenic, since most of the viral genome can be deleted without effect on the transduction efficiency (BARTLETT and SAMULSKI 1997). The principle feasibility of the use of these vectors for p53 gene therapy has recently been shown (QAZILBASH et al. 1997). Problems which have to be overcome here are the frequent contamination with adenoviral particles and producing high titer stocks sufficient for transfection in vivo.

Another virus which has the potential for future gene therapy approaches are the lentiviruses. These viruses also belong to the group of retroviruses, but can infect dividing and non-dividing cells (VERMA and SOMIA 1997). HIV is a well known example of a lentivirus. Editing its sequence renders these constructs non-pathogenic. Furthermore, these vectors can be produced in higher titers compared to retroviruses and they have been shown to induce long term expression.

Besides the engineering of the vector system used, other intriguing methods for delivery of p53 have been published by PHELAN et al. (1998) who reported the use of a construct containing a Herpes Simplex Virus (HSV) tegument protein, VP22, together with the wt-p53 gene. The principle is based on the observation, that the HSV-1 structural protein VP22 has the remarkable property of intercellular transport (ELLIOTT and O'HARE 1997). VP22 protein is exported from the cytoplasm of an infected cell and subsequently imported by neighboring cells, where it accumulates in the nucleus. It was shown that a chimeric protein containing the VP22 in conjunction with the entire p53 coding region produces a 90 kDa fusion protein which retains the abilities of the VP22 protein. Furthermore, it was demonstrated that this protein readily accumulates in the nucleus of non-transfected cells in cell mixture experiments and these cells subsequently undergo cell cycle arrest and induction of apoptosis. The use of an approach like this would hold considerable potential, especially for cancer gene therapy. First, use of a vector containing the VP22 protein would reach cancer cells which had not transfected directly. In addition the use of this protein would probably allow for the design of constantly producing cell lines, which are then given to the patient instead of trying to transduce the cancer cell itself, which would have great advantages with respect to feasibility and biosafety considerations.

Besides these efforts to optimize gene transfer of p53 into the cell, there are also attempts being made to engineer the normal p53 coding sequence on the molecular level in order to obtain an artificial protein with specific functions. CONSEILLER et al. (1998) constructed a p53-derived chimeric tumor suppressor gene (CTS1) with enhanced in vitro apoptotic properties. The construct contains the core domain of p53 (AA 75–325) and the VP16 transcription activation domain at the amino terminus as well as an optimized leucine zipper for homodimerization at the carboxy terminus. First in vitro data indicate that the chimeric protein is able to bind and induce transcription from a p53 response element, and that the chimeric protein is as least as functional as wt-p53 in the induction of cell cycle arrest and apoptosis. It is an advantage that this protein is not inhibited by mutant p53 as well as the mdm-2 protein. Thus, CTS1 could potentially be useful in treating cancers like osteosarcomas, which frequently show amplification of the mdm-2 gene (OLINER et al. 1992). The practicality of this very interesting approach in vivo has to be studied in detail using animal models.

E. Summary

The availability of novel techniques for the rapid and sensitive detection of p53 alterations with high throughput will make the detection of p53 alterations much easier in the near future. This will enable large scale studies in different tumor entities. First clinical data on the feasibility of gene therapy approaches targeting mutant p53 are encouraging, showing that this approach can be performed in the patient without severe side effects. It will now be a matter of larger Phase II and III trials to prove the clinical benefit. Due to the multiple changes that have accumulated in malignant tumors, introducing a wt-p53 gene will probably not be sufficient for cure of the disease. However, the availability of an agent specifically targeting a defect in cancer cells is a promising new approach and will extend our treatment options for multimodal strategies in the fight against malignancies. Future research must focus on the role of this kind of therapy in combination with conventional chemotherapy and in the situation of minimal residual disease. Here it will be important to develop novel kinds of delivery and to achieve sufficient gene transfer, not only locally but also in disseminated tumor cells. It will also be interesting to see whether the newly discovered p53-homologues are equally potent targets for gene therapy, and whether combinations of different gene therapy approaches have increased potential in vivo.

Acknowledgements. Supported in part by grants from the Wilhelm Sander Stiftung (A.N.) and the Deutsche Forschungsgemeinschaft (Ne 310/6-3, A.N.).

References

Anderson WF (1998) Human gene therapy. Nature 392:25–30
Badie B, Drazan KE, Kramar MH, Shaked A, Black KL (1995) Adenovirus-mediated p53 gene delivery inhibits 9L glioma growth in rats. Neurol Res 17:209–216
Barak Y, Juven T, Haffner R, Oren M (1993) mdm2 expression is induced by wild type p53 activity, EMBO J 12:461–468
Bartlett JS, Samulski RJ (1997) Methods for the Construction and Propagation of Recombinant Adeno-Associated Virus Vectors. In: Robbins PD (ed) Gene Therapy Protocols Humana Press. Totowa, New Jersey, p 25
Bayerdorffer E, Neubauer A, Rudolph B, Thiede C, Lehn N, Eidt S, Stolte M (1995) Regression of primary gastric lymphoma of mucosa-associated lymphoid tissue type after cure of Helicobacter pylori infection. MALT Lymphoma Study Group [see comments]. Lancet 345:1591–1594

Bischoff JR, Kirn DH, Williams A, Heise C, Horn S, Muna M, Ng L, Nye JA, Sampson JA, Fattaey A, McCormick F (1996) An adenovirus mutant that replicates selectively in p53-deficient human tumor cells [see comments]. Science 274:373–376

Buckbinder L, Talbott R, Velasco MS, Takenaka I, Faha B, Seizinger BR, Kley N (1995) Induction of the growth inhibitor IGF-binding protein 3 by p53: Nature 377:646–649

Bunz F, Dutriaux A, Lengauer C, Waldman T, Zhou S, Brown JP, Sedivy JM, Kinzler KW, Vogelstein B (1998) Requirement for p53 and p21 to sustain G2 arrest after DNA damage. Science 282:1497–1501

Canman CE, Lim DS, Cimprich KA, Taya Y, Tamai K, Sakaguchi K, Appella E, Kastan MB, Siliciano JD (1998) Activation of the ATM kinase by ionizing radiation and phosphorylation of p53. Science 281:1677–1679

Carbone DP, Minna JD (1994) In vivo gene therapy of human lung cancer using wild-type p53 delivered by retrovirus [editorial; comment]. J Natl Cancer Inst 86:1437–1438

Chao TC, Korsmeyer SJ (1998) BCL-2 Family: Regulators of Cell Death. Annu Rev Immunol 16:395–419

Chin KV, Ueda K, Pastan I, Gottesman MM (1992) Modulation of activity of the promoter of the human MDR1 gene by Ras and p53. Science 255:459–462

Cirielli C, Riccioni T, Yang C, Pili R, Gloe T, Chang J, Inyaku K, Passaniti A, Capogrossi MC (1995) Adenovirus-mediated gene transfer of wild-type p53 results in melanoma cell apoptosis in vitro and in vivo. Int J Cancer 63:673–679

Clayman GL, el-Naggar AK, Lippman SM, Henderson YC, Frederick M, Merritt JA, Zumstein LA, Timmons TM, Liu TJ, Ginsberg L, Roth JA, Hong WK, Bruso P, Goepfert H (1998) Adenovirus Mediated p53 Gene Transfer in Patients with Advanced Recurrent Head and Neck Squamous Cell Carcinoma. J Clin Oncol 16:2221–2232

Clayman GL, el-Naggar AK, Roth JA, Zhang WW, Goepfert H, Taylor DL, Liu TJ (1995) In vivo molecular therapy with p53 adenovirus for microscopic residual head and neck squamous carcinoma. Cancer Res 55:1–6

Conseiller E, Debussche L, Landais D, Venot C, Maratrat M, Sierra V, Tocque B, Bracco L (1998) CTS1: a p53-derived chimeric tumor suppressor gene with enhanced in vitro apoptotic properties. J Clin Invest 101:120–127

de-Kant E, Heide I, Thiede C, Herrmann R, Rochlitz CF (1996) MDR1 expression correlates with mutant p53 expression in colorectal cancer metastases. J Cancer Res Clin Oncol 122:671–675

Eastham JA, Hall SJ, Sehgal I, Wang J, Timme TL, Yang G, Connell CL, Elledge SJ, Zhang WW, Harper JW et al. (1995) In vivo gene therapy with p53 or p21 adenovirus for prostate cancer. Cancer Res 55:5151–5155

el-Deiry WS, Harper JW, O'Connor PM, Velculescu VE, Canman CE, Jackman J, Pietenpol JA, Burrell M, Hill DE, Wang Y et al. (1994) WAF1/CIP1 is induced in p53-mediated G1 arrest and apoptosis. Cancer Res 54:1169–1174

Elliott G, O'Hare P (1997) Intercellular trafficking and protein delivery by a herpesvirus structural protein. Cell 88:223–233

Evan G, Littlewood T (1998) A matter of life and cell death, Science 281:1317–1322

Fujiwara T, Cai DW, Georges RN, Mukhopadhyay T, Grimm EA, Roth JA (1994a) Therapeutic effect of a retroviral wild-type p53 expression vector in an orthotropic lung cancer model [see comments]. J Natl Cancer Inst 86:1458–1462

Fujiwara T, Grimm EA, Mukhopadhyay T, Zhang WW, Owen SL, Roth JA (1994b) Induction of chemosensitivity in human lung cancer cells in vivo by adenovirus-mediated transfer of the wild-type p53 gene. Cancer Res 54:2287–2291

Giaccia AJ, Kastan MB (1998) The complexity of p53 modulation: emerging patterns from divergent signals. Genes Dev 12:2973–2983

Gjerset RA, Turla ST, Sobol RE, Scalise JJ, Mercola D, Collins H, Hopkins PJ (1995) Use of wild-type p53 to achieve complete treatment sensitization of tumor cells expressing endogenous mutant p53. Mol Carcinog 14:275–285

Hall AR, Dix BR, O'Carroll SJ, Braithewaite AW (1998) p53-dependent cell death/apoptosis is required for a productive adenovirus infection. Nat Med 4:1068–1072

Harris MP, Sutjipto S, Wills KN, Hancock W, Cornell D, Johnson DE, Gregory RJ, Shepard HM, Maneval DC (1996) Adenovirus-mediated p53 gene transfer inhibits growth of human tumor cells expressing mutant p53 protein. Cancer Gene Ther 3:121–130

Haupt Y, Maya R, Kazaz A, Oren M (1997) Mdm2 promotes the rapid degradation of p53. Nature 387:296–299

Hawkins DS, Demers GW, Galloway DA (1996) Inactivation of p53 enhances sensitivity to multiple chemotherapeutic agents. Cancer Res 56:892–898

Heise C, Sampson JA, Williams A, McCormick F, Von-Hoff DD, Kirn DH (1997) ONYX-015, an E1B gene-attenuated adenovirus, causes tumor-specific cytolysis and antitumoral efficacy that can be augmented by standard chemotherapeutic agents [see comments]. Nat Med 3:639–645

Hermeking H, Eick D (1994) Mediation of c-Myc-induced apoptosis by p53. Science 265:2091–2093

Hollstein M, Sidransky D, Vogelstein B, Harris CC (1991) p53 mutations in human cancers. Science 253:49–53

Hussain SP, Harris CC (1998) Molecular epidemiology of human cancer: contribution of mutation spectra studies of tumor suppressor genes. Cancer Res 58:4023–4037

Jost CA, Marin MC, Kaelin-WG J (1997) p73 is a human p53-related protein that can induce apoptosis [see comments]. Nature 389:191–194

Kaghad M, Bonnet H, Yang A, Creancier L, Biscan JC, Valent A, Minty A, Chalon P, Lelias JM, Dumont X, Ferrara P, McKeon F, Caput D (1997) Monoallelically expressed gene related to p53 at 1p36, a region frequently deleted in neuroblastoma and other human cancers. Cell 90:809–819

Kastan MB, Zhan Q, el-Deiry WS, Carrier F, Jacks T, Walsh WV, Plunkett BS, Vogelstein B, Fornace-AJ J (1992) A mammalian cell cycle checkpoint pathway utilizing p53 and GADD45 is defective in ataxia-telangiectasia. Cell 71:587–597

Katayose D, Gudas J, Nguyen D, Srivastava S, Cowan K, Seth P (1995) Cytotoxic effects of adenovirus-mediated wild-type p53 protein expression in normal and tumor mammary epithelial cells. Clin Cancer Res 1:889–897

Kidd VJ (1998) Proteolytic Activities that mediate Apoptosis. Annu Rev Physiol 60:533–573

King KL, Cidlowski JA (1998) Cell Cycle Regulation and Apoptosis. Annu Rev Physiol 60:601–617

Kinzler KW, Vogelstein B (1997) Cancer-susceptibility genes. Gatekeepers and caretakers [news; comment]. Nature 386:761, 763

Kirn DH, Hermiston T, McCormick F (1998) Onyx-015: clinical data are encouraging, Nat Med 4:1341–1342

Kirsch DG, Kastan MB (1998) Tumor suppressor p53: implications for tumor development and prognosis, J Clin Oncol 16:3158–3168

Ko LJ, Prives C (1996) p53: puzzle and paradigm. Genes Dev 10:1054–1072

Koshland DE (1993) Molecule of the year [editorial]. Science 262:1953

Kubbutat MH, Jones SN, Vousden KH (1997) Regulation of p53 stability by Mdm2. Nature 387:299–303

Lane DP (1992) Cancer. p53, guardian of the genome [news; comment] [see comments]. Nature 358:15–16

Lane DP, Crawford LV (1979) T antigen is bound to a host protein in SV 40-transformed cells. Nature 278:261–263

Lesoon WL, Kim WH, Kleinman HK, Weintraub BD, Mixson AJ (1995) Systemic gene therapy with p53 reduces growth and metastases of a malignant human breast cancer in nude mice. Hum Gene Ther 6:395–405

Linzer DI, Levine AJ (1979) Characterization of a 54K dalton cellular SV 40 tumor antigen present in SV40-transformed and uninfected embryonal carcinoma cells. Cell 17:43–52

McCurrach ME, Connor TM, Knudson CM, Korsmeyer SJ, Lowe SW (1997) Bax-deficiency promotes drug resistance and oncogenic transformation by attenuating p53-dependent apoptosis. Proc Natl Acad Sci USA 94:2345–2349

Meijerink JP, Mensink EJ, Wang K, Sedlak TW, Sloetjes AW, de-Witte T, Waksman G, Korsmeyer SJ (1998) Hematopoietic malignancies demonstrate loss-of-function mutations of BAX. Blood 91:2991–2997

Meng RD, Shih H, Prabhu NS, George DL, el-Deiry WS (1998) Bypass of abnormal MDM2 inhibition of p53-dependent growth suppression. Clin Cancer Res 4:251–259

Milner J, Medcalf EA (1991) Cotranslation of activated mutant p53 with wild type drives the wild-type p53 protein into the mutant conformation. Cell 65:765–774

Miyashita T, Reed JC (1995) Tumor suppressor p53 is a direct transcriptional activator of the human Bax gene. Cell 80:293–299

Mujoo K, Maneval DC, Anderson SC, Gutterman JU (1996) Adenoviral-mediated p53 tumor suppressor gene therapy of human ovarian carcinoma. Oncogene 12:1617–1623

Nguyen DM, Spitz FR, Yen N, Cristiano RJ, Roth JA (1996) Gene therapy for lung cancer: enhancement of tumor suppression by a combination of sequential systemic cisplatin and adenovirus-mediated p53 gene transfer. J Thorac Cardiovasc Surg 112:1372–1376

Nielsen LL, Dell J, Maxwell E, Armstrong L, Maneval D, Catino JJ (1997) Efficacy of p53 adenovirus-mediated gene therapy against human breast cancer xenografts. Cancer Gene Ther 4:129–138

Nielsen LL, Maneval DC (1998) p53 tumor suppressor gene therapy for cancer, Cancer Gene Ther 5:52–63

Ogawa N, Fujiwara T, Kagawa S, Nishizaki M, Morimoto Y, Tanida T, Hizuta A, Yasuda T, Roth JA, Tanaka N (1997) Novel combination therapy for human colon cancer with adenovirus-mediated wild-type p53 gene transfer and DNA-damaging chemotherapeutic agent. Int J Cancer 73:367–370

Oliner JD, Kinzler KW, Meltzer PS, George DL, Vogelstein B (1992) Amplification of a gene encoding a p53-associated protein in human sarcomas [see comments]. Nature 358:80–83

Osada M, Ohba M, Kawahara C, Ishioka C, Kanamaru R, Katoh I, Ikawa Y, Nimura Y, Nakagawara A, Obinata M, Ikawa S (1998) Cloning and functional analysis of human p51, which structurally and functionally resembles p53. Nat Med 4:839–843

Owen-Schaub LB, Zhang W, Cusack JC, Angelo LS, Santee SM, Fujiwara T, Roth JA, Deisseroth AB, Zhang WW, Kruzel E et al. (1995) Wild-type human p53 and a temperature-sensitive mutant induce Fas/APO-1 expression. Mol Cell Biol 15:3032–3040

Papadopoulos N, Lindblom A (1997) Molecular basis of HNPCC: mutations of MMR genes. Hum Mutat 10:89–99

Phelan A, Elliott G, O'Hare P (1998) Intercellular delivery of functional p53 by the herpesvirus protein VP22. Nat Biotechnol 16:440–443

Polyak K, Xia Y, Zweier JL, Kinzler KW, Vogelstein B (1997) A model for p53-induced apoptosis [see comments]. Nature 389:300–305

Pomerantz J, Schreiber AN, Liegeois NJ, Silverman A, Alland L, Chin L, Potes J, Chen K, Orlow I, Lee HW, Cordon CC, DePinho RA (1998) The Ink4a tumor suppressor gene product, p19Arf, interacts with MDM2 and neutralizes MDM2's inhibition of p53. Cell 92:713–723

Qazilbash MH, Xiao X, Seth P, Cowan KH, Walsh CE (1997) Cancer gene therapy using a novel adeno-associated virus vector expressing human wild-type p53. Gene Ther 4:675–682

Raelson JV, Nervi C, Rosenauer A, Benedetti L, Monczak Y, Pearson M, Pelicci PG, Miller-WH J (1996) The PML/RAR alpha oncoprotein is a direct molecular target of retinoic acid in acute promyelocytic leukemia cells. Blood 88:2826–2832

Rampino N, Yamamoto H, Ionov Y, Li Y, Sawai H, Reed JC, Perucho M (1997) Somatic frameshift mutations in the BAX gene in colon cancers of the microsatellite mutator phenotype. Science 275:967–969

Roth JA, Nguyen D, Lawrence DD, Kemp BL, Carrasco CH, Ferson DZ, Hong WK, Komaki R, Lee JJ, Nesbitt JC, Pisters KM, Putnam JB, Schea R, Shin DM, Walsh GL, Dolormente MM, Han CI, Martin FD, Yen N, Xu K, Stephens LC, McDonnell TJ, Mukhopadhyay T, Cai D (1996) Retrovirus-mediated wild-type p53 gene transfer to tumors of patients with lung cancer [see comments]. Nat Med 2:985–991

Rothmann T, Hengstermann A, Whitaker NJ, Scheffer M, zur Hausen H (1998) Replication of ONYX-015, a potential anticancer adenovirus, is independent of p53 status in tumor cells. J Virol 72:9470–9478

Russell SJ, Ye YW, Waber PG, Shuford M, Schold-SC J, Nisen PD (1995) p53 mutations, O6-alkylguanine DNA alkyltransferase activity, and sensitivity to procarbazine in human brain tumors. Cancer 75:1339–1342

Sakamuro D, Eviner V, Elliott KJ, Showe L, White E, Prendergast GC (1995) c-Myc induces apoptosis in epithelial cells by both p53-dependent and p53-independent mechanisms. Oncogene 11:2411–2418

Sandig V, Brand K, Herwig S, Lukas J, Bartek J, Strauss M (1997) Adenovirally transferred p16INK4/CDKN2 and p53 genes cooperate to induce apoptotic tumor cell death. Nat Med 3:313–319

Scardigli R, Bossi G, Blandino G, Crescenzi M, Soddu S, Sacchi A (1997) Expression of exogenous wt-p53 does not affect normal hematopoiesis: implications for bone marrow purging. Gene Ther 4:1371–1378

Schuler M, Rochlitz CF, Horowitz JA, Schlegel J, Perruchoud AP, Kommoss F, Bolliger CT, Kauczor HU, Dalquen P, Fritz MA, Swanson S, Herrmann R, Huber C (1998) A phase I study of adenovirus-mediated wild-type p53 gene transfer in patients with advanced non-small cell lung cancer. Hum Gene Ther 9:2075–2082

Scolnick DM, Chehab NH, Stavridi ES, Lien MC, Caruso L, Moran E, Berger SL, Halazonetis TD (1997) CREB-binding protein and p300/CBP-associated factor are transcriptional coactivators of the p53 tumor suppressor protein. Cancer Res 57:3693–3696

Seth P, Katayose D, Li Z, Kim M, Wersto R, Craig C, Shanmugam N, Ohri E, Mudahar B, Rakkar AN, Kodali P, Cowan K (1997) A recombinant adenovirus expressing wild type p53 induces apoptosis in drug-resistant human breast cancer cells: a gene therapy approach for drug-resistant cancers. Cancer Gene Ther 4:383–390

Sheard MA, Vojtesek B, Janakova L, Kovarik J, Zaloudik J (1997) Up-regulation of Fas (CD95) in human p53wild-type cancer cells treated with ionizing radiation. Int J Cancer 73:757–762

Shieh SY, Ikeda M, Taya Y, Prives C (1997) DNA damage-induced phosphorylation of p53 alleviates inhibition by MDM2. Cell 91:325–334

Siliciano JD, Canman CE, Taya Y, Sakaguchi K, Appella E, Kastan MB (1997) DNA damage induces phosphorylation of the amino terminus of p53. Genes Dev 11:3471–3481

Steegenga WT, Riteco N, Jochemsen AG, Fallaux FJ, Bos JL (1998) The large E1B protein together with the E4orf6 protein target p53 for active degradation in adenovirus infected cells. Oncogene 16:349–357

Tada M, Matsumoto R, Iggo RD, Onimaru R, Shirato H, Sawamura Y, Shinohe Y (1998) Selective sensitivity to radiation of cerebral glioblastomas harboring p53 mutations. Cancer Res 58:1793–1797

Trink B, Okami K, Wu L, Sriuranpong V, Jin J, Sidransky D (1998) A new p53 homologue. Nat Med 4:747–748

Verma IM, Somia N (1997) Gene therapy – promises, problems and prospects [news]. Nature 389:239–242

Wills KN, Maneval DC, Menzel P, Harris MP, Sutjipto S, Vaillancourt MT, Huang WM, Johnson DE, Anderson SC, Wen SF et al. (1994) Development and characterization of recombinant adenoviruses encoding human p53 for gene therapy of cancer. Hum Gene Ther 5:1079–1088

Woo RA, McLure KG, Lees-Miller SP, Rancourt DE, Lee PWK (1998) DNA-dependent protein kinase acts upstream of p53 in response to DNA damage. Nature 394:700–704

Wotherspoon AC, Doglioni C, Diss TC, Pan L, Moschini A, de-Boni M, Isaacson PG (1993) Regression of primary low-grade B-cell gastric lymphoma of mucosa-associated lymphoid tissue type after eradication of Helicobacter pylori [see comments]. Lancet 342:575–577

Wu GS, Burns TF, McDonald ER, Jiang W, Meng R, Krantz ID, Kao G, Gan DD, Zhou JY, Muschel R, Hamilton SR, Spinner NB, Markowitz S, Wu G, el-Deiry WS (1997) KILLER/DR5 is a DNA damage-inducible p53-regulated death receptor gene [letter]. Nat Genet 17:141–143

Xiong Y, Hannon GJ, Zhang H, Casso D, Kobayashi R, Beach D (1993) p21 is a universal inhibitor of cyclin kinases [see comments]. Nature 366:701–704

Xu M, Kumar D, Srinivas S, Detolla LJ, Yu SF, Stass SA, Mixson AJ (1997) Parenteral gene therapy with p53 inhibits human breast tumors in vivo through a bystander mechanism without evidence of toxicity. Hum Gene Ther 8:177–185

Xu M, Kumar D, Stass SA, Mixson AJ (1998) Gene therapy with p53 and a fragment of thrombospondin I inhibits human breast cancer in vivo. Mol Genet Metab 63:103–109

Yang A, Kaghad M, Wang Y, Gillet E, Fleming MD, Dotsch V, Andrews NC, Caput D, McKeon F (1998) p63, a p53 homolog at 3q27–29, encodes multiple products with transactivating, death-inducing, and dominant-negative activities. Mol Cell 2:305–316

Yang C, Cirielli C, Capogrossi MC, Passaniti A (1995) Adenovirus-mediated wild-type p53 expression induces apoptosis and suppresses tumorigenesis of prostatic tumor cells. Cancer Res 55:4210–4213

Yew PR, Berk AJ (1992) Inhibition of p53 transactivation required for transformation by adenovirus early 1B protein, Nature 357:82–85

Yin C, Knudson CM, Korsmeyer SJ, Van-Dyke T (1997) Bax suppresses tumorigenesis and stimulates apoptosis in vivo. Nature 385:637–640

Zhang J, Cado D, Chen A, Kabra NH, Winoto A (1998) Fas-mediated apoptosis and activation-induced T-cell proliferation are defective in mice lacking FADD/Mort1. Nature 392:296–300

Zhang WW, Fang X, Mazur W, French BA, Georges RN, Roth JA (1994) High-efficiency gene transfer and high-level expression of wild-type p53 in human lung cancer cells mediated by recombinant adenovirus. Cancer Gene Ther 1:5–13

CHAPTER 10
Antioxidants: Protection Versus Apoptosis

Y. DELNESTE, E. ROELANDTS, J.-Y. BONNEFOY, and P. JEANNIN

A. Introduction

Tissue homeostasis is tightly regulated by both proliferation and cell death. These processes are crucial in embryogenesis during the development of the central nervous system (RAFF 1993; NARAYANAN 1997) and the development (VON BOEHMER 1992) and function of the immune system (COHEN and DUKE 1992). The cell death associated with tissue turnover is called apoptosis or "programmed cell death" (PCD) and is distinct from necrosis, which results from tissue or cell injury, hypoxia or hyperthermia. Apoptosis is an active process requiring cell activation and is characterized by particular morphological and biochemical changes (KERR et al. 1972), such as condensation of cytoplasm, membrane blebbing, nucleus segmentation and DNA fragmentation into oligomers of 180–200 bp (WYLLIE et al. 1980).

The crucial role played by apoptosis in regulating a normal homeostasis is illustrated in pathologies which are associated with an excessive (graft rejection, AIDS) or with a deficient (autoimmune disease, cancer) cell death. Autoimmune diseases result from the failure to regulate autoreactive T cells, which can be due to mutations in apoptosis-signaling molecules such as Fas (CD95) (FISHER et al. 1995; RIEUX-LAUCAT et al. 1995). Graft rejection is a consequence of the killing of engrafted cells (KABELITZ 1998). Cancer is characterized by the absence of death of uncontrolled proliferating cells. While tumor cell growth is a multiparameter mechanism, numerous studies have reported alterations of apoptosis-regulating molecules, such as mutations in the p53 tumor suppressor gene, in tumor cells (PFEIFER and DENISSENKO 1998). Moreover, the massive $CD4^+$ T cell depletion in HIV-infected patients is induced by apoptosis (GROUX et al. 1992).

Most of our knowledge of the cellular and molecular mechanisms that regulate apoptosis comes from the study of the immune system: apoptosis is involved in thymic selection (OGASAWARA et al. 1995), peripheral tolerance (WANG and LENARDO 1997) and regulation of the outcome of an immune response (JU et al. 1995). Different external signals can induce apoptosis, including UV radiation, hypoxia, serum deprivation, and physiological inducers such as cytokines, for example, tumor necrosis factor alpha (TNFα) and membrane-associated molecules such as Fas ligand (Fas-L), following interaction with their ligands, TNF-RI (CD120a) and TNF-RII (CD120b), or Fas, respectively. Recent progress has been made in the identification of the intracellular signaling pathways responsible for apoptosis, showing that specific transduction molecules are responsible for the induction (i.e., FLICE; Muzio et al. 1996) or inhibition (i.e., FLIP; Irmler et al. 1997) of the apoptotic signal. Whatever the nature of the stimulus, signaling finally leads to a cascade of catalytic activation of caspases, culminating in apoptosis. Three different and non-exclusive biochemical processes have been suggested to be critical for apoptosis: cytoplasmic proteases (such as those belonging to the ICE family), endonucleases (responsible for the DNA degradation) and oxidative stress.

Several studies have now clearly demonstrated that the intracellular redox status can influence apoptosis: numerous antioxidants, including natural intracellular enzymes, such as superoxide dismutase (SOD) or catalase, as well as chemical compounds with antioxidant properties, such as N-acetyl-L-cysteine (NAC) or dithiothreitol (DTT), can prevent apoptosis of different cell types. Antioxidants are of particular interest since some of them, such as vitamin C or NAC, are poorly toxic and usually used in humans. The aim of this review is: (1) to summarize arguments in favor of the redox regulation of apoptosis, and (2) to report data concerning the protective mechanisms of antioxidants.

B. Apoptosis and the Cellular Redox Status

Oxygen plays a key role in the metabolism of aerobic cells. The generation of highly reactive oxygen species (ROS), such as singlet oxygen (O_2), hydrogen peroxide (H_2O_2), the superoxide anion (O_2^-) and the hydroxyl radical ($OH^.$), is associated with respiration. ROS are important in many physiological processes such as signal transduction. They can act as second messengers (Schreck and Baeuerle 1991a) or in innate immunity (Barja 1993). Indeed, ROS produced by neutrophils and monocytes/macrophages can kill pathogens. Nevertheless, ROS are highly reactive molecules and can, thus, cause extensive damages to macromolecules, including DNA. Different cellular redox systems have thus been created during evolution to protect the intracellular reducing status in the face of the highly oxidizing extracellular environment. These systems involve enzymes, such as glutathione peroxidase, catalase and superoxide dismutase, and chemical compounds (carotenoids, α-tocopherol). Among these, reduced glutathione (GSH) is one of the most important antioxidants in the soluble compartment of the cell while α-tocopherol is mainly located in the membrane.

Since several studies have demonstrated that the production of ROS and the modulation of the intracellular redox status participate in the apoptotic process (reviewed by Buttke and Sandstrom 1994; Powis et al. 1997), different authors have reported that, even if antioxidants protect some cell types against apoptosis, ROS are not involved in the apoptotic process. That oxidative stress is thought to be involved in the apoptotic process results essentially from the observation that physiological (glutathione) and nonphysiological reducing agents (NAC, DTT) can prevent cell death induced by oxidants (H_2O_2) and membrane molecule transducing apoptotic signals, such as Fas or TNF-R. In this section, we summarize the pro and con arguments regarding the role of ROS in apoptosis.

I. Exogenous ROS or Oxidants can Trigger Apoptosis

Numerous data have demonstrated that exogenous ROS or oxidants can induce apoptosis of different cell types and thus argue in favor of a direct role of ROS in cell death. H_2O_2 produced by monocytes/macrophages and neutrophils (oxidative burst) can trigger the death of pathogens as well as bystander cells (SZATROWSKI and NATHAN 1991). Exogenous H_2O_2 induces in vitro apoptosis of different types of cells including tumor cell lines (LENNON et al. 1991), muscle cells (STANGEL et al. 1996), monocytes (LAOCHUMROONVORAPUNG et al. 1996), neurons (KAMATA et al. 1996) and mature effector T cells (ZETTL et al. 1997). H_2O_2-induced cell death can be inhibited by catalase, SOD, or desferrioxamine, and can exert its effect directly or via the generation of hydroxyl radical (OH·) (LI et al. 1997). In addition to H_2O_2, nitric oxide (NO) also induces apoptosis of macrophages and monocytes (ALBINA et al. 1993). In a similar manner, UV and X-ray irradiation induce apoptosis through the generation of ROS. The antineoplastic drugs, doxorubicin (BENCHEKROUN et al. 1993) or ether-linked lipids (DIOMEDE et al. 1994), can induce apoptosis by eliciting the formation of ROS. Oxidizing agents such as diamide, which induces sulfhydryl oxidation, induces apoptosis of T helper lymphocytes at 200 μM and necrosis at 400 μM (SATO et al. 1995). The direct exposure of cells to oxidants increases intracellular levels of Ca^{2+}, depletes ATP, and induces the oxidation of NADPH, glutathione and lipids. In a similar manner, oxidized low density lipoproteins and lipid hydroxyperoxides such as 15-hydroperoxyeicosatetraenoic acid (15HPETE) induce apoptosis (ESCARGUEIL et al. 1992).

II. Apoptosis is Associated with an Alteration of the Redox Status

Different experimental evidence suggests that the generation of ROS can be involved in most types of cell death: (1) antioxidants can protect or delay apoptosis induced by stimuli other than oxidants; and (2) the modulation of endogenous antioxidants regulates cell sensitivity to apoptosis.

In agreement with the hypothesis that apoptosis is associated with a decrease in antioxidant defenses, STEFANELLY et al. (1995) reported that glucocorticoid-induced thymocyte apoptosis is reduced when oxygen tension is lowered below 5%, suggesting that ROS generation could also be implicated in apoptosis induced by a nonoxidative stimulus. Dexamethasone-induced apoptosis is associated with a selective decrease in the mRNAs encoding SOD, catalase, glutathione peroxidase or thioredoxin, the molecules responsible for the antioxidant defense. Primary cultured sympathetic neurons (PC12) die by apoptosis when deprived of nerve growth factor (NGF). Addition of NGF increases the levels of catalase and glutathione peroxidase, suggesting that growth factor withdrawal may involve a down-regulation of antioxidant defenses, resulting in an increase of ROS sensitivity produced during normal metabolism. Pre-apoptotic and apoptotic cells have lower GSH, protein sulfhydryl and α-tocopherol than do normal cells. Inhibition of GSH neo-synthesis using buthionine sulfoxymide (BSO), an irreversible inhibitor of γ-glutamyl cysteine synthetase (HUANG et al. 1988), the enzyme responsible for glutathione synthesis, is unable to induce cell death but renders cells more susceptible to oxidative stress-induced apoptosis (ZHONG et al. 1993). The anti-Fas mAb-induced cell death of Jurkat cells is associated with a rapid efflux of intracellular levels of GSH with no increase of oxidized glutathione (GSSG), and survival is prolonged when cells are treated with GSH. The efflux of GSH level may thus be responsible for a breakdown in the maintenance of a reducing environment (VAN DEN DOBBELSTEEN et al. 1996). Primary cultured sympathetic neurons undergo apoptosis when deprived of NGF. Injection of Cu/Zn SOD, or transfection with the cDNA encoding for these molecules a few hours before deprivation, delays apoptosis (GREENLUND et al. 1995). Transfection with MnSOD protects tumor cells against cytostatic and cytotoxic concentrations of TNFα or IL-1α and against chemical (doxorubicin) and physical (irradiation) apoptotic inducers, suggesting that resistance to apoptosis is associated with the intracellular level of antioxidant defense (HIROSE et al. 1993). In a similar manner, dexamethasone-induced cell death in thymomas is associated with an early decrease in the regulated expression of the primary antioxidant defense enzymes prior to chromatin condensation (BRIEHL et al. 1995).

III. The Antioxidant Activity of the Apoptosis Inhibitor Molecule Bcl-2

One of the most important arguments in favor of ROS involvement in apoptosis comes from the observation that Bcl-2, one of the most potent anti-apoptotic intracellular molecules, has antioxidant activity (HOCKENBERY et al. 1993; Kane et al. 1993). Bcl-2 was originally described associated with the t(14;18) translocation (q32;q21) in B cell lymphomas (KORSMEYER 1992). *Bcl-2* is homologous to *ced-9*, a cell-death gene in the nematode worm *Caenorhabditis elegans*. Interestingly, *ced-9* is part of a bi-cistronic gene co-encoding a protein similar to cytochrome b560 of complex II from the mitochondrial respiratory chain, suggesting that Ced-9 may have redox or ROS-regulatory activities (HENGARTNER and HORVITZ 1994). The evidence for involvement of Bcl-2 in regulating cell death comes from the observation that Bcl-2 knockout mice show apoptosis of thymocytes and spleen cells (VEIS et al. 1993). Due to its homology to Ced-9, Bcl-2 has been also suspected to have redox regulatory properties and HOCKENBERRY et al. (1993) have shown that Bcl-2 is an antioxidant: overexpression of Bcl-2 suppressed lipid peroxidation, which is induced by H_2O_2 or t-butyl hydroperoxide, and protected against H_2O_2- and menadione-induced oxidative apoptosis. In a similar manner, overexpression of Bcl-2 in the GTI-7 neural cell line prevented necrosis resulting from glutathione depletion, which is normally associated with the generation of ROS (KANE et al. 1993). The treatment of cells with TNFα is followed by a decrease of Bcl-2 expression which precedes cell death (CHEN et al. 1995). Moreover, the transfection of cells with Bcl-2 renders breast carcinoma cells totally resistant to TNFα- or Fas-mediated apoptosis (JAATTELA et al. 1995). Similar results were observed with Bcl-X_L, a member of the Bcl-2 family of apoptotic regulatory molecules, which protects WEHI-231 B cells from oxidant-induced apoptotic signals such as serum deprivation or gamma irradiation (FANG et al. 1995).

IV. Are ROS Really Involved in Apoptosis?

While several studies have reported the involvement of ROS in cell death (apoptosis or necrosis), many authors have presented evidence indicating that ROS are not involved in the apoptotic process. Cell death can occur at low oxygen tension where ROS are unlikely to be produced: apoptosis induced by different stimuli, such as anti-Fas mAb, IL-3 withdrawal (JACOBSON and RAFF 1995), dexamethasone and serum deprivation (MUSCHEL et al. 1995), can occur in near-anaerobic conditions. Under these conditions, apoptosis already occurred in response to ROS-generating compounds. Moreover, hypoxia can also induce apoptosis of the T lymphoma cell line WEHY7.1 (MUSCHEL et al. 1995). It is also interesting to note that ROS are not involved in all types of cell death: apoptosis induced via Fas activation does not require the generation of ROS (SCHULZE-OSTHOFF et al. 1994). LEE and SHACTER (1997) reported that Bcl-2 did not protect Burkitt's lymphoma cells against H_2O_2-induced apoptosis although it protected against ionomycin-induced cell death. Together, these results demonstrate that ROS can induce apoptosis but are not strictly required for the process of cell death. Moreover, the potential involvement of ROS in apoptosis can also be dependent on the nature of the target cell. Indeed, NAC protects endothelial cells but not L929 tumor cells from TNFα-mediated cell death (SCHRODER et al. 1993).

Collectively, these data suggest that (1) ROS can be generated as a result of some apoptosis-inducing signals (such as TNFα) but not of others (Fas triggering) (HUG et al. 1994; SCHULZE-OSTHOFF et al. 1994), and (2) ROS are generated in some cell types but not others. As a consequence, it is now widely accepted that apoptosis is a redox-regulated mechanism, explaining why antioxidants can protect against apoptosis induced by several different stimuli.

C. Anti-Apoptotic Properties of Antioxidants: Mechanisms of Action

I. ROS Scavenging and Reducing Activities of Antioxidants

The main physiological function of antioxidants is to scavenge ROS, which can be involved in cell death of some cell types (JACOBSON and RAFF 1995). The production of ROS is one of the intracellular mechanisms induced by TNFα and antioxidants can prevent the TNFα-mediated cell death of different cell types. The antioxidants cysteine (LEE et al. 1995) and catalase (SANDSTROM and BUTTKE 1993) are spontaneously secreted by cells that inhibit apoptosis. Thioredoxin, an important intracellular thiol antioxidant (WOLLMAN et al. 1988) protects glial cells during re-perfusion after ischemia (TOMIMOTO et al. 1993), delays the onset of glucocorticoid-induced apoptosis of thymocytes (SLATER et al. 1995; WOLFE et al. 1994) and protects U937 cells against TNFα-induced apoptosis (MATSUDA et al. 1991). Vitamin E and catalase, two potent antioxidants, prevent dexamethasone-induced apoptosis (BAKER et al. 1996). Oxidative stress may lead to the formation of oxidized lipids in the cell membrane (HALLIWELL et al. 1988), which are potent inducers of apoptosis (SANDSTROM et al. 1994; ESCARGUEIL et al. 1992) and are suspected to be involved in TNFα-mediated cell death (LARRICK and WRIGHT 1990).

The protective scavenging effect of antioxidants has also been suspected for the treatment of some neurodegenerative diseases, particularly in Alzheimer's disease (reviewed by DAVIS 1996). The generation of senile plaques is associated with the cytotoxic properties of β-amyloid (βA4) (YANKNER et al. 1990; BEHL et al. 1992), which induces the production of ROS. While the mechanism responsible for ROS production in βA4-induced apoptosis remains unclear (alteration of antioxidant defenses), it is interesting to note that ROS can be produced by βA4 (HENSLEY et al. 1994). Antioxidants (such as vitamin E) have been shown to prevent βA4-induced cell death (BEHL et al. 1992, 1994; MATTSON and GOODMAN 1995). In a similar manner, the transfection of catalase or glutathione peroxidase protects PC12 cells against βA4-induced apoptosis (SAGARA et al. 1996).

II. Replenishment of Intracellular GSH Levels

Cells must maintain a normal intracellular concentration of GSH, since it participates in numerous important physiological processes such as maintenance of the redox status, DNA and protein synthesis, drug detoxification, amino acid transport, and acts as a cofactor for several enzymes. In a normal situation, the ratio of GSH to GSSG is higher than 20. For example, a normal level of GSH correlates with the capacity of peripheral blood mononuclear cells to enter the cell cycle: low levels of GSH are associated with a decrease in cell cycle progression from G1 to S phase (IWATA et al. 1994). Increasing intracellular levels of GSH by using NAC, a precursor of GSH neosynthesis, or GSH ethyl ester, protects human peripheral blood T cells against Fas-mediated apoptosis (DEAS et al. 1997) as well as protecting against TNFα-mediated cell death of oligodendrocytes and L929 fibroblasts (MAYER and NOBLE 1994). The protective effect was (1) inhibited by BSO, demonstrating that maintaining the concentration of GSH is an important protective pathway against apoptosis, and (2) was observed with all the thiol-containing compounds used (cysteine, captopril, D-penicillamine and 2-mercaptoethanol) but not with non-thiol antioxidants (catalase, vitamin E), suggesting that the protection was not related to the scavenging of ROS. The stimulation of Jurkat cells with an agonistic anti-Fas mAb induces a rapid decrease of intracellular levels of GSH with no increase of GSSG, suggesting that apoptosis is associated with a rapid efflux of GSH (CHIBA et al. 1996). This efflux is responsible for the alteration of the intracellular reducing environment and can thus affect the scavenging of ROS. The survival of cells is prolonged when they are treated with permeable GSH-diethyl esters, which maintain normal intracellular levels of GSH. In a similar manner, cysteine starvation inhibits DNA synthesis and the cytotoxic activity of T cell clones; this mechanism can be mimicked by BSO (LIANG et al. 1991). In addition to protecting T cells against apoptosis, several papers have reported that thiols, and especially NAC, enhance T cell functions and/or T cell growth (LIANG et al. 1991; SMYTH 1991; EYLAR et al. 1993; YIM et al. 1994).

The crucial role played by thiol antioxidants in protecting against apoptosis has been clearly illustrated in HIV patients: these subjects, even when asymptomatic, present low levels of extracellular cystine and cysteine (ECK et al. 1989). Various authors have reported that (1) low levels of GSH (STAAL et al. 1992), in association with increased levels of GSSG (AUKRUST et al. 1995), follow HIV infection and promote HIV replication (STAAL et al. 1990); and (2) that disturbance of glutathione redox status is associated with a selective depletion of native $CD4^+$ T cells (STAAL et al. 1992; AUKRUST et al. 1996). Moreover, GSH deficiency has been associated with impaired survival in HIV disease (ROEDERER et al. 1991). Such alterations of the glutathione redox status have also been recently noted in synovial T cells of patients suffering from rheumatoid arthritis (MAURICE et al. 1997).

III. Thiol Antioxidants Induce the Shedding of Membrane Fas

Numerous studies have shown that antioxidants protect against Fas-mediated apoptosis. However, Fas-mediated apoptosis is not dependent on ROS production (HUG et al. 1994) and can act independently of extracellular Ca^{2+} (ROUVIER et al. 1993), suggesting that the ROS scavenging and metal ion chelating properties of antioxidants are not responsible for this protective effect. Moreover, NAC does not modulate the expression of the anti-apoptotic factor Bcl-2 (unpublished personal observation). Based on the role played by Fas and TNF-R in transducing the apoptotic signal, we have evaluated whether thiol antioxidants may directly affect their expression.

Molecules belonging to the NGF/TNF-R family play a crucial role in transducing an apoptotic signal following binding with their specific ligands. Among these molecules, Fas (CD95) has been extensively studied. Fas is a 48 kDa cell surface glycoprotein expressed by several cell types, including immature thymocytes and activated T cells. Fas transduces a death signal when triggered with an agonistic anti-Fas mAb, or following interaction with Fas-L. The couple Fas–Fas-L is important in maintaining homeostasis within the immune system and in preventing autoimmune diseases (KRAMMER et al. 1994), as indicated by animal models. Indeed, Fas- (lpr) and Fas-L-deficient mice (gld) present an excessive peripheral T cell proliferation and autoimmune disorders. TNFα is a pleiotropic cytokine expressed as a membrane protein (25 kDa); a soluble (17 kDa) form results from the shedding of the membrane form. In addition to proinflammatory properties, TNFα induces cell death of TNF-R expressing cells. TNFα-induced signaling is mediated by two receptors, TNF-RI (p55, CD120a) and TNF-RII (p75, CD120b). Soluble TNF-R are generated by shedding of the membrane forms and neutralize the activity of TNFα.

We have demonstrated that thiols downregulate Fas membrane expression on human T cells (DELNESTE et al. 1996). Fas expression was induced in peripheral blood T cells either by stimulation with anti-CD3 mAb or by culture in a medium containing a low concentration of fetal calf serum. The decrease of Fas expression was dependent on the concentration of NAC (significant with 5 mM and maximal with 20 mM) and was complete by 4 h of incubation. Such an effect was only seen with the sulfhydryl-containing compounds tested (NAC, GSH, L- and D-cysteine, DTT and mercaptopropionic acid, MPA), but not with S-substituted (S-methyl cysteine, methionine) or oxidized thiols (GSSG), or with antioxidants lacking a thiol group (catalase, SOD, desferrioxamine and ascorbic acid). These data demonstrate that the thiol-induced decrease of Fas (1) requires a free SH group, (2) is not associated with the antioxidant properties of thiols and (3) does not require GSH neosynthesis. Interestingly, an NAC-induced decrease of Fas was correlated with the release of the shedded form. Indeed, an immunoreactive form of Fas was detected in the culture supernatants by ELISA and western blotting. As a consequence, thiol-treated T lymphocytes were resistant to anti-Fas mAb-induced cell death. Taken together, thiols protect against Fas-mediated apoptosis via both their own anti-apoptotic properties and their ability to induce the shedding of Fas.

While inducing the shedding of membrane Fas, thiols increase both membrane TNF-RI and TNF-RII expression on activated human T lymphocytes (DELNESTE et al. 1997). All the free thiol-containing compounds tested induced an early dose-dependent increase of membrane TNF-R on activated cells, suggesting that thiols may inhibit an enzyme responsible for their shedding. Thiols also increased the levels of TNF-R mRNA later on, which could account for the late increase of membrane TNF-R expression observed. Thus, it is tempting to speculate that, under particular conditions (i.e., activation of the target cells), thiols may increase the sensitivity to TNFα-induced cell death by increasing TNF-R expression. It is interesting to note that thiols may have opposite effects on the TNF-R-mediated cell death since (1) they increase TNF-R expression and thus the sensitivity to TNFα-induced apoptosis, and (2) they protect target cells against apoptosis. The thiol-mediated regulation of TNF-R expression has also been previously reported: ZANG and AGGARWAL (1994) showed that thiol-modifying reagents such as diamide and iodoacetamide induced the shedding of TNF-RI and TNF-RII from a variety of cell types of both myeloid and epithelial origin.

We have previously reported that thiol antioxidants (GSH, cysteine, NAC, DTT) potentiate the activation-induced membrane TNFα expression on human peripheral blood T cells (DELNESTE et al. 1997). In a similar manner, BAUER et al. (1998) recently showed that the antioxidants DTT and pyrrolidine dithiocarbamate potentiate the expression of Fas-L on phorbol myristate acetate plus ionomycin-stimulated Jurkat cells. Both these studies clearly demonstrate that antioxidants may have opposite effects on the apoptotic process since they can protect target cells, but may increase the killing activity of effector cells.

IV. Thiol Antioxidants can Modulate the Generation of Second Messengers and the Expression-Activation of Transcription Factors

1. Modulation of Signaling Molecules

Antioxidants have been reported to modulate the generation of second messengers and the activation of transcription factors which are involved in the signaling pathways of apoptosis.

The apoptosis-signal regulator kinase (ASK) 1 belongs to the mitogen-activated protein kinase family whose molecules are involved in apoptotic signaling (FANGER et al. 1997): overexpression of ASK1 induces apoptosis of epithelial cells cultured in low serum (ICHIJO et al. 1997). SAITOH et al. (1998) have reported that thioredoxin is a potent inhibitor of ASK1: the treatment of L929 cells with the apoptosis-inducing stimuli H_2O_2 or TNFα activates ASK1 which is inhibited by NAC, suggesting a redox regulation of ASK1. Antioxidants can prevent apoptosis, at least in part, by inhibiting the oxidation-induced dissociation of thioredoxin from ASK1.

As mentioned above, T cells from HIV-infected subjects have impaired biological functions and thiol antioxidants have been reported to protect these cells against activation-induced cell death. In a recent study, STEFANOVA et al. (1996) have shown that, in T cells from HIV-infected patients, an oxidation of the thiol groups is responsible for a conformational alteration of p56[lck], Fyn and ZAP70, three molecules involved in the TCR signaling. The modifications, which impair the T cell functions, can be reversed by antioxidants such as DTT. The authors suggested that modification of the sulfhydryl groups might be related to an alteration of the redox status associated with HIV infection. Interestingly, a similar decrease of CD3ζ chain can be also observed in cancer patients (NAKAGOMI et al. 1993; GUNJI et al. 1994) but can be recovered after treatment with NAC (OTSUJI et al. 1996). Taken together, all these studies demonstrate that the oxidative stress induces T cell dysfunction through reduction of the CD3ζ chain and/or the inactivation of kinases and that these modifications can be reversed by antioxidants.

2. Modulation of Transcription Factors

Antioxidants have been shown to modulate the activity of the transcription factors NF-κB and AP-1 which are involved in the induction of the apoptotic process.

In non-stimulated cells, NF-κB is composed of two heterodimeric molecules (p50 and p75) which form a complex with the inhibitory molecule IκB. Activation induces the phosphorylation and proteolysis of IκB, resulting in its dissociation from NF-κB (BEG et al. 1993). As a consequence, NF-κB is activated and translocates to the nucleus. The involvement of NF-κB in cell death is suggested by different observations: serum starvation, which induces apoptosis of 293 cells, is associated with an activation of NF-κB (GRIMM et al. 1996), and the neurotoxic A beta is a potent activator of NF-κB in primary neurons (KALTSCHMIDT et al. 1997). More recently, a direct role for NF-κB in the TNFα-mediated cell death has been clearly evidenced by using NF-κB-deficient mice in which TNFα-induced apoptosis is impaired (BEG and BALTIMORE 1996). Efficient activation of NF-κB-dependent genes following stimulation with PMA, IL-1 or TNFα requires an appropriate intracellular oxidized redox status (ISRAEL et al. 1992). Physiological concentrations of H_2O_2 induce NF-κB specific DNA binding and transactivating activity in Jurkat cells. The antioxidants cysteine, NAC, β-mercaptoethanol, nordihydroguaiaretic acid (NDGA), vitamin E analogs and α-lipoic acid inhibit the activation of NF-κB (STAAL et al. 1990; SCHRECK et al. 1991b; ISRAEL et al. 1992; SUZUKI et al. 1992; SUZUKI and PACKER 1993). The TPA-induced activation of NF-κB is inhibited by BSO, suggesting that antioxidants increase the activity of NF-κB by inhibiting GSSG formation (MIHM et al. 1995). Moreover, a partial depletion in intracellular GSH inhibits the activation and nuclear translocation of NF-κB in the human T cell line MOLT4 (MIHM et al. 1995). Nevertheless, different studies have reported that oxidizing conditions inhibit the DNA binding of NF-κB which can be recovered after treatment with β-mercaptoethanol (TOLEDANO and LEONARD 1991). In a similar manner, thioredoxin potentiates the expression of a NF-κB-linked reporter gene. All these results demonstrate that NF-κB is controlled at two levels: (1) the activation and nuclear translocation involves ROS and can be inhibited by thiol antioxidants such as NAC, and (2) the DNA binding activity of NF-κB is inhibited by oxidizing agents such as diamide and potentiated by thiol antioxidants (MIHM et al. 1995). While antioxidants can interfere directly with the molecule, others have suspected they can modulate the activity of tyrosine kinase and phosphatases within the NF-κB signal transduction pathway (ANDERSON et al. 1994).

The transcription factor AP-1 is comprised of two molecules, Jun and Fos. The potential involvement of AP-1 in apoptosis results, essentially, from the observation that an AP-1 DNA binding site maps to a negative-response region in the promoter of the oncogene c-myc (SCHRIER and PELTENBURG 1993), the expression of which has been associated with the initiation of T cell hybridoma apoptosis (SHI et al. 1992). The DNA binding and transactivation of AP-1 is induced by H_2O_2 (DEVARY et al. 1991). Moreover, treatment of cells with the antioxidant PDTC and the expression of thioredoxin activate AP-1 (MEYER et al. 1993), which could interfere with the expression of c-myc.

D. Conclusions and Therapeutic Perspectives

Among the different molecules able to protect mammalian cells against apoptosis, antioxidants are one of the most important groups because (1) they protect against a wide variety of apoptosis-inducing signals (chemical, physical and physiological), and (2) they protect different type of cells, irrespective of their function or differentiation status. As such, the in vivo biological properties of antioxidants with regard to cell viability and protection against apoptosis have been extensively reported in the literature. It is important to note that the actual concept is that apoptosis requires an alteration of the intracellular redox status to be effective, which can be reverted by the antioxidants. More recently, several studies, focused on defining more precisely the cellular and molecular mechanisms responsible for the protective effects of antioxidants, showed that antioxidants can modulate the function of different crucial pathways required for the transduction of the apoptotic signal, such as transduction molecules, second messengers and transcription factors.

As a consequence, antioxidants, and more precisely, the thiol antioxidants (due to their low toxicity) have been proposed for the treatment of patients suffering from pathologies associated with a disturbance of the redox status including AIDS (DROGE et al. 1992), cancer, Alzheimer's disease and amyotrophic lateral sclerosis (characterized by a motor neuron death resulting from a mutation encoding for Cu/Zn SOD) (HACK et al. 1997). Interestingly, NAC has been shown to restore a normal level of $CD4^+$ T lymphocytes in HIV patients, suggesting that this antioxidant may be useful in the treatment of AIDS. Taken together, all the in vitro and in vivo data show that antioxidants appear as useful drugs for the treatment of pathologies characterized by an abnormal apoptosis associated with an alteration of the redox status.

References

Albina JE, Cui S, Mateo RB, Reichner JS (1993) Nitric oxide-mediated apoptosis in murine peritoneal macrophages. J Immunol 150:5080–5085

Anderson MT, Staal FJ, Gitler C, Herzenberg LA, Herzenberg LA (1994) Separation of oxidant-initiated and redox-regulated steps in the NF-kappa B signal transduction pathway. Proc Natl Acad Sci USA 91:11527–11531

Aukrust P, Svardal AM, Muller F, Lunden B, Berger RK, Ueland PM, Froland SS (1995) Increased levels of oxidised glutathione in CD4+ lymphocytes associated with disrupted intracellular redox balance in human immunodeficiency virus type 1 infection. Blood 86:258–267

Aukrust P, Svardal AM, Muller F, Lunden B, Nordoy I, Froland SS (1996) Markedly disrupted glutathione redox status in CD45RA+CD4+ lymphocytes in human immunodeficiency virus type 1 infection is associated with selective depletion of this lymphocyte subset. Blood 88:2626–2633

Baker AF, Briehl MM, Dorr R, Powis G (1996) Decreased antioxidant defense and increased oxidant stress during dexamethasone-induced apoptosis: *bcl-2* selectively prevents the loss of catalase activity. Cell Death Diff 3:207–213

Barja G (1993) Oxygen radicals, a failure or a success of evolution? Free Radic Res Commun 18:63–70

Bauer MKA, Vogt M, Los M, Siegel J, Wesselborg S, Schulze-Osthoff K (1998) Role of reactive oxygen intermediates in activation-induced CD95 (APO-1/Fas) ligand expression. J Biol Chem 273:8048–8055

Beg AA, Finco TS, Nantermet PV, Baldwin AS Jr (1993) Tumor necrosis factor and interleukin-1 lead to phosphorylation and loss of I kappa B alpha: a mechanism for NF-kappa B activation. Mol Cell Biol 13:3301–3310

Beg AA, Baltimore D (1996) An essential role for NF-kappaB in preventing TNF-alpha-induced cell death. Science 274:782–784

Behl C, Davis JB, Cole GM, Schubert D (1992) Vitamin E protects nerve cell from amyloid β protein toxicity. Biochem Biophys Res Comm 186:944–950

Behl C, Davis JB, Lesley R, Schubert D (1994) Hydrogen peroxide mediates amyloid β protein toxicity. Cell 77:817–827

Benchekroun MN, Sinha BK, Robert J (1993) Doxorubicin-induced oxygen free radical formation in sensitive and doxorubicin-resistant variants of rat glioblastoma cell lines. FEBS Lett 326:302–305

Briehl MM, Cotgreave IA, Powis G (1995) Downregulation of the antioxidant defense during glucocorticoid-mediated apoptosis. Cell Death Differ 2:41–46

Buttke TM, Sandstrom PA (1994) Oxidative stress as a mediator of apoptosis. Immunol today 15:7–10

Chen M, Quintas J, Fuks S, Thompson C, Kufez DW, Weichselbaum RR (1995) Suppression of *BCL-2* messenger RNA production may mediate apoptosis after ionizing radiation, tumour necrosis factor and ceramide. Cancer Res 55:991–994

Chiba T, Takahashi S, Sato N, Ishii S, Kikuchi K (1996) Fas-mediated apoptosis is modulated by intracellular glutathione in human T cells. Eur J Immunol 26:1164–1169

Cohen JJ, Duke RC (1992) Apoptosis and programmed cell death in immunity. Ann Rev Immunol 10:267–293

Davis JB (1996) Oxidative mechanisms in β-amyloid cytotoxicity. Neurodegeneration 5:441–444

Deas O, Dumont C, Mollereau B, Metivier D, Pasquier C, Bernard-Pomier G, Hirsch F, Charpentier B, Senik A (1997) Thiol-mediated inhibition of FAS and CD2 apoptotic signaling in activated human peripheral T cells. Int Immunol 9:117–125

Delneste Y, Jeannin P, Sebille E, Aubry J-P, Bonnefoy J-Y (1996) Thiols prevent Fas(CD95)-mediated T cell apoptosis by down-regulating membrane Fas expression. Eur J Immunol 26:2981–2988

Delneste Y, Jeannin P, Potier L, Romero P, Bonnefoy J-Y (1997) N-acetyl-L-cysteine exhibits antitumoral activity by increasing tumor necrosis factor α-dependent T-cell cytotoxicity. Blood 90:1124–1132

Devary Y, Gottlieb RA, Lau LF, Karin M (1991) Rapid and preferential activation of the c-jun gene during the mammalian UV response. Mol Cell Biol 11:2804–2811

Diomede L, Piovani B, Re F, Principe P, Colotta F, Modest EJ, Salmona M (1994) The induction of apoptosis is a common feature of the cytotoxic action of ether-linked glycerophospholipids in human leukemic cells. Int J Cancer 57:645–649

Dröge W, Eck H-P, Mihm S (1992) HIV-induced cysteine deficiency and T-cell dysfunction – a rationale for treatment with N-acetylcysteine. Immunol today 13:211–214

Eck HP, Gmünder H, Hartmann M, Petzoldt D, Daniel V, Dröge W (1989) Low concentrations of acid soluble thiol (cysteine) in the blood plasma of HIV-1 infected patients. Biol Chem Hoppe-Seyler 370:101–108

Escargueil I, Negre-Salvayre A, Pieraggi MT, Salvayre R (1992) Oxidized low density lipoproteins elicit DNA fragmentation of cultured lymphoblastoid cells. FEBS Lett 305:155–159

Eylar E, Rivera-Quinones C, Molina C, Baez I, Molina F, Mercado CM (1993) N-acetylcysteine enhances T cell functions and T cell growth in culture. Int Immunol 5:97–101

Fang W, Rivard JJ, Ganser JA, LeBien TW, Nath KA, Mueller DL, Behrens TW (1995) Bcl-x_L rescues WEHI 231 B lymphocytes from oxidant-mediated death following diverse apoptotic stimuli. J Immunol 155:66–75

Fanger GR, Gerwins P, Widmann C, Jarpe MB, Johnson GL (1997) MEKKs, GCKs, MLKs, PAKs, TAKs, and tpls: upstream regulators of the c-Jun amino-terminal kinases? Curr Opin Genet Dev 7:67–74

Fisher GH, Rosenberg FJ, Straus SE, Dale JK, Middleton LA, Lin AY, Strober W, Lenardo MJ, Puck JM (1995) Dominant interfering Fas gene mutations impair apoptosis in a human autoimmune lymphoproliferative syndrome. Cell 81:935–946

Greenlund LJ, Deckwerth TL, Johnson EM Jr (1995) Superoxide dismutase delays neuronal apoptosis: a role for reactive oxygen species in programmed neuronal death. Neuron 14:303–315

Grimm S, Bauer MK, Baeuerle PA, Schulze-Osthoff K (1996) Bcl-2 down-regulates the activity of transcription factor NF-kappaB induced upon apoptosis. J Cell Biol 134:13–23

Groux H, Torpier G, Monte D, Mouton Y, Capron A, Ameisen J (1992) Activation-induced death by apoptosis in CD4+ T cells from human immunodeficiency virus-infected asymptomatic individuals. J Exp Med 175:331–340

Gunji Y, Hori S, Aoe T, Asano T, Ochiai T, Isono K, Saito T (1994) High frequency of cancer patients bearing abnormal assembly of the T cell receptor-CD3 complex in peripheral blood T lymphocytes. Jpn J Cancer Res 85:1189–1192

Hack V, Schmid D, Breitkreutz R, Stahl-Henning C, Drings P, Kinscherf R, Taut F, Holm E, Droge W (1997) Cystine levels, cystine flux, and protein catabolism in cancer cachexia, HIV/SIV infection, and senescence. FASEB J 11:84–92

Halliwell B, Hoult JR, Blake DR (1988) Oxidants, inflammation, and anti-inflammatory drugs. FASEB J 2:2867–2873

Hengartner MO, Horvitz H (1994) C. elegans cell survival gene ced-9 encodes a functional homolog of the mammalian proto-oncogene bcl-2. Cell 25:665–676

Hensley K, Carney JM, Mattson MP, Aksenova M, Harris M, Wu JF, Floyd R, Butterfield DA (1994) A model for β-amyloid aggregation and neurotoxicicty based on free radical generation by the peptide: relevance to Alzheimer's disease. Proc Natl Acad Sci USA 91:3270–3274

Hirose K, Longo DL, Oppenheim JJ, Matsushima K (1993) Overexpression of mitochondrial manganese superoxide dismutase promotes the survival of tumor cells exposed to interleukin-1, tumor necrosis factor, selected anticancer drugs, and ionizing radiation. FASEB J 7:361–368

Hockenbery DM, Oltvai ZN, Yin XM, Milliman CL, Korsmeyer SJ (1993) Bcl-2 functions in an antioxidant pathway to prevent apoptosis. Cell 75:241–251

Huang CS, Moore WR, Meister A (1988) On the active site thiol of gamma-glutamylcysteine synthetase: relationships to catalysis, inhibition, and regulation. Proc Natl Acad Sci USA 85:2464–2468

Hug H, Enari M, Nagata S (1994) No requirement of reactive oxygen intermediates in Fas-mediated apoptosis. FEBS Lett 351:311–313

Ichijo H, Nishida E, Irie K, ten Dijke P, Saitoh M, Moriguchi T, Takagi M, Matsumoto K, Miyazono K, Gotoh Y (1997) Induction of apoptosis by ASK1, a mammalian MAPKKK that activates SAPK/JNK and p38 signaling pathways. Science 275: 90–94

Irmler M, Thome M, Hahne M, Schneider P, Hofmann K, Steiner V, Bodmer JL, Schroter M, Burns K, Mattmann C, Rimoldi D, French LE, Tschopp J (1997) Inhibition of death receptor signals by cellular FLIP. Nature 388:190–195

Israel N, Gougerot-Pocidalo MA, Aillet F, Virelizier JL (1992) Redox status of cells influences constitutive or induced NF-kappa B translocation and HIV long terminal repeat activity in human T and monocytic cell lines. J Immunol 149: 3386–3393

Iwata S, Hori T, Sato N, Ueda-Taniguchi Y, Yamabe T, Nakamura H, Masutani H, Yodoi J (1994) Thiol-mediated redox regulation of lymphocyte proliferation. Possible involvement of adult T cell leukemia-derived factor and glutathione in transferrin receptor expression. J Immunol 152:5633–5642

Jäättela M, Benedict M, Tewari M, Shayman JA, Dixit VM (1995) Bcl-x and Bcl-2 inhibit TNF and Fas-induced apoptosis and activation of phospholipase A2 in breast carcinoma cells. Oncogene 10:2297–2305

Jacobson MD, Raff MC (1995) Programmed cell death and Bcl-2 protection in very low oxygen. Nature 374:814–816

Ju ST, Panka DJ, Cui H, Ettinger R, El-Khatib M, Sherr DH, Stanger BZ, Marshak-Rothstein A (1995) Fas(CD95)/FasL interactions required for programmed cell death after T-cell activation. Nature 373:444–448

Kabelitz D (1998) Apoptosis, graft rejection, and transplantation tolerance. Transplantation 65:869–875

Kaltschmidt B, Uherek M, Volk B, Baeuerle PA, Kaltschmidt C (1997) Transcription factor NF-kappaB is activated in primary neurons by amyloid beta peptides and in neurons surrounding early plaques from patients with Alzheimer disease. Proc Natl Acad Sci USA 94:2642–2647

Kamata H, Tanaka C, Yagisawa H, Hirata H (1996) Nerve growth factor and forskolin prevents H_2O_2-induced apoptosis in PC12 cells by glutathione independent mechanism. Neurosci Lett 212:179–182

Kane DJ, Sarafian TA, Anton R, Hahn H, Gralla EB, Valentine JS, Ord T, Bredesen DE (1993) Bcl-2 inhibition of neural death: decreased generation of reactive oxygen species. Science 262:1274–1277

Kerr JFR, Wyllie AH, Currie AR (1972) Apoptosis: a basic biological phenomenon with wide ranging implications in tissue kinetics. Br J Cancer 26:239–257

Korsmeyer SJ (1992) *Bcl-2*: an antidote to programmed cell death. Cancer Surv 15: 105–118

Krammer PH, Behrmann I, Daniel P, Dhein J, Debatin K-M (1994) Regulation of apoptosis in the immune system. Curr Opin Immunol 6:279–289

Laochumroonvarapung P, Paul S, Elkon KB, Kaplan G (1996) H_2O_2 induces monocyte apoptosis and reduces viability of Mycobacterium avium-M. intracellulare within cultured human monocytes. Infect Immun 64:452–459

Larrick JW, Wright SC (1990) Cytotoxic mechanism of tumor necrosis factor-alpha. FASEB J 4:3215–3223

Lee S-H, Fujita N, Imai K, Tsuro T (1995) Cysteine produced from lymph nodes stromal cells suppresses apoptosis of mouse malignant T-lymphomas. Biochem Biophys Res Commun 213:837–844

Lee Y, Shacter E (1997) Bcl-2 does not protect Burkitt's lymphoma cells from oxidant-induced cell death. Blood 89:4480–4492

Lennon SV, Martin SJ, Cotter TG (1991) Dose-dependent induction of apoptosis in human tumour cell lines by widely diverging stimuli. Cell Prolif 24:203–214

Li PF, Dietz R, van Harsdorf R (1997) Reactive oxygen species induce apoptosis of vascular smooth muscle cell. FEBS Lett 404:249–252

Liang SM, Liang CM, Hargrove ME, Ting CC (1991) Regulation by glutathione of the effect of lymphokines on differentiation of primary activated lymphocytes. Influence of glutathione on cytotoxic activity of CD3-AK. J Immunol 146:1909–1913

Matsuda M, Masutani H, Nakamura H, Miyajima S, Yamauchi A, Yonehara S, Uchida A, Irimajiri K, Horiuchi A, Yodoi J (1991) Protective activity of adult T cell leukemia-derived factor (ADF) against tumor necrosis factor-dependent cytotoxicity on U937 cells. J Immunol 147:3837–3841

Mattson MP, Goodman Y (1995) Different amyloidogenic peptides share a similar mechanism of neurotoxicity involving reactive oxygen species and calcium. Brain Res 676:219–224

Maurice MM, Nakamura H, van der Voort EA, van Vliet AI, Staal FJ, Tak PP, Breedveld FC, Verweij CL (1997) Evidence for the role of an altered redox state in hyporesponsiveness of synovial T cells in rheumatoid arthritis. J Immunol 158:1458–1465

Mayer M, Noble M (1994) N-acetyl-L-cysteine is a pluripotent protector against cell death and enhancer of trophic factor-mediated cell survival in vitro. Proc Natl Acad USA 91:7496–7500

Meyer M, Schreck R, Baeuerle PA (1993) H_2O_2 and antioxidants have opposite effects on activation of NF-κB and AP-1 in intact cells: AP-1 as secondary antioxidant-responsive factor. EMBO J 12:2005–2015

Mihm S, Galter D, Dröge W (1995) Modulation of transcription factor NF kappa B activity by intracellular glutathione levels and by variations of the extracellular cysteine supply. FASEB J 9:246–252

Muschel RJ, Bernhard EJ, Garza L, McKenna WG, Koch CJ (1995) Induction of apoptosis at different oxygen tensions: evidence that oxygen radicals do not mediate apoptotic signaling. Cancer Res 55:995–998

Muzio M, Chinnaiyan AM, Kischkel FC, O'Rourke K, Shevchenko A, Ni J, Scaffidi C, Bretz JD, Zhang M, Gentz R, Mann M, Krammer PH, Peter ME, Dixit VM (1996) FLICE, a novel FADD-homologous ICE/CED-3-like protease, is recruited to the CD95 (Fas/APO-1) death-inducing signaling complex. Cell 85:817–827

Nakagomi H, Petersson M, Magnusson I, Juhlin C, Matsuda M, Mellstedt H, Tanpin J-L, Viviver E, Anderson P, Kiessling R (1993) Decreased expression of the signal transducing ζ chains in tumor-infiltrating T-cells and NK cells of patients with colorectal carcinoma. Cancer Res 53:5610–5612

Narayanan V (1997) Apoptosis in development and disease of the nervous system: 1. Naturally occurring cell death in the developing nervous system. Pediatr Neurol 16:9–13

Otsuji M, Kimura Y, Aoe T, Okamoto Y, Saito Y (1996) Oxidative stress by tumor-derived macrophages suppresses the expression of CD3ζ chain of T-cell receptor complex and antigen-specific T cell responses. Proc Natl Acad Sci USA 93:13119–13124

Pfeifer GP, Denissenko MF (1998) Formation and repair of DNA lesions in the p53 gene: relation to cancer mutations? Environ Mol Mutagen 31:197–205

Powis G, Gasdaska JR, Baker A (1997) Redox signaling and the control of cell growth and death. Adv Pharmacol 38:329–359

Raff MC (1993) Programmed cell death and the control of cell survival: lessons from the nervous system. Science 262:695–700

Rieux-Laucat F, Le Deist F, Hivroz C, Roberts IA, Debatin KM, Fischer A, de Villartay JP (1995) Mutations in Fas associated with human lymphoproliferative syndrome and autoimmunity. Science 268:1347–1349

Roederer M, Staal FJT, Osada H, Herzenberg LA, Herzenberg LA (1991) CD4 and CD8 T cells with high intracellular glutathione levels are selectively lost as the HIV infection progresses. Int Immunol 3:933–937

Rouvier E, Luciani MF, Golstein P (1993) Fas involvement in Ca(2+)-independent T cell-mediated cytotoxicity. J Exp Med 177:195–200

Sagara Y, Dargush R, Klier FG, Schubert DN, Behl C (1996) Increased antioxidant enzyme activity in amyloid beta protein-resistant cells. J Neurosci 16:497–505

Saitoh M, Nishitoh H, Fujii M, Takeda K, Tobiume K, Sawada Y, Kawabata M, Miyazono K, Ichijo H (1998) Mammalian thioredoxin is a direct inhibitor of apoptosis signal-regulating kinase (ASK) 1. EMBO J 17:2596–2606

Sandstrom PA, Buttke TM (1993) Autocrine production of extracellular catalase prevents apoptosis of the human CEM T-cell line in serum-free medium. Proc Natl Acad Sci USA 90:4708–4712

Sandstrom PA, Tebbey PW, Van Cleave S, Buttke TM (1994) Lipid hydroperoxides induce apoptosis in T cells displaying a HIV-associated glutathione peroxidase deficiency. J Biol Chem 269:798–801

Sato N, Iwata S, Nakamura K, Hori T, Mori K, Yodoi J (1995) Thiol-mediated redox regulation of apoptosis. Possible roles of cellular thiols other than glutathione in T cell apoptosis. J Immunol 154:3194–3203

Schreck R, Baeuerle PA (1991a) A role for oxygen radicals as second messengers. Trends Cell Biol 1:39–42

Schreck R, Rieber P, Baeuerle P (1991b) Reactive oxygen intermediates as apparently widely used messengers in the activation of the NF-kappa B transcription factor and HIV-1. EMBO J 10:2247–2258

Schrier PI, Peltenburg LT (1993) Relationship between myc oncogene activation and MHC class I expression. Adv Cancer Res 60:181–246

Schroder H, Warren S, Bargetzi MJ, Torti SV, Torti FM (1993) N-acetyl-L-cysteine protects endothelial cells but not L929 tumor cells from tumor necrosis factor-alpha-mediated cytotoxicity. Naunyn Schmiedebergs Arch Pharmacol 347:664–666

Schulze-Osthoff K, Krammer PH, Dröge W (1994) Divergent signalling via APO-1/Fas and the TNF receptor, two homologous molecules involved in physiological cell death. EMBO J 13:4587–4596

Shi Y, Glynn JM, Guilbert LJ, Cotter TG, Bissonnette RP, Green DR (1992) Role for c-myc in activation-induced apoptotic cell death in T cell hybridomas. Science 257:212–214

Slater AFG, Nobel CSI, Maellaro E, Bustamante J, Kimland M, Orrenius S (1995) Nitrone spin traps and a nitroxide antioxidant inhibit a common pathway of thymocyte apoptosis. Biochem J 306:771–778

Smyth MJ (1991) Glutathione modulates activation-dependent proliferation of human peripheral blood lymphocyte populations without regulating their activated function. J Immunol 146:1921–1927

Staal FJ, Roederer M, Herzenberg LA, Herznberg LA (1990) Intracellular thiols regulate activation of nuclear factor kappa B and transcription of human immunodeficiency virus. Proc Natl Acad Sci USA 87:9943–9947

Staal FJ, Ela SW, Roederer M, Anderson MT, Herzenberg LA, Herzenberg LA (1992) Glutathione deficiency and human immunodeficiency virus infection. Lancet 339:909–912

Staal FJ, Anderson MT, Staal GE, Herzenberg LA, Gitler C, Herzenberg LA (1994) Redox regulation of signal transduction: tyrosine phosphorylation and calcium influx. Proc Natl Acad Sci USA 91:3619–3622

Stangel M, Zettl UK Mix E, Zielasek J, Toyka KV, Hartung HP, Gold R (1996) H_2O_2 and nitric oxide-mediated oxidative stress induce apoptosis in rat skeletal muscle myoblasts. J Neuropathol Exp Neurol 55:36–43

Stefanelli C, Stanic I, Bonavita F, Muscari C, Pignatti C, Rossoni C, Caldarera CM (1995) Oxygen tension influences DNA fragmentation and cell death in glucocorticoid-treated thymocytes. Biochem Biophys Res Commun 212:300–306

Stefanova I, Saville MW, Peters C, Cleghorn FR, Schwartz D, Venzon DJ, Weinhold KJ, Jack N, Bartholomew C, Blattner WA, Yarchoan R, Bolen JB, Horak ID (1996)

HIV infection-induced posttranslational modification of T cell signaling molecules associated with disease progression. J Clin Invest 98:1290–1297

Suzuki YJ, Aggarwal BB, Packer L (1992) Alpha-lipoic acid is a potent inhibitor of NF-kappa B activation in human T cells. Biochem Biophys Res Commun 189: 1709–1715

Suzuki YJ, Packer L (1993) Inhibition of NF-kappa B activation by vitamin E derivatives. Biochem Biophys Res Commun 193:277–283

Szatrowski TP, Nathan CF (1991) Production of large amount of hydrogen peroxide by tumor cells. Cancer Res 51:794–798

Toledano MB, Leonard WJ (1991) Modulation of transcription factor NF-kappa B binding activity by oxidation-reduction in vitro. Proc Natl Acad Sci USA 88: 4328–4332

Tomimoto H, Akiguchi I, Wakita H, Kimura J, Hori K, Yodoi J (1993) Astroglial expression of ATL-derived factor, a human thioredoxin homologue, in the gerbil brain after transient global ischemia. Brain Res 625:1–8

van den Dobbelsteen DJ, Nobel SI, Schlegel J, Cotgreave IA, Orrenius S, Slater AFG (1996) Rapid and specific efflux of reduced glutathione during apoptosis induced by anti-fas/APO-1 antibody. J Biol Chem 271:15420–15427

Veis DJ, Sorenson CM, Shutter JR, Korsmeyer SJ (1993) Bcl-2-deficient mice demonstrate fulminant lymphoid apoptosis, polycystic kidneys, and hypopigmented hair. Cell 75:229–240

Von Boehmer H (1992) Thymic selection: a matter of life and deth. Immunol today 13:454–458

Wang J, Lenardo MJ (1997) Molecules involved in cell death and peripheral tolerance. Curr Opin Immunol 9:818–825

Wolfe JT, Ross D, Cohen GM (1994) A role for metals and free radicals in the induction of apoptosis in thymocytes. FEBS Lett 352:58–62

Wollman EE, d'Auriol L, Rimsky L, Shaw A, Jacquot JP, Wingfield P, Graber P, Dessarps F, Robin P, Galibert F, Bertoglio J, Fradelizi D (1988) Cloning and expression of a cDNA for human thioredoxin. J Biol Chem 263:15506–15512

Wyllie AH, Kerr JFG, Currie AR (1980) Cell death: the significance of apoptosis. Int Rev Cytol 68:251–306

Yankner BA, Duffey LK, Kirschner DA (1990) Neurotrophic and neurotoxic effects of amyloid β protein. Reversal by tachykinin neuropeptides. Science 250:279–281

Yim CY, Hibbs JB Jr, McGregor JR, Galinsky RE, Samlowski WE (1994) Use of N-acetyl cysteine to increase intracellular glutathione during the induction of antitumor responses by IL-2. J Immunol 152:5796–5805

Zettl UK, Mix E, Zielasek J, Stangel M, Hartung HP, Gold R (1997) Apoptosis of myelin-reactive T cells induced by reactive oxygen and nitrogen intermediates in vitro. Cell Immunol 178:1–8

Zhong LT, Sarafian T, Kane DJ, Charles AC, Mah SP, Edwards RH, Bredesen DE (1993) bcl-2 inhibits death of central neural cells induced by multiple agents. Proc Natl Acad Sci USA 90:4533–4537

CHAPTER 11
Reactive Oxygen Species and Apoptosis

G. Bauer, S. Dormann, I. Engelmann, A. Schulz, and M. Saran

A. Introduction

There is increasing evidence for the involvement of reactive oxygen species (ROS) in the regulation of central biological functions. Interaction between certain ROS and the generation of highly reactive ROS at desired locations, as well as their modulation by antioxidants and a variety of enzymes, warrant a hitherto unexpected degree of efficiency and specificity. ROS are involved in triggering and mediating apoptosis under physiological and pathophysiological conditions. This paper summarizes the major interdependencies of ROS and their physiological sources, and critically reviews the data on the evidence for the role of ROS during induction and execution of apoptosis. The focus is on the action of superoxide anions, hydrogen peroxide, hydroxyl radicals, hypochlorous acid, nitric oxide and peroxynitrite. Glutathione represents one of the key elements during the regulation of apoptosis. It balances against ROS created by multiple signaling pathways, enzymatic reactions or mitochondria, and it inhibits sphingomyelinase, the key enzyme for the generation of ceramide. This second messenger is intrinsically interwoven with the generation of ROS and with activation of execution-caspases. Mitochondria are both the target and the source of ROS during induction and execution of apoptosis. The control of the mitochondrial permeability transition pore is therefore of central importance for the regulation of apoptosis. Tumor necrosis factor induces apoptosis through a versatile use of ROS. Similarly, ROS are involved in Apo/Fas-triggered or p53-mediated apoptosis at several distinct and synergistically acting steps. Direct apoptosis induction by TGF-beta depends on the action of ROS. Intercellular and intracellular ROS signaling is the basis for intercellular induction of apoptosis, a recently discovered system for the control of oncogenesis. It is based on specific apoptosis induction in transformed cells by their nontransformed neighbors. Superoxide anions released from transformed cells are the key for specific apoptosis induction. During intercellular signaling, a myeloperoxidase-analogous enzyme converts hydrogen peroxide (generated through dismutation of superoxide anions) into hypochlorous acid. This compound reacts with superoxide anions at the membrane of the transformed cells to form the ultimate apoptosis-inducing hydroxyl radical. The limited diffusion pathway of superoxide anions and the extreme reactivity of hydroxyl radicals ensure that apoptosis induction is restricted to transformed cells. The same signaling principle seems to be used when nitric oxide, a long-ranging signal is converted to the reactive peroxynitrite by superoxide anions. These data indicate that natural antitumor mechanisms utilize similar signaling principles for specific apoptosis induction in transformed cells.

B. Reactive Oxygen Species: Shotgun or Precision Tool?

Oxygen radicals that arise from the disintegration of water after adsorption of ionizing radiation have originally been the main focus of radiation research. After it became clear that some of these species also play important roles in biological systems, the acronym reactive oxygen species (ROS) was introduced to encompass a much wider spectrum of reactive species. The term is now used for short-lived entities such as hydroxyl (·OH), alkoxyl (RO·) or peroxyl (ROO·) radicals, for some radical species of medium lifetime such as superoxide (O_2^-) or the nitroxyl radical (NO·) (also termed nitric oxide) and also includes non-radical end products like hydrogen peroxide (H_2O_2), organic hydroperoxides (ROOH) and hypochlorous acid (HOCl), and in some respect also peroxynitrite, the cross-product of NO· and O_2^-. In a broader sense, those valency states of enzymes that use oxygen or hydrogen peroxide to bring the inactive metal in some activated form (i.e., ferryl-perferryl states of peroxidases, cytochrome P-450 enzyme, ribonucleotide reductase) may also be subsumed under the header ROS even though they do not exactly comply with the idea of being freely diffusible entities. For radiobiologists, the destructive nature of ROS through interaction with cellular macromolecules, especially DNA, seemed to be of central importance. Therefore, it was not astonishing to find out that ROS were used by phagocytic cells for antimicrobial action, an effect that required specific recognition of the target by the phagocyte, but no obvious need for specific and balanced reactions during target destruction. Already in the late 1980s SARAN and BORS (1989) postulated the hypothesis that ROS may act as chemical messengers rather than being merely destructive. This hypothesis was later verified, when it was shown that ROS may also be involved in signaling pathways (WOLIN 1996; SUZUKI et al. 1997a; JORDAN and IYENGAR 1998; LEE et al. 1998). Moreover, they act as specific activators or inactivators of enzymes (YAO et al. 1996; LI et al. 1995; ERMACORA et al. 1992; 1994; WITTUNG and MALMSTROM 1996; SAARI et al. 1992), ion channels (RUPPERSBERG et al. 1991), receptors (COFFER et al. 1995; KNEBEL et al. 1996; HUANG et al. 1996), cytokines (BARCELLOS-HOFF et al. 1996) or other regulatory molecules such as transcription factors (SCHRECK et al. 1991). The basis for specific ROS action lies in the ability of the cells to regulate their synthesis or release (e.g., of O_2^- by NADPH oxidase), to modulate their reactions through specific enzymes (e.g., formation of hydrogen peroxide from superoxide anions through SOD; generation of HOCl from hydrogen peroxide and chloride through myeloperoxidase and related enzymes), and to counterbalance their action through antioxidants (like reduced glutathione) or enzymes (like catalase or SOD). The central secret for specific ROS action in biological systems seems, however, to depend on the right site of synthesis and the controlled conversion of less reactive species with a longer range of action to highly reactive species with a short range of action at the desired location. This principle – relevant for the understanding of the specific role of ROS – can be illustrated by two recent papers dealing with the activation of latent TGF-beta (BARCELLOS-HOFF et al. 1996) or with the delicate balance of ROS during phagocyte antimicrobial action (SARAN et al. 1999).

TGF-beta is involved in a multitude of biological functions. It is released from cells as an inactive complex consisting of a large latency-associated protein (LAP) and the smaller cytokine. A change in conformation of the LAP and subsequent release of the smaller cytokine leads to its activation. Conformational change can be achieved by pH-shock, heat, protease cleavage or interaction with the carbohydrate moiety (summarized in HÄUFEL et al. 1999). The work of BARCELLOS-HOFF et al. (1996) showed that ROS can also mediate specific activation. Their model implies that relatively nonaggressive ROS members, form highly reactive hydroxyl radicals. These may oxidize sulfhydryl groups of cysteines and, thus, lead to a conformational change of the molecule, which is required for activation. Accordingly, site-specific generation of highly reactive and therefore short-ranged ROS represents an efficient and specific modulation of protein conformation with significant regulatory consequences.

Analogous ideas have recently been proposed for the scenario of phagocyte interaction with bacteria (SARAN et al. 1999). The specific task of the phagocyte is to perform an aggressive ROS attack on the microbe without damage to its own cellular membranes. This is achieved by the discharge of superoxide anions through a membrane-associated NADPH oxidase. The superoxide anions within the phagosome may either form hydrogen peroxide or remain for a while as superoxide anions. The concentrations of these two relatively nonaggressive members of the ROS family are too low to induce direct damage of either the microbe or the cell membrane. Concomitantly, myeloperoxidase released into the phagosome binds to bacteria and synthesizes hypochlorous acid using hydrogen peroxide and chloride anions. The binding of myeloperoxidase to the bacterium ensures that HOCl is synthesized where it is needed. HOCl is not the ultimate toxic substance, however. Highly reactive hydrogen radicals are produced through the interaction of hypochlorous acid with superoxide anions, as the phagosome is small enough to allow migration of free superoxide anions from the phagocytic membrane to the microbe. This interplay illustrates a complex and fascinating interaction of different members of the ROS family, based on their different reactivity and range of action, allowing an efficient defense system without the danger of damage to the effector cell.

C. Interdependencies of ROS

Several recent reviews on the chemistry of ROS may be used for further reference (Winterbourn 1995a, 1995b; Hampton et al. 1998; Saran et al. 1998). Here we summarize only the major interdependencies of the various species which are essential for the understanding of the role of ROS during induction of apoptosis.

Superoxide anions are mild oxidants but may also reduce compounds of adequate reduction potential, such as cytochrome c, for example. Their chemical half-life is unusually long for a radical and results in a diffusion path length of a few micrometers, i.e., of the order of magnitude of single cells (Saran and Bors 1994). It is conceivable that superoxide anions pass through cellular membranes after reaction with protons. Superoxide anions dismutate spontaneously or are driven by superoxide dismutase to form hydrogen peroxide, an oxidant that readily interacts with thiols. Hydrogen peroxide and chloride ions serve as substrates for myeloperoxidase, lactoperoxidase or eosinophilic peroxidase to form hypochlorous acid. In the presence of Fe^{++} or Cu^{++} ions, hydrogen peroxide forms hydroxyl radicals through the Fenton reaction (Winterbourn 1995b).

Hydroxyl radicals represent the most reactive ROS, readily causing oxidation of thiols or lipid peroxidation. Their ability to react with the next suitable neighbor molecule results in an extremely short chemical half-life and range of action. However, hydroxyl radicals can readily react with chloride ions and start a cascade of reactions ultimately leading to the formation of chlorine and hypochlorous acid (Saran and Bors 1997; Saran et al. 1997, 1999). Reaction of superoxide anions with hypochlorous acid generates hydroxyl radicals (Ramos et al. 1992; Candeias et al. 1993; Hippeli et al. 1997).

HOCl is an oxidizing and chlorinating agent. Its oxidative attack on proteins is directed against sulfhydryl groups (Hu et al. 1993). During HOCl-mediated cartilage degradation, oligomeric polysaccharides are released from cartilage, N-acetyl side chains are degraded via a chlorinated transient product and an interaction of HOCl restrictively with alanine is measured (Schiller et al. 1995). HOCl has no potency for direct lipid peroxidation (Hu et al. 1993), but hydroxyl radicals derived from HOCl/superoxide anion interaction are powerful lipid peroxidants. HOCl and hydrogenperoxide can interact to form water, protons, chloride and molecular oxygen, thus neutralizing their oxidative potential and their biological effects (Saran et al. 1999).

Nitric oxide, a long-lived radical with a wide range of action is known as a regulator of a variety of biological processes. NO· can cause termination of lipid radical chains by formation of less reactive nitrogen-containing products (Rubbo et al. 1998) but can also form highly reactive radicals through several distinct pathways. NO· plus superoxide anions form peroxynitrite in a diffusion controlled reaction (Saran et al. 1990; Huie and Padmaja 1993). This is a very efficient lipid peroxidant and can cause both nitration or oxidation of proteins (RADI et al. 1991; ISCHIROPOULOS et al. 1992; SQUADRITO and PRYOR 1998). Generation of peroxynitrite from NO· and superoxide anions can be inhibited by superoxide dismutase. NO can be oxidized to the nitrite anion, which is used by myeloperoxidase to form nitrogen dioxide (EISERICH et al. 1998). Nitrylchloride can be formed by direct interaction of nitrite with hypochlorous acid. Interestingly, the reaction of NO with hydrogen peroxide (NAPPI and VASS 1998), as well as the decomposition of peroxynitrons acid (BECKMAN et al. 1990; RICHESON et al. 1998), can yield the highly reactive hydroxyl radical.

For the demonstration of the functional role of ROS several enzymes, antioxidants and radical scavengers have been instrumental. Inhibition of a process by superoxide dismutases (either mitochondrial MnSOD or cytosolic Zn, CuSOD) implies a direct functional role of superoxide anions. Lack of inhibition may indicate that they have no direct role in a given process. However, since SODs catalyze the formation of hydrogen peroxide from superoxide anions, lack of inhibition by SOD may, therefore, alternatively indicate that superoxide anion-derived hydrogen peroxide is the essential member in the chain of reactions. In this case, addition of SOD would enhance the process through acceleration of hydrogen peroxide formation from superoxide anions rather than inhibit it. Glutathione is the central antioxidant, reacting with most of the ROS species except superoxide anions (GILLESSEN et al. 1997). N-acetylcystein, which readily passes cell membranes, is a substrate for GSH, but also acts as an antioxidant itself, and seems to react with HOCl, hydrogen peroxide, hydroxyl radicals but not with superoxide anions (ARUOMA et al. 1989). The involvement of hydroxyl radicals can be either substantiated by prevention of the Fenton reaction through chelating iron and copper ions, or by the addition of hydroxyl radical scavengers like DMSO or mannitol. HOCl can be scavenged by taurine, an amino acid which specifically interacts with HOCl, but not with hydroxyl anions, superoxide anions or hydrogen peroxide (ARUOMA et al. 1988; GRISHAM et al. 1984; SHI et al. 1997). Several specific enzyme inhibitors exist, such as diphenyleneiodonium (DPI) for NADPH oxidase, 4-aminobenzoic acid hydrazide (ABAH) for myeloperoxidase (KETTLE et al. 1995, 1997) or N-omega-nitro-L-arginine methyl ester (L-NAME) or N(G)-monomethyl-L-arginine (L-NMMA) for NO synthetase. Substances that release NO· (like sodium nitroprusside), or NO· plus superoxide anions that instantly form peroxynitrite (like 3-morpholinosydnonimine hydrochloride [SIN-1]), have been very useful in elucidating the role of NO and its products in apoptosis induction.

D. Physiological Sources of ROS

For experimental purposes, radiation is still the classical method of generating and investigating ROS. Within cellular systems, however, the contribution of radicals produced by background environmental radiation is negligible. In contrast, other sources of radicals operate here. They may belong to different classes: (1) processes that liberate ROS as unwanted (but unavoidable) byproducts, e.g., electron leakage of mitochondria, redox cycling of quinoid compounds; (2) processes that generate ROS for teleologically intended purposes, e.g. the NADPH oxidase and myeloperoxidase of phagocytes; (3) processes that, during abnormal episodes of ROS generation, result in pathological processes such as the ischemia/reperfusion syndrome. The main point is that none of these effects can be regarded separately. In particular, pathological processes connected with abnormal levels of free metals and those that occur with increases in hydrogen peroxide levels are intrinsically interwoven through the Fenton reaction. The hydroxyl radicals thus formed may, in turn, enter a chloride-dependent pathway since they have a greater chance to react with abundant chloride ions, initiating a sequence of events that leads to the formation of hypochlorous acid (SARAN and BORS 1997; SARAN et al. 1999). Hypochlorous acid may then cause hydroxyl radical formation after interaction with superoxide anions.

Superoxide anions can be generated by a multitude of systems (SEGAL 1992; McCORD and OMAR 1993; MOHAZZAB and WOLIN 1994; BABIOR 1995; McCORD 1995; DARLEY-USMAR and HALLIWELL 1996; WOLIN et al. 1996; SARAN et al. 1998, 1999). Xanthine and xanthine oxidase yield superoxide anions, a reaction which is often used experimentally. The major sources for superoxide anions in vivo are, however, membrane NADPH oxidases and the mitochondria. Membrane NADPH oxidases are central enzymes for the oxidative burst of phagocytes but are also connected to the function of protooncogenes and oncogenes (SUNDARESAN et al. 1996; IRANI et al. 1997; JORDAN and IYENGAR 1998; DIEKMAN et al. 1994; KNAUS et al. 1991). Superoxide anions are involved in the maintenance of ras-mediated transformation (IRANI et al. 1997). Many signals during apoptotic induction aim at mitochondria and cause hypergeneration and release of superoxide anions after the opening of the permeability transition pore, and the disruption of the mitochondrial membrane potential. Fibroblasts possess a distinct NADPH oxidase on their membrane (MEIER et al. 1989, 1991, 1993; THANNICKAL and FANBURG 1995) which is regulated by cytokines. In addition to an inducible system, cells carry a NADPH oxidase system which is ready to respond to a signal as simple as the touch of an electrode (ARBAULT et al. 1997). This process has been discussed by the authors to mimic membrane interaction of intruding bacteria or viruses, and points out the involvement of NADPH oxidase in a cellular alert system. Superoxide anions can also be produced through cyclooxygenase (MOHAZZAB and WOLIN 1994) or microsomal cytochrome P 450 (JOSEPH and JAISWAL 1998).

Hydrogen peroxide can be formed either through dismutation of superoxide anions (ZULUETA et al. 1995) or directly in enzymatic reactions like the oxidation of glucose by glucose oxidase (CHANCE et al. 1979). As mentioned above, hydroxyl radicals are either formed through the Fenton reaction (KOPPENOL 1993; WALLING 1995; WINTERBOURN 1995b; WARDMAN and

CANDEIAS 1996), through interaction of superoxide anions with hypochlorous acid (RAMOS et al. 1992; CANDEIAS et al. 1993; HIPPELI et al. 1997), through decay of peroxynitrons acid, or by interaction of NO· with hydrogenperoxide (BECKMAN et al. 1990; CROW et al. 1994; RICHESON et al. 1998).

HOCl is synthesized by myeloperoxidase and related enzymes (KETTLE and WINTERBOURN 1997). This molecule has mainly been observed in the context of phagocytic activity. Recent evidence from our laboratory shows that HOCl can induce apoptosis in superoxide anion-producing transformed cells during the control of oncogenesis. Myeloperoxidase and superoxide anions are the central players in this scenario (ENGELMANN et al., in preparation). This allows the speculation that the emerging role of myeloperoxidase in a multitude of diseases (DAUGHERTY et al. 1994; NAGRA et al. 1997; WORLITZSCH et al. 1998; MOHAMMED et al. 1998) is possibly also due to HOCl-mediated apoptotic induction. This may be the basis for future therapeutic concepts.

NO· is synthesized both by a constitutively expressed NO· synthetase, as well as by an inducible enzyme (iNOS). It can be synthesized by a variety of cells in the context of physiological reactions and is involved in antitumor defense mechanisms exerted by macrophages and granulocytes. Although primarily it has regulatory functions for the endothelium, NO· also plays a role in an endothelial defense mechanism against tumor cells (UMANSKY et al. 1997; EDMISTON et al. 1998) which may prevent tumor cells present in the bloodstream from entering tissues through the endothelium. Peroxynitrite, formed from NO· and superoxide anions in a diffusion controlled reaction (HUIE and PADMAJA 1993), may be the ultimate reacting ROS in this system.

E. ROS and Apoptosis

I. ROS-Dependent Apoptosis Under Physiological and Pathophysiological Conditions

In 1987, BISHOP et al. tested whether reactive oxygen species might induce apoptosis. They used the xanthine/xanthine oxidase system (which generates superoxide anions) or the radiomimetic substance t-BOOH, known to cause lipid peroxidation (LANGLEY et al. 1993). Superoxide dismutase, which catalyzes the formation of hydrogen peroxide from superoxide anions, did not inhibit cell death induced by the xanthine/xanthine oxidase system, while catalase, which destroys hydrogen peroxide, attenuated cell death. Hydroxyl radical scavengers gave inconsistent results. Cell death was characterized as necrosis. From our present knowledge, we conclude that the authors did not observe ROS-dependent apoptotic induction, as the concentration of ROS applied was probably too high. Under these conditions, direct damage of the membrane may have led to necrosis or secondary necrosis might have been caused by the fast shut-down of cellular metabolism, masking the apoptotic process. (Similarly, the action of TNF, which induces apoptosis in tumor cells, was originally described as necrosis. This observation even gave the cytokine its name). Based on the inhibitor data, hydrogen peroxide seemed to be the responsible molecule in this study. A few years later, hydrogen peroxide was, indeed, shown to induce apoptosis in blastocysts, establishing the role of ROS in the induction of apoptosis (PARCHMENT 1991; PIERCE et al. 1991). Since then, ROS-dependent apoptosis has been described in physiological processes like morphogenesis during mouse embryogenesis (SALAS VIDAL et al. 1998), regression of the tadpole tail (HANADA et al. 1997) and a multitude of pathological processes like neurodegenerative diseases (JENNER and OLANOW 1996; LUO et al. 1998), Downs syndrome (BUSCIGLIO and YANKNER 1995), atherosclerosis (DIMMELER et al. 1997a), heart disease (FERRAI et al. 1998), pesticide intoxication (BAGCHI et al. 1995), pathological effects of asbestos (BROADDUS et al. 1996), prion disease (KRETZSCHMAR et al. 1997), bacterial meningitis (LEIB et al. 1996) and HIV infection (DOBMEYER et al. 1997). In addition, ROS have been demonstrated to be central triggering and modulating elements during natural antitumor mechanisms such as the action of TNF, intercellular induction of apoptosis (a novel regulatory system for the prevention of tumorigenesis based on the induction of apoptosis in transformed cells by their nontransformed neighbors) (JÜRGENSMEIER et al. 1994b; SCHAEFER et al. 1995; LANGER et al. 1996; BAUER 1996) and apoptotic induction through endothelial cells (UMANSKY et al. 1997; EDMISTON et al. 1998) – a mechanism perhaps especially related to the control of metastasis. These effects of ROS will be discussed later in separate chapters.

II. Evidence for the Role of ROS During Induction and Execution of Apoptosis

The role of ROS in triggering, mediating, and executing apoptosis is no longer questioned today. The following experiments justify this conclusion: certain ROS induced apoptosis, specific antioxidants, or antioxidant enzymes inhibited apoptosis, other apoptosis signal molecules triggered intracellular ROS generation, and antioxidants inhibited their effects. In addition, the modulating effect of the cellular redox state on the efficiency of apoptosis and the interaction of cellular antioxidants with ROS and vice versa teach us a lot about ROS involvement during apoptosis. Thereby the reduction of cellular glutathione levels can sometimes be the cause, sometimes the consequence of ROS-mediated apoptosis. The basic question "Is apoptosis mediated by ROS?" has been incisively answered in the reviews by SARAFIAN and BREDESEN (1994); BUTTKE and SANDSTROM (1994); JACOBSON (1996); CLUTTON (1997). The role of the antioxidant defense has been clearly presented by BRIEHL and BAKER (1996); SLATER et al. (1996).

The purpose of this review is to focus on systems where ROS act at different sites of the apoptosis scenario either sequentially or in parallel, and to demonstrate the well-balanced interaction of ROS during apoptosis induction.

Some selected papers on the functional or causative role of ROS for apoptosis shall be mentioned first. PIERCE et al. (1991) presented data on the apoptosis inducing capacity of hydrogen peroxide in blastocysts which was prevented by catalase. Their paper also states the important finding that the intracellular glutathione level opposes the apoptosis-triggering effect of ROS. Since then many papers have used the generation of ROS as a trigger for apoptosis. Rollet Labelle et al. (1998) studied neutrophils that were subjected to either xanthine-xanthine oxidase (production of superoxide anions) or glucose oxidase (production of hydrogen peroxide) in the presence of various inhibitors. In their system, the presence of SOD had no inhibitory effect, indicating that superoxide anions were not directly involved in the induction of apoptosis (note that this result does not exclude the fact that that superoxide anions may have an indirect apoptosis-inducing effect after having formed hydrogen peroxide). Catalase prevented apoptosis induction by both systems, as well as spontaneous apoptosis, indicating that hydrogen peroxide had a functional role in this process. But hydrogen peroxide did not seem to be the ultimate mediating molecule, as prevention of hydroxyl radical formation through the addition of iron chelators prevented apoptosis. This paper demonstrates the sequence from superoxide anion over hydrogen peroxide and suggests the final highly reactive hydroxyl radical to be the ultimate oxidizing agent during apoptosis induction. The functional role of hydroxyl radicals during apoptosis induction has been elaborated in many systems (RAUEN and DE GROOT 1998; Aoshima et al. 1997; LI et al. 1997a; XU et al. 1997). Superoxide anions generated in abundance can lead to the sequence illustrated by Rollet Labelle et al. (1998); at lower concentrations they may serve other functions without being cytotoxic. LI et al. (1997a) demonstrated that superoxide anions applied in a single exposure were mitogenic, whereas their repeated exposure at high concentrations or the direct generation of hydrogen peroxide induced apoptosis – an impressive example of the differential effects of ROS and their concentration-dependent interaction. But superoxide anions may also have direct roles during apoptosis induction. The study by SUZUKI et al. (1997b) demonstrates that snake venom induces apoptosis in endothelial cells. This effect was prevented when MnSOD had been upregulated before treatment, indicating that superoxide anions were directly functional in this system. In addition, their paper clearly demonstrates that a decrease in intracellular glutathione levels by treatment with BSO accelerated apoptosis induction, indicating the role of antioxidant defense during ROS-mediated apoptosis induction. In patients with familial amyotrophic lateral sclerosis the role of superoxide anions in apoptosis induction has been shown by GHADGE et al. (1997). In their study, the presence of mutant SOD causes a higher intracellular concentration of superoxide anions than that measured in the presence of wild type enzyme. Inhibition of apoptosis by SOD may either indicate that superoxide anions trigger apoptosis without the need for the chemical sequence superoxide – hydrogen peroxide – hydroxyl radical, or that the generation of other reactive mediators has been inhibited. This may, for example, be the formation of peroxynitrite from the interaction of superoxide anions with NO (KELLER et al. 1998) or interaction of superoxide anions with HOCl, yielding highly reactive hydroxyl radicals (ENGELMANN et al., in preparation).

There are numerous examples of the inhibition of apoptosis through antioxidative enzymes. Catalase prevented hydrogen peroxide-induced apoptosis (SANDSTROM and BUTTKE 1993), SOD and catalase inhibited neutrophil apoptosis (OISHI and MACHIDA 1997). As shown before, the use of defined antioxidants can elaborate the sequence of ROS interactions and can describe their role. With this approach it has been shown that lipid hydroperoxides (products of lipid peroxidation which can be induced by hydroxyl radicals or peroxynitrite) can induce ROS production, finally leading to the most likely functional hydroxyl radical (AOSHIMA et al. 1997). This represents a fine example of how the primary reaction of a highly reactive but short-ranged ROS with the cell membrane can cause a sequence of ROS-mediated effects, ending in the production of intracellular hydroxyl radicals which mediate apoptosis. Similarly, hyperthermia causes generation of hydroxyl radicals functional in apoptosis (RAUEN and DE GROOT 1998) as shown by the inhibition of apoptosis through antioxidants. The same is shown for prevention of apoptosis induction by oxidized low density lipoproteins through N-acetylcysteine (DEIGNER 1998). The effects of ROS are inhibited by antioxidants, together with the apoptosis-triggering effects of mediators as diverse as TNF, ceramides, or TGF-beta. Their interaction with ROS will be discussed later in this chapter, as they call for the presentation of rather complicated and interacting signaling pathways in apoptosis. The role of cellular antioxidant defense will also be discussed separately and in more detail. Here shall be mentioned only that reduction of the cellular glutathione concentration through inhibition of its de novo synthesis using BSO causes apoptosis which is prevented by antioxidants – another clear proof of the apoptosis inducing role of ROS (ZUCKER et al. 1997a).

Work in progress in our laboratory demonstrates that HOCl can induce apoptosis in transformed fibroblasts (ENGELMANN et al., in preparation). The basis for this specificity is the production of superoxide anions at the membrane of transformed cells (a step necessary for the maintenance of their transformed state). As in the scenario discussed for phagocyte microbe interaction, the relatively nonreactive HOCl may form highly reactive hydroxyl radicals when confronted with superoxide anions. Apoptosis induction by HOCl can therefore be inhibited by SOD (which destroys the activating superoxide anion), by the specific hypochlorous acid scavenger taurine, and by the hydroxyl radical scavenger DMSO. This finding may be of relevance for the understanding of apoptosis induction by phagocytes. As discussed later, it is the central element during intercellular induction of apoptosis. Extending these findings, ENGELMANN et al. (in preparation) demonstrated that myeloperoxidase (MPO) added to transformed cells caused apoptosis, whereas nontransformed cells remained unaffected. Inhibitor studies revealed that the transformed cells produced enough hydrogen peroxide (which is needed by MPO to form HOCl), as well as sufficient superoxide anions for activation of HOCl to hydroxyl radical generation. Myeloperoxidase-mediated apoptosis of transformed cells was inhibited by SOD, catalase, MPO inhibitors, and scavengers of HOCl as well as of hydroxylradicals.

III. Induction and Inhibition of Apoptosis by NO·

NO·, which is relatively stable and can pass cellular membranes, is involved in a multitude of biological effects such as regulation of the vascular tone, antiplatelet and antileukocyte activity, and modulation of cell growth (LOPEZ-FARRE et al. 1998). It has been implicated in the induction as well as the inhibition of apoptosis. Modulation of apoptosis by NO· can lead to physiological or pathophysiological consequences. NO· is involved in natural tumor defense mechanisms like the action of macrophages or NK cells (BRUENE et al. 1998), control of metastasis by endothelial cells (EDMISTON et al. 1998), and also in the intercellular induction of apoptosis (HEIGOLD et al., in preparation). NO· readily reacts with superoxide anions to form peroxynitrite, a highly reactive molecule (SARAN et al. 1990; HUIE and PADJAMA 1993; SQUADRITO and PRYOR 1998). The role of NO· and its derivative peroxynitrite are discussed separately, though some of the NO· effects are certainly due to peroxynitrite activity.

Several reviews exist on the role of NO· in apoptosis (LOPEZ-FARRE et al. 1998; BRUENE et al. 1998; TURPAEV 1998; ALBINA and REICHNER 1998; XIE and FIDLER 1998; DIMMELER and ZEIHER 1997). NO· synthesis from arginine in macrophages causes apoptosis which can be inhibited by inhibitors of NO· synthetase (SARIH et al. 1993; ALBINA et al. 1993). Interleukin-1-beta induced NO· production in pancreatic cells and chondrocytes activated the apoptotic process (ANKARCRONA et al. 1994; BLANCO et al. 1995). NO·-induced apoptosis in macrophages was paralleled by p53 expression (MESSMER et al. 1994; BRUENE et al. 1995) and was antagonized by protein kinase C and protein kinase A-activating compounds (MESSMER et al. 1995). Later studies revealed evidence of p53-dependent and p53-independent signaling pathways during NO·-mediated apoptosis (MESSMER and BRUENE 1996). NO·-induced apoptosis has been shown in many cell systems such as colonic epithelial cells (SANDOVAL et al. 1995), mesangial cells (MUEHL et al. 1996), endothelial cells (LOPEZ-FARRE et al. 1997), vascular smooth muscle cells (IWASHINA et al. 1998), non-lymphocytic leukemia cells (SHAMI et al. 1998), pancreatic carcinoma cells (GANSAUGE et al. 1998; HAJRI et al. 1998) and neurons (LEIST and NICOTERA 1998). In the human promyeloid leukemia cell line HL-60, apoptosis can be induced by high doses of NO· (JUN et al. 1996) as well as by peroxynitrite (LIN et al. 1995). This illustrates the problem of differentiation between direct NO· effects and effects of its derivative. NO· is involved in antitumor mechanisms exerted by macrophages (SVEINBJORNSSON et al. 1996; CUI et al. 1994) or NK cells (FILEP et al. 1996). Interestingly, it is used by endothelial cells to induce apoptosis in lymphoma cells (UMANSKY et al. 1997) and low metastatic colon carcinoma cells (EDMISTON et al. 1998). The study by EDMISTON et al. (1998) indicates that endothelium-tumor cell interaction may be an important control step in the prevention of metastasis and, mechanistically, is based on apoptosis induction. NO· is of central importance in their system but cannot adequately induce apoptosis when the tumor cells are tested separately. Generation of superoxide anions is necessary in parallel, indicating that peroxynitrite is the effective molecule in apoptosis induction. As discussed in more detail later, superoxide anions produced by tumor cells might be the key to their own destruction by endothelium-derived NO· by formation of peroxynitrite and subsequent apoptosis induction. The same effect can be achieved when an inducible NO· synthetase gene is expressed in murine melanoma cells (XIE et al. 1995).

NO· can be scavenged by reduced glutathione (ZHAO et al. 1997). Inactivation of GSH-dependent peroxidase has been discussed to contribute to NO·-mediated apoptosis (ASAHI et al. 1995), a finding that indicates that lipid peroxidation may be triggered by NO·. NO· upregulates the expression of the Apo/Fas receptor (FUKUO et al. 1996) and may thus enhance apoptosis induction by this receptor mediated apoptotic pathway. Nitric oxide-mediated Apo/Fas-dependent apoptosis required activation of caspases (CHLICHLIA et al. 1998). NO· was shown to inhibit mitochondrial cytochrome oxidase and thereby respiration (RICHTER 1997). Triggering of mitochondrial permeability transition (a step used by many apoptosis inducers) is efficiently used by NO· for apoptosis induction (HORTELANO et al. 1997). NO· triggers disruption of the mitochondrial transmembrane potential which is followed by hyperproduction of ROS. These and apoptogenic factors released from mitochondria control the following execution of apoptosis. Inhibition of NO·-mediated apoptosis by Bcl-2 (MESSMER et al. 1996; XIE et al. 1996) and downregulation of Bcl-2 during NO-triggered apoptosis (XIE et al. 1997; TAMATANI et al. 1998; BROCKHAUS and BRUENE 1998) fit into this scenario, as Bcl-2 is a key element for the control of the mitochondrial megachannel.

NO· has been shown to possess both an apoptosis-inducing and an apoptosis-inhibitory effect (SHEN et al. 1998). In some systems this may depend solely on its concentration (SHEN et al. 1998). NO· inhibits TNF- (SHEN al. 1998) and LPS-mediated apoptosis (CENEVIVA et al. 1998). It represents a survival factor for T lymphocytes (SCIORATI et al. 1997) and inhibits Apo/Fas-mediated apoptosis (DIMMELER et al. 1998; HEBESTREIT et al. 1998; MANNICK et al. 1997). Apoptosis inhibition by NO· can be achieved by two different strategies. One is based on the induction of heatshock proteins by NO· (KIM et al. 1997a). This is redox-regulated and requires low concentrations of reduced GSH. As soon as sHsps are expressed, however, they cause a rise of GSH (ARRIGO 1998) which blocks ROS-dependent effects in the apoptosis signaling cascade and neutralizes the remaining NO·. The second strategy is rather direct: NO· causes nitrosylation of caspases and thus directly interferes with the execution of apoptosis (KIM et al. 1997b; LI et al. 1997b; TENNETI et al. 1997; DIMMELER et al. 1997b).

IV. Peroxynitrite: An Efficient Apoptosis Inducer

The highly reactive peroxynitrite is formed by the interaction of NO· with superoxide anions in a diffusion controlled way (SARAN et al. 1990; HUIE and PADMAJA 1993; SQUADRITO and PRYOR 1998). Its chemistry and role for apoptosis induction have been reviewed (SZABO and OHSHIMA 1997; SQUADRITO and PRYOR 1998; REITER 1998; TURPAEV 1998). Under experimental conditions, peroxynitrite is generated either by substances like SIN-1 (3-morpholinosydnonimine hydrochloride) which release NO· and superoxide anions simultaneously, or by the interaction of NO· with superoxide anions derived from cellular superoxide sources (PACKER et al. 1996). SHARPE and COOPER (1998) described a superoxide anion-independent way of peroxynitrite formation through NO and cytochrome c interaction, leading to nitroxylanions (NO–) which can be oxidized by molecular oxygen to form peroxynitrite. Peroxynitrite can pass membranes (DENICOLA et al. 1998) and is scavenged by glutathione (CUZZOCREA et al. 1998). The biological role of superoxide anions for peroxynitrite formation and the role of peroxynitrite for subsequent apoptosis induction was elegantly demonstrated by KELLER et al. (1998). The authors show that overexpression of MnSOD suppresses peroxynitrite generation, lipid peroxidation, mitochondrial dysfunction, and apoptosis. An increased cellular concentration of glutathione peroxidase compensated for the increased hydrogen peroxide concentration caused by the action of SOD and thus prevented hydrogen-peroxide-dependent apoptosis (which otherwise would have masked the specific effect demonstrated). The work presented by GONZALES et al. (1998) or NOACK et al. (1998) leads to the same conclusion. Whereas overexpression of SOD inhibited peroxynitrite formation, downregulation of SOD (using antisense nucleotides) allowed peroxynitrite formation through increase of available superoxide anions (TROY et al. 1996). The same scenario is activated during experimental induction of colitis (SEO et al. 1995). Whereas NO· synthetase is induced and causes NO· production, SOD is downmodulated by the inducing drug 2,4,6-trinitrobenzenesulfonic acid. As a result, peroxynitrite is formed and causes tissue damage. To prevent peroxynitrite formation, NO· can inhibit superoxide production in neutrophils (RODENAS et al. 1998) – a mechanism that could allow direct NO· effects without parallel peroxynitrite-induced apoptosis.

IONNIDIS et al. (1998) demonstrated that NO· exhibited low cytotoxicity for endothelial cells, whereas peroxynitrite was highly toxic – a finding that leads to speculation that, in other systems of NO-mediated apoptosis induction, peroxynitrite might have been the ultimately responsible molecule. For example, apoptosis induction in HL-60 cells has been reported for both NO· (JUN et al. 1996) and peroxynitrite (LIN et al. 1995). Trophic factor deprivation of neuronal cells (i.e., deprivation of exogenous survival factors) causes apoptosis (ESTEVEZ et al. 1998) which can be blocked either by inhibitors of NO· synthesis or scavengers of superoxide anions, indicating peroxynitrite formation and its functional role. Apoptosis induction by peroxynitrite in neuronal cells may be the basis for its role in diseases like multiple sclerosis (Cross et al. 1998), amyotrophic lateral sclerosis (Liu 1996), Alzheimers disease (Van Dyke 1997), and Parkinson's disease (Jenner and Olanow 1996). Peroxynitrite also induces apoptosis in thymocytes (Virag et al. 1998a) and pulmonary cells (Gow et al. 1998). It may be involved in the pathogenesis of asthma (Saleh et al. 1998), rheumatic disease (Carson and Tan 1995), and atherosclerosis (Dusting et al. 1998). During cardiac allograft rejection, cardiac myocyte apoptosis seems to be induced by peroxynitrite because iNOS is expressed and nitrated myocyte proteins can be detected (Szabolcs et al. 1998). LPS-challenged neutrophils, monocytes, and lymphocytes produce peroxynitrate and thus contribute to the increased concentration of peroxynitrite during endotoxic shock (Gagnon et al. 1998).

Peroxynitrite seems to have different modes of chemical reactions. Decomposition of peroxynitrite (Beckman et al. 1990; Richeson et al. 1998) can yield hydroxyl radicals – effective oxidants for proteins and involved in lipid peroxidation, like peroxynitrite itself. Peroxynitrite can oxidize the essential zinc-thiolated moiety of enzymes (Crow et al. 1995) as well as cause nitration or oxidation of tyrosine residues (Macmillan Crow et al. 1998; Zhang et al. 1998; Yamakura et al. 1998; Roberts et al. 1998). Inactivation of MnSOD through peroxynitrite-mediated oxidation and nitration of tyrosines (Macmillan Crow et al. 1998) represents an interesting regulatory pathway to enhance peroxynitrite formation by the increase of superoxide anion concentration. Glutathione peroxidase represents another target that is inactivated through peroxynitrite-dependent oxidation (Padmaja et al. 1998). As this enzyme plays a crucial role in the inhibition of apoptosis (through removal of hydrogen peroxide and lipid peroxides, both involved in apoptosis induction), its inactivation might be one of the ways in which peroxynitrite induces apoptosis. As peroxynitrite increases the degradation of proteins by proteasomes (Grune et al. 1998), it might trigger apoptosis by removing inhibitors of the apoptosis signal pathways (operationally defined as endogenous survival factors (Dormann et al. 1999)). Degradation of endogenous survival factors (molecules controlling a constitutively expressed apoptosis machinery) through the action of peroxynitrite represents a challenging idea waiting for experimental investigation. Interaction of peroxynitrite with mitochondria, causing decreased mitochondrial potential and subsequent hyperproduction and release of ROS as well as apoptogenic factors, may be the mechanism where peroxynitrite action meets the activity of other intracellular apoptosis signals. In accordance with this assumption are the findings that: (i) peroxynitrite causes an increase of ROS (Lin et al. 1997a); (ii) peroxynitrite causes activation of caspase-3 (Lin et al. 1998); (iii) peroxynitrite causes cleavage of poly ADP-ribose polymerase (Szabo 1996; Virag et al. 1998b); and (iv) these effects are blocked by Bcl-2 (Melkova et al. 1997; Lin et al. 1997b), which acts at the site of the mitochondrial megachannel. Richter (1998) showed that peroxynitrite and NO· have differential effects on mitochondria during induction of apoptosis.

The combination of the long-lived, far-ranging signal molecule NO· (without a direct apoptosis-inducing effect at low concentrations) with the relatively nonreactive superoxide anion (with its limited diffusion area) yields formation of the reactive peroxynitrite. This allows precise apoptosis induction in cells that release superoxide anions. This fascinating signaling strategy can be illustrated in the case of certain natural antitumor mechanisms; a transformed cell, generating superoxide anions and approaching a NO-releasing endothel, will suddenly encounter peroxynitrite formation close to and on its membrane (the site of superoxide anion generation), causing its destruction without endangering the endothel, which sends out NO but is not necessarily reached by superoxide anions.

V. Glutathione: Key Element for the Regulation of Apoptosis

Glutathione serves two major functions during the regulation of apoptosis. It balances against ROS created by multiple signaling pathways, enzymatic reactions, or mitochondria and it inhibits sphingomyelinase, the key enzyme for the generation of ceramide, a second messenger which is intrinsically interwoven with the generation of ROS and with activation of execution-caspases.

The metabolism of glutathione has been reviewed by MEISTER and ANDERSON (1983) and MEISTER (1988). Decrease of intracellular glutathione has been shown to be an early event during apoptosis (BEAVER and WARING 1995; MACHO et al. 1997). The functional and causative role of glutathione depletion for induction and execution has been proven by several groups who demonstrated that experimental glutathione depletion causes apoptosis, being able to be inhibited by antioxidants (thus in turn proving the role of ROS in this process) (ZUCKER et al. 1997a; RATAN et al. 1994a; DHANBHOORA and BABSON 1992), enhancing the sensitivity of cells for other apoptosis inducers (CHRISTIE et al. 1994; DEAS et al. 1997; CHIBA et al. 1996; ZUCKER et al. 1997b) or abrogating resistance against apoptosis induction (CHIBA et al. 1996). As expected, the augmentation of intracellular glutathione inhibited apoptosis (CHIBA et al. 1996).

The intracellular level of glutathione determines whether cells die from necrosis or apoptosis (FERNANDES and COTTER 1994). A decrease in intracellular reduced glutathione must not necessarily indicate its oxidation, but may be due to its active extrusion (GHIBELLI et al. 1995; VAN DEN DOBBELSTEEN et al. 1996), a mechanism that enhances ROS-dependent intracellular effects and allows ceramide generation. GHIBELLI et al. (1998) showed that cells can be rescued from apoptosis when extrusion of glutathione is inhibited. The same effect is achieved when inhibitors of macromolecular synthesis shunt cysteine from protein to glutathione synthesis (RATAN et al. 1994b), when transaldolase is downregulated (BANKI et al. 1996), or when small stress proteins are induced (ARRIGO 1998). The finding that virus infection causes glutathione extrusion (CIRIOLO et al. 1997; SCHWARZ 1996) allows the speculation that this event is the switch for induction of apoptosis of virus-infected cells. Thus glutathione extrusion may represent a sort of emergency trigger, causing apoptotic self-destruction of potentially hazardous cells through the action of ROS. Glutathione works as a redox sensor (MARCHETTI et al. 1997) and as such controls apoptosis at two central steps: mitochondrial function and sphingomyelinase activity (LIU and HANNUN 1997; LIU et al. 1998). Mitochondria are the source for massive ROS production and for the release of apoptogenic factors like cytochrome c or AIF, a protease involved in activation of execution caspases. Sphingomyelinase is the key enzyme for the regulation of the ceramide second messenger pathway.

VI. Mitochondria: Target and Source for ROS During Apoptosis Induction

The impressive work of several groups during the last few years has shown that mitochondria are the central element for the regulation of the execution phase of apoptosis (KROEMER et al. 1995, 1997; KROEMER 1997; MIGNOTTE and VAYSSIERE 1998). The mitochondrial permeability transition pore seems to be the main switch that is activated during the induction phase of apoptosis through various stimuli (HIRSCH et al. 1997a; BERNARDI 1996; MACHO et al. 1998). In the context of our review, ICE-1-like caspases, as well as ROS, NO·, and ceramides, are the most important ones. Interaction of ICE-like caspases with the permeability transition pore is not inhibited by Bcl-2, but by crmA (a viral antiapoptotic gene), whereas the activity of ROS and ceramide is efficiently counteracted by Bcl-2 (SUSIN et al. 1997; MARZO et al. 1998), thus explaining the antiapoptotic activity of Bcl-2 in many apoptosis pathways (ZAMZAMI et al. 1998a,b). As a consequence of increased gating of the pore, the mitochondrial transmembrane potential is disrupted, and the mitochondria release factors involved in caspase activation as well as ROS (KROEMER et al. 1997; ZAMZAMI et al. 1995; HIRSCH et al. 1997b). As a consequence glutathione is depleted.

Regarding the role of ROS for apoptosis, it is important to point out that mitochondria are both the target and the source of ROS (RICHTER et al. 1996; KROEMER 1997; BACKWAY et al. 1997). Oxidation of the mitochondrial megachannel pore represents a central event for ROS-dependent regulation of apoptosis (CROMPTON and ANDREEVA 1993; PETRONILLI et al. 1994a,b; WEIS et al. 1994; BERNARDI 1996; CONSTANTINI et al. 1996; MACHO et al. 1998; MARZO et al. 1998). Oxidation of at least two vicinal thiols increases the gating potential of the pore (PETRONILLI et al. 1994a,b). NO· has also been shown to induce apoptosis via triggering of mitochondrial permeability transition (HORTELANO et al. 1997). It seems worthwhile to consider that NO· might have formed peroxynitrite after contact with superoxide anions leaking from intact mitochondria and that peroxynitrite was the actually pore oxidizing agent. This scenario would explain why NO· entered the cell without interacting with other thiols and was able to induce specific oxidation of pore thiols. Oxidation of the pore causes disruption of the mitochondrial membrane potential, hypergeneration and release of superoxide anions, as well as release of apoptogenic factors. There is controversy over whether ROS released from mitochondria play a role for the regulation of apoptosis, or whether caspase-3 activation through mitochondria-derived proteases and the resultant steps are sufficient for execution of apoptosis (ZAMZAMI et al. 1996). The existing literature does not permit clarification of this matter. If we oversimplify induction and execution of apoptosis as unidirectional and highly synchronized events, caspase activation by other proteases may be the rate-limiting step for further events and ROS may be secondary. If, however, we imagine nonsynchronized and initially weak apoptosis stimuli (e.g., through suboptimal TNF action or other death receptor pathways) to cause release of ROS by opening just one mitochondrial megachannel, this could initiate apoptosis through either oxidizing the pores of further mitochondria directly or activating the ceramide pathway (through GSH oxidation) (Fig. 1). Ceramide would then act in the same way as ROS and thus synergize its effect. This scenario of multiple interactions of ROS with mitochondria, release of ROS from mitochondria, generation of ceramides, and their synergistic action with ROS links the execution phase of apoptosis with the induction phase for the sake of signal amplification with the final effect of maximal release of mitochondrial proteases.

VII. Ceramides: First Class Second Messengers

Ceramides represent effective inducers of apoptosis (OBEID et al. 1993; KOLESNICK et al. 1994; HAIMOVITZ-FRIEDMAN et al. 1994). Generation of ceramides is redox-regulated and provides an initial signal transmitter from exogenous ROS as well as a signal amplifier within cells. Generation of ceramides occurs at the cell membrane where sphingomyelins are cleaved by sphingomyelinases (HAIMOVITZ-FRIEDMAN et al. 1994). Mitochondria are the central target structure of ceramides where they directly interact with the mitochondrial permeability transition pore (DECAUDIN et al. 1998), causing a decrease of the mitochondrial membrane potential and release of ROS (QUILLET-MARY et al. 1997) and apoptogenic factors from mitochondria. Ionizing radiation as well as receptor-mediated apoptosis inducers like TNF-alpha or Apo/Fas utilize the ceramide pathway for apoptosis induction (KOLESNICK et al. 1994). Defects in the sphingomyelin pathway cause resistance to radiation (BRUNO et al. 1998; MICHAEL et al. 1997; CHMURA et al. 1997; SANTANA et al. 1996). These findings prove the functional role of ceramides for apoptosis induction and indicate that the interaction of radiation-derived ROS with cellular membranes is the major cause of radiation-induced apoptosis. Activation of sphingomyelinase seems to be the rate-limiting step. This enzyme is inhibited by glutathione (LIU and HANNUN 1997; LIU et al. 1998). General or local depletion of glutathione through oxidation or extrusion therefore represents the initial step for ceramide generation. Intracellular ROS, generated by caspase activated mitochondria seems to be the mediator used by TNF or

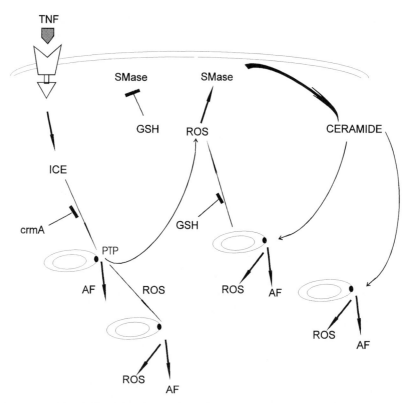

Fig. 1. Intracellular ROS-mediated signaling during apoptosis induction. TNF-triggered activation of an ICE-1-like caspase (inhibitable by crmA but not Bcl-2) causes opening of the mitochondrial permeability transition pore (PTP), depolarization of mitochondrial membrane potential, hypergeneration and release of ROS, in parallel to the release of apoptogenic factors (AF) like cytochrome c and AIF. ROS released from the first mitochondrium can either oxidatively activate the permeability transition pore of other mitochondria and thus enhance the primary signal or activate sphinglomyelinase (SMase) through local depletion of glutathione. As a result of sphinglomyelinase activation, ceramides are generated which activate the mitochondrial transition pore of further mitochondria. The model demonstrates that a first ROS-independent apoptosis-inducing effect can be multiplied by ROS. The effects are nonsynchronous, possibly repetitive and do not fit into a simple categorization of induction and execution phase

Apo/Fas receptor signaling for ceramide generation (BRENNER et al. 1998; SUSIN et al. 1997). As an early consequence of receptor activation, ICE-1-like proteases activate the mitochondrial permeability transition pore (in a crmA-sensitive but Bcl-2-insensitive way), causing mitochondrial membrane depolarization, release of apoptogenic factors, hypergeneration, and release of ROS. These may oxidize glutathione and thus contribute to further activation of sphingomyelinase, leading to ceramide generation. Ceramide in turn acts at the mitochondrial permeability transition pore and accelerates induction of mitochondrial dysfunction, with the consequence of maximal ROS release and appearance of apoptogenic factors.

F. Tumor Necrosis Factor: Apoptosis Induction Through Versatile Use of ROS

TNF receptor activation causes a complex cascade of intracellular protein interaction (reviewed by WALLACH 1997; DARNAY and AGGARWAL 1997) terminating in protein synthesis-independent cytotoxic as well as protein-synthesis-dependent protective mechanisms. The latter utilize NF-κB, a redox-regulated transcription factor involved both in apoptotic and antiapoptotic signaling, dependent on the cell system.

The protective effect of the antioxidant N-acetylcysteine on TNF-induced apoptosis pointed to the functional role of ROS during this process (TALLEY et al. 1995; COSSARIZZA et al. 1995). The role of mitochondria-derived ROS was soon established and superoxide anion production defined as the primary reactive species produced by mitochondria of TNF-treated cells (SCHULZE-OSTHOFF et al. 1992, 1993; HENNET et al. 1993) . The connection between the TNF-activated death receptors and mitochondria through ICE-1-like enzymes (inhibited by crmA but not by Bcl-2) has been established (SLOWIK et al. 1997; SUSIN et al. 1997); the consequences (opening of the mitochondrial megachannel, disruption of the mitochondrial membrane potential, hypergeneration and release of superoxide anions in parallel to apoptogenic factors) have been discussed in the previous chapters. ROS released from mitochondria most probably are the cause of glutathione depletion, sphingomyelinase activation, and ceramide generation (LIU et al. 1998) with the consequences already discussed. TNF receptor interaction thus represents an elaborate example of a complex network of ROS effects which trigger, enhance, and cause apoptosis. The TNF story is not only fascinating due to the versatile use of mechanisms to enhance ROS generation, but also due to the parallel induction of antiapoptotic mechanisms (WONG and GOEDDEL 1989; WONG et al. 1989), using partially the same effector molecules, namely ROS. These activate NF-κB, the redox-sensitive transcription factor already mentioned which is involved in protection against apoptosis (VAN ANTWERP et al. 1996; BEG and BALTIMORE 1996) as well as in apoptosis induction (MARINOVICH et al. 1996). Induction of MnSOD (WONG et al. 1989) represents one of the examples of antiapoptotic responses induced by TNF. The protective effect of this enzyme can be explained by removal of superoxide anions from the cycle of ROS-induced, ROS, and ceramide-mediated effects. The effect of MnSOD either indicates that superoxide anions have direct effects (which is difficult to conceive as superoxide anions do not oxidize glutathione directly) or that reaction products with superoxide anions are involved in the effects measured. It seems worthwhile to test whether catalase is induced in parallel to SOD, allowing inactivation of hydrogen peroxide produced from superoxide anions through the action of SOD.

G. Apo/Fas-Mediated Apoptosis: ROS Involved in Synergistic Pathways

The Apo/Fas system of apoptosis induction (more recently termed CD 95 pathway) represents a receptor controlled apoptosis system, involved in many physiological and pathophysiological processes. Reports on the role of ROS in this system are conflicting and therefore more interesting. In total, a complex system of ROS function in synergistically acting signaling pathways is emerging. Whereas the first studies on ROS and Apo/Fas came to the conclusion that ROS are not involved in Apo/Fas mediated apoptosis induction (HUG et al. 1994) and that the Apo/Fas pathway is different in this respect from TNF-triggered apoptosis (SCHULZE-OSTHOFF et al. 1994), later studies indicate that ROS do play a role during Apo/Fas-mediated apoptosis induction (UM et al. 1996; GULBINS et al. 1996, 1997; RADRIZZANI et al. 1997; CHIBA et al. 1996). The latter conclusions are based on inhibition of Apo/Fas-mediated apoptosis induction by antioxidants and on the demonstration that ROS generation during Apo/Fas-mediated apoptosis has a functional role. The study by GULBINS et al. (1996) demonstrates that interference with ras-mediated superoxide anion production interferes with Apo/Fas-triggered apoptosis. In addition it has been shown that Apo/Fas activity causes glutathione extrusion (VAN DEN DOBBELSTEEN et al. 1996), a process which will accelerate ROS-dependent steps during apoptosis induction. ROS have been shown to be involved in the induction of both Fas ligand (HUG et al. 1997; BAUER et al. 1998) and receptor (DELNESTE et al. 1996), as well as during the central apoptosis-inducing signaling cascade (GULBINS et al. 1996; SUSIN et al. 1997). It is also known that Apo/Fas triggered apoptosis aims at the destabilization of mitochondria with the well known consequences of membrane potential breakdown, release of apoptosis-regulating factors, and further release of ROS. Ceramides also play a role in this complex system (SUSIN et al. 1997). Though there are conflicting results, the overall picture is a network with ROS acting at several steps in parallel, causing a synergistic effect on apoptosis induction. Reasons for discrepancies between different groups may depend on the different cell systems used and on the experimental strategies. In addition, there seem to exist multiple Apo/Fas-dependent apoptosis pathways (SCAFFIDI et al. 1998).

H. p53-Mediated Apoptosis: ROS Action Through Several Subsequent Steps

The p53 system of apoptosis induction represents an excellent example to demonstrate that, during ROS-dependent apoptosis, ROS may act at several subsequent steps, in different and specific ways. Let us assume radiation induced ROS generation inside or outside a cell, leading to DNA damage. At this step, ROS are the causative agents and the degree of the damage induced by it will be monitored by the p53 system, leading either to cell cycle arrest and repair, or induction of apoptosis. Interestingly, p53 is a redox-controlled molecule itself (HAINAUT and MILNER 1993; SUN and OBERLEY 1996; RAINWATER et al. 1995). Apoptosis induction by p53 causes a downstream activation of ROS which is functional during the induction of apoptosis (JOHNSON et al. 1996) . The impressing work by POLYAK et al. (1997) clarifies that p53 activation as a first step causes induction of cellular enzymes involved in ROS generation. These may interact with the mitochondrial megachannel, causing mitochondrial dysfunction, decrease of mitochondrial potential, release of cytochrome c and proteases involved in caspase activation as well as release of mitochondrial ROS. Based on our knowledge of other signaling pathways, it may be assumed that primary ROS generation can activate ceramide synthesis and thus establish a second signaling loop, aiming at the same central structure: the mitochondrion.

The work by CAELLES et al. (1994) indicates the existence of a parallel protein synthesis-independent apoptosis pathway induced by p53. A role for ROS in this pathway has not been elucidated so far.

I. TGF-Beta: Central Roles for ROS

Apoptosis induction by TGF-beta is related to the action of ROS in many ways. As outlined before, TGF-beta activation can be controlled by ROS. TGF-beta and ROS are central players during intercellular induction of apoptosis, a process directed against transformed cells and involving TGF-beta as well as ROS action at different levels. TGF-beta also directly induces apoptosis in various cell systems such as hepatocytes (GRESSNER et al. 1997; INAYAT et al. 1997; MULLAUER et al. 1996; OBERHAMMER et al. 1992; SANCHEZ et al. 1996), hepatoma cells (GRESSNER et al. 1997), tracheal epithelial cells (ANTOSHINA and OSTROWSKI 1997), glial cells (XIA et al. 1997; MARUSHIGE and MARUSHIGE 1994), leukemic B cell precursors (BUSKE et al. 1997), prostatic epithelial cells (HSING et al. 1996), gastric cancer cells (YANAGIHARA et al. 1992; YAMAMOTO et al. 1996), colon adenoma cells (WANG et al. 1995) and ovarian carcinoma cells (LAFRON et al. 1996). Direct induction of apoptosis by TGF-beta seems to be mediated by the action of reactive oxygen species, as it can be inhibited by antioxidants (LAFRON et al. 1996; SANCHEZ et al. 1996). This finding is in line with the ability of TGF-beta to induce an increase in cellular ROS (THANNICKAL et al. 1993, 1995; DAS and FANBURG 1991; OHBA et al. 1994), either by inducing or activating ROS producing enzymes like NADH oxidase (THANNICKAL et al. 1995) or by decreasing the concentration of antioxidant enzyme systems like catalase or glutathione peroxidase (KAYONAKI et al. 1994; ISLAM et al. 1997). The exact signaling pathway of direct apoptosis induction by TGF-beta is not completely understood, but as it is inhibited by Bcl-2, induction of mitochondrial dysfunction followed by ROS release may be one of the key events. Inactivation of endogenous survival factors may be the final step during direct TGF-beta mediated apoptosis induction and may as well be the case for increased sensitivity of TGF-beta pretreated cells for other apoptosis stimuli. The work by SANCHEZ et al. (1997) indicates that TGF-beta induces expression of proteins involved both in increase of ROS and decrease of reduced glutathione. Induction of p53 through the action of TGF-beta (TERAMOTO et al. 1998) connects the action of direct TGF-beta induced ROS production with the multiple ROS effects during p53-mediated apoptosis.

ROS, Apoptosis and Tumorigenesis

I. Intercellular Induction of Apoptosis: Elimination of Transformed Cells Through Diverse Extracellular and Intracellular ROS-Dependent Signaling Steps

Coculture of transformed and nontransformed fibroblasts causes specific elimination of transformed cells (HÖFLER et al. 1993; JÜRGENSMEIER et al. 1994; BAUER 1996). TGF-beta (all isoforms) as well as FGF are central regulatory molecules in this system (JÜRGENSMEIER et al. 1994; ECKERT and BAUER 1998): either added exogenously (to compensate the dilution of endogenous TGF-beta under cell culture conditions) or as shown for TGF-beta, derived from the transformed cells themselves (where they are involved in an autocrine TGF-beta loop to maintain the transformed state (WEHRLE et al. 1994; HACKENJOS et al. 1996)) the cytokines induce specific effects of nontransformed cells directed against their transformed neighbors. Elimination of the transformed cells is due to the induction of apoptosis, as demonstrated by membrane blebbing, chromatin condensation, nuclear fragmentation, and DNA strand breaks detectable by the TUNEL reaction (JÜRGENSMEIER et al. 1994; PANSE et al. 1997; BECK et al. 1997). As shown by clonal analysis, all cells within a population of nontransformed cells are able to mediate apoptosis induction (PICHT et al. 1995). Transformed fibroblasts are regularly sensitive for intercellular induction of apoptosis, no matter what the originally transforming principle had been (JÜRGENSMEIER et al. 1994; BECK et al. 1997; PANSE et al. 1997). Cells transformed by viruses, oncogenes, chemical carcinogens, UV light plus TGF-beta treatment, or spontaneously were equally sensitive for intercellular induction of apoptosis, indicating that sensitivity is a regular feature of transformed cells, which makes them accessible to this natural control mechanism (BAUER 1996). Sensitivity is causally related to the transformed state, as revertants lost sensitivity (BECK et al. 1997). Moreover, cells transformed by an inducible ras oncogene showed sensitivity as long as ras was expressed (SCHWIEGER et al., submitted), cells transiently transformed by the combined action of TGF-beta and EGF exhibited sensitivity as long as they showed the transformed phenotype (HÄUFEL et al., submitted), and fusion products between transformed and nontransformed cells lost both sensitivity and the transformed state (WILMSMEYER and BAUER, in preparation). Intercellular induction of apoptosis has been discussed to represent a hitherto unrecognized control step during oncogenesis (BAUER 1996). Tumor development should therefore require resistance against this mechanism (BAUER 1995, 1996,

1997). In line with this assumption ex vivo tumor cells were found to be resistant against intercellular induction of apoptosis, whereas in vitro transformed cells not challenged with the defense-mechanism of an organism were sensitive (ENGELMANN and BAUER, submitted). This finding indicates that tumor cells must express resistance mechanisms against intercellular induction of apoptosis during tumor development. This idea is further discussed by ENGELMANN and BAUER (submitted). p53 plays no role during intercellular induction of apoptosis, as transformed cells from p53 null/null mice were as sensitive as transformed cells from p53-positive controls and nontransformed cells from p53-negative animals were as efficient in apoptosis-induction as controls (HIPP and BAUER 1997). It was soon realized that antioxidants block intercellular induction of apoptosis (JÜRGENSMEIER et al. 1994b; SCHAEFER et al. 1995). The use of tissue culture inserts allowed distinct phases to be defined during intercellular induction of apoptosis which could be tested independently of each other for the involvement of ROS (LANGER et al. 1996). Phase one is the interaction of TGF-beta or FGF with nontransformed cells. Cells pretreated with either cytokine for two days exert their apoptosis inducing effect on transformed cells even if the cytokines have been removed. This shows that TGF-beta or FGF have induced a cellular program necessary for apoptosis induction by the nontransformed effector cells. The induction of this program in nontransformed cells can be blocked by antioxidants and therefore seems to depend on the action of ROS. Coculture of cytokine-pretreated nontransformed cells and transformed cells represents phase two. If antioxidants or hydroxyl radical scavengers are present early in this step, apoptosis induction in transformed cells can be substantially inhibited. This points to a role of ROS during interaction of nontransformed and transformed cells and during apoptosis induction in transformed cells. ROS-mediated processes in transformed cells are substantiated by the finding that a decrease of intracellular glutathione in transformed cells enhances their apoptosis during intercellular induction of apoptosis. Recent experiments reveal that signaling between nontransformed cells and transformed cells is mediated by a complex interaction of ROS (HERDERNER et al., in preparation; ENGELMANN et al., in preparation). The model which takes into account inhibitor data as well as the knowledge of diffusion ranges of different ROS is based on the synthesis of hypochlorous acid and its subsequent interaction with superoxide anions (Fig. 2). TGF-beta or FGF seem to induce the release of a myeloperoxidase analogous enzyme from nontransformed cells. The peroxidase generates HOCl in the vicinity of transformed cells, utilizing superoxide anion derived hydrogen peroxide and chloride (and therefore, the reaction can be blocked by 4-aminobenzoic acid hydrazide, a specific inhibitor of myeloperoxidase and signaling is abrogated by the specific HOCl scavenger taurine). Superoxide anions synthesized at the membrane of the transformed cell interact with HOCl to yield the highly reactive hydroxyl radical (and therefore intercellular signaling is blocked by superoxide dismutase as well as by hydroxyl radical scavengers). Formation of HOCl in the close vicinity of transformed cells and the small dif-

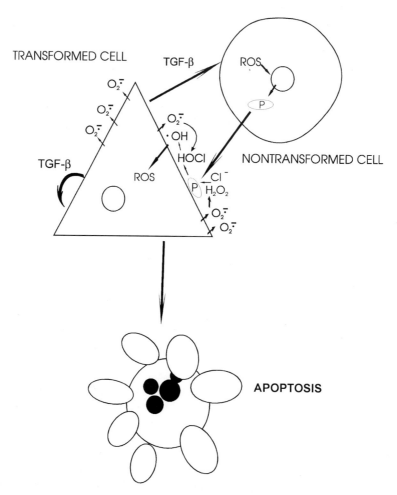

Fig. 2. Signaling during intercellular induction of apoptosis. TGF-beta (or FGF) utilized by transformed cells (*triangle*) for the maintenance of their transformed state induces nontransformed cells (in a ROS-dependent way) to express or activate a peroxidase (enzymatically analogous to MPO but structurally different). This enzyme is released and synthesizes HOCl, utilizing hydrogen peroxide (derived from superoxide anions produced at the membrane of transformed cells) and abundant chloride anions. HOCl and further superoxide anions generate highly reactive hydroxyl anions at the membrane of the transformed cell. Lipid peroxidation by hydroxyl radicals is the first step to transmit a ROS signal into the transformed cells, resulting in apoptosis. Nontransformed cells (without superoxide anion production) are not challenged by HOCl

fusion pathway of superoxide anion derived exclusively from transformed cells ensure that the ultimate signal (the hydroxyl radical) is generated directly at the membrane of the transformed target cell. Lipid peroxidation by hydroxyl radicals seems to be the critical step occurring at the cell membrane. It can be mimicked by other lipid peroxidants like tertiary butylhydroxy peroxide, which causes apoptosis in our cell system as well. As Bcl-2 inhibits intercellular induction of apoptosis (JÜRGENSMEIER et al. 1997a), mitochondrial permeability transition seems to play a functional role during apoptosis induction in transformed cells. This points to possible intracellular roles of ROS or ceramides or both. Endogenous survival factors (operationally defined control elements of the apoptosis machinery) seem to be the targets for endogenous ROS (DORMANN et al. 1999). Their inactivation releases the apoptosis machinery from negative control and leads to the onset of cell death.

Intercellular induction of apoptosis thus represents a well defined system of specific ROS-dependent steps, both intracellular and intercellular in nature. Superoxide anion production by transformed cells performs an outstanding role in this scenario. It is the basis for hydrogen peroxide production, necessary as substrate for myeloperoxidase, and it is the critical radical that reacts with hypochlorous acid to form hydroxyl radicals. The low diffusion range of superoxide anions, the location of the peroxidase close to the transformed cells, and the generation of reactive hydroxyl radicals directly at the membrane of the transformed cells ensure specific apoptosis induction in transformed cells. This well-balanced set of interaction of different members of the ROS family resembles the scenario described for phagocyte-microbe interaction (SARAN et al. 1999). Recent experiments in our laboratory are in favor with the idea that macrophage/tumor cell interaction utilizes the same efficient chemistry and the same strategy of interaction of ROS. These findings on analogous ROS chemistry utilized by different natural antitumor mechanisms allows the hypothesis that resistance against one of the mechanisms may automatically imply resistance against the other. This represents the negative view with respect to tumor formation. The positive view is based on the idea that unraveling of resistance mechanisms of tumor cells and their manipulation towards sensitivity may have the potential to render them sensitive for several natural antitumor mechanisms and may thus have therapeutic potential in the future, when combined with classical tumor treatment.

II. NO-Mediated Control of Tumorigenesis

NO· is used by macrophages and granulocytes for antitumor defense. It may be speculated that peroxynitrite formed through the interaction of NO· with superoxide anions abundant in the vicinity of these cells is the ultimate apoptosis inducer. In addition to these classical NO-utilizing systems, another system with an efficient NO-based antitumor defense mechanism has been characterized recently. Endothelial cells induce apoptosis in tumor cells through the action of NO·. The paper by EDMISTON et al. (1998) shows that induction of apoptosis in colon carcinoma cells of low metastatic potential through NO· derived from endothelial cells is inhibited by SOD, pointing to superoxide dependent peroxynitrite formation. The biological importance of the endothelial system is obvious: it is directed against migrating tumor cells.

The paper by Edmiston indicates that high metastatic potential implies resistance against this system. Intercellular induction of apoptosis as described in the preceding chapter and endothelial cell-dependent apoptosis induction seem to act in concert but at different levels: whereas intercellular induction of apoptosis seems to inhibit newly transformed cells (unless they possess resistance mechanisms that are the basis for further tumor formation), endothelial cell-dependent processes are more aggressive and have the potential to induce apoptosis in tumor cells. Their efficiency ends when a high metastatic potential is acquired. The elucidation of resistance mechanisms against these natural antitumor systems bears exciting diagnostic and therapeutic potential.

III. Sensitivity of Transformed Cells Against Natural Antitumor Mechanisms

Sensitivity of transformed cells for intercellular induction of apoptosis, a process with several ROS-mediated steps, depends on the production of superoxide anions, leading to hydrogen peroxide formation, which is the basis for HOCl synthesis by an extracellular myeloperoxidase analogous enzyme released from TGF-beta treated nontransformed cells. Superoxide production is causally related to the maintenance of the transformed state (YAN et al. 1996; IRANI et al. 1997; JÜRGENSMEIER et al. 1997b) and thus fulfills an interesting double function for the transformed cell: maintenance of the transformed state as well as elimination of transformed cells. Sensitivity for NO-mediated apoptosis through peroxynitrite formation depends on the same principle and therefore both pathways can act synergistically (Fig. 3). As shown in recent model experiments (HEIGOLD et al., in preparation), nontransformed fibroblasts were insensitive to NO· but sensitive to apoptosis induction by peroxynitrite, whereas transformed cells were sensitive to both agents. Apoptosis induction in transformed cells through NO· was inhibited by SOD, indicating that cell-derived superoxide anions are required to form the ultimate inducer peroxynitrite. Cells with an inducible ras oncogene were sensitive to NO· as long as ras was expressed; apoptosis induction was inhibited by SOD. These data demonstrate the link between oncogene expression, superoxide anion generation followed by the expression of the transformed state, as well as induction of processes directed against the transformed cell.

The novel concept for intercellular ROS signaling during the control of oncogenesis depends on long-lived species with relatively low reactivity and wide range of action (like hypochlorous acid or NO·). These interact with the short-ranging superoxide anion and yield hydroxyl radicals or peroxynitrite – molecules that are extremely reactive, short-lived and short-ranging. This trick allows the efficient monitoring of superoxide anion-producing transformed cells and their specific apoptosis induction.

Selectivity of TNF against transformed cells seems to be different and only relative. It is not based on a selective induction process as in the case of NO·

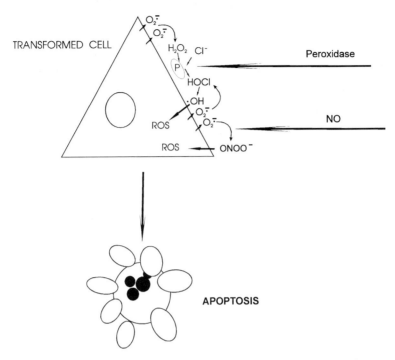

Fig. 3. Superoxide anion production by transformed fibroblasts is the basis for specific recognition by natural antitumor mechanisms. Superoxide anion generation (as a consequence of oncogene activation and NADPH oxidase activity) allows HOCl synthesis through peroxidase (released from TGF-beta-treated nontransformed fibroblasts or derived from phagocytes, i.e., myeloperoxidase). The unspecific signal molecule HOCl is converted to a highly reactive hydroxyl radical at the membrane of the transformed cell through interaction with superoxide anions. This step warrants both efficiency and specificity of peroxidase-mediated antitumor effects. NO· released from endothelial cells, macrophages, granulocytes, NK cells, or fibroblasts represents a nontoxic signal molecule which may be converted to peroxynitrite through interaction with superoxide anions derived from transformed cells. This scheme demonstrates one of the central principles for ROS-mediated signaling discussed in the text – conversion of a long-ranging, nonreactive signal molecule into a more reactive and extremely short-ranging effector molecule at the desired site. This strategy seems to be utilized by different natural antitumor mechanisms (intercellular induction of apoptosis, phagocytes, endothelial cells). Superoxide anions derived from transformed cells are the key elements in this mechanism. Their short range of action ensures that the ultimate signal is generated close to the transformed cell. In this way specific apoptosis induction in transformed cells is warranted

or HOCl, but reflects a differential response of transformed and nontransformed cells. As shown by SCHULZ AND BAUER (in preparation), TNF in an oxidative pathway downmodulates endogenous survival factors in both cell types, demonstrating that TNF signaling is functional in both. However, as nontransformed cells possess higher concentrations of endogenous survival factors as transformed ones, downmodulation after TNF action is not complete and therefore the apoptosis machinery is still inhibited. In transformed cells the lower initial concentration of survival factors is completely destroyed and apoptosis starts. These data demonstrate that the decrease in survival factor concentration after oncogenic transformation has a direct biological relevance during control of tumorigenesis. In the case of HOCl- or NO-driven apoptosis, the differential concentration of survival factors has a modulating effect on the efficiency of the reaction, though it is not the decisive step.

These findings open the way for a better understanding of ROS-dependent signaling pathways involved in the processes of transformation, tumor formation, and metastasis on one side and natural antitumor mechanisms on the other. The knowledge of these mechanisms may enable therapeutic interference in the future.

Acknowledgements. We are grateful to all members of our group who established the experimental evidence for the concept of intercellular induction of apoptosis. We thank our families and friends for support, discussion, and encouragement. The work of our group (presented essentially in the chapter on intercellular induction of apoptosis) was supported by the Deutsche Krebshilfe and the Deutsche Forschungsgemeinschaft.

References

Albina JE, Cui S, Mateo RB, Reichner JS (1993) Nitric oxide-mediated apoptosis in murine peritoneal macrophages. J Immunol 150:5080–5085

Albina JE, Reichner JS (1998) Role of nitric oxid in mediation of macrophage cytotoxicity an apoptosis. Cancer and Metastasis 17:39–53

Ankarcrona M, Dypbukt JM, Bruno B, Nicotera P (1994) Interleukin-1-beta–induced nitric oxide production activates apoptosis in pancreatic RINm5F cells. Exp Cell Res 213:172–177

Antoshina E, Ostrowski LE (1997) TGF-beta-1 induces growth arrest and apoptosis but not ciliated cell differentiation in rat tracheal epithelial cell cultures. In Vitro Cellular and Developmental Biology 33:212–217

Aoshima H, Satoh T, Sakai N, Yamada M, Enokido Y, Ikeuchi T, Hatanaka H (1997) Generation of free radicals during lipid hydroperoxide-triggered apoptosis in PC12h cells. Biochem. Biophys. Acta 1345:35–42

Arbault S, Pantano P, Sojic N, Amatore C, Best Belpomme M, Sarasin A, Vuillaume M (1997) Activation of the NADPH oxidase in human fibroblasts by mechanical intrusion of a single cell with an ultramicroelectrode. Carcinogenesis 18:569–574

Arrigo AP (1998) Small stress proteins: chaperones that act as regulators of intracellular redox state and programmed cell death. Biol Chem 379:19–26

Aruoma OI, Halliwell B, Hoey BM, Butler J (1988) The antioxidant action of taurine, hypotaurine and their metabolic precursors. Biochemical Journal 256:251–256

Aruoma OI, Halliwell B, Hoey BM, Butler J (1989) The antioxidant action of N-acetylcysteine. Its reaction with hydrogen peroxide, hydroxyl radical, superoxide and hypochlorous acid. Free Radical Biology and Medicine 6:593–598

Asahi M, Fujii J, Suzuki K, Seo HG, Kuzuya T, Hori M, Tada M, Fujii S, Taniguchi N (1995) Inactivation of glutathione peroxidase by nitric oxide: implication for cytotoxicity. J Biol Chem 270:21035–21039

Babior BM (1995) Mechanism of activation of the respiratory burst oxidase. In: Davies KJA, Ursini F (eds) The oxygen paradox. Padova. CLEUP Univ Press, pp 749–756

Backway KL, Mcculloch EA, Chow S, Hedley DW (1997) Relationships between the mitochondrial permeability transition and oxidative stress during ara-C toxicity. Cancer Res 57:2446–2451

Bagchi D, Bagchi M, Hassoun EA, Stohs SJ (1995) In vitro and in vivo generation of reactive oxygen species, DNA damage and lactate dehydrogenase leakage by selected pesticides. Toxicology 104:129–140

Banki K, Hutter E, Colombo E, Gonchoroff NJ, Perl A (1996) Glutathione levels and sensitivity to apoptosis are regulated by changes in transaldolase expression. J Biol Chem 271:32994–33001

Barcellos Hoff MH, Dix TA (1996) Redox-mediated activation of latent transforming growth factor-beta-1. Molecular Endocrinology 10:1077–1083

Bauer G (1995) Resistance to TGF-β-induced elimination of transformed cells is required during tumor progression. Int J Oncol 6:1227–1229

Bauer G (Invited Review) (1996) Elimination of transformed cells by normal cells: a novel concept for the control of carcinogenesis. Histol Histopathol 11:237–255

Bauer G (1997) Interference of papilloma viruses with p53-dependent and -independent apoptotic pathways: clues to viral oncogenesis (Review-Hypothesis). Oncology Reports, 4, 273–275

Bauer MKA, Vogt M, Los M, Siegel J, Wesselborg S, Schulze Osthoff K (1998) Role of reactive oxygen intermediates in activation-induced CD95 (Apo/Fas) ligand expression. J Biol Chem 273:8040–8055

Beaver JP, Waring P (1995) A decrease in intracellular glutathione concentration precedes the onset of apoptosis in murine thymocytes. Eur J Cell Biol 68:47–54

Beck E, Schäfer R, Bauer G (1997) Sensitivity of transformed fibroblasts for intercellular induction of apoptosis is determined by their transformed phenotype. Exp Cell Res 234:47–56

Beckman JS, Beckman TW, Chen J, Marshall PA, Freeman BA (1990) Apparent hydroxyl radical production by peroxynitrite: implications for endothelial injury from nitric oxide and superoxide. Proc Natl Acad Sci USA 87:1620–1624

Beg AA, Baltimore D (1996) An essential role for NF-kappaB in preventing TNF-alpha-induced cell death. Science 274:782–784

Bernardi P (1996) The permeability transition pore: control points of a cyclosporin A-sensitive mitochondrial channel involved in cell death. Biochem. Biophys Acta 1275:5–9

Bishop CJ, Rzepcyk CM, Stenzel D, Anderson K (1987) The role of reactive oxygen metabolites in lymphocyte-mediated cytolysis. J Cell Science 87:473–482

Blanco FJ, Ochas RL, Schwarz H, Lotz M (1995) Chondrocyte apoptosis induced by nitric oxide. Am J Pathol 146:75–85

Brenner B, Ferlinz K, Grassme H, Weller M, Koppenhoefer U, Dichgans J, Sandhoff K, Lang F, Gulbins E (1998) Fas/CD95/Apo-I activates the acidic sphingomyelinase via caspases. Cell Death Differ. 5:29–37

Briehl MM, Baker AF (1996) Modulation of the antioxidant defence as a factor in apoptosis. Cell death and Differentiation 3:63–70

Broaddus VC, Yang L, Scavo LM, Ernst JD, Boylan AM (1996) Asbestos induces apoptosis in human and rabbit pleural mesothelial cells via reactive oxygen species. J Clin Investigation 98:2050–2059

Brockhaus F, Bruene B (1998) U937 apoptotic cell death by nitric oxide-Bcl-2 down-regulation and caspase activation. Exp Cell Res 238:33–41

Bruene B, Messmer UK, Sandau K (1995) The role of nitric oxide in cell injury. Toxicol Letters 82–83:233–237

Bruene B, Vonknethen A, Sandau KB (1998) Nitric oxide and its role in apoptosis. Eur J Pharmacol 351:261–272

Bruno AP, Laurent G, Averbeck D, Demur C, Bonnet J, Bettaieb A, Levade T, Jaffrezou JP (1998) Lack of ceramide generation in TF-1 human myeloid leukemic cells resistant to ionizing radiation. Cell Death Diff 5:172–182

Busciglio J, Yankner BA (1995) Apoptosis and increased generation of reactive oxygen species in Down's syndrome neurons in vitro. Nature 378:776–779

Buske C, Becker D, Feuring Buske M, Hannig H, Wulf G, Schaefer C, Hiddenmann W, Woermann B (1997) TGF-beta inhibits growth and induces apoptosis in leukemic B cell precursors. Leukemia 11:386–392

Buttke TM, Sandstrom PA (1994) Oxidative stress as a mediator of apoptosis. Immunology today 15:7–10

Caelles C, Helmberg A, Karin M (1994) p53-dependent apoptosis in the absence of transcriptional activation of p53-target genes. Nature 370:220–223

Candeias LP, Patel KB, Stratford MRL, Wardmann P (1993) Free hydroxyl radicals are formed on reaction between the neutrophil-derived species superoxide anion and hypochlorous acid. FEBS 333:151–153

Carson DA, Tan EM (1995) Apoptosis in rheumatic disease. Bulletin on the Rheumatic Diseases 44:1–3

Ceneviva GD, Tzeng E, Hoyt DG, Yee E, Gallagher A, Engelhardt JF, Kim YM, Billiar TR, Watkins SA, Pitt BR (1998) Nitric oxide inhibits lipopolysaccharide-induced apoptosis in pulmonary artery endothelial cells. Am J Physiol 19:L717–L728

Chance B, Sies H, Boveris A (1979) Hydroperoxide metabolism in mammalian organs. Physiol. Rev. 59:527–605

Chiba T, Takahashi S, Sato N, Ishii S, Kikuchi K (1996) Fas-mediated apoptosis is modulated by intracellular glutathione in human T cells. Eur J Immunol 26:1164–1169

Chlichlia K, Peter ME, Rocha M, Scaffidi C, Bucur M, Krammer PH, Schirrmacher V, Umansky V (1998) Caspase activation is required for nitric oxide-mediated, CD95(Apo-1/FAS)-dependent and independent apoptosis in human neoplastic lymphoid cells. Blood 91:4311–4320

Chmura SJ, Nodzenski E, Beckett MA, Kufe DW, Quintans J, Weichselbaum RR (1997) Loss of ceramide production confers resistance to radiation-induced apoptosis. Canc Res 57:1270–1275

Christie NA, Slutsky AS, Freeman BA, Tanswell AK (1994) A critical role for thiol, but not ATP, depletion in 95% 02-mediated injury of preterm pneumocytes in vitro. Arch. Biochem. Biophys 313:131–138

Ciriolo MR, Palamara AT, Incerpi S, Lafavia E, Bue MC, De Vito P, Garaci E, Rotilio G (1997) Loss of GSH, oxidative stress, and decrease of intracellular pH as sequential steps in viral infection. J Biol Chem 272:2700–2708

Clutton S (1997) The importance of oxidative stress in apoptosis. British Medical Bulletin 53:662–668

Coffer PJ, Burgering BMT, Peppelenbosch MP, Bos JL, Kruijer W (1995) UV activation of receptor tyrosine kinase activity. Oncogene 11:561–569

Constantini P, Chernyak BV, Petronilli V, Bernardi P (1996) Modulation of the mitochondrial permeability transition pore by pyridine nucleotides and dithiol oxidation at two separate sites. J Biol Chem 271:6746–6751

Cossarizza A, Franeschi C, Monti D, Salvioli S, Bellesia E, Rivabene R, Biondo L, Rainaldi G, Tinari A, Malorini W (1995) Protective effect of N-acetylcysteine in tumor necrosis factor alpha-induced apoptosis in U937 cells: the role of mitochondria. Exp Cell Res 220:232–240

Crompton M, Andreeva L (1993) On the involvement of a mitochondrial pore in reperfusion injury. Basic Res Cardiol 88:513–523

Cross AH, Manning PT, Keeling RM, Schmidt RE, Misko TP (1998) Peroxynitrite formation within the central nervous system in active multiple sclerosis. J Neuroimmunology 88:45–56

Crow JP, Beckman JS, Mccord JM (1995) Sensitivity of the essential zinc-thiolate moiety of yeast alcohol dehydrogenase to hypochlorite and peroxynitrite. Biochemistry 34:3544–3552

Crow JP, Spruell C, Chen J, Gunn C, Ischiropoulos H, Tsai M, Smith CD, Radi R, Koppenol WH, Beckman JS (1994) On the pH-dependent yield of hydroxyl radical products from peroxynitrite. Free Rad Biol Med 16:331–338

Cui S, Reichner JS, Matero RB, Albina JE (1994) Activated murine macrophages induce apoptosis in tumor cells through nitric oxide-dependent or-independent mechanism. Cancer Res 54:2462–2467

Cuzzocrea S, Zingarelli B, Oconnor M, Salzman AL, Szabo C (1998) Effect of L-Buthionine-(S,R)-sulphoximine, an inhibitor of gamma-glutamylcysteine synthetase on peroxynitrite- and endotoxic shock-induced vascular failure. British J Pharamacol 123:525–537

Darley Usmar V, Halliwell B (1996) Blood radicals. Reactive nitrogen species, reactive oxygen species, transition metal ions, and the vascular system. Pharmaceutical Research 13:649–662

Darnay BG, Aggarwal BB (1997) Early events in TNF signaling: a story of associations and dissociations. J Leukocyte Biology 61:559–566

Das SK, Fanburg BL (1991) TGF-ß1 produces a "prooxidant" effect on bovine pulmonary artery endothelial cells in culture. Am J Physiol 261:L249–L254

Daugherty A, Dunn JL, Rateri DL, Heinecke JW (1994) Myeloperoxidase, a catalyst for lipoprotein oxidation, is expressed in human atherosclerotic lesions. J Clin Invest 94:437–444

Deas O, Dumont C, Mollereau B, Metivier D, Pasquier C, Bernard-Pomier G, Hirsch F, Charpentier B, Senik A (1997) Thiol-mediated inhibition of Fas and CD2 apoptoic signaling in activaated human peripheral T cells. Int Immun 9:117–125

Decaudin D, Marzo I, Brenner C, Kroemer G (1998) Mitochondria in chemotherapy-induced apoptosis: a prospective novel target of cancer therapy. Int J Oncol 12: 141–152

Deigner HP (1998) Apoptosis caused by oxidized LDL is manganese superoxide dismutase an p53 dependent. FASEB Journal 12:461–4677

Delneste Y, Jeannin P, Sebille E, Aubry JP, Bonnefoy JY (1996) Thiols prevent Fas (CD95)-mediated T cell apoptosis by down-regulating membrane Fas expression. Europ J Immunol 26:2981–2988

Denicola A, Souza JM, Radi R (1998) Diffusion of peroxynitrite across membranes. Proc Natl Acad Sci USA 95:3566–3571

Dhanbhoora CM, Babson JR (1992) Thiol depletion induces lethal cell injury in cultured cardiomyocytes. Arch Biochem Biophys 293:130–139

Diekman D, Abo A, Johnston C, Segal AW, Hall A (1994) Interaction of Rac with p67-phox and regulation of phagocytic NADPH oxidase activity. Science 265: 531–533

Dimmeler S, Haendeler J, Galle J, Zeiher AM (1997a) Oxidized low-density lipoprotein induces apoptosis of human endothelial cells by activation of CPP32-like proteases: a mechanistic clue to the "response to injury" hypothesis. Circulation 95: 1760–1763

Dimmeler S, Haendeler J, Nehls M, Zeiher AM (1997b) Suppression of apoptosis by nitric oxide via inhibition of interleukin-1-beta-converting enzyme (ICE)-like and cysteine protease protein (CPP)-32-like proteases. J Exp Med 185:601–607

Dimmeler S, Haendeler J, Sause A, Zeiher AM (1998) Nitric oxide inhibits apo-1/Fas-mediated cell death. Cell Growth & Differ 9:415–422

Dimmeler S, Zeiher AM (1997) Nitric oxide and apoptosis: another paradigm for the double-edged role of nitric oxide. Nitric Oxide 1:275–281

Dobmeyer TS, Findhammer S, Dobmeyer JM, Klein SA, Raffel B, Hoelzer D, Helm EB, Kabelitz D, Rossol R (1997) Ex vivo induction of apoptosis in lymphocytes is mediated by oxidative stress: role for lymphocyte loss in HIV infection. Free Radical Biology Medicine 22:775–785

Dormann S, Schwieger A, Hanusch J, Häufel T, Engelmann I, Bauer G (1999) Intercellular induction of apoptosis through modulation of endogenous survival factor concentration: a review. Anticancer Res 19:105–112

Dusting GJ, Fennessy P, Yin ZL, Gurevich V (1998) Nitric oxide in atherosclerosis – vascular protector or villain. Clinical and Experimental Pharmacology and Physiology 25 [Suppl S]:S34–S41

Eckert S, Bauer G (1998) TGF-ß isoforms and fibroblast growth factor exhibit analogous indirect antioncogenic activity through triggering of intercellular induction of apoptosis. Anticancer Res 18:45–52

Edmiston KH, Shoji Y, Mizoi T, Ford R, Nachman A, Jessup JM (1998) Role of nitric oxide and superoxide anion in elimination of low metastatic human colorectal carcinomas by unstimulated hepatic sinusoidal endothelial cells. Cancer Res 58:1524–1531

Eiserich JP, Hristova M, Cross CE, Jones AD, Freeman BA, Halliwell B, Van der Vliet A (1998) Formation of nitric oxide-derived inflammatory oxidants by myeloperoxidase in neutrophils. Nature 391:393–397

Ermacora MR, Delfino JM, Cuenoud B, Schepartz A, Fox RO (1992) Conformation-dependent cleavage of staphylococcal nuclease with a disulfide-linked iron chelate. Proc. Natl Acad Sci USA 89:6383–6387

Ermacora MR, Ledmann DW, Hellinga HW, Hsu GW, Fox RO (1994) Mapping staphylococcal nuclease conformation using an EDTA-Fe derivative attached to genetically engineered cysteine residue. Biochemistry 33:13625–13641

Estevez AG, Spear N, Manuel SM, Radi R, Henderson CE, BArbeito L, Beckman JS (1998) Nitric oxide and superoxide contribute to motor neuron apoptosis induced by trophic factor deprivation. J Neuroscience 18:923–931

Fernandez RS, Cotter TG (1994) Apoptosis or necrosis: intracellular levels of glutathione influence mode of cell death. Biochem Pharm 48:675–681

Ferrai R, Agnoletti L, Comini L, Gaia G, Bachetti T, Cargnoni A, Ceconi C, Curello S, Visioli O (1998) Oxidative stress during myocardial ischemia and heart failure. European Heart Journal 19 (Suppl. B):B2–B11

Filep JG, Baron C, Lachance S, Perreault C, Chan JSD (1996) Involvement of nitric oxide in target-cell lysis and DNA fragmentation induced by murine natural killer cells. Blood 87:5136–5143

Fukuo K, Hata S, Suhara T, Nakahashi T, Shinto Y, Tsujimoto Y, Morimoto S, Ogihara T (1996) Nitric oxide induces upregulation of Fas and apoptosis in vascular smooth muscle. Hypertension 27:823–826

Gagnon C, Leblond FA, Filep JG (1998) Peroxynitrite production by human neutrophils, monocytes and lymphocytes challenged with lipopolysaccharide. FEBS Letters 431:107–110

Gansauge S, Nussler AK, Beger HG, Gansauge F (1998) Nitric oxide-induced apoptosis in human pancreatic carcinoma cell lines is associated with a G(1)-arrest an increase of the cyclin-dependent kinase inhibitor P21 (WAF1/CIP1) Cell Growth Differ 9:611–617

Ghadge GD, Lee JP, Bindokas VP, Jordan J, Ma L, Miller RJ, Roos RP (1997) Mutant superoxide dismutase-1-linked familial amyotrophic lateral sclerosis: molecular mechanisms of neuronal cell death and protection. J Neuroscience 17:8756–8766

Ghibelli L, Coppola S, Rotilio G, Lafavia E, Maresca V, Ciriolo MR (1995) Non-oxidative loss of glutathione in apoptosis via GSH extrusion. Biochem Biophys Res Comm 216:313–320

Ghibelli L, Fanelli C, Rotilio G, Lafavia E, Coppola S, Colussi C, Civitarealel P, Ciriolo MR (1998) Rescue of cells from apoptosis by inhibition of active GSH extrusion. FASEB J 12:479–486

Gillissen A, SchaerlingB, Jaworska M, Bartling A, Rasche K, Schultze Werninghaus G (1997) Oxidant scavenger function of ambroxol in vitro: a comparison with N-acetylcysteine. Research in Experimental Medicine 196:389–398

Gonzalez Zulueta M, Ensz LM, Mukhina G, Lebovitz RM, Zwacka RM, Engelhardt JF, Oberley LW, Dawson VL, Dawson TM (1998) Manganese superoxide dismutase protects nNOS neurons form NMDA and nitric oxide mediated neurotoxicity. J Neuroscience 18:2040–2055

Gow AJ, Thom SR, Ischiropoulos H (1998) Nitric oxide and peroxynitrite-mediated pulmonary cell death. Am J Physiol 274:L112–L118

Gressner AM, Lahme B, Mannherz HG, Polzar B (1997) TGF-beta-mediated hepatocellular apoptosis by rat and human hepatoma cells and primary rat hepatocytes. Journal of Hepatology 26:1079–1092

Grisham MB, Jefferson MM, Melton DF, Thomas EL (1984) Chlorination of endogenous amines by isolated neutrophils: ammonia-dependent bactericidal cytotoxic and cytolytic effects of the chloramines. J Biol chem 259:10404–10413

Grune T, Blasig IE, Sitte N, Roloff B, Haseloff R, Davies KJA (1998) Peroxynitrite increases the degradation of aconitase and other cellular proteins by proteasome. J Biol Chem 273:10857–10862

Gulbins E, Brenen B, Schlottmann K, Welsch J, Heinle H, Koppenhoefer U, Linderkamp O, Coggeshall KM, Lang F (1996) Fas-regulated programmed cell death is mediated by a Ras-regulated O_2^- synthesis. Immunology 89:205–212

Gulbins E, Welsch J, Lepple Wienhuis A, Heinle A, Lang F (1997) Inhibition of Fas-induced apoptotic cell death by osmotic cell shrinkage. Biochem. Biophys Res Communications 236:517–521

Hackenjos K, Langer C, Zabel S, Bauer G (1996) Transformed cells trigger induction of their own apoptosis in coculture with normal cells. Oncology Reports 3:27–31

Haimovitz-Friedman A, Kan C-C, Ehleiter D, Persaud RS, Mcloughlin M, Fuks Z, Kolesnick RN (1994) Ionizing radiation acts on cellular membranes to generate ceramide and initiate apoptosis. J Exp Med 180:525–535

Hainaut P, Milner J (1993) Redox modulation of p53 conformation and sequence-specific DNA binding in vitro. Cancer Research 53:4469–4473

Hajri A, Metzger E, Vallat F, Coffy S, Flatter E, Evrard S, Marescaux J, Aprahamian M (1998) Role of nitric oxide in pancreatic tumour growth – in vivo and in vitro studies. British J Cancer 78:841–849

Hampton MB, Kettle AJ, Winterbourn CC (1998) Inside the neutrophil phagosome: oxidants, myeloperoxidase, and bacterial killing. Blood 92:3007–3017

Hanada H, Kashiwagi A, Takehara Y, Kanno T, Yabuki M, Sasaki J, Inoue M, Utsumi K (1997) Do reactive oxygen species underlie the mechanism of apoptosis in the tadpole tail? Free Radical Biology Medicine 23:294–301

Häufel T, Dormann S, Hanusch J, Schwieger A, Bauer G (1999) Three distinct roles for TGF-beta during intercellular induction of apoptosis: a review. Anticancer Res 19:105–112

Hebestreit H, Dibbert B, Balatti I, Braun D, Schapowal A, Blaser K, Simon HU (1998) Disruption of Fas receptor signaling by nitric oxide in eosinophils. J Exp Med 187:415–425

Hennet T, Richter C, Peterhans E (1993) Tumor necrosis factor alpha induces superoxide generation in mitochondria of L929 cells. Biochem J 289:587–592

Hipp ML, Bauer G (1997) Intercellular induction of apoptosis in transformed cells does not depend on p53. Oncogene 15:791–797

Hippeli S, Rohnert U, Koske D, Elstner EF (1997) OH-radical-type reactive oxygen species derived from superoxide and nitric oxide: a sensitive method for their determination and differentiation. Zeitschrift für Naturforschung 52:564–570

Hirsch T, Marzo I, Kroemer G (1997a) Role of the mitochondrial permeability transition pore in apoptosis. Bioscience Reports 17:67–76

Hirsch T, Marchetti P, Susin SA, Dallaporta B, Zamzami N, Marzo I, Geuskens M, Kroemer G (1997b) The apoptosis-necrosis paradox: apoptogenic proteases activated after mitochondrial permeability transition determine the mode of cell death. Oncogene 15:1573–1581

Höfler P, Wehrle I, Bauer G (1993) TGF-ß induces an inhibitory effect of normal cells directed against transformed cells. Int J Cancer 54:125–130

Hortelano S, Dallaporta B, Zamzami N, Hirsch T, Susin SA, Marzo I, Bosca L, Kroemer G (1997) Nitric oxide induces apoptosis via triggering mitochondrial permeability transition. FEBS Letters 410:373–377

Hsing AY, Kadomatsu K, Bonham MJ, Danielpour D (1996) Regulation of apoptosis induced by transforming growth factor-beta-1 in nontumorigenic and tumorigenic rat prostatic epithelial cell lines. Cancer Research 56:5146–5149

Hu ML, Louise S, Cross CE, Motchnik P, Halliwell B (1993) Antioxidant protection against hypochlorous acid in human plasma. J Lab Clin Med 121:257–262

Huang RP, Wu JX, Fan Y, Adamson ED (1996) UV activates growth factor receptors via reactive oxygen intermediates. J Cell Biol 133:211–220

Hug H, Enari M, Nagat S (1994) No requirement for reactive oxygen intermediates of Fas-mediated apoptosis. FEBS Letters 351:311–313

Hug H, Strand S, Grambihler A, Galle J, Hack V, Stremmel W, Krammer PH, Galle PR (1997) Reactive oxygen intermediates are involved in the induction of CD95 ligand mRNA expression b cytostatic drugs in hepatoma cells. J Biol Chemistry 272:28191–28193

Huie RE, Padmaja S (1993) The reaction of NO with superoxide. Free Rad Res Comm 18:195–199

Inayat Hussain SH, Couet C, Cohen GM, Cain K (1997) Processing/activation of CPP32-like proteases is involved in transforming growth factor beta-1-induced apoptosis in rat hepatocytes. Hepatology 25:1516–1526

Ioannidis I, Batz M, Kirsch M, Korth HG, Sustmann R, Degroot H (1998) Low toxicity of nitric oxide against endothelial cells under physiological oxygen partial pressures. Biochem Journal 329:425–430

Irani K, Xia Y, Zweier JL, Sollott SJ, Der CJ, Fearon ER, Sundaresan M, Finkel T, Goldschmidt-Clermon PJ (1997) Mitogenic signaling mediated by oxidants in ras-transformed fibroblasts. Science 275:1649–1652

Ischiropoulos H, Zhu J, Tsai M, Martin JC, Smith CD, Beckman JS (1992) Peroxynitrite-mediated tyrosine nitration catalyzed by superoxide dismutase. Arch Biochem Biophys 298:431–437

Islam KN, Kayanoki Y, Kaneto H, Suzuki K, Asahi M, Fujii J, Taniguchi N (1997) TGF-beta-1 triggers oxidative modifications and enhances apoptosis in HIT cells through accumulation of reactive oxygen species by suppression of catalase and glutathione peroxidase. Free Radical Biology and Medicine 22:1007–1017

Iwashina M, Shichiri M, Marumo F, Hirata Y (1998) Transfection of inducible nitric oxide synthase gene causes apoptosis in vascular smooth muscle cells. Circulation 98:1212–1218

Jacobson D (1996) Reactive oxygen species and programmed cell death. TIBS 21:83–86

Jenner P, Olanow CW (1996) Oxidative stress and the pathogenesis of Parkinsons's disease. Neurology 47:S161–S170

Johnson TM, Yu ZX, Ferrans VJ, Lowenstein RA, Finkel T (1996) Reactive oxygen species are downstream mediators of p53-dependent apoptosis. Proc Natl Acad Sci USA 93:11848–11852

Jordan JD, Iyengar R (1998) Ras, superoxide and signal transduction. Biochemical Pharmacology 55:1339–1346

Joseph P, Jaiswal AK (1998) NAD(P)H:quinone oxidoreductase 1 reduces the mutagenicity of DNA caused by NADPH:P450 reductase-activated metabolites of benzo(a)pyrene quinones. British Journal of Cancer 77:709–719

Jun CD, Lee DK, Chun YH, Yuon DW, Park SK, Song JH, Lee MS, Choi HS, Han EJ, Park YH, Yuon CH, Chung HT (1996) High-dose nitric oxide induces apoptosis in HL-60 human myeloid leukemia cells. Exp Mol Medicine 28:101–108

Jürgensmeier JM, Viesel E, Höfler P, Bauer G (1994a) TGF-β-treated normal fibroblasts eliminate transformed fibroblasts by induction of apoptosis. Cancer Research 54:393–398

Jürgensmeier JM, Höfler P, Bauer G (1994b) TGF-β-induced elimination of transformed fibroblasts by normal cells: independence of cell-to-cell contact and dependence on ROS. International Journal of Oncology 5:525–531

Jürgensmeier J, Bauer G (1997a) Interference of Bcl-2 with intercellular control of carcinogenesis. Int J Cancer 71:698–704

Jürgensmeier J, Panse J, Schäfer R, G. Bauer (1997b) Reactive oxygen species as mediators of the transformed phenotype. Int J Cancer 70:587–589

Kayanoki Y, Fujii J, Suzuki K, Kawata S, Matsuzawa Y, Taniguchi N (1994) Suppression of antioxidative enzyme expression by transforming growth factor beta-1 in rat hepatocytes. J Biol Chem 269:15488–15492

Keller JN, Kindy MS, Holtsberg FW, St Clair DK, Yen HC, Germayer A, Steiner SM, Brucekeller AJ, Hutchins JB, Mattson MP (1998) Mitochondrial manganese superoxide dismutase prevents neural apoptosis and reduces ischemic brain injury – suppression of peroxynitrite production, lipid peroxidation, and mitochondrial dysfunction. J Neuroscience 18:687–697

Kettle AJ, Gedye CA, Hampton MB, Winterbourn CC (1995) Inhibition of myeloperoxidase by benzoic acid hydrazides. Biochemical Journal 308:559–563

Kettle AJ, Gedye CA, Winterbourn CC (1997) Mechanisms of inactivation of myeloperoxidase by 4-aminobenzoic acid hydrazide. Biochem Journal 321:503–508

Kettle AJ, Winterbourn CC (1997) Myeloperoxidase: a key regulator of neutrophil oxidant production. Redox Report 3:3–15

Kim YM, De Vera ME, Wathins SC, Billar SC (1997a) Nitric oxide protects cultured rat hepatocytes from tumor necrosis factor-alpha-induced apoptosis by inducing heat shock protein 70 expression. J Biol Chem 272:1402–1411

Kim YM, Talanian RV, Billiar TR (1997b) Nitric oxide inhibits apoptosis by preventing increases in caspase-3-like activity via two distinct mechanism. J Biol Chem 272:31138–31148

Knaus UG, Heyworth PG, Evans T, Curnutte JT, Bokoch GM (1991) Regulation of phagocyte oxygen radical production by the GTP-binding protein rac 2. Science 254:1512–1515

Knebel A, Rahmsdorf HJ, Ullrich A, Herrlich P (1996) Dephosphorylation of receptor tyrosine kinases as target of regulation by radiation, oxidants or alkylating agents. EMBO Journal 15:5314–5325

Kolesnick RN, Haimovitz-Friedman A, Fuks Z (1994) The sphingomyelin signal transduction pathway mediates apoptosis for tumor necrosis factor, Fas and ionizing radiation. Biochem Cell Biol 72:471–474

Koppenol WH (1993) The Centennial of the Fenton Reaction. Free Rad Biol Med 15:645–651

Kretzschmar HA, Giese A, Brown DR, Herms J, Keller B, Schmidt B, Groschup M (1997) Cell death in prion disease. J Neural Transm [Suppl] 50:191–210

Kroemer G (1997) The proto-oncogene Bcl-2 and its role in regulating apoptosis. Nat Med 3:614–620

Kroemer G, Petit P, Zamzami N, Vayssiere JL, Mignotte B (1995) The biochemistry of programmed cell death. FASEB J 9:1277–1287

Kroemer G, Zamzami N, Susin SA (1997) Mitochondrial control of apoptosis. Immunol Today 18:44–51

Lafron C, Mathieu C, Guerrin M, Pierre O, Vidal S, Valette A (1996) Transforming growth factor beta-1-induced apoptosis in human ovarian carcinoma cells: protection by the antioxidant N-acetylcysteine and Bcl-2. Cell Growth Differ 7:1095–1104

Langer C, Jürgensmeier JM, Bauer G (1996) Reactive oxygen species act both at TGF-β-dependent and -independent steps during induction of apoptosis of transformed cells by normal cells. Exp Cell Res 222:117–124

Langley RE, Palayoor ST, Coleman CN, Bump EA (1993) Modifiers of radiation-induced apoptosis. Radiation Res 136:320–326

Lee SL, Wang WW, Fanburg BL (1998) Superoxide as an intermediate signal for serotonin-induced mitogenesis. Free Radical Biology Medicine 24:855–858

Leib SL, Kim YS, Chow LL, Sheldon RA, Tauber MG (1996) Reactive oxygen intermediates contribute to necrotic and apoptotic neuronal injury in an infant rat

model of bacterial meningitis due to group B streptococci. J Clin Investigation 98:2632–2639
Leist M, Nicotera P (1998) Apoptosis, excitotoxicity, and neuropathology. Exp Cell Res 239:183–201
Li S, Nguyen TH, Schoneich C, Borchardt RT (1995) Aggregation and precipitation of human relaxin induced by metal-catalyzed oxidation. Biochemistry 34:5762–5772
Li PF, Dietz R, Vonharsdorf R (1997a) Differential effect of hydrogen peroxide and superoxide anion on apoptosis and proliferation of vascular smooth muscle cells. Circulation 96:3602–3609
Li JR, Billiar TR, Talanian RV, Kim YM (1997b) Nitric oxide reversibly inhibits seven members of the caspase family via S-nitrosylation. BBRC 240:419–424
Lin KT, Xue JY, Lin MC, Spokas EG, Sun FF, Wong PYK (1998) Peroxynitrite induces apoptosis of HL-60 cells by activation of a caspase-3 family protease. Am J Physiol Cell Physiol 43:855–860
Lin KT, Xue JY, Nomen M, Spur B, Wong PYK (1995) Peroxynitrite-induced apoptosis in HL-60 cells. J Biol Chem 270:16487–16490
Lin KT, Xue JY, Sun FF, Wong PYK (1997a) Reactive oxygen species participate in peroxynitrite-induced apoptosis in HL-60 cells. Biochem Biophys Res Communications 230:115–119
Lin KT, Xue JY, Wong PYK (1997b) Bcl-2 blocks peroxynitrite-induced apoptosis in HL-60 cells, an association with reactive oxygen species. Inflamm Res 46 Suppl 2:S157–S158
Liu B, Andrieu Abadie N, Levade T, Zhang P, Obeid LM, Hannun YA (1998) Glutathione regulation of neutral sphingomyelinase in tumor necrosis factor-alpha-induced cell death. J Biol Chem 273:11313–11320
Liu B, Hannun YA (1997) Inhibition of the neutral magnesium-dependent sphingomyelinase by glutathione. J Biol Chem 272:16281–16287
Liu D (1996) The roles of free radicals in amyotrophic lateral sclerosis. J Mol Neuroscience 7:159–167
Lopez Farre A, Rodriguezfeo JA, Demiguel LS, Rico L, Casado S (1998) Role of nitric oxide in the control of apoptosis in the microvasculature. Int J Biochem Cell Biol 30:1095–1106
Lopez Farre A, De Miguel LS, Caramelo C, Gomez Macias J, Garcia R, Mosquera JR, De Frutos T, Millas I, Rivas F, Echezarreta G, Casado S (1997) Role of nitric oxide in autocrine control of growth and apoptosis of endothelial cells. Am J Physiol 272:H760–H768
Luo Y, Umegaki H, Wang X, Abe R, Roth GS (1998) Dopamine induces apoptosis through an oxidation-involved SAPK/JNK activation pathway. J Biol Chem 273:3756–3764
Macho A, Blazquez MV, Navas P, Munoz E (1998) Induction of apoptosis by vanilloid compounds does not require de novo gene transcription and activator protein 1 activity. Cell Growth Diff 9:277–286
Macho A, Hirsch T, Marzo I, Marchetti P, Dallaporta B, Susin SA, Zamzami N, Kroemer G (1997) Glutathione depletion is an early and calcium elevation is a late event of thymocyte apoptosis. J Immun 158:4612–4619
Macmillan Crow LA, Crow JP, Thompson JA (1998) Peroxynitrite-mediated inactivation of manganese superoxide dismutase involves nitration and oxidation of critical tyrosine residues. Biochemistry 37:1613–1622
Mannick JB, Miao XQ, Stamler JS (1997) Nitric oxide inhibits Fas-induced apoptosis. J Biol Chem 272:24125–24128
Marchetti P, Decaudin D, Macho A, Zamzami N, Hirsch T, Susin SA, Kroemer G (1997) Redox regulation of apoptosis: impact of thiol oxidation status on mitochondrial function. Eur J Immun 27:289–296
Marinovich M, Viviani B, Corsini E, Ghilardi F, Galli CL (1996) NF-kappaB activation by triphenyltin triggers apoptosis in HL-60 cells. Exp Cell Res 226:98–104

Marushige K, Marushige Y (1994) Induction of apoptosis by transforming growth factor beta-1 in glioma and trigeminal neurinoma cells. Anticancer Research 14: 2419–2424

Marzo I, Brenner C, Zamzami N, Susin SA, Beutner G, Brdiczka D, Remy R, Xie ZH, Reed JC, Kroemer G (1998) The permeability transition pore complex: a target for apoptosis regulation by caspases and Bcl–2-related proteins. J Exp Med 187:1261–1271

McCord JM (1995) Superoxide radical: controversies, contradictions, and paradoxes. Proc Soc Exp Biol Med 209:112–117

McCord JM, Omar BA (1993) Sources of free radicals. Toxicol Ind Health 9:23–37

Meier B, Radeke HH, Selle S, Younes M, Sies H, Resch K, Habermehl GG (1989) Human fibroblasts release reactive oxygen species in response to interleukin-1 or tumor necrosis factor-alpha. Biochemical Journal 263:539–546

Meier B, Cross AR, Hancock JT, Kaup FJ, Jones OTG (1991) Identification of a superoxide-generating NADPH oxidase system in human fibroblasts. Biochem Journal 275:241–246

Meier B, Jesaitis AJ, Emmendoerffer A, Roesler J, Quinn MT (1993) The cytochrome B-558 molecules involved in the fibroblast and polymorphonuclear leucocyte superoxide-generating NADPH oxidase systems are structurally and genetically distinct. Biochem Journal 289:481–486

Meister A (1988) Gutathion metabolism and its selective modification. J Biol Chem 263:17205–17208

Meister A, Anderson ME (1983) Glutathione. Ann Rev Biochem 52:711–760

Melkova Z, Lee SB, Rodriguez D, Esteban M (1997) Bcl-2 prevents nitric oxide-mediated apoptosis and poly(ADP-ribose) polymerase cleavage. FEBS Letters 403:273–278

Messmer UK, Bruene B (1996) Nitric oxide-induced apoptosis: p53-dependent and p53-independent signaling pathways. Biochem J 319:299–305

Messmer UK, Ankarcrona M, Nicotera P, Bruene B (1994) p53 expression in nitric oxide-induced apoptosis. FEBS Letters 355:23–26

Messmer UK, Lapetina EG, Bruene B (1995) Nitric oxide-induced apoptosis in RAW 264.7 macrophages is antagonized by protein kinase C- and protein kinase A-activating compounds. Mol Pharmacol 47:757–765

Messmer UK, Reed JC, Bruene B (1996) Bcl-2 protects macrophages from nitric oxide-induced apoptosis. J Biol Chem 271:20192–20197

Michael JM, Lavin MF, Wattters DJ (1997) Resistance to radiation-induced apoptosis in Burkitt's lymphoma cells is associated with defective ceramide signaling. Canc Res 57:3600–3605

Mignotte B, Vayssiere JL (1998) Mitochondria and apoptosis. Eur J Biochem 252:1–15

Mohammed JR, Mohammed BS, Pawluk LJ, Bucci DM, Baker NR, Davis WB (1988) Purification and cytotoxic potential of myeloperoxidase in cystic fibrosis sputum. J Lab Clin Med 112:711–720

Mohazzab HKM, Wolin MS (1994) Sites of superoxide anion production detected by lucigenin in calf pulmonary artery smooth muscle. Am J Physiol 267:L815–L822

Muehl H, Sandau K, Bruene B, Briner VA, Pfeilschifter J (1996) Nitric oxide donors induce apoptosis in glomerular mesangial cells, epithelial cells and endothelial cells. Eur J Pharmacol 317:137–149

Mullauer L, Grasl Kraupp B, Bursch W, Schulte Hermann R (1996) Transforming growth factor beta-1-induced cell death in preneoplastic foci of rat liver and sensitization by the antiestrogen tamoxifen. Hepatology 23:840–847

Nagra RM, Becher B, Tourtellotte WW, Antel JP, Gold D, Paladino T, Smith RA, Nelson JR, Reynolds WF (1997) Immunohistochemical and genetic evidence of myeloperoxidase involvement in multiple sclerosis. J Neuroimmunol 78:97–107

Nappi AJ, Vass E (1998) Hydroxyl radical formation resulting from the interaction of nitric oxide and hydrogen peroxide. Biochem Biophys Acta 1380:55–63

Noack H, Lindenau J, Rothe F, Asayama K, Wolf G (1998) Differential expression of superoxide dismutase isoforms in neuronal and glial compartments in the course of excitotoxically mediated neurodegeneration – relation to oxidative and nitrergic stress. Glia 23:285–297

Obeid LM, Linardic CM, Karolak LA, Hannun YA (1993) Programmed cell death induced by ceramide. Science 259:1769–1771

Oberhammer FA, Pavelka M, Sharma S, Tiefenbacher R, Purchio AF, Bursch W, Schulte Hermann R (1992) Induction of apoptosis in cultured hepatocytes and in regressing liver by transforming growth factor beta-1. Proc Natl Acad Sci USA 89:5408–5412

Ohba M, Shibanuma M, Kuroki T, Nose K (1994) Production of hydrogen peroxide by transforming growth factor-β1 and its involvement in induction of egr-1 in mouse osteoblastic cells. J Cell Biol 126:1079–1088

Oishi K, Machida K (1997) Inhibition of neutrophil apoptosis by antioxidants in culture medium. Scand. J Immunol 45:21–27

Packer MA, Porteous CM, Murphy MP (1996) Superoxide production by mitochondria in the presence of nitric oxide forms peroxynitrite. Biochemistry and Molecular Biology International 40:527–534

Padmaja S, Squadrito GL, Pryor WA (1998) Inactivation of glutathione peroxidase by peroxynitrite. Archives of Biochemistry and Biophysics 349:1–6

Panse J, Hipp ML, Bauer G (1997) Fibroblasts transformed by chemical carcinogens are sensitive for intercellular induction of apoptosis: implications for the control of oncogenesis. Carcinogenesis 18:259–264

Parchment RE (1991) Programmed cell death apoptosis in murine blastocysts: extracellular free radicals, polyamines and other cytotoxic agents. In vivo 5:493–500

Petronilli V, Nicolli A, Costantini P, Colonna R, Bernardi P (1994a) Regulation of the permeability transition pore, a voltage-dependent mitochondrial channel inhibited by cyclosporin A. Biochimica Biophysica Acta 1187:255–259

Petronilli V, Constantini P, Scorrano L, Colonna R, Passamonti S, Bernardi P (1994b) The voltage sensor of the mitochondrial permeability transition pore is tuned by the oxydation-reduction state of vicinal thiols-Increase of the gating potential by oxidants and its reversal by reducing agents. J Biol Chem 269:16638–16642

Picht G, Hundertmark N, Schmitt C P, Bauer G (1995) Clonal analysis of the effect of TGF-β on the apoptosis-inducing activity of normal cells. Exp Cell Research 218:71–78

Pierce GB, Parchment RE, Lewellyn AL (1991) Hydrogen peroxide as a mediator of programmed cell death in the blastocyst. Differentiation 46:181–186

Polyak K, Xia Y, Zweier JL, Kinzler KW, Vogelstein B (1997) A model for p53-induced apoptosis. Nature 389:300–305

Quillet-Mary A, Jaffrezou JP, Mansat V, Bordier C, Naval J, Laurant G (1997) Implication of mitochondrial hydrogen peroxide generation in ceramide-induced apoptosis. J Biol Chem 272:21388–21395

Radi R, Beckman JS, Bush KM, Freeman BA (1991) Peroxynitrite-induced membrane lipid peroxidation: the cytotoxic potential of superoxide and nitric oxide. Arch Biochem Biophys 288:481–484

Radrizzani M, Accornero P, Delia D, Kurrle R, Colombo MP (1997) Apoptosis induced by HIV-gp120 in a Th1 clone involves the generation of reactive oxygen intermediates downstream CD95 triggering. FEBS letters 411:87–92

Rainwater R, Parks D, Anderson ME, Tegtmeyer P, Mann K (1995) Role of cysteine residues in regulation of p53 function. Mol Cell Biol 15:3892–3903

Ramos CL, Pou S, Britigan BE, Cohen MS, Rosen GM (1992) Spin trapping evidence for myeloperoxidase-dependent hydroxyl radical formation by human neutrophils and monocytes. J Biol Chem 267:8307–8312

Ratan RR, Murphy TH, Baraban JM (1994a) Oxidative stress induces apoptosis in embryonic cortical neurons. J Neurochem 62:376–379

Ratan RR, Murphy TH, Baraban JM (1994b) Macromolecular synthesis inhibitors prevent oxidative stress-induced apoptosis in embryonic cortical neurons by shunting cysteine from protein synthesis to glutathione. J Neurosci 14:4385–4392

Rauen U, Degroot H (1998) Cold-induced release of reactive oxygen species as a decisive mediator of hypothermia injury to cultured liver cells. Free Radicals Biol Medicine 24:1316–1323

Reiter RJ (1998) Oxidative damage in the central nervous system – protection by melatonin. Progress in Neurobiology 56:359–384

Richeson CE, Mulder P, Bowry VW, Ingold KU (1998) The complex chemistry of peroxynitrite decomposition – new insights. J Am Chem Soc 120:7211–7219

Richter C (1997) Reactive oxygen and nitrogen species regulate mitochondrial Ca-2+ homeostasis and respiration. Bioscience Reports 17:53–66

Richter C (1998) Nitric oxide and its congeners in mitochondria – implications for apoptosis. Environmental Health Perspectives 106 Suppl 5:1125–1130

Richter C, Schweizer M, Cossarizza A, Franceschi C (1996) Control of apoptosis by the cellular ATP level. FEBS Letters 378:107–110

Roberts ES, Lin HL, Crowley JR, Vuletich JL, Osawa Y, Hollenberg PF (1998) Peroxynitrite-mediated nitration of tyrosine and inactivation of the catalytic activity of cytochrome P450 2B1. Chemical Research in Toxicology 11:1067–1074

Rodenas J, Mitjavila MT, Carbonell T (1998) Nitric oxide inhibits superoxide production by inflammatory polymorphonuclear leukocytes. Am J Physiol 274:C827–C830

Rollet Labelle E, Grange MJ, Elbim C, Marquetty C, Gougerotpocidalo MA, Pasquier C (1998) Hydroxyl radical as a potential intracellular mediator of polymorphonucleas neutrophil apoptosis. Free Radical Biology and Medicine 24:563–572

Rubbo H (1998) Nitric oxide and peroxynitrite in lipid peroxidation. Medicina 58: 361–368

Ruppersberg JP, Stocker M, Pongs O, Heinemann SH, Frank R, Koenen M (1991) Regulation of fast inactivation of cloned mammalian Ik(A) channels by cysteine oxidation. Nature 352:711–714

Saari H, Sorsa T, Lindy O, Suomalainen K, Halinen S, Konttinen YT (1992) Reactive oxygen species as regulators of human neutrophil and fibroblast interstitial collagenases. Int J Tissue Reactions 14:113–120

Salas Vidal E, Lomeli H, Castro Obregon S, Cuervo R, Escalante Alcade D, Covarrubias L (1998) Reactive oxygen species participate in the control of mouse embryonic cell death. Exp Cell Res 238:136–147

Saleh D, Ernst P, Lim S, Barnes PJ, Giaid A (1998) Increased formation of the potent oxidant peroxynitrite in the airways of asthmatic patients is associated with induction of nitric oxide synthase – effect of inhaled glucocorticoid. FASEB Journal 12: 929–937

Sanchez A, Alvarez AM, Benito M, Fabregat I (1996) Apoptosis induced by transforming growth factor-beta in fetal hepatocyte primary cultures: involvement of reactive oxygen intermediates. J Biol Chem 271:7416–7422

Sanchez A, Alvarez AM, Benito M, Fabregat I (1997) Cycloheximide prevents apoptosis, reactive oxygen species production, and glutathione depletion induced by transforming growth factor beta in fetal rat hepatocytes in primary culture. Hepatology 26:935–943

Sandoval M, Liu X, Oliver PD, Zhang XJ, Clark DA, Miller MJS (1995) Nitric oxide induces apoptosis in a human colonic epithelial cell line T84. Mediators of Inflammation 4:248–250

Sandstrom PA, Buttke TM (1993) Autocrine production of extracellular catalase prevents apoptosis of the human CEM T-cell line in serum-free medium. Proc Natl Acad Sci USA 90:4708–4712

Santana P, Pena LA, Haimovitz-Friedman A, Martin S, Green D, Mcloughlin M, Cordon-Cardo C, Schuchman EH, Fuks Z, Kolesnick R (1996) Acid sphingomyelinase-deficient human lymphoblasts and mice are defective in radiation-induced apoptosis. Cell 86:189–199

Sarafian TA, Bredesen DE (1994) Invited commentary. Is apoptosis mediated by reactive oxygen species? Free Rad Res 21:1–8

Saran M, Bors W (1989) Oxygen radicals as chemical messengers: a hypothesis. Free Rad Res Comm 7:213–220

Saran M, Bors W (1994) Signaling by O2- and NO: how far can either radical, or any specific reaction product transmit a message under in vivo conditions? Chemico-Biological interactions 90:35–45

Saran M, Bors W (1997) Radiation chemistry of physiological saline reinvestigated: evidence that chloride-derived intermediates play a key role in cytotoxicity. Radiation Res 147:70–77

Saran M, Michel C, Bors W (1990) Reaction of NO with O_2^-<?>$^-$ Implication for the action of endothelium-derived relaxing factor (EDRF). Free Rad Res Comm 10:221–226

Saran M, Winkler C, Fellerhoff B (1997) Hydrogen peroxide protects yeast cells from inactivation by ionizing radiation: a radiobiological paradox. Int J Rad Biol 72: 745–750

Saran M, Michel C, Bors W (1998) Radical functions in vivo: a critical review of current concepts and hypotheses. Zeitschrift für Naturforschung 53c:210–227

Saran M, Beck-Speier I, Fellerhoff B, Bauer G (1999) Phagocytic killing of microorganisms by radical processes: consequences of the reaction of hydroxyl radicals with chloride yielding chlorine atoms. Free Radical Biology and Medicine 26:482–490

Sarih M, Souvannavong V, Adam A (1993) Nitric oxide induces macrophage death by apoptosis. BBRC 191:503–508

Scaffidi C, Fulda S. Srinivasan A, Friesen C, Li F, Tomaselli KJ, Debatin KM, Krammer PH, Peter ME (1998) Two CD95 (Apo/Fas) signaling pathways. EMBO Journal 17:1675–1687

Schaefer D, Jürgensmeier J, Bauer G (1995) Catechol interferes with TGF-β-induced elimination of transformed cells by normal cells: implications for the survival of transformed cells during carcinogenesis. Int J Cancer 60:520–526

Schiller J, Arnhold J, Arnold K (1995) NMR studies of the action of hypochlorous acid on native pig articular cartilage. Eur J Biochem 233:672–676

Schreck R, Riever P, Baeuerle PA (1991 Reactive oxygen intermediates as apparently widely used messengers in activation of the NF-kappaB transcription factor and HIV-1. EMBO J 10:2247–2258

Schulze-Osthoff K, Beyaert R, Vandevoorde V, Haegeman G, Fiers W (1993) Depletion of the mitochondrial electron transport abrogates the cytotoxic and gene-inductive effects of TNF. EMBO J 12:3095–3104

Schulze-Osthoff K, Bakker AC, Vanhaesebroeck B, Beyaert R, Jacob WA, Fiers W (1992) Cytotoxic activity of tumor necrosis factor is mediated by early damage of mitochondrial functions. J Biol Chem 267:5317–5323

Schulze-Osthoff K, Krammer PH, Droege W (1994) Divergent signaling via Apo-1/Fas and the TNF receptor, two homologous molecules involved in physiological cell death. EMBO J 13:4587–4596

Schwarz KB (1996) Oxidative stress during viral infection: a review. Free Radic Biol Medic 21:641–649

Sciorati C, Rovere P, Ferrarini M, Heltai S, Manfredi AA, Clementi E (1997) Autocrine nitric oxide modulates CD95-induced apoptosis in gamma-deltaT lymphocytes. J Biol Chem 272:23211–23215

Segal AW (1992) Composition and function of the NADPH oxidase of phagocytic cells with particular reference to redox components located within the plasma membrane. In: Cochrane CG, Giombrone MA (eds) Biological oxidants: generation an injurious consequences. San Diego. Academic Press; Cell Mol Mechan Inflamm 4:1–20

Seo HG, Takata I, Nakamura M, Tatsumi H, Suzuki K, Fujii J, Taniguchi N (1995) Induction of nitric oxide synthase and concomitant suppression of superoxide dismutases in experimental colitis in rats. Archives of Bichemistry and Biophysics 324:41–47

Shami PJ, Sauls DL, Weinberg JB (1998) Schedule and concentration-dependent induction of apoptosis in leukemia cells by nitric oxide. Leukemia 12:1461–1466

Sharpe MA, Cooper CE (1998) Reactions of nitric oxide with mitochondrial cytochrome c – a novel mechanism for the formation of nitroxyl anion and peroxynitrite. Biochem Journal 332:9–19

Shen YH, Wang XL; Wilcken DEL (1998) Nitric oxide induces and inhibits apoptosis through different pathways. FEBS Letters 433:125–131

Shi X, Flynn DC, Porter DW, Leonard SS, Vallyathan V, Casranova V (1997) Efficacy of taurine-based compounds as hydroxyl radical scavengers in silica induced peroxidation. Annals of Clinical and Laboratory Science 27:365–374

Slater AFG, Stefan C, Nobel I, van den Dobbelsteen DJ, Orrenius S (1996) Intracellular redox changes during apoptosis. Cell Death and Differentiation 3:57–62

Slowik MR, Min W, Ardito T, Karsan A, Kashgarian M, Pober JS (1997) Evidence that tumor necrosis factor triggers apoptosis in human endothelial cells by interleukin-1-converting enzyme-like protease-dependent and -independent pathways. Lab Invest 77:257–267

Squadrito GL, Pryor WA (1998) Oxidative chemistry of nitric oxide – the roles of superoxide, peroxynitrite, and carbon dioxide. Free Radical Biology Medicine 25:392–403

Sun Y, Oberley LW (1996) Redox regulation of transcriptional activators. Free Radical Biology Medicine 21:335–348

Sundaresan M, Yu ZX, Ferrans VJ, Sulciner DJ, Gutkind JS, Irani K, Goldschmidt Clermont PJ, Finkel T (1996) Regulation of reactive-oxygen-species generation in fibroblasts by rac1. Biochemical Journal 318:379–382

Susin SA, Zamzami N, Castedo M, Daugas E, Wang HG, Geley S, Fassy F, Reed JC, Kroemer G (1997) The central Executioner of apoptosis: multiple connections between protease activation and mitochondria in Fas/APO-1/CD95-and ceramide-induced apoptosis. J Exp Med 186:25–37

Suzuki YJ, Forman HJ, Sevanian A (1997a) Oxidants as stimulators of signal transduction. Free Radical Biology and Medicine 22:269–285

Suzuki K, Nakamura M, Hatanaka Y, Kayanoki Y, Tatsumi H, Taniguchi N (1997b) Induction of apoptotic cell death in human endothelial cells treated with snake venom: implication of intracellular reactive oxygen species and protective effects of glutathione and superoxide dismutase. J Biochemistry 122:1260–1264

Sveinbjornsson B, Olsen R, Seternes OM, Seljelid R (1996) Macrophage cytotoxicity against murine Meth a sarcoma involves nitric oxide-mediated apoptosis. BBRC 223:643–649

Szabo C (1996) DNA strand breakage and activation of poly-ADP ribosyltransferase: a cytotoxic pathway triggered by peroxynitrite. Free Radical Biol Medicine 21:855–869

Szabo C, Ohshima H (1997) DNA damage induced by peroxynitrite: subsequent biological effects. Nitric Oxide 1:373–385

Szabolcs MJ, Ravalli S, Minanov O, Sciacca RR, Michler RE, Cannon PJ (1998) Apoptosis and increased expression of inducible nitric oxide synthase in human allograft rejection. Transplantation 65:804–812

Talley AK, Dewhurst S, Perry SW, Dollard SC, Gummuluru S, Fine SM, New D, Epstein LG, Gendelman H, Gelbard HA (1995) Tumor necrosis factor alpha-induced apoptosis in human neuronal cells: protection by the antioxidant N-acetylcysteine and the genes Bcl-2 and crmA. Mol Cell Biol 15:2359–2366

Tamatani M, Ogawa S, Niitsu Y, Tohyama M (1998) Involvement of Bcl-2 family and caspase-3-like protease in no-mediated neuronal apoptosis. J Neurochem 71:1588–1596

Tenneti L, D´Emilia DM, Lipton SA (1997) Suppression of neuronal apoptosis by S-nitrosylation of caspases. Neuroscience Letters 236:139–142

Teramoto T, Kiss A, Thorgeirsson SS (1998) Induction of p53 and Bax during TGF-beta-1 initiated apoptosis in rat liver epithelial cells. Biochem Biophys Res Commun 251:56–60

Thannickal VJ, Fanburg BL (1995) Activation of an H-2O-2-generating NADH oxidase in human lung fibroblasts by transforming growth factor beta-1. J Biol Chem 270:30334–30338

Thannickal VJ, Hassoun PM, White AC, Fanburg BL (1993) Enhanced rate of H_2O_2 release from bovine pulmonary endothelial cells induced by TGF-beta-1. American Journal of Physiology 265:L622–L626

Troy CM, Derossi D, Pronchiantz A, Greene LA, Shelanski ML (1996) Downregulation of Cu/Zn superoxide dismutase leads to cell death via the nitric oxide-peroxynitrite pathway. J Neuroscience 16:253–261

Turpaev KT (1998) Nitric oxide in intercellular communication. Mol Biol 32:475–484

Um HD, Orenstein JM, Wahl SM (1996) Fas mediates apoptosis in human monocytes by a reactive oxygen intermediate-dependent pathway. J Immunol 156:3469–3477

Umansky V, Bucur M, Schirrmacher V, Rocha M (1997) Activated endothelial cells induced apoptosis in lymphoma cells: role of nitric oxide. Int J Oncol 10:465–471

Van Antwerp DJ, Martin SJ, Kafri T, Green DR, Verma IM (1996) Suppression of TNF-alpha-induced apoptosis by NF-kappaB. Science 274:787–789

Van Den Dobbelsteen DJ, Nobel CSJ, Schlegel J, Cotgreave IA, Orrenius S, Slater AFG (1996) Rapid and specific efflux of reduced glutathione during apoptosis induced by anti-Fas/APO-1 antibody. J Biol Chem 271:15420–15427

Van Dyke K (1997) The possible role of peroxynitrite in Alzheimer's disease: a simple hypothesis that could be tested more thoroughly. Medical Hypotheses 48:375–380

Virag L, Scott GS, Cuzzocrea S, Marmer D, Salzman AL (1998a) Peroxynitrite-induced thymocyte apoptosis – the role of caspases and poly(ADP-ribose) synthetase (PARS) activation. Immunology 94:345–355

Virag L, Salzman AL, Szabo C (1998b) Poly (ADP-ribose) synthetase activation mediates mitochondrial injury during oxidant-induced cell death. J Immunol 161:3753–3759

Wallach D (1997) Cell death induction by TNF: a matter of self control. TIBS 22:107–109

Walling C (1995) Fenton's reagent revisited. Accts Chem Res 8:125–131

Wang CY, Eshleman JR, Willson JKV, Markovitz S (1995) Both transforming growth factor beta and substrate release are inducers of apoptosis in a human colon adenoma cell line. Cancer Research 55:5101–5105

Wardman P, Candeias LP (1996) Fenton chemistry: an introduction. Radiation Res 145:523–531

Wehrle I, Jakob A, Höfler P, Bauer G (1994) Transformation of murine fibroblasts by UV light and TGF-β: establishment of an autocrine TGF-ß loop. Int J Oncology 5:1341–1346

Weis M, Kass GEN, Orrenius S (1994) Further characterization of the events involved in mitochondrial Ca-2+ release and pore formation by prooxidants. Biochem Pharm 47:2147–2156

Weller M, Frei K, Groscurth P, Krammer PH, Yonekawa Y, Fontana A (1994) Anti-Fas-APO-1 antibody-mediated apoptosis of cultured human glioma cells: induction and modulation of sensitivity by cytokines. J Clin Invest 94:954–964

Winterbourn CC (1995a) Free radical toxicology and antioxidant defence. Clin Exp Pharmacol Physiol 22:877–880

Winterbourn CC (1995b) Toxicity of iron and hydrogen peroxide: the Fenton reaction. Toxicology letters 82/83:969–974

Wittung P, Malmstrom BG (1996) Redox-linked conformational changes in cytochrome c oxidase. FEBS letters 388:47–49

Wolin MS (1996) Reactive oxygen species and vascular signal transduction mechanisms. Microcirculation 3:1–17
Wong GH, Goeddel DV (1989) Induction of manganous superoxide dismutase by tumor necrosis factor: possible protective mechanism. Science 242:941–943
Wong GHW, Elwell JH, Oberley LW, Goeddel DV (1989) Manganous superoxide dismutase is essential for cellular resistance to cytotoxicity of tumor necrosis factor. Cell 58:923–931
Worlitzsch D, Herberth G, Ulrich M, Doering D (1998) Catalase, myeloperoxidase and hydrogen peroxide in cystic fibrosis. Europ Respiratory Journal 11:377–383
Xiao BG, Bai XF, Zhang GX, Link H (1997) Transforming growth factor beta-1 induces apoptosis of rat microglia without relation to Bcl-2 oncoprotein expression. Neuroscience Letters 226:71–74
Xie K, Huang S, Dong Z, Juang SH, Gutman M, Xie QW, Nathan C, Fidler IH (1995) Transfection with the inducible nitric oxide synthase gene suppresses tumorigenicity and abrogates metastasis by K-1735 murine melanoma cells. J Exp Med 181:1333–1343
Xie K, Huang S, Wang Y, Beltran PJ, Juang SH, Dong Z, Reed JC, Mcdonnel TJ, McConkey DJ, Fidler IJ (1996) Bcl-2 protects cells from cytokine-induced nitric-oxide-dependent apoptosis. Cancer Immunology, Immunotherapy 43:109–115
Xie K, Wang Y, Huang S, Xu L, Bielenberg D, Salas T, Mcconkey DJ, Jiang W, Fidler IJ (1997) Nitric oxide-mediated apoptosis of K-1735 melanoma cells is associated with downregulation of Bcl-2. Oncogene 15:771–779
Xie KP, Fidler IJ (1998) Therapy of cancer metastasis by activation of the inducible nitric oxide synthase. Cancer and Metastasis 17:55–75
Xu Y, Nguyen Q, Lo DC, Czaja MJ (1997) c-myc-dependent cell apoptosis results from oxidative stress and not a deficiency of growth factors. J Cell Physiol 170:192–199
Yamakura F, Taka H, Fujimura T, Murayama K (1998) Inactivation of human manganese-superoxide dismutase by peroxynitrite is caused by exclusive nitration of tyrosine 34 to 3-nitrotyrosine. J Biol Chem 273:14085–14089
Yamamoto M, Maehara Y, Sakaguchi Y, Kusumoto T, Ichiyoshi Y, Sugimachi K (1996) Transforming growth factor-beta-1 induces apoptosis in gastric cancer cells through a p53-independent pathway. Cancer 77:1628–1633
Yan T, Oberley LW, Zhong W, St Clair DK (1996) Manganese-containing superoxide dismutase overexpression causes phenotypic reversion in SV 40-transformed human lung fibroblasts. Cancer Res 56:2864–2871
Yanagihara K, Tsumuraya M (1992) Transforming growth factor beta-1 induces apoptotic cell death in human gastric carcinoma cells. Cancer Research 52:4042–4045
Yao Y, Yin D, Jas GS, Kuczera K, Williams TD, Schoneich C, Squier TC (1996) Oxidative modification of a carboxyl-terminal vicinal methionine in calmodulin by hydrogen peroxide inhibits calmodulin-dependent activation of the plasma membrane Ca-ATPase. Biochemistry 35:2767–2787
Zamzami N, Marchetti P, Castedo M, Decaudin D, Macho A, Hirsch T, Susin SA, Petit PX, Mignotte B, Kroemer G (1995) Sequential Reduction of mitochondrial Transmembrane potential and generation of reactive oxygen species in early programmed cell death. J Exp Med 182:367–377
Zamzami N, Susin SA, Marchetti P, Hirsch T, Gomez-Monterrey I, Castedo M, Kroemer G (1996) Mitochondrial control of nuclear Apoptosis. J Exp Med 183:1533–1544
Zamzami N, Marzo I, Susin SA, Brenner C, Larochette N, Marchetti P, Reed J, Kofler R, Kroemer G (1998a) The thiol crosslinking agent diamide overcomes the apoptosis-inhibitory effect of Bcl-2 by enforcing mitochondrial permeability transition. Oncogene 16:1055–1063
Zamzami N, Brenner C, Marzo I, Susan SA, Kroemer G (1998b) Subcellular and submitochondrial mode of action of Bcl-2-like oncoproteins. Oncogene 16:2265–2282

Zhang HW, Squadrito GL, Pryor WA (1998) The reaction of melatonin with peroxynitrite – formation of melatonin radical cation and absence of stable nitrated products. Biochem. Biophys Res Commun 251:83–87

Zhao Z, Francis CE, Welch G, Loscalzo J, Ravid K (1997) Reduced glutathione prevents nitric oxide-induced apoptosis in vascular smooth muscle cells. Biochim Biophys 1359:143–152

Zucker B, Hanusch J, Bauer G (1997a) Glutathione depletion in fibroblasts is the basis for induction of apoptosis by endogenous reactive oxygen species. Cell Death and Differ 4:388–395

Zucker B, Bauer G (1997b) Intercellular induction of apoptosis of transformed cells is modulated by their intracellular glutathione concentration. Int J Oncol 10:141–146

Zulueta JJ, Yu FS, Hertig IA, Thannickal VJ, Hassoun PM (1995) Release of hydrogen peroxide in response to hypoxia-reoxygenation: role of an NAD(P)H oxidase-like enzyme in endothelial cell plasma membrane. Am. J. Respir. Cell Mol Biol 12: 41–49

CHAPTER 12
Clearance of Apoptotic Lymphocytes by Human Kupffer Cells. Phagocytosis of Apoptotic Cells in the Liver: Role of Lectin Receptors and Therapeutic Advantages

L. DINI

A. Introduction

This chapter (see also SAVILL and BEBB, Chap. 6, this volume,) deals with the removal of apoptotic cells. The engulfment of cells undergoing apoptosis can be considered a specialized form of phagocytosis, playing a major role in the general tissue homeostasis in physiological and pathological conditions. Phagocytic recognition of apoptotic cells is less well understood than the death program itself, but an increasing number of recent studies are highlighting its importance. A particular aspect of phagocytosis of apoptotic cells will be considered: the Kupffer-cell-mediated removal of apoptotic lymphocytes.

I. Apoptotic Cells: Fast Food for Phagocytes

Apoptosis in vivo is followed almost inevitably by rapid uptake into adjacent phagocytic cells (SAVILL et al. 1993; SAVILL 1997). Condemned cells are swiftly identified and engulfed by phagocytes. The fact that "free" or "nonphagocytosed" dying cells are rarely observed in vivo because of their swift removal partly explains why apoptosis has been only recently identified as a frequent physiological event.

Apoptotic cell removal by phagocytes is a key factor of the program of events associated with this type of cell death in diverse processes: during development favoring the remodeling of embryonic tissue, during physiological situations like thymic involution, for the maintaining of the normal tissue homeostasis, during pathological conditions and resolution of inflammatory response (MEAGHER et al. 1992; HASLETT et al. 1994; FADOK et al. 1998b; SAVILL 1998). The fact that dead cells are ingested by neighboring ones during development suggests that this process serves as a fundamental homeostatic role in multicellular organisms (CLARKE 1990; ELLIS et al. 1991a,b; NISHIKAWA et al. 1998). Investigations of cell death in the nematode *Coenorhabditis elegans* and mutations that affect this process have been particularly enlightening (ELLIS et al. 1991b). Cells that die are phagocytosed not by specialized phagocytes, which are absent from this simple invertebrate, but by neighboring cells. Six mutants that perturb engulfment have been reported (ELLIS et al. 1991b).

Phagocyte recognition of "apoptotic self" is also essential in protecting tissues from inflammatory injury due to leakage of noxious contents from dying cells and possibly limiting the development of auto-immune responses (REN and SAVILL 1998). Unlike other receptor-mediated phagocytic responses of macrophages, ingestion of apoptotic neutrophils does not lead to release of pro-inflammatory mediators (MEAGHER et al. 1992). In fact, the phagocytosis of apoptotic neutrophils, in contrast to Fc-receptor-mediated phagocytosis (RAVETCH 1994) and immunoglobulin G-opsonized apoptotic cells, actively inhibits the production of interleukin-1beta (IL-1β), IL-8, IL-10, granulocyte macrophage colony-stimulating factor (G-MCSF), and tumor necrosis factor-alpha (TNFα), as well as leukotriene C4 and tromboxane B2, by human monocyte-derived macrophages (FADOK et al. 1998b). In contrast, production of transforming growth factor-beta 1 (TGFβ1), prostaglandin E2 or PAF results in inhibition of lipopolysaccharide (LPS)-stimulated cytokine production (FADOK et al. 1998b). Leukocyte recruitment is apparently restricted to situations in which phagocytic capacity is exceeded and apoptotic cells become secondarily necrotic before clearance (OGASAWARA et al. 1993).

The final intracellular fate of intact ingested cells undergoing apoptosis is the lysosomal enzyme destruction. However, little is known about signaling events downstream of apoptotic cell binding to specific receptors. Recently LIU and HENGARTNER (1998) cloned the *ced-6* gene from *C. elegans* that is required for engulfment of apoptotic cells. It encodes a protein with a phosphotyrosine-binding domain and appears to be an adaptor molecule that functions within a specific signal-transduction pathway.

But what are the mechanisms underlining the phagocytosis of apoptotic cells? Recent data indicate that apoptotic cells are marked for disposal by mechanisms which remain poorly understood. Investigations employing a variety of cell types and species imply that changes of the plasma membrane could include surface sugar and charge changes, and exposure of phosphatidylserine (PS) leads to recognition by uncharacterized phagocyte receptors (SAVILL et al. 1993; HART et al. 1996; SAVILL 1997; FADOK et al. 1998a). Although several systems of recognition on the surface of the phagocyte have been proposed to trigger or execute the apoptotic engulfment, the nature of the molecules involved and their molecular roles are still ill defined. Available data have identified candidate phagocyte molecules for restraining apoptotic cells (i.e., lectins, thrombospondin (TPS), CD14, scavenger receptors), transmembrane signaling for phagocytosis ($\alpha_v\beta_3$, CD36, ABC1, an ATP binding Cassette transporter, CED-6) and cytoskeletal reorganization (CED-5) (SAVILL et al. 1990, 1992a; FADOK et al. 1992a,b, 1998a; DINI et al. 1993, 1996; FLORA and GREGORY 1994; REN et al. 1995; LUCIANI and CHIMINI 1996; DEVITT et al. 1998; LIU and HENGARTNER 1998; SAVILL 1998; WU and HORVITZ 1998). These aberrant exposures, as well as several independent mechanisms, allow for the recognition of apoptotic cells by different phagocyte populations and by non-phagocytic cells such as fibroblasts and epithelial cells (SAVILL et al. 1989, 1990). Therefore, individual phagocytes might employ parallel or redundant phagocytic receptor systems. It is conceivable that the several systems of recognition on the surface of the phagocyte proposed to trigger or execute the apoptotic engulfment may act sequentially, each recognizing cells at different stages of the death program. Indeed, data from the literature indicate that macrophages have evolved distinct mechanisms for safe recognition of late apoptotic neutrophils, complicating attempts to clarify this mechanism in vivo (SAVILL 1998). A full understanding of this complexity will require definition of recognition mechanisms which operate *in vivo* in higher organisms. In fact, an active phagocytosis of apoptotic cells and bodies exerted by the hepatic sinusoidal cells is observed in vivo during the massive liver involution generated by a single injection of lead nitrate (DINI et al. 1996a).

B. Recognizing Death: Phagocytosis of Apoptotic Cells in the Liver

I. Liver Apoptosis

Apoptosis is considered a process whereby organisms eliminate "unwanted" (damaged, precancerous, or excessive) cells. However, apoptosis is also the complement of mitosis, and in concert with it determines maintenance, growth, or involution of tissue (GERSCHENSON and ROTELLO 1991). Although apoptosis occurs at a negligible rate in the normal liver, a variety of physiological conditions, diseases, and xenobiotic treatments can cause this form of cell death. Regression of the liver during starvation is accompanied by an enhanced rate of apoptosis (BURSH et al. 1992). Cell loss through apoptosis has also been detected in liver during physiological cellular renewal, in cellular depletion after the "overshoot" of cell regeneration of animals subjected to partial hepatectomy (TESSITORE et al. 1989), and after stimulation with mitogens or hyperplasia-inducing treatments (COLUMBANO et al. 1985; BURSCH et al. 1986). Moreover, apoptosis is also induced by stressful stimuli and by unfavorable environmental conditions (COLUMBANO et al. 1985; BURSCH et al. 1992; GRASL KRAUPP et al. 1994; LEDDA-COLUMBANO et al. 1996). Accordingly, a large number of toxins produce hepatocyte apoptosis.

In the liver, like other organs, the apoptotic process can be divided into four phases, the first three being: an induction phase, the nature of which depends on the specific death-inducing signals; an effector phase, during which the "central executioner" is activated and the cell becomes committed to die; and a degradation phase, during which cell acquires the biochemical and morphological features of endstage apoptosis. In this cascade of events, the "point of no return" would be the step at which the cell becomes irreversibly committed to the loss of essential cellular functions. The fourth, and last phase, is the engulfment of the dead corpse by macrophages and other "occasional" phagocytes. The apoptotic cells within an organ are not, however, easily detectable. Attempts to detect apoptotic cells in clinical samples are rarely successful. A hypothesis is that apoptotic cells are cleared from the circulation by phagocytosis before they become detectable by conventional morphological or cytometric methods (DURRIEU et al. 1998). DNR-treated K562 cells were eliminated by phagocytes while apoptosis was never observed by any of the above methods (DURRIEU et al. 1998).

Phagocytosis, one of the peculiar functions of the liver, is beautifully operated by the sinusoidal cells (i.e., endothelial and Kupffer cells) (SMEDSRØD et al. 1990, 1994; TOTH and THOMAS 1992). Endothelial and Kupffer cells have many specific functions that are essential for the preservation of homeostasis in liver under several conditions and the endocytosis is pivotal for this role. Endocytosis, and particularly receptor-mediated endocytosis that is a major route for protein or glycoconjugate ligand transport into liver cells, is not only essential for the removal of plasma proteins but also of particulate material from the blood such as apoptotic cells and/or bodies, that are produced at the end point of the apoptotic process (DINI et al. 1996a). Due to their location in the sinusoids, and combined with the fact that they represent the majority of the body's fixed macrophages, Kupffer cells are predominant participants in this process. They are the first cells of the mononuclear phagocyte system to come into contact with particulate and immunoreactive materials coming from the blood, potentially noxious like apoptotic cells. However, the functions of these cells include not only the phagocytosis of foreign particles (JONES and SUMMERFIELD 1982) but also the removal of endotoxin (RUITER et al. 1981), tumor cells (ROOS et al. 1978), and liposomes (ROERDINK et al. 1981), the presentation of antigens mediating immune responses (RIFAI and MANNIK 1984), the metabolism of lipoproteins (VAN BERKEL et al. 1992), and the secretion of mediators such as oxygen-derived free radicals, nitrogen intermediates, several cytokines and arachidonate metabolites (SHIRATORY et al. 1993). Many of their phagocytic activities are mediated by specific receptors: carbohydrate-specific receptors (DINI and KOLB-BACHOFEN 1989), receptors for fibronectin and receptors for surface-bound fragments of C3 (WARDLE 1987; KEMPKA et al. 1990) that enable Kupffer cells to bind and endocytose denatured proteins and lipids (NENSTER et al. 1992; VAN BERKEL et al. 1992) and glycoproteins (STEER and CLARENBURG 1979), opsonized foreign particles (KOLB-BACHOFEN 1992), bacteria, yeasts, and viruses (KIRN et al. 1982), apoptotic bodies (DINI et al. 1993), and immune complexes (WARDLE 1987).

II. Hepatic Lectin-Like Receptors

Among the several alternative mechanisms reported for removal of apoptotic cells, that are mainly related to the cell type and system used, it has been reported that in the liver recognition and phagocytosis of apoptotic cells are operated by means of hepatic lectin-like receptors (DINI et al. 1996a). The first demonstration that the asialoglycoprotein receptor (ASGPR) (likely in cooperation with other carbohydrate receptors) is involved in the phagocytosis of apoptotic hepatocytes by healthy ones was performed on newborn hepatocyte cultures induced to undergo apoptosis by hormonal treatments (DINI et al.

1992). The apoptotic bodies, floating in the culture supernatants, were removed by the hepatocytes. The idea that the apoptotic cell surface might expose normally masked sugar residues, rendering them available for interaction with lectin-like receptors on hepatocytes, was supported by the ability of the specific receptor antibodies and sugar moieties to block their binding and uptake by the living liver cells. Therefore hepatocyte recognition and internalization of apoptotic cells is due to the exposition of several glycans, in particular galactose/N-acetyl-galactosamine, on the surface of apoptotic cells (DINI et al. 1992). The presence of galactose/N-acetyl-galactosamine, mannose/N-acetylglucosamine on the surface of apoptotic hepatocytes was observed on cells derived both from the supernatant of the cultures as well as isolated from livers of rat treated to induce apoptosis in vivo (DINI et al. 1992).

In the liver the clearance of galactose-terminated particles from the circulation is performed by a galactose-specific uptake mechanism on Kupffer cells. This receptor shows a high affinity for particulate ligands that expose galactose groups, like desialylated erythrocyte (KOLB-BACHOFEN et al. 1982). It is worth noting that liver endothelial cells also reveal galactose-specific receptors on their surface (DINI et al. 1993) for receptor-mediated endocytosis of circulating modified glycoproteins and for engulfment of large-sized materials (STEFFAN et al. 1986). Moreover, liver endothelial and Kupffer cells take up a wide range of molecules with a net negative charge by the so-called scavenger receptor (VAN BERKER et al. 1992) and with mannose- and N-acetylglucosamine residues by lectin-like receptors. The presence of receptors that specifically interacted with mannose- and N-acetylglucosamine-terminated glycoproteins on sinusoidal liver cells was first described by STEER and CLARENBURG (1979). Liver endothelial cells are the primary site for uptake of these glycoproteins (HUBBARD et al. 1979). Although this receptor has been identified on Kupffer cells, it contributes to a much lower degree (sixfold lower) to the uptake of various mannose-exposing ligands from the circulation than with the endothelial cells (PRAANING-VAN DALEN et al. 1987).

The above reported data shows that, due to exposing of several normally masked glycans on the surface of dead cells, all the main three liver cell types possess receptors that can potentially recognize apoptotic cells (MORRIS et al. 1984; DUVALL et al. 1985; DINI et al. 1992; HALL et al. 1994). Therefore, liver cells are predictable actors in the recognition and subsequent engulfing of apoptosing cells, probably by means of specific carbohydrate-receptors.

Modulation of cell surface molecules has been reported for cells undergoing the process of apoptosis in different experimental conditions (EMOTO et al. 1997; SAVILL 1998) but very little is known about receptor molecules on dying cells or on neighboring healthy ones. On the surface of non-apoptotic liver cells (i.e., hepatocytes, Kupffer cells, endothelial cells), the expression of ASGP-R, galactose-specific receptor, and mannose-specific receptor is modulated (enhanced or decreased) during the entire process of apoptosis, induced in vivo by administration of a potent liver mitogen, lead nitrate (DINI et al. 1993, 1995). The number and distribution of binding sites is receptor and cell-type dependent during the days following the metal injection. However, the intensity and the persistence of the modulation are specific for the different liver cell types, thus indicating different (time and modality) involvement for hepatocytes, Kupffer cells, and endothelial cells during the process of apoptosis. It is worth mentioning that a relationship of carbohydrate receptor expression to the differentiated and/or metabolic state of liver cells has been well documented. The mechanism(s) responsible for this regulation has not yet been completely clarified, even though post-translational modulations are indicated (Massimi et al. 1996).

Irrespective of the liver cell type, galactose and mannose receptors cooperate for the removal of apoptotic cells: decrement of galactose binding sites are paralleled by mannose binding sites overexpression. In this way, carbohydrate specific receptors are always expressed in great amounts on the cell surface. The meaning of all the above-mentioned changes has to be better understood. To this end we are currently studying the modification of hepatic membrane composition in relation to apoptosis.

Hepatic membrane composition may be under the control of mitochondria. A single intravenous injection of lead nitrate was able to lower the activity of the mitochondrial tricarboxylate carrier and the lipogenic enzymes as well as modify the lipid mitochondrial composition, but leaving unaltered the ultrastructure of the mitochondria (Dini et al. 1999). In particular, the reduced activities of cytosolic lipogenic enzymes could suggest a putative mitochondrial control of apoptotic membrane alterations through the tricarboxylate carrier (Dini et al. 1999) In fact, besides other functions, the tricarboxylate carrier plays an important role in fatty acid biosynthesis since it catalyzes the transport of acetyl-CoA, condensed with oxaloacetate in the form of citrate, from mitochondria to the cytosol of the cell, where lipogenesis occurs. Interestingly, in a recent paper Castedo et al. (1995) has shown that the mitochondrial transmembrane potential disruption leads to phosphatidylserine exposure on the plasma membrane, thus causing alterations of the surface that will facilitate the phagocytic recognition and removal of cells en route to apoptosis.

The use of an in vivo model of induction of apoptosis in the liver (Columbano et al. 1985) highlights the role of lectin-like receptors (in particular galactose- and mannose-specific receptors) in the recognition of dead cells (Dini et al. 1993). During the metabolic disorder of the liver, generated by lead nitrate treatment, sinusoidal liver cells (i.e., Kupffer and endothelial cells) activate phagocyte apoptotic hepatocytes and circulating apoptotic cells by using both galactose and mannose-specific receptors, as suggested by inhibition uptake experiments. In particular, Kupffer cells at five and fifteen days from the lead nitrate injection are very active in internalizing apoptotic cells (two- to threefold the control), but phagosomes containing apoptotic hepatocytes are often seen inside the cytoplasm of parenchymal cells and endothelial cells. The ability of endothelial liver cells to recognize and internalize apoptotic cells and/or bodies (maintained even after isolation and cultivation) has been already reported (DINI et al. 1995; DINI and CARLÀ 1998a) and it is in line with the capacity of the hepatic sinusoidal wall to interact with particulate materials (WARDLE 1987; DINI and KOLB-BACHOFEN 1989; KOLB-BACHOFEN 1992) and to operate as a protective barrier for the systemic circulation (TOTH and THOMAS 1993). Interestingly, apoptotic lymphocytes are retained by the sinusoids in a heterogeneous distribution: apoptotic cells in the periportal tract are double those in the perivenous region (DINI and CARLÀ 1998a). The reason should be found in the differences existing between periportal and centrilobular endothelial cells regarding the fenestration pattern (MORIN et al. 1984) and to the uneven expression of galactose and mannose-specific receptors (ROCHA et al. 1993).

Although the mannose receptor-mediated endocytosis is a characteristic of the endothelial cells as a whole, the uneven distribution down the length of the sinusoidal pathway of the mannose receptor (ROCHA et al. 1993) suggests that this function occurs preferentially in the periportal segment (ASUMENDI et al. 1996). Mannose receptor expression on the liver endothelium is upregulated by IL-1 and is associated with increased removal of apoptotic cells and tumor cell adhesion (VIDAL-VANACLOCHA et al. 1994; DINI et al. 1995). The ability to recognize apoptotic lymphocytes has therefore been related to the amount of carbohydrate receptors expressed on the cell surface (DINI and CARLÀ 1998a).

Summarizing, multiple data are in favor of the involvement of hepatic carbohydrate receptors in the apoptotic cell and/or body clearance: (i) the cell surface of dead hepatocytes expresses great amounts of galactose/*N*-acetylgalactosamine/mannose residues; (ii) hepatocytes, Kupffer, and endothelial cells express on their cell surface the carbohydrate receptor systems; (iii) these receptors are modulated differently during the in vivo onset of apoptosis; (iv) during in vivo onset of apoptosis hepatocytes, Kupffer and endothelial cells show large phagosome containing apoptotic bodies; (v) LPS and IL1β stimulation of endothelial cells markedly enhances the phagocytosis of apoptotic lymphocytes, probably by increasing the carbohydrate receptors expressed on the cell surfaces; (vi) the removal of apoptotic cells is reduced by about 70% by addition of specific saccharide.

III. Kupffer Cells Phagocytic Activity

To accomplish phagocytosis of apoptotic cells, the recognition process must be followed by internalization (Figs. 1–3). This latter phenomenon needs cytoplasmic movements that generate fine filamentous processes immediately adjacent to the particle, in which the cytoskeleton plays a major role (WATANABE 1988). Since endocytosis is a multistep process that includes cellular movements, in particular the extension of pseudopodia, a decrease in ruffling movements of the pseudopodia of Kupffer cells indicates an inhibition of phagocytic capacity (WATANABE et al. 1990). During some pathological conditions of the liver (such as adenoma and cirrhotic nodules) Kupffer cells pos-

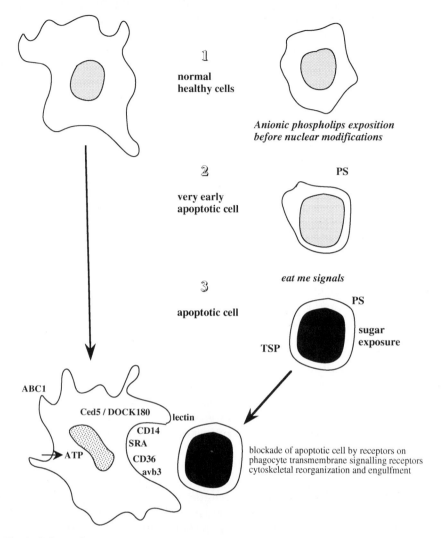

Fig. 1. Schematic representation of the current understanding of the uptake pathway of apoptotic cells by phagocytes. The molecular modifications of the plasma membrane during apoptosis that lead to swift recognition represents those of human lymphocytes. The very early stages of apoptosis (2) are characterized by modification of the membrane lipid asymmetry and external exposition of phosphatidylserine; nuclear modifications are not yet visible; (3) The late stages of apoptosis are characterized by chromatin compacting, round shape, and eventual production of membrane-bounded apoptotic bodies. Normally hidden sugar residues are exposed on the extracellular face of the plasma membrane; (4) membrane modifications of apoptotic cells are necessary for their bound by receptors on phagocytes followed by later internalization. The "phagocyte" represents a combination of features that have been attributed to different cell types capable of phagocytosing apoptotic cells, including human Kupffer cells. These adhesions allow to interact with signaling pathway that allow apoptotic cells to reach their final fate within the phagocytes. To engulf the apoptotic cells cytoskeletal reorganizations are also necessary. Abbreviations: PS, phosphatidylserine; TPS, thrombospondin; CD36/integrin/$\alpha_v\beta_3$, vitronectin receptor; ABC1, ATP binding cassette transporter; Ced5, Coenorabditis elegans gene; DOCK 180, adaptor/signaling molecules; SRA, scavenger receptor class A

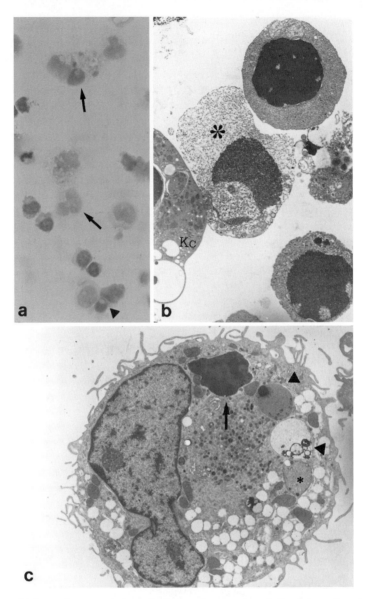

Fig. 2a–c. Light and transmission electron micrographs of the interaction between apoptotic lymphocytes and cultured Kupffer cells (KC) at different interval times. **a** Apoptotic lymphocytes when incubated with Kupffer cells at 37°C for 5, 10, 15, 30, or 60 min are promptly bound (*arrowhead*) and phagocytosed (*arrow*). ×800. **b** An apoptotic lymphocyte (*), whose chromatin aggregates into dense masses and the nucleus is displaced to one edge of the cell, adhering closely to the plasma membrane of a human Kupffer cell (Kc) at 5 min of incubation. Within 5 min of coculture almost all the apoptotic lymphocytes are bound to the plasma membrane of Kupffer cells, while after 10 min of incubation the majority of apoptotic cells are internalized by the Kupffer cells, thus suggesting a very rapid mechanism of recognition. **c** At longer times of cocultivation, phagosomes containing dark material, which represent residues of the partially digested apoptotic lymphocytes, are visible inside Kupffer cells (*arrow*). ×5000. **c** Two phagolysosomes containing apoptotic lymphocytes remnants with still recognizable nuclear dense masses (*arrowheads*). Secondary lysosomes resulting from degradation of phagocytosed apoptotic cells are also visible (*asterisk*). ×6500

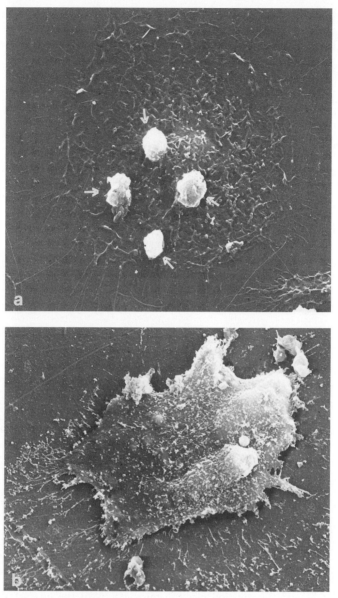

Fig. 3a,b. Scanning transmission electron micrographs of cocultures of apoptotic lymphocytes and human Kupffer cells. **a** Human Kupffer cells are characterized by prominent membrane ruffling with microvilli of variable length accompanied by numerous pseudopodia when cultured in normal condition. Conversely, apoptotic cells are recognized by their round, smooth surface that is a consequence of the disappearance of microvilli during the apoptotic process (*arrows*). Apoptotic lymphocytes added to the culture medium adhere to the surface of the Kupffer cells. ×10,000. **b** A few minutes later Kupffer cells, that are very active in phagocytosis, have completely internalized the apoptotic lymphocytes. After 15 min of coculture round protusions (representing the internalized apoptotic lymphocytes) are often visible inside the cells. When Kupffer cells were incubated with the carbohydrate-specific receptor inhibitors (i.e., sugars or modified glycoproteins) before and during the incubation with apoptotic lymphocytes, their phagocytic activity was dramatically reduced. The addition of healthy lymphocytes to the Kupffer cell cultures does not result in the recognition and internalization of the blood cells. ×7000 sessing a flattened shape and few or no pseudopodia have been described as hypoactive (BURT et al. 1993), while during the process of activation both the number and length of the surface projections of the Kupffer cells usually increased. Conversely, activated Kupffer cells (i.e., LPS, $Pb(NO_3)_2$, cytokine stimulation) show an enhanced phagocytic capacity toward apoptotic cells. Observations of phagocytosed particles have led to the proposal of several possible mechanisms through which internalization is achieved (SWANSON and BAER 1995). One model, "zippering", requires the sequential recruitment of cell-surface receptors on the extending pseudopodia into positions in which they can interact with appropriate ligands. Thus internalization of the particle requires sequential interactions between receptor and ligands in addition to those responsible for initial binding. A second model, "triggering", suggests that initial attachment is itself sufficient to initiate phagocytosis. The multitude of different receptors that have been implicated in apoptotic cell uptake could be consistent with the "zippering" mechanism, which requires sequential receptor recruitment (PLATT et al. 1998). It is worth noting that the state of the phagocyte is also particularly important in the apoptotic recognition (SAVILL et al. 1993). The particular mechanism employed by macrophages and/or other amateur phagocytes may be regulated by external influences. The exposure of human monocyte-derived macrophages to granulocyte-macrophage colony stimulating factor (GM-CSF), a cytokine known to be present at inflammation sites, increased the recognition of apoptotic human neutrophils (SAVILL et al. 1993). Cytokines implicated in repair of injured tissue (i.e., transforming growth factor, TGF-β; platelet-derived growth factor, PDGF) and those involved in the initiation of inflammation (i.e., interferon gamma, IFN-γ, interleukin-1, IL-1 and tumor necrosis factor-a TNF-α) also stimulated TPS-dependent recognition of apoptotic neutrophils (SAVILL et al. 1993). IL-5 modulates macrophage phagocytosis of apoptotic eosinophils (STERN et al. 1992). LPS and IL1β upregulate the mannose receptor expression of liver cells and consequently the phagocytic activity of sinusoidal cells (DINI et al. 1995).

Kupffer cells represent an useful tool for the studies of phagocytosis: they can be used in situ or can be isolated from the livers (different vertebrates including man) and maintained in suspension or in adhesion cultures (NEAUD et al. 1995). The binding and the internalization activity of sugar-exposing ligands is present in situ as well as in vitro, but the amount of both binding sites and internalized ligands is dramatically different in the three experimental models, thus suggesting that different physiological states can be induced by different experimental conditions (DINI et al. 1998b). However, the relative capacity of internalization is almost unchanged in the different systems if the rate of binding to internalization is considered.

The altered morphology of isolated and cultured Kupffer cells, with few and shorter microvilli and pseudopodia compared to the in situ cells (whose traditional image is "stellate" due to the presence of microvillous projections, blebs, etc.), could be one of the reasons for the reduced phagocytic activity of these cells, probably caused by decreasing the number of specific receptors on their surfaces and by decreasing the number and the length of pseudopodia.

A parallel decrement in carbohydrate receptor expression, phagocytosis of apoptotic cells, and microvillous projections is found. Therefore, Kupffer cells phagocytosis of apoptotic cells mediated by specific receptors are dependent on the extent to which these receptors are expressed and, in turn, on the physiological state of the cells. In fact, it has been described several times that the expression of the galactose-specific receptor is sensitive to the physiological and pathological condition of the cells (MASSIMI et al. 1995).

C. Human Kupffer Cells Removal of Apoptotic Lymphocytes

I. Lymphocyte Cell Surface Modifications

In vivo, apoptotic lymphocytes are recognized and phagocytosed by macrophages (Kupffer cells included) well before the final stages of DNA degradation and cell lysis (PRADHAN et al. 1997). The recognition process is apparently triggered by modifications of the cell surface. On the surface of the apoptotic lymphocytes, fewer varieties of potential ligand have so far emerged, the leading contender being PS, closely followed by carbohydrate changes; other possibilities remain, for the present on the sidelines. What is most disappointing at the time of writing is that no macrophage receptor has yet been linked definitively to a ligand in an apoptotic lympocytes.

Lymphocytes, like almost all other cell types, once induced to apoptosis by different apoptosing stimuli (mild hyperthermia, oxidative stress, chx, etc.) develop characteristic apoptotic morphological features that in turn depend on the specific biochemical events involved in the dead process (KUMAR 1995). Asymmetric distribution of phospholipids across the bilayer of lymphocytes plasma membrane (maintained by an ATP-dependent aminophospholipid translocase and dissipated by activation of a non-specific lipid flipsite) is lost as part of the program of cell death, by down regulation of the translocase and activation of the non-specific lipid flipsite. As a consequence, PS is exposed on the cell surface. In cells in which apoptosis is induced through the Fas system, such as HeLa cells (SHIRATSUCHI et al. 1998), T lymphocytes under activation-induced death (BRUNNER et al. 1995; DHEIN et al. 1995), acute lymphocytic leukemia cell lines treated with an anti-cancer drug, doxorubicin (FRIESEN et al. 1996), and influenza virus-infected cultured cells, PS externalization preceded other apoptotic events (STUART et al. 1998). Cells which have lost membrane asymmetry are recognized by macrophages (McEVOY et al. 1986; SCHLEGEL and WILLIAMSON 1987; PRADHAN et al. 1994), but it is still being debated whether PS externalization is sufficient for phagocytosis induction. PS externalization independent of apoptosis caused by N-ethylmaleimide treatment leads to PS-mediated phagocytosis and externalized PS by itself induces apoptosing cell phagocytosis before plasma membrane permeability increased

(SHIRATSUCHI et al. 1998). Moreover, that PS exposure has functional consequences is demonstrated by the ability of artificial lipid vesicles containing PS to inhibit enhanced phagocytosis of apoptotic lymphocytes by macrophages. Understanding the mechanisms that govern membrane lipid sidedness, including those that promote a collapse of phospholipid asymmetry, seems essential to the comprehension of the disease states in which this unwanted PS exposure, or lack of PS exposure, is observed (KUYPERS 1998).

However, other signals besides PS are also involved in recognition of apoptotic lymphocytes. Studies with other inhibitors indicate that macrophages also utilize integrin-mediated and lectin-like recognition systems, although each is restricted to either unactivated or activated macrophages, thus indicating that the signals for recognition of apoptotic lymphocytes are complex and involve multiple recognition systems (SCHLEGEL et al. 1996). During our studies, aimed at characterizing modifications of lymphocytes cell surface during the apoptotic process, we found that the glucidic residues of glycoproteins of plasma membrane were substantially changed in the apoptotic lymphocytes compared to normal cells (FALASCA et al. 1996). In particular our binding experiments, using four different fluorescent conjugate-lectins (Concanavalin-A, *Phaseolus limensis*, *Ricinus communis*, and *Ulex europaeus*) with different hapten sugar specificity, indicate that a relevant amount of desialylated glycans are exposed on the surface of apoptotic cells. The membranes of apoptotic lymphocytes express increased amount of *N*-acetyl-galactosamine, D-galactose, and mannose residues when compared with normal ones. In fact normal and apoptotic cells express the same amount of fucose residues. The same findings were confirmed at the ultrastructural level by labeling apoptotic lymphocytes with gold particles conjugated lectins (ConA-Au_{17} and PHA-Au_{17}) that resulted in labeling as small aggregates distributed all over the cell surface of apoptotic cells.

Interestingly, from our studies of cell surface glycoconjugates between normal and apoptotic lymphocytes isolated from different species (i.e., human, rat), it turns out that cell surface modifications of lymphocytes undergoing apoptosis are related to the species. In fact *Dolichos biflorus* (DBA) (*N*-acetyl-D-galactosamine) binding is detectable only on rat apoptotic lymphocytes while *Limulus polyphemus* (LPA) (*N*-acetyl-D-galactosamine, *N*-acetyl-D-glucosamine, *N*-acetylneuramic acid) binds on human apoptotic lymphocytes. Moreover, PS, whose exposition precedes sugar modifications (personal communication), is also differently expressed on dying rat and human lymphocytes. Rat apoptotic lymphocytes exhibit a higher intensity of Annexin V-FITC binding than human ones. These differences are attributed to the different rate of removal and internalization by murine sinusoidal liver cells (DINI 1999). In addition, time course of cell surface glycoconjugates modifications during apoptosis show that normally masked sugar residues are exposed sequentially.

All the above-mentioned plasma membrane changes correlate with the fact that apoptosis is accompanied by water loss, shrinkage of the cell, and enzymatic fixation of the membrane that leads to peculiarities in the antigenic make-up of the apoptotic cell membrane. CARBONARI et al. (1994), using differential light scattering analysis and identifying specific changes of apparent density of the same surface antigens, discriminated between viable, apoptotic, and necrotic lymphocytes. We do not know how these modifications of the cell surface carbohydrates could occur. Probably they are due to the exposure of new membranes derived from the fusion of endoplasmic reticulum or Golgi vesicles during the onset of apoptosis, or they may also be due to a possible desialylation process that causes the exposure of normally masked residues (MORRIS et al. 1984). This latter mechanism is responsible for the removal of aged erythrocytes by the liver (KOLB et al. 1981).

II. Kupffer Cells Recognition and Phagocytosis of Apoptotic Lymphocytes

Kupffer cells isolated from human liver biopsies recognize and phagocyte in a very efficient manner lymphocytes undergoing apoptosis, induced by different stimuli (heat-shock 43°C; cycloheximide), but not normal living ones (FALASCA et al. 1996). That this recognition is mediated by the carbohydrate specific receptors is strongly suggested by the contemporary presence of the galactose- and mannose-specific receptors on human Kupffer cells and the sugar residues on apoptotic lymphocytes. The hepatic removal of apoptotic cells, proposed in rats (DINI et al. 1996a) is therefore extended to human Kupffer cells. The atypical exposure of sugars is one of the molecular signals for the recognition of apoptotic lymphocytes by Kupffer cells. Phagocytosis is inhibited by sugar cocktail (glucose, N-acetyl-galactosamine, methyl-mannopyranoside, fucose) or, to a lesser extent, by desialylated glycoproteins (lactosylated bovine serum albumin, asialofetuin), but not by unmodified glycoproteins (fetuin, bovine serum albumin). The use of single compounds or modified glycoproteins never reaches the level of inhibition achieved by the sugar cocktail, thus suggesting cooperation among galactose- and mannose-specific receptors. However, the use of diverse molecular mechanisms by human Kupffer cells in the removal of apoptotic cells different from those we assayed cannot be excluded.

The multiple receptor ligand interactions (galactose and mannose) required for recognition and binding of apoptotic lymphocytes is a clever way for safe phagocytosis of blood circulating dead cells. Moreover, the fact that the same receptor systems for the recognition of apoptotic cells are shared among the different liver cells (DINI et al. 1996) suggests a differential involvement of liver cells in this activity. We propose that, while hepatocytes accomplish the selective removal of neighboring dying cells, Kupffer cells mediate the clearance of circulating apoptotic cells, which escape the removal by neighboring cells or derive from other body and/or cell districts. It is worth noting that the liver is the specialized site where T cells, undergoing apoptosis in vivo are eliminated (HUANG et al. 1994). However, the molecular mechanisms that control the accumulation and apoptosis of activated T cells in the liver are still unknown (HUANG et al. 1994).

The recognition of the apoptotic lymphocytes once added to human Kupffer cell cultures is a very rapid process, being almost entirely completed within a few minutes of incubation. Apoptotic cells immediately adhere to Kupffer cells and are detected as dark material inside large phagosomes (Figs. 2 and 3). Kupffer cells were never able to bind and internalize non-apoptotic lymphocytes when added to the cultures, even at the longer incubation times. In addition, recognition of apoptotic rat lymphocytes was significantly reduced compared to those of human apoptotic lymphocytes. It is of note that, in vivo as well as in vitro, Kupffer cells phagocyte apoptotic lymphocytes faster than endothelial liver cells, which internalize apoptotic cells only after long times of incubation. This fact suggests that liver cells are sequentially recruited for the removal of apoptotic cells. In particular, it could be speculated that in vivo phagocytosis of apoptotic cells by endothelial cells is restricted to the situation in which, due to the high number of circulating apoptotic cells, phagocytic capacity of Kupffer cells is exceeded.

It has been repeatedly claimed in this review that, to signal their "edible" status, cells undergoing apoptosis exhibit qualitatively and quantitatively cell surface modifications (including PS and sugar expositions) that are generated in a complex and evolving pathway. However, PS and sugar residues are not the unique key signal for the removal of apoptotic lymphocytes: in fact, when Kupffer cells are incubated with apoptotic U937 cells, the recognition was impaired in spite of the exposition of both PS and sugar residues. The "signal" that discriminates between apoptotic lymphocytes and U937 cells is far from being familiar. It is tempting only to speculate that a swift recognition of apoptotic cells is lost for cell lines and, conversely, it is an important phenomenon in vivo to prevent the inflammatory response.

It should also be borne in mind that the pathways to activation of apoptosis can be different and there may be many triggers for a suicide pathway even in a single cell type (DINI et al. 1996b; COBB et al. 1996). It could therefore be hypothesized that different signals coming from the environment (i.e., exposure to or withdrawal of a hormone or a growth factor, as in thymus atrophy after glucocorticoid administration, the response of cell damage to antitumoral drugs, oxidative stress, and heat shock) could determine the expression of different markers on the cell surface of dying cells to signal their presence.

D. Concluding Remarks and Future Perspectives

The importance of the phagocytosis of dying cells as a process in itself, rather than simply as the endpoint of programmed cell death, is finally being acknowledged; hence it is beginning to receive more attention and research

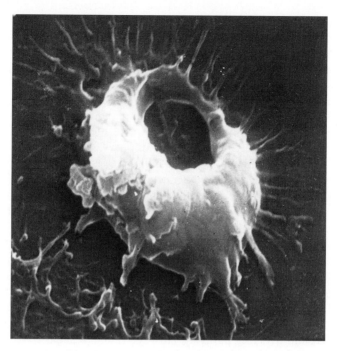

Fig. 4. Scanning transmission electron micrograph of human Kupffer cell incubated with apoptotic lymphocytes. The cell is showing an electron microscopy artifact of preparation that, however, is self explicating the reorganization of the cell during phagocytosis. The "donut" shape of the cell is due, after the glutaraldehyde fixation, to the loss of the lymphocyte before its definitive engulfment. ×25,000

effort. The previous brief discussion of the recognition and ingestion of apoptotic cells by Kupffer cells and by other liver cells shows clearly that human liver macrophages are active participants in the removal of apoptotic cells and that this removal is swift and efficient without eliciting an inflammatory/immune response (Fig. 4). Indeed, phagocytosis of apoptotic cells is not simply passively non-inflammatory but is actively anti-inflammatory (Voll et al. 1997; Fadok et 1998b). A macrophage receptor, CD 14, that is involved in the recognition and non-phlogistic removal of apoptotic cells (Devitt et al. 1998) was known as receptor for the bacterial endotoxin, lipopolysaccharide, which macrophage after binding elicits pro-inflammatory responses. Conversely, at least one unfavorable effect on the phagocytosis of apoptotic cells has been reported in the development of AIDS (Kornbluth 1994). Although apoptosis is often assumed to be a biological dead end, linear, unintegrated retroviral DNA survives apoptosis in avian leukosis virus systems. The viral DNA in apoptotic debris might spontaneously transfect macrophages that are avidly phagocytosing apoptosing cells, and thus lead to the production of new virions. Such a hypothetical accessory infection pathway may explain why anti-

HIV cytotoxic cells are unable to clear this virus from the body (KORNBLUTH 1994).

The presence of multiple molecular mechanism(s) involved in the recognition of apoptotic cells could probably be explained by the sequential recognition of cells at different stages of the apoptotic program and the existence of regional specialization in the recognition process. Cell clearance in vivo might depend upon more than one type of phagocyte, each developing a single mechanism as was found in the inflamed glomerulus where apoptotic neutrophils can be taken up by both macrophages and glomerular mesangial cells (SAVILL et al.1992b; MEAGHER et al. 1992). Possibly a cell undergoing apoptosis displays multiple signals of its status so that the probability of its removal increases and consequently the margin of safety for the whole organism is increased. For example, fibroblasts recognize apoptotic neutrophils via a vitronectin interaction but an additional involvement of a lectin-like mechanism was suggested by the inhibitory effects of mannose and fucose (HALL et al. 1994). Therefore, selection of one or more mechanisms for recognition of apoptotic cells by a particular cell type might depend upon the species, the lineage of the apoptotic cell, or the nature of the phagocyte involved (DINI 1999).

Other peculiarities are emerging in the complex field of the recognition mechanisms of apoptotic cells. In fact, cooperation in the removal of dead cells is restricted not only to the use of more than one cell surface receptor exposed on the phagocytic cells, but also to cooperation among different cellular type sharing the same receptor system for the recognition and removal of apoptotic cells. This fact is well illustrated in the liver where both hepatocytes, Kupffer as well as endothelial cells, operate the plasma clearance of apoptotic cells generated during the involuting phase of liver hyperplasia induced by a single injection of lead nitrate by means of a sugar recognition mechanism (DINI et al. 1993, 1995). These data, together with the fact that the phagocytic activity in endothelial cells can be enhanced in macrophage-depleted rats and that IL-1 induces in vitro overexpression of mannose-specific receptors on endothelial cells, suggest a cooperation with Kupffer cells in phagocytosis.

It is worthwhile to note that the study of the mechanisms of the phagocytosis during the process of apoptosis it is not merely a speculative exercise, since defects of phagocytosis of apoptotic cells might have deleterious consequences for neighboring healthy cells. The logical consideration of the importance of phagocytosis leads to thoughts on the contribution of defective clearance as a factor in the pathogenesis of inflammatory diseases. The relevance of phagocytosis to the dysregulation of the immune system that underlies specific pathological conditions requires examination: for example, whether compromising the capability to ingest apoptosing cells contributes to autoantibody production (BOTTO et al. 1998; HERMANN et al. 1998).

Further investigations of the molecular mechanisms of recognition and ingestion of apoptotic cells will be important for the identification of the target structures present on apoptotic cells and for a better understanding of the fate of apoptotic cells. This in turn may allow manipulation of phagocyte responses to apoptotic cell stimuli and the development of novel therapeutic strategies (for example, during tissue repair) as an effective anti-inflammatory and immunosuppressive strategy. Moreover, the investigation of the potential therapeutic use in administering agents to enhance, specifically, phagocytic clearance of apoptotic cells to remove unwanted cells (i.e., malignancy, targeted by apoptosis inducing treatments), should lead to the development of new therapeutics to overcome diseases for which effective medical treatment is not yet available.

References

Asumendi A, Alvarez A, Martinez I, Smedsrod B, Vidal-Vanaclocha F (1996) Hepatic sinusoidal endothelium heterogeneity with respect to mannose receptor activity is interleukin 1 dependent. Hepatology 23:1521–1529

Botto M, Dell'Agnola C, Bygrave AE, Thompson EM, Cook HT, Petry F, Loos M, Pandolfi PP, Walport MJ (1998) Homozygous C1q deficiency causes glomerulonephritis associated with multiple apoptotic bodies. Nat Genet 19:56–59

Brunner T, Mogil RJ, LaFace D, Yoo NJ, Mahboubi A, Echeverri F, Martin SJ, Force WR, Lynch DH, Ware CF, Green DR (1995) Cell-autonomous Fas (CD95)/ Fas-ligand interaction mediates activation-induced apoptosis in T-cell hybridomas. Nature 373:441–444

Bursch W, Dusterberg B, Schulte-Hermann R (1986) Growth, regression and cell death in rat liver as related to tissue levels of the hepatomitogen cytoproterone acetate. Arch Toxicol 59:221–227

Bursch W, Oberhammer F, Schlte-Hermann R (1992) Cell death by apoptosis and its protective role against disease. Trends Pharmacol Sci 13:245–251

Burt AD, Le Bail B, Balabaud C, Bioulac-Sage P (1993) Morphological investigation of sinusoidal cells. Semin Liver Dis 13:21–38

Carbonari M, Cibati M, Cherchi M, Sbarigia D, Pesce AM, Dell'Anna L, Modica A, Fiorilli M (1994) Detection and characterization of apoptotic peripheral blood lymphocytes in human immunodeficiency virus infection and cancer chemotherapy by a novel flow immunocytometric method. Blood 83:1268–1277

Castedo M, Macho A, Zamzami N, Hirsh T, Marchetti P, Uriel J, Kroemer G (1995) Mitochondrial perturbations define lymphocytes undergoing apoptotic depletion in vivo. Eur J Immunol 25:3277–3284

Clarke P (1990) Developmental cell death: morphological diversity and multiple mechanisms. Anat Embryol 181:195–213

Cobb JP, Hotchkiss RS, Karl IE, Buchman TG (1996) Mechanisms of cell injury and death British Journal of Anaesthesia. 77:3–10

Columbano A, Ledda-Columbano GM, Coni P, Faa G, Liguori C, Santacruz G, Pani G (1985) Occurrence of cell death (apoptosis) during the involution of liver hyperplasia. Lab Invest 52:670–677

Devitt A, Moffatt OD, Raykundalia C, Capra JD, Simmons DL, Gregory CD (1998) Human CD 14 mediates recognition and phagocytosis of apoptotic cells. Nature 392:505–508

Dhein J, Walczak H, Baumler C, Debatin KM, Kramer PH (1995) Autocrine T-cell suicide mediated by APO-1/. Nature 373:438–441

Dini L, Kolb-Bachofen V (1989) Preclustered receptor arrangement is a prerequisite for galactose-specific clearance of large particulate ligands in rat liver. Exp Cell Res 184:235–240

Dini L, Autuori F, Lentini A, Oliverio S, Piacentini M (1992) The clearance of apoptotic cells in the liver is mediated by the asialoglycoprotein receptor. FEBS Lett 296:174–178

Dini L, Falasca L, Lentini A, Mattioli P, Piacentini M, Piredda L, Autuori F (1993) Galactose-specific receptor modulation related to the onset of apoptosis in rat liver. Europ J Cell Biol 61:329–337

Dini L, Lentini A, Diez Diez G, Rocha M, Falasca L, Serafino L, Vidal-Vanaclocha F (1995) Phagocytosis of apoptotic bodies by liver endothelial cells. J Cell Sci 108:967–973

Dini L, Ruzittu M, Falasca L (1996a) Recognition and phagocytosis of apoptotic cells. Scanning microscopy 10:239–252

Dini L, Coppola S, Ruzittu M, Ghibelli L (1996b) Multiple pathways for apoptotic nuclear fragmentation. Exp Cell Res 223:340–347

Dini L, Carlà EC (1998) Hepatic sinusoidal endothelium heterogeneity with respect to the recognition of apoptotic cells. 240:388–393

Dini L, Ruzittu M, Carlà EC, Falasca L (1998) Relationship between cellular shape and recepto-mediated endocytosis: an ultrastructural and morphometric study in rat Kupffer cells. Liver 18:99–109

Dini L, Giudetti AM, Ruzittu M, Gnoni GV, Zara V (1999) Citrate carrier and lipogenic enzyme activities in lead nitrate-induced proliferative and apoptotic phase in rat liver. Biochim Cell Mol Intern 47(4):607–614

Dini L (1999) Endothelial liver cell recognition of apoptotic peripheral blood lymphocytes. Biochemical Society Transactions 26:635–637

Durrieu F, Belloc F, Lacoste L, Dumain P, Chabrol J, Dachary-Prigent J, Morjani H, Boisseau MR, Reiffers J, Bernard P, Lacombe F (1998) Caspase activation is an early event in anthracycline-induced apoptosis and allows detection of apoptotic cells before they are ingested by phagocytes. Exp. Cell Res 240:165–175

Duvall E, Wyllie AH, Morris RG (1985) Macrophage recognition of cells undergoing programmed cell death (apoptosis). Immunology 56:351–358

Ellis RE, Jacobson DM, Horvitz HR (1991a) Genes required for the engulfment of cell corpses during programmed cell death in *Caenorhabditis elegans*. Genetics 129:79–94

Ellis RE, Yuan J, Horvitz HR (1991b) Mechanisms and functions of cell death. Annu Rev Cell Biol 7:663–698

Emoto K, Toyama-Sorimachi N, Karasuyama H, Inoue K, Umeda M (1997) Exposure of phosphatidylethanolamine on the surface of apoptotic cells. Exp Cell Res 232:430–434

Fadok VA, Savill JS, Haslett C, Bratton DL, Doherty DE, Campbell PA, Henson PM (1992a) Different populations of macrophages use either the vitronectin receptor or the phosphatidylserine receptor to recognize and remove apoptotic cells. J Immunol 149:4029–4035

Fadok VA, Voelker DR, Campbell PA, Cohen JJ, Bratton DL, Henson PM (1992b) Exposure of phosphatidylserine on the surface of apoptotic lymphocytes triggers specific recognition and removal by macrophages. J Immunol 148:2207–2216

Fadok VA, Bratton DL, Frasch SC, Warner ML, Henson PM (1998a) The role of phosphatidylserine in recognition of apoptotic cells by phagocytes. Cell Death Differ 5:551–562

Fadok VA, Bratton DL, Konoval A, Freed PW, Westcott JY, Henson PM (1998b) Macrophages that have ingested apoptotic cells in vitro inhibit proinflammatory cytokine production through autocrine/paracrine mechanisms involving TGF-β, PGE2, and PAF. J Clin Invest 101:890–898

Falasca L, Bergamini A, Serafino A, Balabaud C, Dini L (1996) Human Kupffer cell recognition and phagocytosis of apoptotic peripheral blood lymphocytes. Exp Cell Res 224:152–162

Flora PK, Gregory CD (1994) Recognition of apoptotic cells by human macrophages: inhibition by a monocyte/macrophage-specific monoclonal antibody. Eur J Immunol 24:2625–2632

Friesen C, Herr I, Krammer PH, Debatin KM (1996) Involvement of the CD95 (APO-1/Fas) receptor/ligand system in drug-induced apoptosis in leukemia cells. Nature Med 2:574–577

Gerschenson LE, Rotello RJ (1991) Apoptosis and cell proliferation are terms of the growth equation. In: Apoptosis. The molecular basis of cell death. Cold Spring Harbor Laboratory Press, p 175

Grasl Kraupp B, Bursch W, Ruttkay Nedecky B, Wagner A, Lauer B, Schulte-Hermann R (1994) Food restriction eliminates preneoplastic cells through apoptosis and antagonizes carcinogenesis in rat liver. Proc Natl Acad Sci USA 91:9995–9999

Hall SE, Savill JS, Henson PM, Haslett C (1994) Apoptotic neutrophils are phagocytosed by fibroblasts with participation of the fibroblast vitronectin receptor and involvement of a mannose/fucose-specific lectin. J Immunol 153: 3218–3227

Hart SP, Haslett C, Dransfield I (1996) Recognition of apoptotic cells by phagocytes. Experientia 52:950–956

Haslett C, Savill JS, Whyte MKB, Stern M, Dransfield I, Meagher LC (1994) Granulocyte apoptosis and the control of inflammation. Phil Trans R Soc London Bbiol Sci 345:327–333

Herrmann M, Voll RE, Zoller OM, Hagenhofer M, Ponner BB, Kalden JR (1998) Impaired phagocytosis of apoptotic cell material by monocyte-derived macrophages from patients with systemic lupus erythematosus. Arthritis Rheum 41:1241–1250

Huang L, Soldevila G, Leeker M, Flavell R, Crispe N (1994) The liver eliminates T cells undergoing antigen-triggered apoptosis in vivo. Immunity 1:741–749

Hubbard AL, Wilson G, Ashwell G, Stukenbrok H (1979) An electron microscopic autoradiographic study of the carbohydrate recognition system in rat liver. I distribution of 125I-ligands among the liver cell types. J. Cell Biol 83:47–64

Jones EA, Summerfield JA (1982) Kupffer cells In: Arias I, Popper H, Schacter D, Shafritz DA (eds), The liver: biology and pathobiology, Raven Press, New York, p 507

Kempka G, Roos P, Kolb-Bachofen V (1990) A membrane-associated form of C-reactive protein is the galactose-specific particle receptor on rat Kupffer cells. J Immunol 144:1004–1009

Kirn A, Gut JP, Gendrault JL (1982) Interaction of viruses with sinusoidal cells. In: Popper H, Schaffner F (eds) Progress in liver diseases. New York: Grune & Stratton, p 377

Kolb H, Friedrick E, Suss R (1981) Lectin mediates homing of neuraminidase-treated erythrocytes to the liver as revealed by scintigraphy. Hoppe-Seyler's Z Physiol Chem 362:1609–1614

Kolb-Bachofen V, Schlepper-Schafer J, Vogell W (1982) Electron microscopic observations of the hepatic microscopic evidence for an asailoglycoprotein receptor on Kupffer cells: localization of lectin mediated endocytosis. Cell 29:859–866

Kolb-Bachofen V (1992) A review on the biological properties of C-reactive protein. Immunobiol 183:133–145

Kornbluthh RS (1994) Significance of T cell apoptosis for macrophages in HIV infection. J Leukoc Biol 56:247–256

Kumar S (1995) ICE-like proteases in apoptosis. Trends Biochem Sci 20:198–202

Kuypers FA (1998) Phospholipid asymmetry in health and disease Curr Opin Hematol 5:122–131

Ledda-Columbano GM, Shinozuka H, Katyal SL, Columbano A (1996) Cell proliferation, cell death and hepatocarcinogenesis. Cell Death Differ 3:17–22

Liu QA, Hengartner MO (1998) Candidate adaptor protein CED-6 promotes the engulfment of apoptotic cells in *C. elegans*. Cell 93:961–972

Luciani MF, Chimini G (1996) The ATP binding cassette transporter ABCD1, is required for the engulfment of corpses generated by apoptotic cell death. EMBO J 15:226–235

Martin SJ, Reutelingsperger CPM, McGahon AJ, Rader JA, van Schie RCA, LaFace DM, Green DR (1995) Early redistribution of plasma membrane phosphatidylserine is a general feature of apoptosis regardless of the initiating stimulus: inhibition by overexpression of Bcl-2 and Abl. J Exp Med 182:1545–1556

Massimi M, Conti Devirgiliis, Kolb-Bachofen V, Dini L (1995) Independent modulation of galactose-specific receptor expression in rat liver cells. Hepatology 22:1819–1828

Massimi M, Falasca L, Felici A, Dini L, Conti Devirgiliis L (1996) Expression of the asialoglycoprotein receptor in cultured rat hepatocytes is modulated by cell density. Bioscience Reports 16:477–484

McEvoy L, Williamson P, Schlegel RA (1986) Membrane phospholipid asymmetry as a determinant of erytrocyte recognition by macrophages. Proc Natl Acad Sci USA 83:3311–3315

McMurchie EJ, Raison JK (1979) Membrane lipid fluidity and its effect on the activation energy of membrane-associated enzymes. Biochim Biophys Acta 554:364–374

Meagher LC, Savill JS, Baker A, Fuller R, Haslett C (1992) Phagocytosis of apoptotic neutrophils does not induce macrophage release of thromboxane B2. J Leuk Biol 52:269–273

Morin O, Patry P, and Lafleur L (1984) Heterogeneity of endothelial cells of adult rat liver as resolved by sedimentation velocity and flow cytometry. J Cell Physiol 119:327–334

Morris RG, Hargreaves AD, Duvall E, Wyllie AH (1984) Surface changes in thymocytes undergoing apoptosis. Am J Path 115:426–436

Nenseter MS, Gudmundsenn O, Roos N, Maelandsmo G, Drevon CA, Berg T (1992) The role of liver endothelial and Kupffer cells in clearing low density lipoprotein from blood in hypercholestrerolemic rabbits. J lipid Res 33:867–877

Neaud V, Dubuisson L, Balabaud C, Bioulac-Sage P (1995) Ultrastructure of human Kupffer cells maintained in culture. J Submicrosc Cytol Pathol 27:161–170

Nishikawa A, Murata E, Akita M, Kaneko K, Moriya O, Tomita M, Hayashi H (1998) Roles of macrophages in programmed cell death and remodelling of tail and body muscle of Xenopus laevis during metamorphosis. Histochem Cell Biol 109:11–17

Ogasawara J, Wtanabe-Fukunaga R, Adachi M, Matsuzawa A, Kasugai T, Kitamura Y, Itoh N, Suda T, Nagata S (1993) Lethal effect of the anti-Fas antibody in mice. Nature 364:806–809

Platt N, Pedro da Silva R, Gordon S (1998) Recognizing death: the phagocytosis of apoptotic cells. Trends in Cell Biology 8:365–372

Pradhan D, Williamson P, Schlegel RA (1994) Phosphatidylserine vesicles inhibit phagocytosis of erythrocytes with a symmetric transbilayer distribution of phospholipids. Mol Membr Biol 11:181–187

Pradhan D, Krahling S, Williamson P, Schlegel RA (1997) Multiple systems for recognition of apoptotic lymphocytes by macrophages. Mol Biol Cell 8:767–778

Praaning-van Dalen DP, de Leeuw AM, Brouwer A, Knook DL (1987) Rat liver endothelial cells have a greater capacity than Kupffer cells to endocytose N-acetylglucosamine- and mannose- terminated glycoproteins. Hepatology 7:672–679

Ravetch JV (1994) Fc receptors: rubor redux. Cell 78:553–560

Ren V, Silverstein RL, Allen J, Savill J (1995) CD36 gene transfer confers capacity for phagocytosis of cells undergoing apoptosis. J Exp Med 181:1857–1862

Ren V, Savill J (1998) Apoptosis: the importance of being eaten. Cell Death Differ 5:563–568

Rifai A, Mannik M (1984) Clearance of circulating IgA immune complexes is mediated by a specific receptor on Kupffer cells in mice. J Exp Med 160:125–137

Rocha M, Lentini A, Asumendi A, Falasca L, Autuori F, Dini L, Vidal-Vanaclocha F (1993) In situ and in vitro correlation between mannose receptor expression and fenestration pattern in endothelial cells selected from different zones of liver lobule. In Knook D, Wisse E (ed) Cells of the hepatic sinusoid, vol 5. Kupffer Cell Foundation, Leiden, The Netherlands, p 470

Roos E, Dingemans KP, Van de Pavert IV, Van den Bergh-Weerman MA (1978) Mammary-carcinoma cells in mouse liver: infiltration of liver tissue and interaction with Kupffer cells. Brit J Cancer 38:88–99

Roerdink F, Dijkstra J, Hartman G, et al. (1981) The involvement of parenchymal, Kupffer and endothelial liver cells in the hepatic uptake of intravenously injected liposomes. Effects of lanthanum and gadolinium salts. Biochim Biophys Acta 677:79–89

Ruiter DJ, Van der Meulen J, Brouwer A, Hummel MJ, Mauw BJ, Van der Ploeg JC, Wisse E (1981) Uptake by liver cells of endotoxin following its intravenous injection. Lab Invest 45:38–45

Savill JS, Henson PM, Haslett C (1989) Phagocytosis of aged human neutrophils by macrophages is mediated by a novel "charge-sensitive" recognition mechanism. J Clin Invest 84:1518–1527

Savill J, Dransfield L, Hogg N, Haslett C (1990) Vitronectin receptor-mediated phagocytosis of cells undergoing apoptosis. Nature 343:170–173

Savill J, Hogg N, Ren Y, Haslett C (1992a) Thrombospondin cooperates with CD36 and the vitronectin receptor in macrophage recognition of neutrophils undergoing apoptosis. J Clin Invest 90:1513–1522

Savill J, Smith J, Sarraf C, Ren Y, Abbott F, Ress A (1992b) Glomerular mesangial cells and inflammatory macrophages ingest neutrophilis undergoing apoptosis. Kidney Int 42:924–936

Savill J, Fadok V, Henson P, Haslett C (1993) Phagocyte recognition of cells undergoing apoptosis Immunol Today 14:131–136

Savill JS (1997) Recognition and phagocytosis of cells undergoing apoptosis. Br Med Bull 53:491–508

Savill JS (1998) Phagocytic docking without shocking Nature 392:442–443

Schlegel RA, Williamson P (1987) Membrane phospholipid organization as a determinant of blood cell-reticuloendothelial cell interactions. J Cell Physiol 132:381–384

Schlegel RA, Callahan M, Krahling S, Pradhan D, Williamson P (1996) Mechanisms for recognition and phagocytosis of apoptotic lymphocytes by macrophages. Adv Exp Med Biol 406:21–28

Shiratsuchi A, Osada S, Kanazawa S, Nakanishi Y (1998) Essential role of phosphatidylserine externalization in apoptosing cell phagocytosis by macrophages. Biochem Biophys Res Commun 246:549–555

Shiratori Y, Tanaka M, Kawase T, Shiina S, Komatsu Y, Omata M (1993) Quantification of sinusoidal cell function in vivo. Semin LiverDis 13:39–49

Smedsrød B, Pertoft H, Gustafson S, Laurent TC (1990) Scavenger functions of the liver endothelial cell. Biochem J 266:313–327

Smedsrød B, Deblaser PJ, Braet F, Lovisetti P, Vanderkerken K, Wisse E, Geerts A (1994) Cell biology of liver endothelial and Kupffer cells. Gut 35:1509–1516

Steer CJ, Clarenburg R (1979) Unique distribution of glycoprotein receptors on parenchymal and sinusoidal cells of rat liver. J Biol Chem 254:4457–4461

Steffan AM, Gendrault JL, McCuskey RS, McCuskey PA, Kirn A (1986) Phagocytosis, an unrecognized property of murine endothelial liver cells. Hepatology 6:830–836

Stern M, Meagher L, Savill J, Haslett C (1992) Apoptosis in human eosinophils. Programmed cell death in the eosinophil leads to phagocytosis by macrophages and is modulated by IL-5. J Immunol 148:3543–3549

Stuart MC, Damoiseaux JG, Frederik PM, Arends JW, Reutelingsperger CP (1998) Surface exposure of phosphatidylserine during apoptosis of rat thymocytes precedes nuclear changes. 76:77–83

Swanson JA, Baer SC (1995) Phagocytosis by zippers and triggers. Trends Cell Biol 5:89–93
Tessitore L, Valente G, Bonelli G, Costelli P, Baccino FM (1989) Regulation of cell turnover in the livers of tumor bearing rats: occurrence of apoptosis. Int J Cancer 44:697–700
Toth CA, Thomas P (1992) Liver endocytosis and Kupffer cells. Hepatology 16:255–266
Van Berkel TJC, De Rijke JB, Kruijt JK (1992) Recognition of modified lipoprotein by various scavenger receptors on Kupffer and endothelial liver cells. In: Windler E, Greten H, eds. Hepatic endocytosis of lipids and proteins. Munchen, FRG: Zuckschwerdt, p 443
Vidal-Vanaclocha F, Amezaga C, Asumendi A, Kaplanski G, Dinarello CA (1994) Interleukin-1 receptor blockade reduces the number and size of murine B16 melanome hepatic metastasis. Cancer Res 54:2667–2672
Voll RE, Hermann M, Roth EA, Stach C, Kalden JR, Girkontaite I (1997) Immunosuppressive effects of apoptotic cells. Nature 390:350–351
Wardle EM (1987) Kupffer cells and their function. Liver 7:63–70
Watanabe S (1988) Calmodulin antagonists inhibit the phagocytic activity of cultured Kupffer cells. Lab Invest 59:214–218
Watanabe S, Hirose M, Ueno T et al. (1990) Integrity of the cytoskeletal system is important for phagocytosis by Kupffer cells. Liver 10:249–254
Wu YC, Horvitz HR (1998) *C. elegans* phagocytosis and cell-migration protein CED-5 is similar to human DOCK 180. Nature 392:501–504

CHAPTER 13
Drug-Induced Apoptosis of Skin Cells and Liver

M. Neuman, R. Cameron, N. Shear, and G. Feuer

A. Prevalence of Drug-Induced Apoptosis

A variety of man-made and naturally occurring chemicals can induce apoptosis in a number of cell types (Cameron and Feuer, Chap. 1, this volume; Pessayre et al., Chap. 3, this volume). Therapeutic agents which can cause apoptosis include glucocorticoids and a number of chemotherapeutic drugs including bleomycin, cisplatin, cytosine arabinoside, doxorubicin, methotrexate, nitrogen mustard, and vincristine (Cameron and Feuer, Chap. 1, this volume). We have been studying the process of the induction of apoptosis by selected drugs *in vitro* and *in vivo*. The chemotherapeutic drug methotrexate induces apoptosis in skin cells and in liver cells in vitro and, in addition, apoptosis of hepatocytes was observed in liver biopsies of patients treated with methotrexate for psoriasis. In a series of further studies, we also examined the drugs acetaminophen and valproic acid for their apoptotic inducing effects on hepatocytes in vitro.

B. Methotrexate-Induced Apoptosis

Methotrexate is an antimetabolite which binds to the enzyme dihydrofolate reductase. Methotrexate acts by inhibiting the synthesis of purine and pyrimidine nucleotides and appears to exert its toxicity by means of DNA strand breakage in cells of the liver and skin (Sano et al. 1991). The mechanism of methotrexate toxicity to hepatocytes has been studied by a number of groups (Vonen and Morland 1984; Muller et al. 1997, 1998; Los et al. 1997; Rashid et al. 1999). It was suggested from these studies that one mechanism of apoptosis induction in hepatocytes is associated with the CD95 receptor ligand interaction. Methotrexate is known to up-regulate CD95 receptors. Methotrexate-induced apoptosis of hepatocytes was also shown to be mediated by caspases (Los et al. 1997). In our studies, we investigated the effect of methotrexate in normal neonatal primary skin cells, epidermal skin cells of the line A431, normal human primary hepatocytes, and human HepG2 cells. The presence of cytokines and the level of cytotoxicity in apoptosis were examined as well as cytoviability and glutathione content. Transmission electron microscopy was used and we attempted to quantify the differences in morphology found in electron micrographs from liver biopsies of patients with methotrexate toxicity. We also examined the effect of methotrexate in combination with ethanol. We concluded that, at lower doses, methotrexate or ethanol will not cause cellular apoptosis, although ethanol produces oxidative stress which can then promote methotrexate-induced apoptosis.

I. Apoptosis of Hepatocytes *In Vivo*

Patients receiving methotrexate therapy, usually for the treatment of psoriasis, are known to be at risk of liver disease including steatosis, hepatic fibrosis, and even cirrhosis (GILBERT et al. 1990; WHITING-O'KEEFE et al. 1991).

We have studied liver changes in a group of 20 patients with psoriasis undergoing chronic methotrexate therapy by light and electron microscopy. In six patients, liver biopsy was performed using morphometric analysis. On each grid prepared for electron microscopy, a minimum of 500 hepatocytes were examined in each case. Magnification for the electromicrographs for morphometry was set at 2500× in each case in order to make relative comparisons between the patients exposed to methotrexate and a group of 51 control liver biopsies representing a group of patients with antibodies to hepatitis C virus with normal histology and no liver pathology. Quantitation was made of the number and size of lipid vesicles, size of mitochondria, number of apoptotic cells and of apoptotic bodies. Random photomicrographs were taken. In addition, the length and axial ratio of mitochondria and lipid droplets were measured. For each cell, the numerical density or number of lipid droplets per cell was quantified. The ultrastructural changes seen in the methotrexate treated patients were very striking compared to the controls with normal histology: (a) steatosis of both macrovesicular and microvesicular type involving 25%–75% of all hepatocytes in the six patients examined; (b) marked dilatations of the smooth endoplasmic reticulum (SER) compared to controls; (c) proliferation and microvesiculation of endoplasmic reticulum; (d) diffuse mitochondrial changes with increases in size and paracrystalline inclusions; and (e) the presence of scattered apoptotic hepatocytes (Fig. 1) as compared to control liver tissues which showed no apoptotic cells. These ultrastructural changes, including microvesicular steatosis, mitochondrial changes, and proliferation and dilatation of the SER, are not specific for methotrexate but represent characteristic responses of the liver to drug toxicity (PHILLIPS et al. 1987; FEUER and DE LA IGLESIA 1996). A much wider group of drugs, however, such as chemotherapeutic drugs, seem to cause apoptosis in hepatocytes (CAMERON and FEUER, Chap. 1, this volume).

II. Apoptosis of Hepatocytes *In Vitro*

1. Initial Studies

Human hepatocytes derive from two sources, namely human hepatoblastoma cells or HepG2 cells (Fig. 2) were obtained from the Wistar Institute, Philadelphia, PA, and human normal primary hepatocytes were obtained from donor

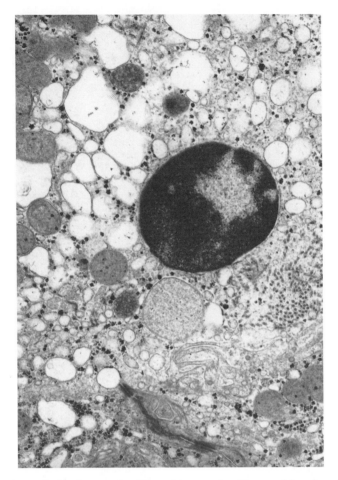

Fig. 1. Electron photomicrograph of liver biopsy of a 43-year-old male patient with psoriasis at two years post-treatment with methotrexate which shows an apoptotic hepatocyte nucleus, ×8400

livers (Fig. 3). These cells were used to analyze the *in vitro* toxicity to human hepatocytes of drugs such as methotrexate, methotrexate plus ethanol, acetaminophen, and valproic acid. Our previous studies, and those by others, had shown that specific molecules had a critical effect on drug-induced hepatotoxicity *in vitro* including cytokines such as TNFα, glutathione (SHEAR et al. 1995), and the effect of a sublethal and almost subtoxic level of ethanol in combination with a drug like methotrexate *in vitro*.

2. Effects of Co-exposures of Hepatocytes with Methotrexate and Ethanol

Methotrexate added alone at a dose of 10 mmol/L concentration caused some hepatocytes to enlarge and simulate mild steatosis with the appearance of lipid

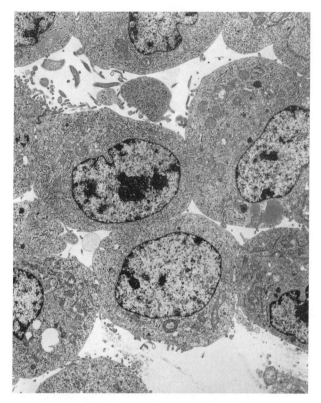

Fig. 2. Electron photomicrograph of untreated HepG2 cells *in vitro* showing normal looking hepatocytic nuclei and cytoplasmic organelles, ×4200

droplets. Similarly, the addition of 40mmol/L ethanol to hepatocytes for 24h in vitro showed few, if any, differences compared to control cells. We had previously found that a dose of 80mmol/L ethanol to similar cells for 24h had induced a number of toxic effects (Fig. 4) including changes in mitochondria, SER, and accumulation of abundant lipid vesicles (NEUMAN et al. 1996). The addition of this subtoxic dose of ethanol of 40mmol/L with 10mmol/L of methotrexate in a combined form for 24h in vitro with hepatocytes resulted in a number of toxic manifestations including increases in numbers of lipid droplets, enlargement of the SER, and changes in mitochondria with a reduction in the number of mitochondrial cristae. Similar effects were further accentuated if an additional dose of the same combination of ethanol and methotrexate were added for an additional 24h. There was an additional three-fold increase in the number of lipid vesicles, further ballooning of endoplasmic reticulum, and further alterations in mitochondria. In addition, many hepatocytes became apoptotic as evidenced by the dense aggregations of nuclear chromatin. Image analysis of hepatocytes exposed to the ethanol and methotrexate in combination showed that these cells were much larger at

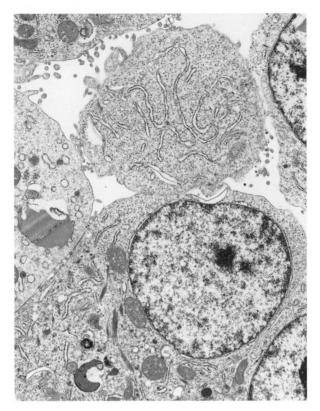

Fig. 3. Electron photomicrograph of untreated human hepatocytes in primary cultures showing features of normal hepatocytes, ×4200

6025 ± 345 microns compared to controls exposed only to plain media which were 4425 ± 525 microns in size. In addition, electron microscopic morphometry showed the hepatocytes exposed to methotrexate plus ethanol had a threefold increase in the length of mitochondria, a 2.5× increase in size diameter of lipid droplets, and a twofold increase in the number of lipid droplets per cell compared to control untreated hepatocytes in vitro.

III. Apoptosis of Skin Cells *In Vitro*

Methotrexate has been a commonly used and effective drug in the treatment of psoriasis, a skin condition which involves the formation of scaly and itchy plaques on the skin. HEENEN et al. (1998) had found that keratinocytes from psoriatic plaques were resistant to apoptosis. Psoriatic plaques had also been shown by WRONE-SMITH et al. (1997) to overexpress Bcl-x_L, an apoptosis-inhibiting protein. Methotrexate may serve to reduce the hyperplasia characteristic of psoriatic skin by means of the induction of apoptosis in keratinocytes (HEENEN et al. 1998). SNYDER (1988) had proposed that the mech-

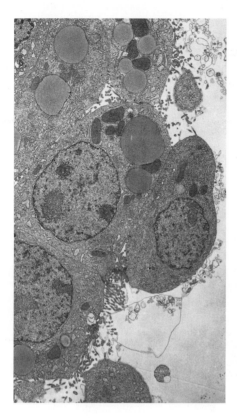

Fig. 4. Electron photomicrograph of HepG2 cells exposed to 80mmol/l ethanol for 24h in vitro which shows steatosis and megamitochondria, ×4200

anism of methotrexate toxicity involved the depletion of cellular deoxynucleoside triphosphate pools which affected the DNA excision repair process in cultured human fibroblasts. This same effect on DNA synthesis can lead to a deoxynucleotide pool imbalance and subsequently to apoptosis. Skin cells which were studied were obtained from two sources: one source was skin obtained of healthy neonates and the second were cultured skin cells of the epidermal cell line A431, obtained from Wistar Institute, Philadelphia, PA. When keratinocytes of the A431 cell line were exposed to a similar combination of 40mmol/L ethanol and 10mmol/L methotrexate for two doses over 48h in culture, multiple apoptotic skin cells were evident (Fig. 5), similar to what was seen with the hepatocytes *in vitro*.

C. Acetaminophen-Induced Apoptosis of Hepatocytes and Skin Cells In Vitro

Exposure of acetaminophen *in vitro* is cytotoxic to human hepatocytes, particularly when there is depletion of glutathione. Protection against aceta-

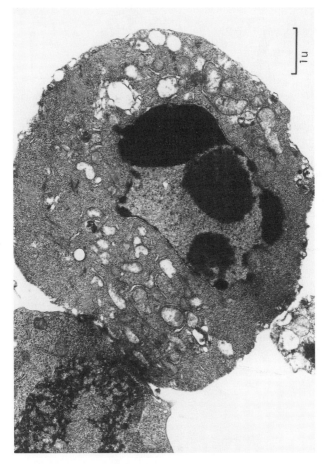

Fig. 5. Electron photomicrograph of A431 skin cells exposed to subtoxic doses of ethanol of 40 mmol/L together with methotrexate at 10 mmol/L dose *in vitro* for 24 h shows apoptosis of an A431 skin cell, ×8400

minophen hepatotoxicity, therefore, could be induced by agents such as *N*-acetylcysteine. Acetylcysteine acts in a manner similar to glutathione by preventing the binding of the toxic metabolite of acetaminophen to liver cell macromolecules. Glutathione substrates are depleted in the process of detoxification of acetaminophen and can be replenished by sulfhydryl compounds from the diet or by cystine-containing drugs such as *N*-acetylcysteine. The glutathione S transferase reaction is central to the detoxification of acetaminophen. Apoptosis was observed in hepatocytes in vivo when high doses of acetaminophen were administered to ICR mice. DNA fragmentation began at 2 h post treatment and extended to 24 h. The morphologic appearance of apoptosis, namely the nuclear condensations, began as early as 2–6 h after exposure to acetaminophen. We have shown similar responses *in vitro* (Fig. 6).

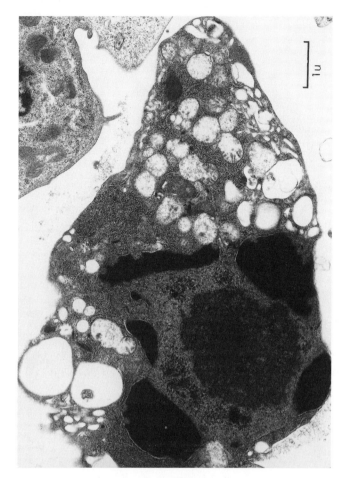

Fig. 6. Electron photomicrograph of A431 skin cells exposed to subtoxic dose of ethanol of 40mmol/L combined with acetaminophen also at a subtoxic dose *in vitro* for 24h and it shows apoptosis of an A431 skin cell, ×8400

D. Valproic Acid-Induced Apoptosis of Hepatocytes In Vitro

Valproic acid is a drug frequently used in the treatment of epilepsy. This drug has excellent therapeutic effects in the treatment of several forms of epilepsy but has been linked in rare cases to severe and fatal hepatotoxicity (ZIMMERMAN 1982). Anti-convulsants such as valproic acid are typically present with idiosyncratic hepatotoxicity, and with valproic acid the risk of fatal hepatotoxicity has been rare, being reported in one study as 1 in 50,000 (DREIFUSS et al. 1989). This study also reported that 90% of patients with valproic acid induced fatal hepatic failure were below the age of 20.

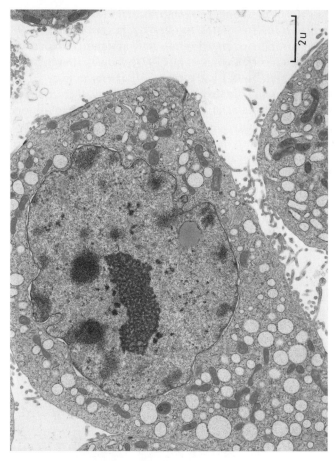

Fig. 7. Electron photomicrograph of HepG2 liver cells exposed to valproic acid alone *in vitro* for 24h which shows diffuse microvesicular steatosis, ×4200

Various studies have elucidated possible mechanisms of hepatotoxicity (ZIMMERMAN 1982; TAKEUCHI et al. 1988; KASSAHUN et al. 1994; JURIMA-ROMET et al. 1996; ZIMMERMAN and ISHAK 1996). One significant factor derived from these studies is the production of the toxic metabolite 4-en-valproate, which is the favored metabolite when the metabolism of valproic acid is shifted from the usual β-oxidation to ω-oxidation. Induction of cytochrome P450 activity favors the shift towards this type of metabolism of valproic acid. The reactive metabolites formed by this pathway then bind to macromolecules, deplete glutathione, and inhibit fatty acid metabolism, resulting in hepatic microvesicular steatosis (Fig. 7). Patients using valproic acid had low levels of the cofactors carnitine, coenzyme A, and acetyl-coenzyme, which are necessary for the β-oxidation of fatty acids. The carnitine deficiency may predispose these patients to hepatoxicity because of increasing serum fatty acid levels which then promote the shift of metabolism of valproic acid towards the pathway which generates reactive intermediates. Studies by TAKEUCHI et al. (1988) showed that the administration of DL-carnitine and albumin reduced valproic acid hepatotoxicity. Studies by FISHER et al. (1994) showed that the toxicity of valproic acid and its metabolites had a range of toxicity in liver slices from adult or weanling rats but similar toxicities in slices derived from human livers. A study by JURIMA-ROMET et al. (1996) found that levels of glutathione were critical to valproic acid toxicity to rat hepatocytes in vitro and found a protective effect of anti-oxidants such as vitamins C and E.

We have shown that valproic acid hepatotoxicity is enhanced *in vitro* by inducers of cytochrome p4502E1 (NEUMAN et al. 1999). Normal human hepatocytes *in vitro*, when treated with a combination of valproic acid and 40mmol/L ethanol for 24h, show apoptosis (Fig. 8). Cells treated with valproic

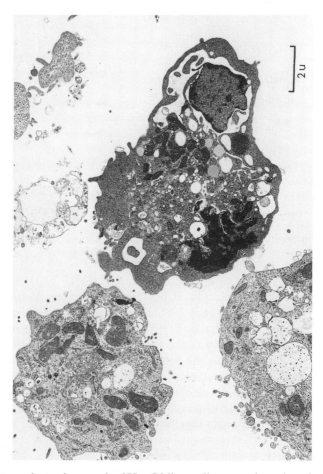

Fig. 8. Electron photomicrograph of HepG2 liver cells exposed to ethanol at 40mmol/L dose and valproic acid in combination in vitro for 24h which shows apoptosis of a HepG2 liver cell, ×8400

acid alone, however, showed only microvesicular steatosis without apoptosis (Fig. 7). In contrast, liver cells exposed only to 40mmol/L ethanol without valproic acid showed only mild steatosis without apoptosis.

E. Conclusion

We have been able to show in a series of *in vitro* studies using skin cells and human liver cells that a variety of different drugs are able to induce apoptosis in hepatocytes and skin cells, including methotrexate, acetaminophen, and valproic acid. The addition of tissue culture environments which add specific metabolic stresses to these cells, such as induction of specific cytochrome P450s or depletion of glutathione, have been shown to enhance the induction of apoptosis *in vitro* for skin cells and for human liver cells. Intracellular ATP levels in human T-cell lines have been shown by EGUCHI et al. (1997) to be critical in directing the process of cell death so that cells undergoing apoptosis can be driven towards necrosis in ATP-depleting conditions. Apoptosis of hepatocytes was also observed in liver biopsies of patients treated with methotrexate for psoriasis. In summary, it has been possible to undertake mechanistic studies of the induction of apoptosis of human skin cells and human liver cells *in vitro*.

Acknowledgements. We thank Marie Maguire for her excellent work in the preparation of this manuscript, and Gady Katz for his expert technical assistance on all aspects of these studies.

References

Anderson GD, Acheampong AA, Wilensky AJ, Levy RH (1992) Effect of valproate dose on formation of hepatotoxic metabolites. Epilepsia 33:736–742
Cameron RG, Blendis LM (1996) Clinical studies and role of necrosis in hepatotoxicity. In: Cameron RG, Feuer G, de la Iglesia FA (eds) Drug-induced hepatotoxicity. Springer, Berlin Heidelberg New York, pp 25–42
Coulter DL (1984) Carnitine deficiency: a possible mechanism for valproate hepatotoxicity. Lancet 1:689–695
Dreifuss FE, Santilli N (1986) Valproic acid hepatic fatalities: analysis of United States cases. Neurology 39:201–207
Eguchi Y, Shimizu S, Tsujimoto Y (1997) Intracellular ATP levels determine cell death fate by apoptosis or necrosis. Cancer Res 57:1835–1840
Feuer G, de la Iglesia FA (1996) Subcellular biochemical and pathological correlated in experimental models of hepatotoxicity. In: Cameron RG, Feuer G, de la Iglesia FA (eds) Drug-induced hepatotoxicity. Springer, Berlin Heidelberg New York, pp 43–74
Fisher RL, Saniuk JT, Nau H, Gordolfi AJ, Brendel K (1994) Comparative toxicity of valproic acid and its metabolites in liver slices. Toxicol Vitro 8:371–379
Gilbert SC, Klintmalin G, Mentor A, Silverman A (1990) Methotrexate-induced cirrhosis requiring liver transplantation. Arch Intern Med 150:889–891
Grewal KK, Racz WJ (1993) Intracellular calcium disruption as a secondary event in acetaminophen-induced hepatotoxicity. Can J Physiol Pharmacol 71:26–33

Harman AW, Kyle ME, Serroni A, Farber JL (1991) The killing of cultured hepatocytes by N-actetyl-p-benzoquinoneimine (NAPQI) as a model of the cytotoxicity of acetaminophen. Biochemical Pharmacology 41:1111–1117

Heenen M, Laporte M, Noel JC, de Graef C (1998) Methotrexate induces apoptotic cell death in human keratinocytes. Arch Dermatol Res 290:240–245

Jurima-Romet M, Tang W, Huang HS, Whitehouse LW (1996) Cytotoxicity of unsaturated metabolites of valproic acid and protection by vitamins C and E in glutathione-depleted rat hepatocytes. Toxicology 112:69–85

Kassahun K, Farrell K, Abbott F (1991) Identification and characterization of the glutathione and N-acetylcystine conjugates of toxic metabolites of valproic acid. Drug Metab Disp 19:525–535

Larrauri A, Fabra R, Gomez-Lechon MJ, Trullenque R, Castell JV (1987) Toxicity of paracetamol in human hepatocytes. Comparison of the protective effects of sulfhydryl compounds acting as glutathione precursors. Molecular Toxicology 1:301–311

Lee WM (1995) Drug-induced hepatotoxicity. New Engl J Med 333:1118–1127

Los M, Herr I, Friesen C, Fulda S, Schulze-Osthoff K, Debatin KM (1989) Cross-resistance of CD-95 and drug-induced apoptosis as a consequence of deficient activation of caspases (ICE/Ced-3-proteases). Blood 90:3118–3129

Mitchell JR, Jollow DJ, Potter WZ, David DC, Gillette JR, Brodie BB (1973) Acetaminophen-induced hepatic necrosis. I. Role of drug metabolism. J Pharmacol Exp Ther 187:185–194

Muller M, Strand S, Hug H, Heinemann EM, Walczak H, Hofmann WJ, Stremmel W, Krammer PH, Galle P (1997) Drug-induced apoptosis in hepatoma cells is mediated by the CD95 (APO/Fas) receptor/ligand system and involves activation of wild-type p53. J Clin Invest 99:403–413

Muller M, Wilder S, Bannasch D, Israeli D, Lehlbach K, Li-Weber M, Friedman S, Galle P. Stremmel W, Oren M, Krammer P (1998) p53 activates the CD95 (APO-1/Fas) gene in response to DNA damage by anticancer drugs. J Exp Med 188:2033–2045

Neuman MG, Cameron RG, Shear NH, Bellantani S, Tiribelli C (1995) Effect of tauroursodeoxycholic and ursodeoxycholic acid on ethanol-induced cell injuries in the human HepG2 cell line. Gastroenterology 109:555–563

Phillips MJ, Poucell S, Patterson J, Valencia P (1987) The liver: an atlas and text of ultrastructural pathology. Raven, New York

Rashid A, Wu T, Huang C, Chen C, Lin H, Yang S, Lee F, Diehl A (1999) Mitochondrial proteins that regulate apoptosis and necrosis are induced in mouse fatty liver. Hepatology 29:1131–1138

Ray SD, Murnaw VR, Raje RR, Fariss MW (1996) Protection of acetaminophen-induced hepatocellular apoptosis and necrosis by cholesteryl hemisuccinate pretreatment. J Pharmacol Exp Ther 279:1470–1483

Saile B, Knittel T, Matthes N, Schott P, Ramadori G (1997) CD95/CD95-L mediated apoptosis of the hepatic stellate cell. Am J Path 151:1265–72

Sano H, Kubota M, Kasai Y et al. (1991) Increased methotrexate-induced DNA strand breaks and cytotoxicity following mutational loss of thymidine kinase. Int J Cancer 48:92–95

Shear NH, Malkjewicz IM, Klein D, Koren G, Randor S, Neuman MG (1995) Acetaminophen-induced toxicity to human epidermoid cell line A 431 and hepatoblastoma cell line Hep G2, in vitro, is diminished by silymarin. Skin Pharmacol 8:279–291

Snyder RD (1988) Consequences of the depletion of cellular deoxynucleoside triphosphate pools on the excision-repair process in cultured human fibroblasts. Mutat Res 200:193–199

Takeuchi T, Sugimoto T, Nishida N, Kobayashi Y (1988) Protective effect of D,L-carnitine on valproate-induced hyperammonemia and hypoketonemia in primary cultured rat hepatocytes. Biochem Pharmacol 37:2255–2260

Vonen B, J Morland (1984) Isolated rat hepatocytes in suspension: potential hepatotoxic effects of six different drugs. Arch Toxic 56:33–37

Whiting-O'Keefe QE, Fye KH, Sack KD (1991) Methotrexate and histologic hepatic abnormalities: a meta-analysis. Am J Med 90:711–716

Wrone-Smith T, Mitra R, Thompson C, Jasty R, Castle V, Nickoloff B (1997) Keratinocytes derived from psoriatic plaques are resistant to apoptosis compared with normal skin. Am J Pathol 151:1321–29

Zimmerman HJ, Ishak KG (1982) Valproate-induced hepatic injury, analysis of 23 fatal cases. Hepatology 2:591–597

Zimmerman HJ, Ishak KG (1996) Antiepileptic drugs. In: Cameron RG, Feuer G, de la Iglesia FA (eds) Drug-induced hepatotoxicity. Springer, Berlin Heidelberg New York, pp 637–662

CHAPTER 14
Apoptosis and Eosinophils

H.-U. SIMON

A. Introduction

Apoptosis is the most common form of physiologic cell death. It is essential for organ developments during embryogenesis. After development completion, a multicellular organism must renew many lineages. For instance, red and white blood cells are constantly generated from hematopoietic progenitor cells. Therefore, physiological cell death is a necessary process to maintain correct cell numbers.

It is also clear that apoptosis is regulated by survival factors. Whereas most of these factors act on many cells of different lineage, only some are specific. For instance, interleukin-5 (IL-5) appears to be a specific survival factor for eosinophils, at least within the human system (BAGLEY et al. 1997). Therefore, and not surprisingly, eosinophilia and high IL-5 expression have often been associated, especially in chronic allergic disorders such as bronchial asthma and atopic dermatitis. Moreover, the phenomenon of delayed eosinophil apoptosis has been demonstrated in nasal polyposis (SIMON et al. 1997a) and atopic dermatitis (WEDI et al. 1997). In addition, glucocorticoids appear to exert their effects in bronchial asthma in part due to the induction of eosinophil apoptosis (WOOLLEY et al. 1996). These data suggest that dysregulated apoptosis of inflammatory cells such as eosinophils may represent an important pathogenic mechanism in chronic allergic responses.

In this chapter we will summarize our current knowledge about the regulation of eosinophil apoptosis and discuss the importance of these findings for the inflammatory process in allergic disorders.

B. Characteristics and Measurements of Apoptotic Eosinophils

Apoptosis is characterized by morphologic changes in the dying cell. This is also true in the case of eosinophils. The most readily observed morphologic features involve the nucleus, where the chromatin becomes extremely condensed before a complete collapse of the nucleus, can be observed. Second, a loss of cell volume is clearly detectable. Figure 1A shows these two morphologic changes of apoptotic cells in eosinophils. In necrosis there are no changes

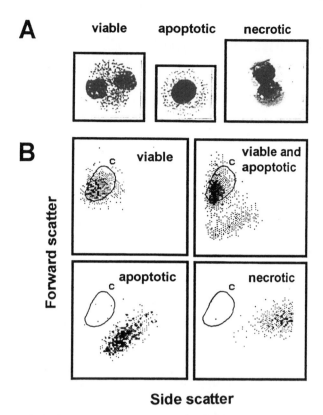

Fig. 1A,B. Morphologic features of human eosinophils undergoing apoptosis in vitro. **A** Cells were stained with Giemsa-May-Grünwald. Apoptosis is associated with compaction of the nuclear chromatin. Moreover, the cell volume decreases in apoptotic cells. In contrast, in necrosis, no change of nuclear morphology occurs. In addition, necrosis is associated with an increase of the cell volume because it is a lytic process. **B** Forward light scatter (FS) vs side light scatter (SS) analysis using a flow cytometer. Apoptosis is associated with a shift of the high FS/low SS population to the low FS/high SS population. In contrast, necrosis is not associated with low FS

of the nucleus. Moreover, necrosis is characterized by rapid cell swelling and lysis. Therefore, the cell volume is increased.

These morphologic differences between apoptotic and necrotic eosinophils also allow the determination of the form of cell death by flow cytometric analysis using light scatter measurements. Viable eosinophils display relatively high forward light scatter (FS) and relatively low side light scatter (SS). Culturing of purified eosinophils is associated with the appearance of a second, clearly separated cell subpopulation with low FS and high SS. Staining of the cells with a fluorescent dye demonstrates that the latter subpopulation represents apoptotic cells. Therefore, induction of eosinophil apoptosis results in a clear shift of the high FS/low SS population to the low FS/high SS population (Fig. 1B).

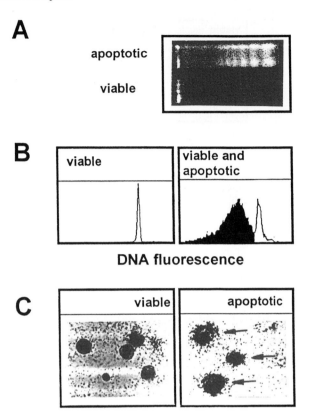

Fig. 2A–C. Different methods for detecting DNA fragmentation, a hallmark of apoptotic cells, in eosinophils. **A** Extracted DNA of purified eosinophils is analyzed by agarose gel electrophoresis. A typical DNA ladder is seen in apoptotic, but not in viable cells. **B** DNA of purified eosinophils is stained by propidium iodide. DNA fragmentation is analyzed by flow cytometry. Fragmented DNA (*black*) can be discriminated from normal, high-molecular weight DNA (*narrow peak*). **C** In situ detection of apoptotic eosinophils in nasal polyp tissues. DNA fragmentation is detected by radioactive (^{35}S-dATP) in situ labeling. Eosinophils can also be identified by immunohistochemistry using an eosinophil specific mAb (e.g., anti-ECP mAb)

Besides morphologic changes, DNA fragmentation is another hallmark of apoptotic cells. There are many different techniques to analyze DNA fragmentation. The classical technique is DNA electrophoresis. This technique proves whether internucleosomal fragmentation has occurred, which is visualized by the appearance of a ladder pattern on gel electrophoresis (Fig. 2A). However, although this method is specific, it does not give quantitative information about the amount of apoptosis. Another way to analyze DNA fragmentation is based on the observation that cellular DNA of apoptotic cells is less stainable with fluorescent dyes. Measurements are performed by flow cytometry. The advantage of this technique is that apoptotic cells can be mea-

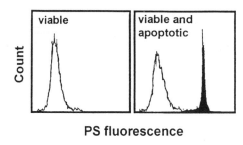

Fig. 3. Apoptosis is associated with phosphatidylserine (PS) redistribution. PS is normally confined to the inner plasma membrane leaflet. In contrast, PS appears on the external leaflet in apoptotic eosinophils. Annexin V is a PS-binding protein and can be used to detect apoptotic cells (*right panel, black*)

sured quantitatively as a hypodiploid cell population (Fig. 2B). A further technique to detect DNA fragmentation is the Terminal deoxynucleotidyl Transferase (TdT) uridine triphosphate (UTP) Nick End Labeling (TUNEL) method. Using this technique, free 3' OH-ends of DNA fragments are labeled with FITC-dUTP (purified blood eosinophils, analysis by flow cytometry) or ^{35}S-dATP (tissue eosinophils, analysis by light microscopy, Fig. 2C).

Apoptotic cells are removed by phagocytosis by neighboring cells, while retaining their intact plasma membrane. Several kinds of structural changes of the plasma membrane have been identified that lead to phagocyte recognition. For instance, cell surface exposure of phosphatidylserine (PS) is one such event that can be easily monitored using FITC-conjugated Annexin V and flow cytometric analysis (Fig. 3).

C. Role of Delayed Eosinophil Apoptosis for the Development of Eosinophilia in Allergic Tissues

Previously published work suggested that, in allergic inflammation, eosinophils specifically adhere to the endothelium and migrate into tissues with the help of eosinophil-specific chemoattractants (ROTHENBERG 1998). However, these mechanisms alone cannot explain the selective accumulation of eosinophils in allergic inflammation (SIMON 1998a). This view is strongly supported by observations in vivo. For instance, after antigen challenge, an increased initial, nonspecific recruitment of inflammatory cells, including neutrophils, has been reported in murine and human models (KOH et al. 1993; LUKACS et al. 1995; RICHARDS et al. 1996; TERAN et al. 1997).

Therefore, we suggested, based on well-documented in vitro studies (YAMAGUCHI et al. 1991; HER et al. 1991; STERN et al. 1992), an additional mechanism, namely the specific inhibition of eosinophil apoptosis by cytokines (SIMON and BLASER 1995). We have recently demonstrated that there is indeed a dramatic increase in the life span of eosinophils due to delayed apoptosis in nasal polyp compared to control nasal tissues (SIMON et al. 1997a). IL-5 is most likely responsible for this phenomenon, since lymphocytes, mast cells, and eosinophils themselves were found to express high amounts of IL-5 protein. Moreover, treatment of the eosinophilic-infiltrated tissue with neutralizing anti-IL-5 antibody induced eosinophil apoptosis and decreased tissue eosinophilia (SIMON et al. 1997a). Therefore, IL-5 appears to be a key cytokine within allergic inflammatory sites, and inhibition of this cytokine may represent an attractive approach to treat allergic disorders in the future.

D. Role of Tyrosine Kinases Activation in Cytokine-Mediated Antiapoptosis

The growth and differentiation of eosinophils are critically regulated by the three hematopoietins IL-3, IL-5, and GM-CSF. All three cytokines have overlapping functions on eosinophils. The action of IL-5 is specific for eosinophils whereas that of IL-3 and GM-CSF is not. IL-3 and GM-CSF also affect the growth and differentiation of other granulocytes and macrophages. The mechanisms of their overlapping functions are explained by the composition of their receptor complexes. All three cytokines have ligand-specific α receptor subunits but they share a common β (βc) subunit. The latter is considered the most important signaling receptor for these hematopoietins. The βc receptor is physically associated with the tyrosine kinases Lyn (PAZDRAK et al. 1995a; YOUSEFI et al. 1996), Jak1 (OGATA et al. 1998), and Jak2 (PAZDRAK et al. 1995b; VAN DER BRUGGEN et al. 1995; SIMON et al. 1997b; OGATA et al. 1998) (Fig. 4). This physical association occurs in basal conditions without growth factor stimulation. Stimulation of eosinophils with IL-5 or other hematopoietins results in tyrosine phosphorylation of these and other kinases. There are also reports of activation of Fes, Btk, and Fyn by the hematopoietins in myeloid cell lines. Whether these kinases are activated in eosinophils is unknown at this time.

Jak kinases have a propensity to tyrosine phosphorylate and activate the Stat family of nuclear factors. Indeed, IL-5 activates Stat1 (PAZDRAK et al. 1995b; VAN DER BRUGGEN et al. 1995), Stat3 (CALDENHOVEN et al. 1995), and Stat5 (MUI et al. 1995) nuclear factors (Fig. 4). The activation of other tyrosine kinases results in the propagation of signals via a number of downstream signaling pathways including the Ras-Raf-MAP kinase (PAZDRAK et al. 1995a), the PI-3 kinase-c-akt (COFFER et al. 1998), and other pathways. The propagation of signals via these pathways is facilitated by adapter proteins such as Shc, Grb2, and GTPase-activating proteins, e.g., Sos, as well as tyrosine phosphatases such as SHP-1 and SHP-2. Indeed, in eosinophils, IL-5 has been shown to activate not only Shc and Grb2 (BATES et al. 1998), but also SHP-2 (PAZDRAK et al. 1997) (Fig. 4).

TGF-β is a pleiotropic immunoregulatory cytokine that, for instance, antagonizes the effects of IL-5 on eosinophils (ALAM et al. 1994). In addition to blocking the antiapoptotic effects of IL-5, it also inhibits eosinophil degranulation and cytokine production. The mechanisms of this inhibitory effect of

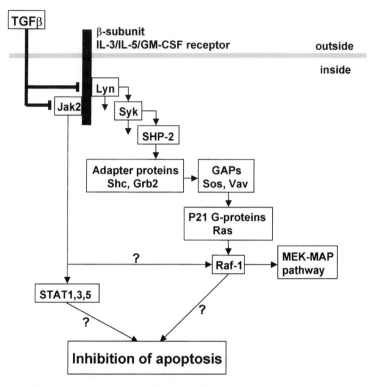

Fig. 4. Simplified scheme suggesting how cytokine-mediated antiapoptotic signals are transduced in eosinophils. Both Lyn- and Jak2-initiated pathways appear to be essential for antiapoptosis. The MEK-MAP pathway does not seem to be involved in the regulation of apoptosis, but might be important for eosinophil secretion

TGF-β is unknown. It has been demonstrated that TGF-β blocks tyrosine phosphorylation of Jak2 and Lyn tyrosine kinases (PAZDRAK et al. 1995c) (Fig. 4). Furthermore, it inhibits the activation of ERK MAP kinase and Stat1 nuclear factor. However, the signaling molecules mediating these effects have not yet been identified. Tyrosine phosphatases have been studied but do not seem to be involved. It is possible that TGF-β activates some of the newly-described inhibitors of tyrosine kinases, which subsequently mediate its inhibitory effects (HELDIN et al. 1997).

E. The MEK-ERK MAP Kinase Pathway Does Not Mediate Antiapoptotic Signals Initiated Via the IL-5 Receptor

The importance of signaling molecules in the antiapoptotic effect of IL-5 has been investigated. Specific depletion of Lyn (YOUSEFI et al. 1996; PAZDRAK et al. 1998), Syk (YOUSEFI et al. 1996), SHP-2 (PAZDRAK et al. 1997), and Raf-1 (PAZDRAK et al. 1998) by antisense oligodeoxynucleotides results in complete abrogation of the antiapoptotic effects of IL-5 and other hematopoietins. Similarly, the inhibition of Jak2 by the specific inhibitor AG490 also abrogates the effects of IL-5 (SIMON et al. 1997b; PAZDRAK et al. 1998). The results suggest that these signaling molecules are involved in propagating antiapoptotic signals provided by IL-5. Interestingly, although ERK is activated by IL-5, the inhibition of ERK activation by the MEK inhibitor PD98059 has minimal effects on eosinophil survival (R. ALAM, personal communication). Likewise, the specific inhibitor of p38 MAP kinase SB202589 does not block the survival-promoting effect of IL-5. These results are quite startling and provocative since it has been described that Raf-1 is critical for eosinophil survival (PAZDRAK et al. 1998). At this point, we believe, that the survival-promoting signal provided by Raf-1 is not propagated via the MEK-ERK MAP kinase pathway as the existing dogma would imply, but rather contributes to the phosphorylation of BAD (WANG et al. 1996), a Bcl-2- and Bcl-x_L-associated protein.

Since Bcl-x_L but not Bcl-2 is significantly expressed in eosinophils (DIBBERT et al. 1998), the phosphorylation of BAD via Raf-1 may enable Bcl-x_L to homodimerize (WANG et al. 1996), thereby exerting its antiapoptotic effects in this cellular system. Moreover, there are reports on Raf-1 activation by Jak2 kinase (XIA et al. 1996). Thus, it is possible that signals from Lyn, Jak2, Syk, and SHP-2 converge on Raf-1 to mediate activation and functional compartmentalization of Bcl-x_L. In this scenario, signals are also transduced via the MEK-ERK pathway. However, the latter pathway appears to be redundant for survival, although it is likely to be important for other cellular functions.

F. The Effects of Glucocorticoids on Eosinophil Apoptosis

Glucocorticoids have been used for decades as clinical tools to suppress both the immune response and the process of inflammation. However, only recently, we have begun to understand the molecular mechanisms of the effects of glucocorticoids. For instance, administration of glucocorticoids to patients with eosinophilia results in a marked decline in the number of circulating eosinophils (ROTHENBERG 1998). The reduction of eosinophil numbers appears to be due to the induction of eosinophil apoptosis (WOOLLEY et al. 1996).

What is the mechanism of the induction of eosinophil apoptosis by glucocorticoids in vivo? Asthma and other allergic disorders are characterized by T cell activation (MCFADDEN and GILBERT 1992). T cells produce cytokines, among them eosinophil survival factors such as IL-5. It is now clear that glucocorticoids suppress the transcription of the IL-5 and other cytokine genes. This inhibition of transcription is the consequence of inhibition of the potent

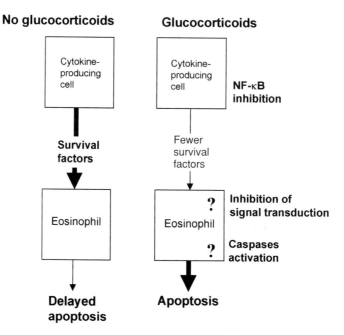

Fig.5. The effects of glucocorticoids in eosinophilic inflammation. Glucocorticoid therapy is often associated with a reduction of eosinophil numbers due to the induction of apoptosis. Probably the most important mechanism responsible for this observation is the decreased expression of eosinophil survival factors due to NF-κB inhibition. Other possible mechanisms might be inhibition of signal transduction pathways initiated by survival factors and/or direct induction of apoptosis in eosinophils

inflammatory transcription factor NF-κB (SCHWIEBERT et al. 1996). Therefore, one possible mechanisms of how eosinophil apoptosis can be mediated by glucocorticoids is the reduced expression of eosinophil survival factors (Fig. 5).

There is, however, the possibility that glucocorticoids could act directly on eosinophils. Indeed, there is experimental evidence that glucocorticoids inhibit the activity of eosinophil survival factors (WALLEN et al. 1991). Such inhibition of survival signals may be the consequence of disruption of signaling pathways (BAUS et al. 1996), although this has not formally been demonstrated in eosinophils. Recently, it has also been observed that glucocorticoids may also directly induce eosinophil apoptosis (MEAGHER et al. 1996) (Fig. 5).

G. Role of CD95 Ligand/CD95 Molecular Interactions in the Regulation of Eosinophil Apoptosis

It is now clear that eosinophils do not only undergo apoptosis in the absence of survival factors, but can also be triggered to die via specific surface death receptors. One of these death receptors expressed by eosinophils is CD95

(Fas/APO-1) (MATSUMOTO et al. 1995; TSUYUKI et al. 1995; DRUILHE et al. 1996; HEBESTREIT et al. 1996). The ligand of CD95 (CD95L, FasL, APO-1L) is highly expressed by activated T cells (GREEN and WARE 1997). Thus, the same cells that produce eosinophil survival factors also express at least one death factor for eosinophils. Interestingly, activation of the CD95-mediated apoptotic pathway in eosinophils occurs even in the presence of eosinophil survival factors (MATSUMOTO et al. 1995; TSUYUKI et al. 1995). Therefore, the newly discovered additional possibility of actively inducing eosinophil apoptosis makes sense: CD95L/CD95 molecular interactions may serve to limit eosinophil expansion independently from eosinophil hematopoietin expression within inflammatory sites.

Some of the intracellular signaling mechanisms initiated by survival and death signals have recently been identified in eosinophils. However, these studies revealed that the story is less simple than we thought. There is not only a passive and an active way to induce apoptosis in eosinophils. Moreover, not only cytokine-mediated delayed apoptosis but also CD95L-induced eosinophil death can be counterregulated, as discussed in greater detail below.

H. Nitric Oxide, but Not Eosinophil Hematopoietins, Mediates CD95 Resistance

Tissue eosinophils within inflammatory sites may not always undergo apoptosis following CD95 stimulation (HEBESTREIT et al. 1996). This phenomenon, also called CD95 resistance, could result in an unlimited expansion of eosinophils. Indeed, in nasal polyp tissues, where CD95 resistance has been observed, an extraordinary infiltration of eosinophils is usually observed. Thus, CD95 resistance is of pathophysiological relevance in chronic eosinophilic disorders and, therefore, CD95 signal transduction studies in eosinophils appear to be important.

The mechanisms of CD95 resistance has generated great interest in other cellular systems as well. Previously published data have provided evidence that mutations (FISHER et al. 1995; RIEUX-LAUCAT et al. 1995) as well as splicing variants that lack intracellular (CASCINO et al. 1996) or transmembrane (CHENG et al. 1994; SIMON et al. 1996) parts of the death receptor are associated with nonfunctional Fas receptors. Furthermore, lack of cell activation or costimulation via antigen (ROTHSTEIN et al. 1995) or cytokine (FOOTE et al. 1996) receptors appears to decrease susceptibility to CD95-mediated apoptosis. High levels of Bcl-2 (ITOH et al. 1993), Bcl-x_L (BOISE and THOMPSON 1997), viral (THOME et al. 1997) or cellular (IRMLER et al. 1997) FLIP, ALG-3 (LACANA et al. 1997), or IL-1β (TATSUTA et al. 1996) may also contribute to the development of CD95 resistance. In addition, the Abl kinase has been identified as a negative regulator of CD95-initiated signaling events (MCGAHON et al. 1995). Thus, CD95 resistance may often be associated with unwanted cell expansion associated with disease.

What is the mechanism(s) of CD95 resistance in eosinophils? The observation that eosinophil apoptosis following CD95 activation can be induced even in the presence of IL-5 or GM-CSF makes it unlikely that eosinophil survival factors account for this phenomenon. Meanwhile, we have recently observed that nitric oxide (NO) prevents CD95-mediated apoptosis in eosinophils (HEBESTREIT et al. 1998). This striking protective effect of NO appears to be of pathophysiological relevance since increased concentrations of NO are present within allergic inflammatory sites (BARNES and LIEW 1995).

I. Role of Sphingomyelinase-Mediated Pathways in CD95 Signaling

Activation of CD95 leads to stimulation of a proteases cascade, which, when started, is irreversible (NAGATA 1997; THORNBERRY and LAZEBNIK 1998). These proteases belong to the interleukin-1-converting enzyme (ICE) family of cysteine proteases, now called caspases (ALNEMRI et al. 1996), and appear to be directly responsible for the induction of apoptosis. However, other signaling events involving tyrosine phosphorylation (EISCHEN et al. 1994), sphingomyelinase (SMase)-ceramide (CIFONE et al. 1993), and Ras-Raf-1-MAP kinases (GOILLOT et al. 1997) pathways might be equally important.

The availability of a natural inhibitor (NO) of Fas receptor signaling allowed us to determine the roles of several biochemical events for the induction of apoptosis following death receptor activation in eosinophils. Ceramide, generated by activated SMase, triggers apoptosis in response to CD95 activation (CIFONE et al. 1993; GULBINS et al. 1995; TEPPER et al. 1995) and many other death stimuli (HAIMOVITZ-FRIEDMAN et al. 1994). We found that NO blocks the death signal distal to SMase. In contrast, activation of SMase was abrogated when the tetrapeptide YVAD was used to block caspases activity. Therefore, a caspase, such as caspase 8, appears to be proximal to SMase activation (Fig. 6).

Moreover, CD95 activation and ceramide induce activation of another, alternative MAP kinase pathway, resulting in Jun kinase (JNK) stimulation. JNK activation has been shown to be critical for induction of apoptosis in many systems. We observed that JNK is also activated following CD95 crosslinking in eosinophils. In contrast, JNK activation is completely blocked in the presence of NO. These findings suggest that JNK activation is also necessary for CD95-mediated apoptosis in eosinophils, and NO may act at the level of, or proximal to, JNK activation to prevent eosinophil apoptosis (Fig. 6).

The observation that it is possible to block activation of SMase by using the YVAD inhibitor suggests that there is some caspases activation even in the presence of NO. We hypothesize at this point that the generation of ceramide and subsequent JNK activation may represent a signaling event responsible for amplification of the proteolytic cascade. Therefore, disruption

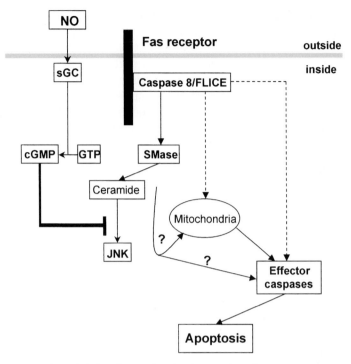

Fig. 6. A proposed model showing molecular interactions between caspases and sphinomyelinase (SMase) pathways in CD95 (Fas receptor, APO-1) -mediated apoptosis in eosinophils. In this model, caspases (FLICE, caspase 8) is, at least following stimulation, physically associated with the death receptor. Activation of this caspase (and, perhaps, other initiator caspases) appears to be essential for SMase activation. SMase generates ceramide and activates JNK that can be blocked by second messengers of NO (cGMP). Ceramide and/or further distal located signaling molecules activate either directly or indirectly (via mitochondria) effector caspases (e.g., CPP32, caspase 3) leading to apoptotic death

of ceramide-induced signals prevents further caspases activation. This idea is further supported by previously published reports demonstrating that the central effector caspase, caspase 3, is not only a target of initiator caspases (ENARI et al. 1996; THORNBERRY and LAZEBNIK 1998), but also of MAP kinase- and JNK-signaling pathways (GOILLOT et al. 1997; YANG et al. 1997). This process may involve cytochrome C release from mitochondria (Fig. 6).

Taken together, activation of eosinophils via the CD95 molecule in the presence of NO leads to an immediate but limited activation of caspases able to degrade only a limited number of substrates. Probably, these substrates can be replaced without any damage to the cell. Obviously, under condition of NO presence, CD95-mediated caspase 8 activation is unable to activate effector caspases to induce apoptosis either directly or indirectly via the mitochondria (Fig. 6). In contrast, in the absence of NO, ceramide-mediated amplification of the proteolytic cascade takes place and the apoptotic process initiated via CD95 proceeds, causing irreversible damage to the cells.

Therefore, we have learned from the eosinophil system that the apoptosis signal triggered by death receptors can be modulated by intracellular mechanisms, at least in those cases where caspase 8 needs the mitochondrial amplification loop to activate effector caspases (GREEN 1998). Thus, the described data are relevant for the understanding of both the pathophysiological role of NO, a secretory product released in increased amounts within chronic eosinophilic inflammatory responses, and basic mechanisms of how the cell death machinery works.

J. Role of Tyrosine Kinase Activation in CD95 Signaling

Tyrosine phosphorylation has been shown to be involved in CD95 transmembrane signaling in many cellular systems (EISCHEN et al. 1994), although it is still controversial whether tyrosine kinase activation is necessary for CD95-mediated apoptosis (LATINIS and KORETZKY 1996). However, a role of tyrosine phosphorylation is supported by the observation that expression of SHP-1 is a prerequisite for CD95-induced apoptosis in several lymphoid cell lines (SU et al. 1995). Moreover, another tyrosine phosphatase, FAP-1, has been shown to associate with CD95 and to exert a negative influence on CD95 signal transduction (SATO et al. 1995).

We have recently demonstrated that tyrosine phosphorylation is an important event involved in CD95 transmembrane signal transduction in human and mouse eosinophils (SIMON et al. 1998b). CD95 is physically associated with a number of tyrosine-phosphorylated proteins, as shown by co-immunoprecipitation studies. Moreover, phosphorylation of both tyrosine residues within the intracellular part of the human CD95 molecule has recently been demonstrated (GRADL et al. 1996). Interestingly, these two tyrosine residues are also present in the amino acid sequence predicted for the murine CD95 cDNA, consistent with the suggestion that these tyrosine residues are important for signal transduction. Furthermore, tyrosine kinase blockers inhibited CD95-mediated apoptosis in both human and mouse eosinophils in vitro, and prevented, at least partially, CD95-mediated resolution of eosinophilic inflammation in a mouse in vivo model of lung eosinophilia (SIMON et al. 1998b). Taken together, these results strongly implicate tyrosine kinase activation as likely involved in the death response following CD95 crosslinking in eosinophils.

The demonstration of tyrosine phosphorylation of CD95, which does not itself contain an intrinsic kinase activity, suggests that a cytoplasmic tyrosine kinase is associated with the receptor. We have identified Lyn as an important tyrosine kinase which transduces death signals via CD95 in eosinophils (SIMON et al. 1998b). These data are in agreement with previously published work demonstrating a reduced susceptibility to CD95-mediated death in B cells from Lyn-deficient mice (WANG et al. 1996). Thus, Lyn emerges as a signaling molecule capable of inducing two mutually exclusive cellular functions, cell survival and cell death. Similar observations have been previously reported for other signaling molecules such as ceramide (KOLESNICK and FUKS 1995) or Ras (SATOH et al. 1991). It is possible that Lyn facilitates pro-apoptotic activities when a concurrent activation of the Jak-STAT pathway does not occur, as seen following activation of IL-3/IL-5/GM-CSF receptors (SIMON et al. 1997b; PAZDRAK et al. 1998). Moreover, the data confirm previously published work suggesting that Lyn may represent a common element involved in granulocyte signaling following activation with widely different agonists (GAUDRY et al. 1995).

While it is clear that CD95 activation results in increased tyrosine phosphorylation, a requirement of tyrosine kinase activation for CD95-mediated apoptosis is controversial, especially in T cells (LATINIS and KORETZKY 1996). In contrast, tyrosine phosphorylation appears to modulate the functional death response in eosinophils, neutrophils, and B cells. One possible explanation for this discrepancy could be that the levels of CD95 surface expression seem to be critical for the efficacy of the death signal (CLEMENT and STAMENKOVIC 1994). Since the levels of CD95 surface expression are relatively low in eosinophils and neutrophils compared to T cells (HEBESTREIT et al. 1996), it is possible that in granulocytes tyrosine kinase activation is also required for optimal signal transduction via CD95. In contrast, an optimal interaction between CD95 and second messengers molecules (e.g., caspase 8) might already be present in activated T cells and therefore the activation of tyrosine kinases might be not functionally relevant in these systems. Thus, in this model, the role of tyrosine kinase activation could be to decrease the threshold of needed receptor molecules per cell and/or caspase 8 intracellular activity for induction of apoptosis in granulocytes and B cells. This assumption is supported by the fact that tyrosine kinase activation appears to be independent from caspases since the caspase inhibitor, YVAD, completely blocked Fas receptor-mediated death, but did not abrogate the activation of Lyn in eosinophils.

K. Concluding Remarks

Since eosinophils are prominent in allergic inflammation, investigators became interested in how these cells accumulate in tissues and about their role within the inflammatory cascade. There is increasing evidence from several laboratories that eosinophil numbers are regulated in vivo, not only by eosinophil production in the bone marrow, but also by the amount of eosinophil apoptosis. Moreover, it has been directly demonstrated that eosinophil apoptosis is delayed in allergic inflammatory sites, and that this mechanism contributes to the expansion of these cells in tissue.

Inhibition of eosinophil apoptosis can be achieved by at least two mechanisms – increased expression of eosinophil survival factors and disruption of death signals. There are many urgent questions to be answered in the near future. For instance, how long do eosinophils act as effector cells in allergic inflammation when they do not undergo apoptosis? Which other death receptors (in addition to CD95) are expressed in eosinophils? What is their function? Which caspases play a role? What are the differences in the apoptosis regulation between eosinophils and neutrophils? How does a cell decide its outcome when it receives survival and death signals at the same time, a situation that very likely occurs in vivo? Clearly, there is much more to learn about eosinophil apoptosis.

Acknowledgements. Work of the author's laboratory is supported by the Swiss National Science Foundation (32–49210.96), OPO Foundation (Zurich), EMDO Foundation (Zurich), Saurer Foundation (Zurich), Foundation for Scientific Research (Zurich), and the Silva Casa Foundation (Bern).

References

Alam R, Forsythe P, Stafford S, Fukuda Y (1994) Transforming growth factor β abrogates the effects of hematopoietins on eosinophils and induces their apoptosis. J Exp Med 179:1041–1045
Alnemri ES, Livingston DJ, Nicholson DW, Salvesen G, Thornberry NA, Wong WW, Yuan J (1996) Human ICE/CED-3 protease nomenclature. Cell 87:171
Bagley CJ, Lopez AF, Vadas MA (1997) New frontiers for IL-5. J Allergy Clin Immunol 99:725–728
Barnes PJ, Liew FY (1995) Nitric oxide and asthmatic inflammation. Immunol Today 16:128–130
Bates ME, Busse WW, Bertics PJ (1998) Interleukin 5 signals through Shc and Grb2 in human eosinophils. Am J Respir Mol Cell Biol 18:75–83
Baus E, Andris F, Dubois PM, Urbain J, Leo O (1996) Dexamethasone inhibits the early steps of antigen receptor signaling in activated T lymphocytes. J Immunol 156:4555–4561
Boise LH, Thompson CB (1997) Bcl-xL can inhibit apoptosis in cells that have undergone Fas-induced protease activation. Proc Natl Acad Sci USA 94:3759–3764
Caldenhoven E, van Dijk T, Raaijmakers JA, Lammers JW, Koenderman L, de Groot RP (1995) Activation of the STAT3/acute phase response factor transcription factor by interleukin-5. J Biol Chem 270:25778–25784
Cascino I, Papoff G, De Maria R, Testi R, Ruberti G (1996) Fas/APO-1 (CD95) receptor lacking the intracytoplasmic signaling domain protects tumor cells from Fas-mediated apoptosis. J Immunol 156:13–17
Cheng J, Zhou T, Liu C, Shapiro JP, Brauer MJ, Kiefer MC, Barr PJ, Mountz JD (1994) Protection from Fas-mediated apoptosis by a soluble form of the Fas molecule. Science 263:1759–1762
Cifone MG, De Maria R, Roncaioli P, Rippo MR, Azuma M, Lanier LL, Santoni A, Testi R (1993) Apoptotic signaling through CD95 (Fas/APO-1) activates an acidic sphingomyelinase. J Exp Med 182:1545–1556
Clement MV, Stamenkovic I (1994) Fas and tumor necrosis factor receptor-mediated cell death: similarities and distinctions. J Exp Med 180:557–567
Coffer PJ, Schweizer RC, Dubois GR, Maikoe T, Lammers JW, Koenderman L (1998) Analysis of signal transduction pathways in human eosinophils activated by chemoattractants and T-helper 2-derived cytokines interleukin-4 and interleukin-5. Blood 91:2547–2557
Dibbert B, Daigle I, Braun D, Schranz C, Weber M, Blaser K, Zangemeister-Wittke U, Akbar AN, Simon HU (1998) Role for Bcl-x$_L$ in delayed eosinophil apoptosis

mediated by granulocyte-macrophage colony-stimulating factor and interleukin-5. Blood 92:778–783
Druilhe A, Cai Z, Hailé S, Chonaib S, Petrolani M (1996) Fas-mediated apoptosis in cultured human eosinophils. Blood 87:2822–2830
Eischen CM, Dick CJ, Leibson PJ (1994) Tyrosine kinase activation provides an early and requisite signal for Fas-induced apoptosis. J Immunol 153:1947–1954
Enari M, Talanian RV, Wong WW, Nagata S (1996) Sequential activation of ICE-like and CPP32-like proteases during Fas-mediated apoptosis. Nature 380:723–726
Fisher GH, Rosenberg FJ, Straus SE, Dale JK, Middelton LA, Lin AY, Strober W, Lenardo MJ, Puck JM (1995) Dominant interfering Fas gene mutations impair apoptosis in a human autoimmune lymphoproliferative syndrome. Cell 81:935–946
Foote LC, Howard RG, Marshak-Rothstein A, Rothstein TL (1996) IL-4 induces Fas resistance in B cells. J Immunol 157:2749–2753
Gaudry M, Gilbert C, Barabé F, Poubelle PE, Naccache PH (1995) Activation of Lyn is a common element of the stimulation of human neutrophils by soluble and particulate agonists. Blood 86:3567–3574
Goillot E, Raingeaud J, Ranger A, Tepper RI, Davis RJ, Harlow E, Sanchez I (1997) Mitogen-activated protein kinase-mediated Fas apoptotic signaling pathway. Proc Natl Acad Sci USA 94:3302–3307
Gradl G, Grandison P, Lindridge E, Wang Y, Watson J, Rudert, F (1996) The CD95 (Fas/APO-1) receptor is phosphorylated in vitro and in vivo and constitutively associates with several cellular proteins. Apoptosis 1:131–140
Green DR, Ware CF (1997) Fas-ligand: privilege and peril. Proc Natl Acad Sci USA 94:5986–5990
Green DR (1998) Apoptotic pathways: the road to ruin. Cell 94:695–698
Gulbins E, Bissonnette R, Mahboubi A, Martin S, Nishioka W, Brunner T, Baier G, Baier-Bitterlich G, Byrd C, Lang F, Kolesnick R, Altman A, Green D (1995) Fas-induced apoptosis is mediated via a ceramide-initiated RAS signaling pathway. Immunity 2:341–351
Haimovitz-Friedman A, Kan C, Ehleiter D, Persaud R, McLoghlin M, Fuks Z, Kolesnick RN (1994) Ionizing radiation acts on cellular membranes to generate ceramide and initiate apoptosis. J Exp Med 180:525–535
Hebestreit H, Yousefi S, Balatti I, Weber M, Crameri R, Simon D, Hartung K, Schapowal A, Blaser K, Simon HU (1996) Expression and function of the Fas receptor on human blood and tissue eosinophils. Eur J Immunol 26:1775–1780
Hebestreit H, Dibbert B, Balatti I, Braun D, Schapowal A, Blaser K, Simon HU (1998) Disruption of Fas receptor signaling by nitric oxide in eosinophils. J Exp Med 187:415–425
Heldin CH, Miyazomo K, ten Dijke P (1997) TGF-β signaling from cell membrane to nucleus through SMAD proteins. Nature 390:465–471
Her E, Frazer J, Austen KF, Owen WF Jr (1991) Eosinophil hematopoietins antagonize the programmed cell death of eosinophils. Cytokine and glucocorticoid effects on eosinophils maintained by endothelial cell-conditioned medium. J Clin Invest 88:1982–1987
Irmler M, Thome M, Hahne M, Schneider P, Hofmann K, Steiner V, Bodmer JL, Schröter M, Burns K, Mattmann C, Rimoldi D, French LE, Tschopp J (1997) Inhibition of death receptor signals by cellular FLIP. Nature 388:190–195
Itoh N, Tsujimoto Y, Nagata S (1993) Effect of Bcl-2 on Fas antigen-mediated cell death. J Immunol 151:621–627
Koh YY, Dupuis R, Pollice M, Albertine KH, Fish JE, Peters SP (1993) Neutrophils recruited to the lungs by segmental antigen challenge display a reduced chemotactic response to leukotriene B4. Am J Respir Cell Mol Biol 8:493–499
Kolesnick R, Fuks Z (1995) Ceramide: a signal for apoptosis or mitogenesis? J Exp Med 181:1949–1952

Lacana E, Ganjei JK, Vito P, D'Adamio L (1997) Dissociation of apoptosis and activation of IL-1β-converting enzyme/Ced-3 proteases by ALG-2 and the truncated Alzheimer's gene ALG-3. J Immunol 158:5129–5135

Latinis KM, Koretzky GA (1996) Fas ligation induces apoptosis and Jun kinase activation independently of CD45 and Lck in human T cells. Blood 87:871–875

Lukacs NW, Strieter RM, Kunkel SL (1995) Leukocyte infiltration in allergic airway inflammation. Am J Respir Cell Mol Biol 13:1–6

Matsumoto K, Schleimer RP, Saito H, Iikura Y, Bochner BS (1995) Induction of apoptosis in human eosinophils by anti-Fas antibody treatment in vitro. Blood 86:1437–1443

McFadden ER, Gilbert IA (1992) Asthma. N Engl J Med 327:1928–1937

McGahon AJ, Nishioka WK, Martin SJ, Mahboubi A, Cotter TG, Green DR (1995) Regulation of the Fas apoptotic cell death pathway by Abl. J Biol Chem 270:22625–22631

Meagher LC, Cousin JM, Seckl JR, Haslett C (1996) Opposing effects of glucocorticoids on the rate of apoptosis in neutrophilic and eosinophilic granulocytes. J Immunol 156:4422–4428

Mui AL, Wakao H, O'Farrell AM, Harada N, Miyajima A (1995) IL-3, GM-CSF and IL-5 transduce signals through two Stat5 homologs. EMBO J 14:1166–1175

Nagata S (1997) Apoptosis by death factor. Cell 88:355–365

Ogata N, Kuro T, Yamada A, Koike M, Hanai N, Ishikawa T, Takatsu K (1998) Jak2 and Jak1 constitutively associate with an interleukin-5 (IL-5) receptor α and βc subunit, respectively, and are activated upon IL-5 stimulation. Blood 91:2264–2271

Pazdrak K, Schreiber D, Forsythe P, Justement L, Alam R (1995a) The signal transduction mechanism of IL-5 in eosinophils: the involvement of Lyn tyrosine kinase and the ras-raf 1-MEK-MAP kinase pathway. J Exp Med 181:1827–1834

Pazdrak K, Stafford S, Alam R (1995b) The activation of the Jak-STAT1 signaling pathway by IL-5 in eosinophils. J Immunol 155:397–402

Pazdrak K, Justement L, Alam R (1995c) Mechanism of inhibition of eosinophil activation by transforming growth factor-β. Inhibition of Lyn, MAP, Jak2 kinases and STAT1 nuclear factor. J Immunol 155:4454–4458

Pazdrak K, Adachi T, Alam R (1997) SHPTP2/SHP2 tyrosine phosphatase is a positive regulator of the interleukin-5 receptor signal transduction pathways leading to the prolongation of eosinophil survival. J Exp Med 186:561–568

Pazdrak K, Olszewska-Pazdrak B, Stafford S, Garofalo RP, Alam R (1998) Lyn, Jak2 and Raf-1 kinases are critical for the anti-apoptotic effect of interleukin-5, whereas only Raf-1 kinase is essential for eosinophil activation and degranulation. J Exp Med 188:421–429

Richards IM, Kolbasa KP, Hatfield CA, Winterrowd GE, Vonderfecht SL, Fidler SF, Griffin RL, Brashler JR, Krzesicki RF, Sly LM, Ready KA, Staite ND, Chin JE (1996) Role of very late activation antigen-4 in the antigen-induced accumulation of eosinophils and lymphocytes in the lungs and airway lumen of sensitized brown Norway rats. Am J Respir Cell Mol Biol 15:172–183

Rieux-Laucat F, Le Deist F, Hivroz C, Roberts IAG, Debatin KM, Fischer A, de Villartay JP (1995) Mutations in Fas associated with human lymphoproliferative syndrome and autoimmunity. Science 268:1347–1349

Rothenberg ME (1998) Eosinophilia. N Engl J Med 338:1592–1600

Rothstein TL, Wang JKM, Panka DJ, Foote LC, Wang Z, Stanger B, Cui H, Ju ST, Marshak-Rothstein A (1995) Protection against Fas-dependent Th1-mediated apoptosis by antigen receptor engagement in B cells. Nature 374:163–165

Sato T, Irie S, Kitada S, Reed JC (1995) FAP-1: a protein tyrosine phosphatase that associates with Fas. Science 268:411–415

Satoh T, Nakafuku M, Miyajima A, Kaziro Y (1991) Involvement of ras p21 protein in signal transduction pathways from interleukin 2, interleukin 3, and granulocyte/macrophage colony-stimulating factor, but not from interleukin-4. Proc Natl Acad Sci USA 88:3314–3318

Schwiebert LA, Beck LA, Stellato C, Bickel CA, Bochner BS, Schleimer RP (1996) Glucocorticosteroid inhibition of cytokine production: relevance to antiallergic actions. J Allergy Clin Immunol 97:143–152

Simon HU, Blaser K (1995) Inhibition of programmed eosinophil death: a key pathogenic event for eosinophilia? Immunol Today 16:55–55

Simon HU, Yousefi S, Dommann-Scherrer CC, Zimmermann DR, Bauer S, Barandun J, Blaser K (1996) Expansion of cytokine-producing T cells associated with abnormal Fas expression and hypereosinophilia. J Exp Med 183:1071–1082

Simon HU, Yousefi S, Schranz C, Schapowal A, Bachert C, Blaser K (1997a) Direct demonstration of delayed eosinophil apoptosis as a mechanism causing tissue eosinophilia. J Immunol 158:3902–3908

Simon HU, Yousefi S, Dibbert B, Levi-Schaffer F, Blaser K (1997b) Anti-apoptotic signals of granulocyte-macrophage colony-stimulating factor are transduced via Jak2 tyrosine kinase in eosinophils. Eur J Immunol 27:3536–3539

Simon HU (1998a) Eosinophil apoptosis in allergic diseases – an emerging new issue. Clin Exp Allergy 28:1321–1324

Simon HU, Yousefi S, Dibbert B, Hebestreit H, Weber M, Branch DR, Blaser K, Levi-Schaffer F, Anderson GP (1998b) Role for tyrosine phosphorylation and Lyn tyrosine kinase in Fas receptor-mediated apoptosis in eosinophils. Blood 92:547–557

Stern M, Meagher L, Savill J, Haslett C (1992) Apoptosis in human eosinophils. Programmed cell death in the eosinophils leads to phagocytosis by macrophages and is modulated by IL-5. J Immunol 148:3543–3549

Su X, Zhou T, Wang Z, Yang P, Jope RS, Mountz JD (1995) Defective expression of hematopoietic cell protein tyrosine phosphatase (HCP) in lymphoid cells blocks Fas-mediated apoptosis. Immunity 2:353–362

Tatsuta T, Cheng J, Mountz JD (1996) Intracellular IL-1β is an inhibitor of Fas-mediated apoptosis. J Immunol 157:3949–3957

Tepper CG, Jayadev S, Liu B, Bielawska A, Wolff R, Yonehara S, Hannun YA, Seldin MF (1995) Role of ceramide as an endogenous mediator of Fas-induced cytotoxicity. Proc Natl Acad Sci USA 92:8443–8447

Teran LM, Campos MG, Begishvilli BT, Schröder JM, Djukanovic R, Shute JK, Church MK, Holgate ST, Davies DE (1997) Identification of neutrophil chemotactic actors in bronchoalveolar lavage fluid of asthmatic patients. Clin Exp Allergy 27:396–405

Thome M, Schneider P, Hofmann K, Fickenscher H, Meinl E, Neipel F, Mattmann C, Burns K, Bodmer JL, Schröter M, Scaffidi C, Krammer PH, Peter ME, Tschopp J (1997) Viral FLICE-inhibitory proteins (FLIPs) prevent apoptosis induced by death receptors. Nature 386:517–521

Thornberry NA, Lazebnik Y (1998) Caspases: enemies within. Science 281:1312–1316

Tsuyuki S, Bertrand C, Erard F, Trifilieff A, Tsuyuki J, Wesp M, Anderson GP, Coyle AJ (1995) Activation of the Fas receptor on lung eosinophils leads to apoptosis and the resolution of eosinophilic inflammation of the airways. J Clin Invest 96:2924–2931

van der Bruggen T, Caldenhoven E, Kanters D, Coffer P, Raaijmakers JA, Lammers JW, Koenderman L (1995) Interleukin-5 signaling in human eosinophils involves JAK2 tyrosine kinase and STAT1α. Blood 85:1442–1448

Wallen N, Kita H, Weiler D, Gleich GJ (1991) Glucocorticoids inhibit cytokine-mediated eosinophil survival. J Immunol 147:3490–3495

Wang HG, Rapp UR, Reed JC (1996) Bcl-2 targets the protein kinase Raf-1 to mitochondria. Cell 87:629–638

Wang J, Koizumi T, Watanabe T (1996) Altered antigen receptor signaling and impaired Fas-mediated apoptosis of B cells in Lyn-deficient mice. J Exp Med 184:831–838

Wedi B, Raap U, Lewrick H, Kapp A (1997) Delayed eosinophil programmed cell death in vitro: a common feature of inhalant allergy and extrinsic and intrinsic dermatitis. J Allergy Clin Immunol 100:536–543

Woolley KL, Gibson PG, Carty K, Wilson AJ, Twaddell SH, Woolley MJ (1996) Eosinophil apoptosis and its resolution of airway inflammation in asthma. Am J Respir Crit Care Med 154:237–243

Xia K, Mukhopadhyay NK, Inhorn RC, Barber DL, Rose PE, Lee RS, Narsimhan RP, D'Andrea AD, Griffin JD, Roberts TM (1996) The cytokine-activated tyrosine kinase JAK2 activates Raf-1 in a p21ras-dependent manner. Proc Natl Acad Sci USA 93:11681–11686

Yamaguchi Y, Suda T, Ohta S, Tominaga K, Miura Y, Kasahara T (1991) Analysis of the survival of mature eosinophils: interleukin-5 prevents apoptosis in mature human eosinophils. Blood 78:2542–2547

Yang X, Khosravi-Far R, Chang HY, Baltimore D (1997) Daxx, a novel Fas-binding protein that activates JNK and apoptosis. Cell 89:1067–1076

Yousefi S, Hoessli DC, Blaser K, Mills GB, Simon HU (1996) Requirement of Lyn and Syk tyrosine kinases for the prevention of apoptosis by cytokine in human eosinophils. J Exp Med 183:1407–1414

CHAPTER 15
Thymocyte and B-Cell Death Without DNA Fragmentation

T. Itoh, M. Nakamura, H. Yagi, H. Soga, and T. Ishii

A. Introduction

In mammals there are several cell renewal systems including the epidermis, the intestinal epithelium, and blood cells, in which a number of cells are generated every day, while a similar number of cells are lost due to cell death. In the hematopoietic system, for example, most blood cells, once they mature, die at various intervals with different life-spans for each lineage cells; mature neutrophils die within 2–3 days, whereas denucleated mature erythrocytes are totally discarded every 120 days by splenic or liver macrophages. Among hematopoietic lineages, most lymphoid cells are also short-lived, indicating they have rather short life-spans, though some of them are long-lived memory cells. The most characteristic aspect of the lymphoid cell fate is their repertoire generating mechanism. In both T and B lymphocytes, their extensively diversified repertoire is characteristically produced by their enormous proliferating activities and by consequent massive cell death, leaving only a minor population with an appropriately selected repertoire specificity. In T lymphocytes, the site for repertoire generation is the thymus, and for B lymphocytes the bone marrow and the germinal center. In this chapter we carefully examine in situ cell death of thymocytes and B cells at the germinal center and discuss their cell death mechanism.

B. Functional and Structural Characteristics of the Thymus

I. Differentiation of Thymocytes

Since the thymus is well known for its production of a tremendous amount of dead cells, the organ has been considered to be a useful model for investigation of programmed cell death or apoptosis.

Hematopoietic progenitors migrate from the bone marrow, via the blood vessels, and enter the thymus. Upon entering the thymus, they extensively proliferate at the subcapsular region of the thymus, gradually change the location toward the deep cortex after the cessation of the cell division, and through a series of complex selection processes they proceed along the differentiation pathway, finally reaching the vessels at the cortico-medullary junction,

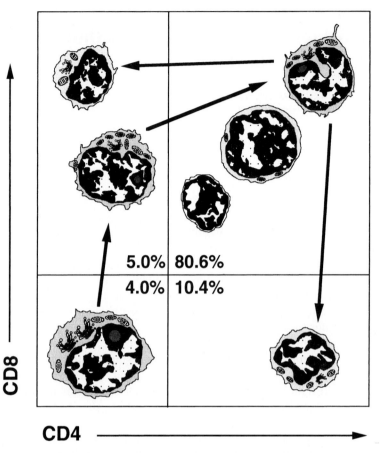

Fig. 1. A schematic representation of thymocyte differentiation defined by CD4 and CD8 in flow cytometry. The most immature thymocytes do not bear either of the surface antigens, then they express both of them (double positive thymocytes), and finally a small number of double positive thymocytes differentiate into CD4 (ca. 10%) or CD8 single positive (ca. 5%) mature thymocytes. The majority of double positive thymocytes undergo cell death within 3–4 days

from where they presumably leave the thymus for the periphery. At first they do not bear any of the markers specific for mature T lymphocytes including CD4 and CD8, but gradually acquire the expression of both. On the other hand, they begin to express T cell receptor (TCR), and then thymocytes lose either of the coreceptors (CD4 or CD8) (Fig. 1). Some of them become capable of expressing TCR at high intensity (TCR^{hi}), and ultimately only those thymocytes that are allowed to mature (positively selected; single positive for CD4 or CD8, and TCR^{hi}) will leave the thymus for peripheral lymphatic tissues after the establishment of self tolerance (VON BOEHMER 1988, 1992).

The thymic microenvironment is thus divided into several compartments: the subcapsular region, where massive proliferation of thymocytes takes place; the cortex, where nearly all thymocytes cease cell division and thymocyte differentiation as well as critical selection might occur; the cortico-medullary junction, with abundant venules with relatively wide lumen through which a number of thymocytes are considered to leave for the periphery; and the medulla, the site for the accumulation of mature thymocytes (RITTER and CRISPE 1994). Among these regions, the cortex is the most important in terms of thymocyte differentiation, selection, and/or death.

II. Thymocyte Selection

There may be at least two types of selection processes that might be inevitably related to thymocyte death. One such easily comprehensible type of selection is the deletion of self-reactive cells which is generated during the enormous repertoire formation (COHEN 1991; WILLIAMS 1994). Another type would be abortive generation of aberrant nonfunctional (non-selected) thymocytes with inappropriate TCR (COHEN 1991). Under a physiological condition, i.e., in the normal thymus, to produce immunologically competent T cell populations which are single-positive for CD4 or CD8 with high TCR, the progenitor population, first of all, likely generates a huge number of progenies, which include a small subset of thymocytes ($CD4^+8^+$) with a potential for productive differentiation, as well as a much larger subset of aberrant nonfunctional thymocytes ($CD4^+8^+$). This takes place, on the basis of probability, completely at random; the progenitor population first generates progenies without any apparent bias, and then the progenies undergo biased selection. A minor population has been considered to have the capability of interacting surrounding stromal cells with appropriate ligands for TCR and coreceptors, thus transducing positive signals to thymocytes and eventually leading to positive selection, whereas nonfunctional abortive thymocytes do not receive any triggers from the microenvironment, resulting in a large number of non-selected thymocytes which soon die (COHEN 1991; RITTER and CRISPE 1994; VON BOEHMER 1988, 1992). Among a small positively selected population, a still smaller number would become self-reactive by chance; it is hypothesized that these cells are actively deleted in the thymus by a so far undefined process.

III. Thymocyte Death

As a result of the selection process of thymocytes to generate a large diversity of TCR repertoire, an extremely large fraction of thymocytes die within the thymus, even without leaving it (COHEN 1991). As discussed above, the number of dead cells resulting from aberrant nonfunctional thymocytes in the normal thymus is much larger than that from the extremely tiny population of self-reactive thymocytes.

Table 1. Reported definition of various cell deaths

Types of cell death	DNA fragmentation	Chromatin condensation	Membrane integrity	Phagocytosis	Enzymes involved
Apoptosis	+	Peripheral	+	Later	Caspases (cysteine proteases), endonuclease
Pyknosis	–	Heavy, overall	+	Later	Not reported
Necrosis	–	–	–	–	Not reported

As the thymus generates a huge number of dead cells, the organ has long been regarded as one of the best organs to perform investigations of "apoptosis." Since the notion of apoptosis was first introduced into this field, numerous studies have been reported on thymocyte death.

A couple of papers published more than ten years ago prompted us to undertake an investigation on thymocyte death (COHEN and DUKE 1984; KIZAKI et al. 1989); in the reports, death of thymocytes was induced in vitro and detected by a ladder pattern by electrophoresis, but the detection of thymocyte death in situ (or in vivo) was not presented. Quite surprisingly, even fresh thymocytes (immediately after the suspension was made) have frequently been used as a negative control for the ladder formation (COHEN and DUKE 1984; COSSARIZZA et al. 1994; KIZAKI et al. 1989; WALKER et al. 1994). The ladder formation in electrophoresis is one means to detect DNA fragmentation, which has often been recognized as the major hallmark of apoptosis (Table 1) (ARENDS and WYLLIE 1991; COHEN 1991; GOLDSTEIN et al. 1991; RAFF 1992). We therefore decided to carry out experiments to investigate in situ thymocyte death carefully and extensively under physiological conditions. Until the time we started the series of experiments, only a few reports had been presented on thymocyte death in situ (SURH and SPRENT 1994).

First of all we examined thymocyte death in situ by the terminal deoxynucleotidyl transferase dUTP-biotin nick end labeling (TUNEL) method (GAVRIELI et al. 1992). This procedure allows us to detect DNA fragmentation on frozen sections. Only a few thymocytes (less than 1%) could be detected by the TUNEL method in frozen sections (Fig. 2 and see also Fig. 8) (NAKAMURA et al. 1995). This finding was entirely different from what a number of investigators had repeatedly postulated until then, i.e., thymocytes die by apoptosis, by definition, which should accompany DNA fragmentation (COHEN 1991; WILLIAMS 1994). Since dead thymocytes detected in our studies by the TUNEL method have often been found to form clusters, we simultaneously performed staining of TUNEL and histochemistry to detect DNA fragmentation and acid phosphatase (ACP) (BARKA and ANDERSON 1962) (Fig. 3). Acid phosphatase is an enzyme representative for lysosomes; lysosomes are characteristic organelles for macrophages. All TUNEL positive nuclei (pigmented brown by diaminobenzidine precipitation) were observed entirely overlapping with the red spots (where pararosaniline precipitated in acid phosphatase

Fig. 2. A micrograph of the normal mouse thymus stained by the TUNEL method. Only a few (less than 1%) thymocytes, scattered throughout the cortex, are stained. The *bar* indicates 100 µm

reaction (BARKA and ANDERSON 1962)); DNA fragmentation was detected only within macrophages (NAKAMURA et al. 1995). This finding was further confirmed by TUNEL and Mac-2 double immunofluorescence staining (Fig. 4). All green fluorescence signals (TUNEL staining) were encircled by red signals (Mac-2 staining), suggesting that all DNA fragmented nuclei were present only inside Mac-2-positive phagocytes (macrophages). In plain transmission electron microscopy, macrophages ingesting several nuclei at various stages of digesting processes were in fact discerned (Fig. 5) (NAKAMURA et al. 1995).

Up to this point, though by circumstantial evidence, we confidently came to the conclusion that, under the normal physiological condition, nearly all (dying) thymocytes die by pyknosis, but not by apoptosis (NAKAMURA et al. 1995). "Pyknosis," a type of cell death mainly defined by morphology, is characterized by heavy overall chromatin condensation, conventionally considered to occur in thymocytes (Table 1) (AREY 1974). To obtain more conclusive evidence, we applied the TUNEL method at the electron microscopic level. As shown in Fig. 6, positive signals for TUNEL were only detected in the nuclei present inside the phagocytes (macrophages); no matter how morphologically

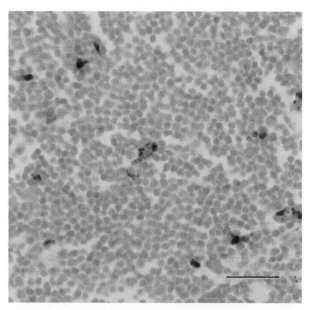

Fig. 3. The normal mouse thymus simultaneously stained for TUNEL and ACP. Positive staining for the TUNEL method (thymocytes with fragmented DNA) is completely overlapping with ACP histochemical staining (macrophages). The *bar* indicates 20 µm

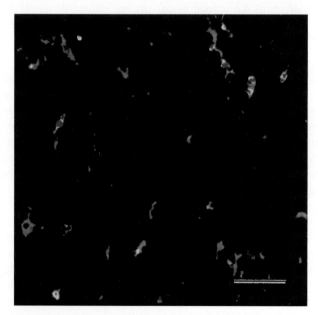

Fig. 4. The normal mouse thymus subjected to double fluorescence staining with TUNEL (Texas Red) and Mac-2 (FITC). All TUNEL positive thymocytes are observed within Mac-2 positive cells. The *bar* indicates 35 µm

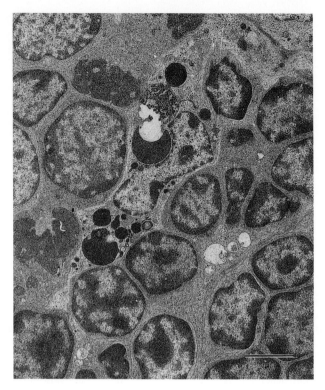

Fig. 5. Transmission electron micrograph of a macrophage of the mouse thymus ingesting several dead cells (thymocytes). The *bar* indicates 2 μm

obvious the evidence for cell death is, i.e., by their pyknotic nuclei and by their extremely small cell size, the cells, unless they are phagocytosed, were TUNEL negative even when in the close vicinity of phagocytes. These results collectively indicate that thymocytes in the normal thymus do not die by apoptosis: thymocytes do not show DNA fragmentation, even though they are apparently dead before being phagocytosed (Ishii et al. 1997; Nakamura et al. 1995). DNA fragmentation was only detected in the nuclei which had been ingested by macrophages. Indeed, in transmission electron microscopy (TEM), relatively abundant (ca. 10%) extremely small (smaller than red blood cells), heavily chromatin-condensed pyknotic thymocytes could be seen, which should be considered to be dead (Fig. 7) (Nakamura et al. 1995). One can perform the TUNEL method by flow cytometry (Kishimoto et al. 1995; Ojeda et al. 1992). Figure 8 shows a representative result of the normal mouse thymocytes by TUNEL flow cytometry. Virtually no TUNEL positive thymocytes could be detected by flow cytometry (Ishii et al. 1997). Flow cytometry also demonstrated the presence of about 10% small dead cells at the time of cell preparation (Fig. 9); these apparently dead cells were negative for TUNEL staining.

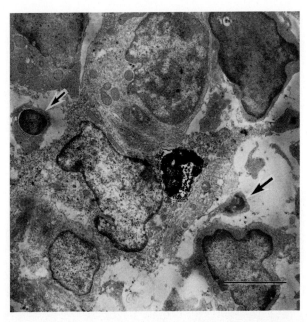

Fig. 6. Electron micrograph of the mouse thymus stained by TUNEL method. At the center of this micrograph, one macrophage can be seen. Only an engulfed cell (a dead thymocyte) by this macrophage is stained heavily black (positive for TUNEL), whereas small pyknotic cells adhering to the macrophage, even though they are obviously dead by their morphological features (extremely small cell size and pyknotic nuclei), are negative for TUNEL staining (*arrows*). The *bar* indicates 2 μm

It is well established that corticosteroids induce rapid and massive cell death in thymocytes (CLAMAN 1972; COWAN and SORENSON 1964). It has also been demonstrated that, in vitro, corticosteroids caused apoptosis in thymocytes; DNA fragmentation could be detected by electrophoresis as a ladder pattern (CLARKE et al. 1993; COHEN and DUKE 1984; PERANDONES et al. 1993; WYLLIE 1980). Accordingly, we next set up experiments to determine whether immediate and extensive thymocyte death induced by in vivo injection of corticosteroids is really apoptosis or not. Figure 10 shows the TUNEL staining of the mouse thymus 2h and 4h after steroid injection. Positive cells increased greatly in number, and apparently formed slightly larger clusters, all of which were colocalized with ACP positive cells (Fig. 11), again indicating that cells with DNA fragmentation were all phagocytosed. TEM showed that pyknotic cells became prominent 2h and 4h after the steroid injection (Fig. 12). TUNEL electron microscopy and TUNEL flow cytometry could not detect TUNEL-positive (free) cells even after the steroid treatment, when clusters of TUNEL-positive cells within macrophages became prominent in frozen sections of the steroid-treated thymuses (Figs. 13 and 14). All these findings strongly suggest that even in the case of steroid treatment, thymocytes die by pyknosis, not by

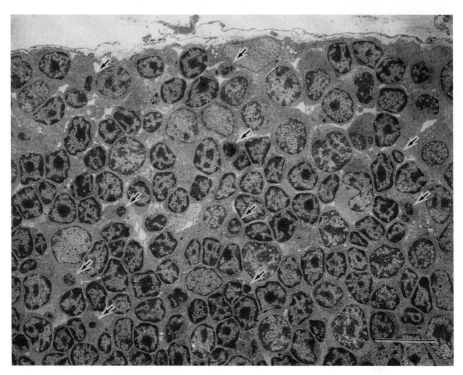

Fig. 7. A plain transmission electron micrograph of the normal mouse thymus. Small, heavily chromatin-condensed pyknotic thymocytes (*arrows*), apparently dead but not yet phagocytosed, are frequently (nearly 10% in this micrograph) observed. The *bar* indicates 7 μm

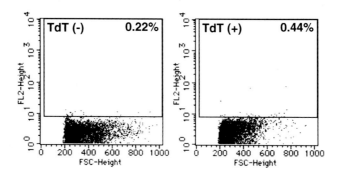

Fig. 8. TUNEL flow cytometry of the normal mouse thymocytes. TUNEL positive thymocytes are not documented by this method beyond the detection limit

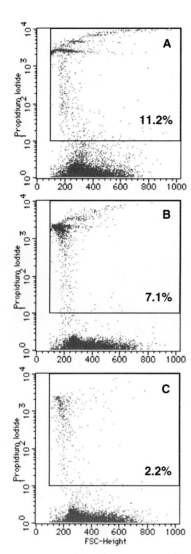

Fig. 9A–C. Flow cytometric analysis of the normal mouse thymocytes: **A** at the time of preparation of the suspension, if the centrifugation is not carried out, about 10% of small propidium iodide staining-positive dead cells are detected, consistent with the electron microscopic finding (see Fig. 7) that about 10% of dead cells are present in the normal thymus. The centrifugation (washing) is regularly performed to eliminate the dead cells for better presentation of the data; **B** two washes; **C** five washes

typical apoptosis which has to be accompanied by DNA fragmentation (UEDA and SHAH 1994; WYLLIE 1980).

Discrepancies would be pointed out. The most important point in our study is that we performed the entire experiments in vivo, not in vitro. Most studies on thymocyte death so far reported were set up in vitro (COHEN et al. 1992; COHEN and DUKE 1984; KIZAKI et al. 1989; MCCONKEY et al. 1989; PEITSCH et al. 1993). It would be desirable to carry out investigations in vivo as much as possible, especially if we find discrepancies between in vivo and in vitro studies. Second, we carried out the TUNEL method at the electron micro-

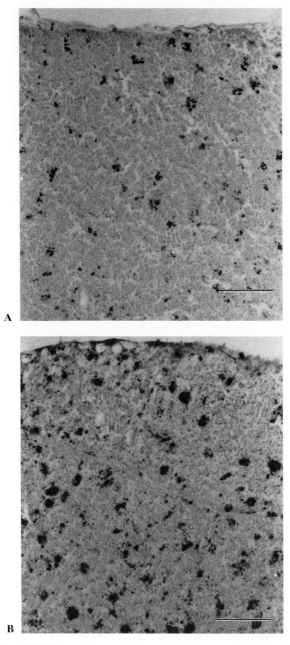

Fig. 10A,B. The TUNEL staining of the mouse thymus treated with corticosteroid (hydrocortisone sodium phosphate, 250mg/kg): **A** 2h after treatment; **B** 4h after treatment. TUNEL positive cells are significantly prominent compared to the normal thymus, and they aggregate to form clusters. The number of TUNEL positive cells increases with time, and the size of aggregation also becomes larger with time. The *bar* indicates 100μm

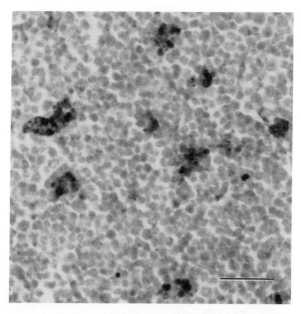

Fig. 11. The corticosteroid-treated mouse thymus simultaneously stained for TUNEL and ACP (2 h after treatment). TUNEL positive thymocytes are completely overlapping with positive cells for ACP histochemical staining. The *bar* indicates 20 µm

scopic level. It has been impossible to determine by the light microscope alone whether all pyknotic cells conventionally detectable in normal thymus are truly phagocytosed or not, and consequently, whether they are TUNEL-positive or not. To address these issues, we undertook in vivo investigation of frozen sections of the mouse thymus by the TUNEL method at an electron microscopic level. As a result, it was clearly demonstrated that, under normal condition (NAKAMURA et al. 1995) and under a condition of steroid administration (ISHII et al. 1997; NAKAMURA et al. 1997), most thymocytes die by pyknosis, not by typical apoptosis, which, by definition, must be accompanied by DNA fragmentation (PEITSCH et al. 1993; WYLLIE et al. 1980) prior to being phagocytosed by macrophages.

C. Functional and Structural Characteristics of the Germinal Center

I. Affinity Maturation of B Cells

In mammals, B lymphocytes undergo the first half of differentiation in the bone marrow to the stage of mature B lymphocyte with surface expression of IgM and IgD, but at this stage in the bone marrow they have not yet encoun-

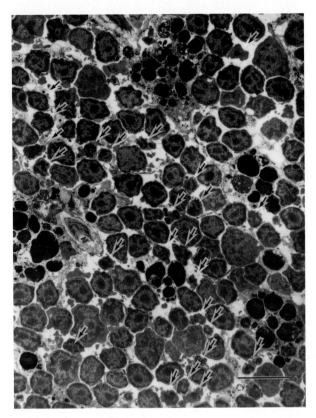

Fig. 12. A plain transmission electron micrograph of the corticosteroid-treated mouse thymus 2h after the treatment. Small, pyknotic thymocytes (*arrows*), apparently dead but not yet phagocytosed, are substantially observed. The *bar* indicates 9 μm

tered foreign antigens. They emigrate from the bone marrow thereafter to the spleen, the lymph node, the tonsil, and the Peyer's patch, where they mature into terminally differentiated functional antibody-forming B cells or memory B cells. The second half of the B cell differentiation process, after they encounter foreign antigens and with the help of T cells (COHEN 1991; JACOBSON et al. 1974), that takes place in the germinal center of the spleen, the lymph node, the tonsil, or the Peyer's patch, can be regarded as "the fine tuning process". Meanwhile, B lymphocytes, already having rearranged the immunoglobulin genes and having expressed IgM and IgD on the surface, further mutate ("hypermutate") (JACOB et al. 1991; KALLBERG et al. 1994) their immunoglobulin genes into those producing and expressing immunoglobulins with higher affinity ("affinity maturation"; also called "somatic mutation") for the antigens they have recently encountered (LIU et al. 1992; PASCUAL et al. 1994).

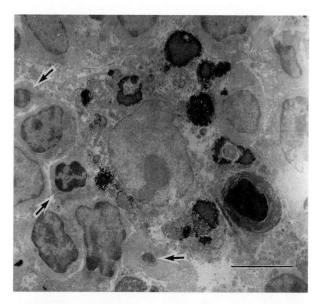

Fig. 13. An electron micrograph of the corticosteroid-treated mouse thymus stained by the TUNEL method. A macrophage ingesting abundant dead cells can be seen. Phagocytosed cells (dead thymocytes) are stained heavily black (positive for TUNEL), whereas small pyknotic apparently dead cells present in the close vicinity of the macrophage are negative for TUNEL staining (*arrows*). The *bar* indicates 3 μm

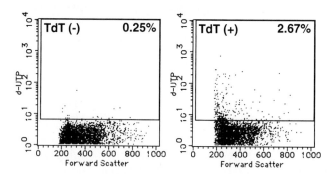

Fig. 14. TUNEL flow cytometry of the corticosteroid-treated mouse thymocytes (4 h after treatment). In contrast to the finding with the transmission microscope (prominent pyknotic dead cells in a substantial amount), only an extremely small fraction of TUNEL positive thymocytes could be detected (2%)

The germinal center is a well-organized site for the affinity maturation (Fig. 15) (LIU and ARPIN 1997; LIU et al. 1992). Conventionally, the germinal center is divided into three compartmentalized regions – the dark region, the light region, and the follicular mantle. At the dark region, "virgin" mature B

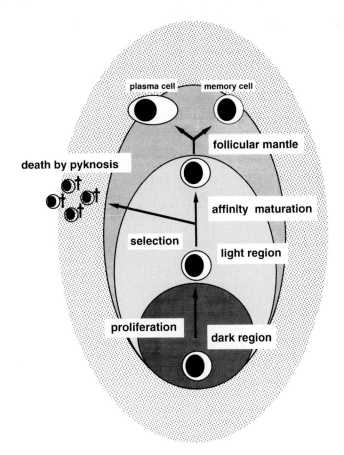

Germinal Center

Fig. 15. Diagram illustrating functional compartments of the typical germinal center in the lymph node. After the antigenic stimulation, oligoclonal B lymphocytes proliferate greatly in the dark region, then go through the selection process in the light region via the interaction with FDC, and ultimately terminally differentiate into plasma cells or memory B cells. Non-selected, abortive nonfunctional B cells are destined to die by pyknosis

cells, after they meet the antigen, proliferate; at the light region, B cells, after they cease to proliferate, go through the selection process; finally, at the follicular mantle, only positively selected B cells are allowed to differentiate further into antibody-forming cells or memory B cells. The germinal center can thus be defined as the microenvironment for antibody formation; alternatively, it could be regarded as the site for terminal differentiation for B cells.

II. B-Cell Selection

At the dark region, virgin mature B cells proliferate to make a large number of B cell clones with a diversified repertoire of B cell receptors (BCR = immunoglobulins). From the basal to the apical light region, multiclonal B cells are subjected to the selection process via the interaction with follicular dendritic cells (FDC), more exactly, through the interaction of newly formed (through the hypermutation process) BCR with the antigen adsorbed on the surface of FDC. B cells expressing BCR able to receive positive signals through the interaction with the antigen would further terminally differentiate into antibody-forming B cells or memory B cells. On the other hand, those expressing aberrant BCR unable to interact with the antigen on FDC ("abortive" nonfunctional clones) could not proceed along the differentiation pathway, ultimately resulting in cell death. Self-reactive clones might be generated during these processes, but the number should be extremely low when one assumes that the process of the somatic mutation takes place randomly.

III. B-Cell Death

As discussed above, when the selection process of B cells in the germinal center was carefully examined, the process of B-cell selection turned out to be remarkably similar to the selection process of thymocytes as described earlier in this chapter, i.e., both T cell and B cells first proliferate, undergo selection processes, and, to generate a relatively minor population of positively selected competent cells, the majority of them are left behind without any interaction with critical ligand molecules in the selecting microenvironments. Accordingly, we analyzed the cell death of the mouse germinal center with the same procedure used to analyze thymocyte death (NAKAMURA et al. 1996). We hypothesized that the major dying population of the germinal center (aberrant nonfunctional clones), almost identically to thymocytes, would die by pyknosis, not by apoptosis. Since "apoptosis" by definition requires DNA fragmentation prior to being phagocytosed, it is absolutely necessary, if the relevant cell death is postulated to be apoptosis, to demonstrate fragmented DNA in nuclei of cells apparently dead but not yet phagocytosed.

The results presented in Figs. 16 and 17 clearly revealed that TUNEL positive nuclei formed clusters in the light region, and that all TUNEL positive nuclei were surrounded by positive reaction for ACP (NAKAMURA et al. 1996), being in good accordance with thymocyte death. In the germinal center, fragmented DNA were detected only within macrophages by TUNEL electron microscopy (Fig. 18).

These observations on the cell death of the germinal center indicated that almost all dying cells in the germinal center die by pyknosis (NAKAMURA et al. 1996). They never showed typical apoptosis pattern of DNA fragmentation prior to processing of dead cells by other cells (phagocytosis). Altogether,

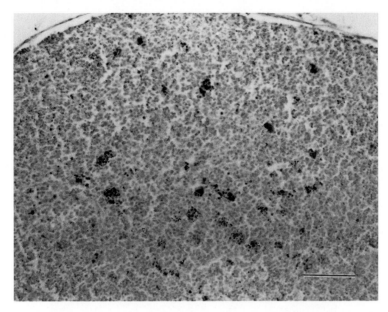

Fig. 16. TUNEL staining of the germinal center of the draining lymph node of the mouse immunized with sheep red blood cells (SRBC). Positive aggregates can be observed, mainly in the basal light region. The *bar* indicates 100 μm

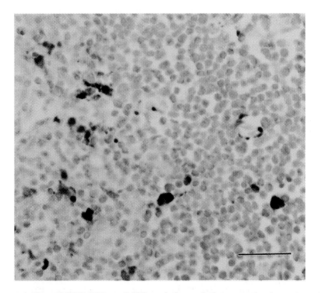

Fig. 17. Micrographs of TUNEL + ACP staining of the germinal center of the draining lymph node of the mouse immunized with SRBC. Colocalization of TUNEL positive cells with ACP positive cells are clearly demonstrated. The *bar* indicates 20 μm

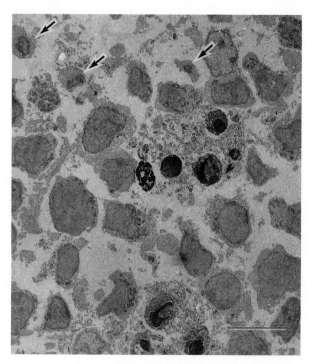

Fig. 18. Electron micrograph of the germinal center of the lymph node of the SRBC-injected mouse stained by the TUNEL method. Two macrophages ingesting dead cells can be seen. Phagocytosed cells (dead thymocytes) are stained heavily black (positive for TUNEL), whereas small pyknotic apparently dead cells (*arrows*) unphagocytosed and in the close vicinity of the macrophage, are negative for TUNEL staining. The *bar* indicates 5 μm

findings are completely identical to those obtained from the cell death analysis on thymocytes.

D. Summary

The study of thymocyte death in the thymus and B-cell death of the germinal center demonstrated common features of cell death (Fig. 19).

Both cells are destined to make a large repertoire of antigen receptors. To generate a repertoire, the first step (actually, really the first step for thymocytes and the first step for the second half of the differentiation process of B cells) for them to go through is cell division to produce a relatively large population of progenies. Simultaneously, they rearrange or mutate their receptor genes; thymocytes rearrange TCR genes, and B cells hypermutate rearranged BCR genes. Through interaction with molecules (counter receptors/ligands) expressed by stromal cells in the microenvironment, only thymocytes or B cells

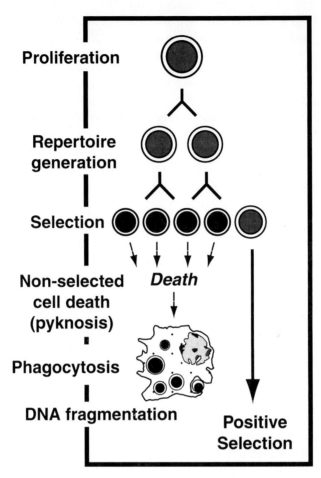

Fig. 19. Common features of cell death mechanisms of the thymus and the germinal center. In both cell death processes: first thymocytes or B lymphocytes proliferate to generate the repertoire diversity; second they are subjected to selection processes based on the receptor repertoire – during the selection processes, some are positively selected while the majority of them are left behind without receiving any stimuli for further differentiation; finally the cells, unable to interact with the surrounding microenvironment, have to advance the death program (pyknosis). Dead cells are to be phagocytosed some times later, and their nuclei do not show signs of DNA fragmentation until they are phagocytosed

with receptors capable of receiving appropriate positive signals from surrounding environment would be positively selected (FARR et al. 1985; KOSCO et al. 1992). Two types of cells which would not have been positively selected might be driven to death; self-reactive clones and aberrant nonfunctional clones. The number of cells of the latter population is far greater than that of the former, as discussed earlier. Nearly all dead cells detected both in the thymus and the germinal center are abortively generated nonfunctional cells. As demonstrated, these dead cells do not show DNA fragmentation, despite the fact that some of them display apparent morphological features of cell death, prior to phagocytosis by professional phagocytes. Thymocytes and B cells of the germinal center showed DNA fragmentation only after they are phagocytosed by macrophages, thus strongly indicating that the cell death observed with thymocytes and B cells of the germinal center could not be regarded as typical apoptosis.

E. Prospects

Recently we also examined the interdigital tissues of the limb buds of developing mice, whose PCD has been known for typical apoptosis with DNA fragmentation. To our surprise it was also found that virtually all TUNEL positive cells were phagocytosed by either surrounding mesenchymal cells of the same lineage as dead cells or professional phagocytes bearing characteristic markers for macrophages (our unpublished observations).

Findings presented in this study, together with our recent observations with interdigital tissue, strongly indicate that there might be in vivo other types of cell death than apoptosis. Thymocyte death (WYLLIE 1993), B-cell death (MANGENEY et al. 1991), and moreover cell death in the interdigital tissue (GARCIA-MARTINEZ et al. 1993), all of which have long been attributed to apoptosis, never showed DNA fragmentation when carefully examined. Therefore, although it has been widely accepted that in apoptosis the cell severs self DNA into oligonucleosomal subunits long before the dead cell is ingested by phagocytes, it is necessary either to redefine the apoptosis or to reexamine the cell death in vivo; it is essential to determine whether prior DNA fragmentation should be a diagnostic sign for apoptosis, whether cell death without DNA fragmentation (before phagocytosis) should be designated as apoptosis (in this case, it would be extremely difficult to distinguish conventional cell death pyknosis or necrosis from special case death apoptosis), or whether we should essentially change our definition of apoptosis.

Genes and enzymes have been extensively investigated in relation to apoptosis, and some of the candidates are now strongly suspected to be involved in apoptosis of various kinds of cells including mammalian cultured cells (MIURA et al. 1993; VAUX and STRASSER 1996) and those of nematodes (ELLIS and HORVITZ 1986; VAUX and STRASSER 1996). One such gene, ced-3, originally isolated in *Caenorhabditis elegans* (ELLIS and HORVITZ 1986), has been found to be homologous to the gene encoding interleukin-1β converting enzyme (ICE) of higher vertebrates (YUAN et al. 1993). Recently, these genes were collectively renamed as caspases (ALNEMRI et al. 1996; JACOBSON et al. 1997). However, while trying to accumulate evidence on these death-related genes, controversial findings have also been reported. In either case of caspase-1 (ICE) knockout mice (KUIDA et al. 1995) or caspase-3 (CPP32) knockout mice (KUIDA et al. 1996), thymocyte death was regularly observed, indicating that these enzymes, although generally considered essentially to be involved in apoptosis, do not play fundamental roles in thymocyte "apoptosis." Accordingly, we should point out that the term "apoptosis" has become inappropriate in its original definition. From the results of death-related gene-knockout mice, either the concept that these death-related genes are deeply involved in apoptosis is incorrect, or if one assumes that the hypothesis for the death-related enzymes is correct, then, one must say that thymocytes, under normal conditions, do not undergo apoptosis.

In any case, the first thing we should do is determine precisely whether or not the cell deaths in vivo in various organs or tissues really demonstrate DNA fragmentation, as one of their earliest signs prior to phagocytosis, not as later signs due to the degradation process by phagocytes.

Acknowledgements. We express our deepest gratitude to Dr R. Suzuki, Shionogi Pharmaceutical Company, for his encouragement and helpful discussions throughout the study. We greatly appreciate the expert technical assistance of Mr M. Ito and Mr Y. Suzuki. We also thank Ms K. Omori for her secretarial assistance, and Mr D. Arakawa for reviewing the manuscript. This work was in part supported by a Grant-In-Aid for Scientific Research from the Ministry of Education, Science and Culture, Japan (07407066 to TI, 09770001 to HY), and by The Funds for Comprehensive Research on Long Term Chronic Diseases from the Ministry of Health and Welfare of Japan (to TI).

References

Alnemri ES, Livingston DJ, Nicholson DW, Salvesen G, Thornberry NA, Wong WW, Yuan J (1996) Human ICE/Ced-3 protease nomenclature. Cell 87:171
Arends MJ, Wyllie AH (1991) Apoptosis: mechanism and roles in pathology. Intern Rev Exp Pathol 32:223–254
Arey LB (1974) Human histology, 4th edn. WB Saunders, Philadelphia
Barka T, Anderson PJ (1962) Histochemical methods using hexazonium pararosaniline as coupler. J Histochem Cytochem 10:741–753
Claman HN (1972) Corticosteroids and lymphoid cells. New Engl J Med 287:388–397
Clarke AR, Purdie CA, Harrison DJ, Morris RG, Bird CC, Hooper ML, Wyllie AH (1993) Thymocyte apoptosis induced by p53-dependent and independent pathways. Nature 362:849–852
Cohen GM, Sun X-M, Snowden RT, Dinsdale D, Skilleter DN (1992) Key morphological features of apoptosis may occur in the absence of internucleosomal DNA fragmentation. Biochem J 286:331–334
Cohen JJ (1991) Programmed cell death in the immune system. Adv Immunol 50:55–85
Cohen JJ, Duke RC (1984) Glucocorticoid activation of a calcium dependent endonuclease in thymocyte nuclei leads to cell death. J Immunol 132:38–42
Cossarizza A, Kalashnikova G, Grassilli E, Chiappelli F, Salvioli S, Capri M, Barbieri D, Troiano L, Monti D, Franceschi C (1994) Mitochondrial modifications during rat thymocyte apoptosis: a study at the single cell level. Exp Cell Res 214:323–330
Cowan WK, Sorenson GD (1964) Electron microscopic observations of acute thymic involution produced by hydrocortisone. Lab Invest 13:353–370

Ellis HM, Horvitz R (1986) Genetic control of programmed cell death in the nematode *C. elegans*. Cell 44:817–829

Farr AG, Anderson SK, Marrack P, Kappler J (1985) Expression of antigen-specific, major histocompatibility complex-restricted receptors by cortical and medullary thymocytes in situ. Cell 43:543–550

Garcia-Martinez V, Macias D, Ganan Y, Garcia-Lobo JM, Francia MV, Fernandez-Teran MA, Hurle JM (1993) Internucleosomal DNA fragmentation and programmed cell death (apoptosis) in the interdigital tissue of the developing chick leg bud. J Cell Sci 106:201–208

Gavrieli Y, Sherman Y, Ben-Sasson SA (1992) Identification of programmed cell death in situ via specific labeling of nuclear DNA fragmentation. J Cell Biol 119:493–501

Goldstein, Ojcius DM, Young JD-E (1991) Cell death mechanisms of the immune system. Immunol Rev 121:29–65

Ishii T, Nakamura M, Yagi H, Soga H, Kayaba S, Gotoh T, Satomi S, Itoh T (1997) Glucocorticoid-induced thymocyte death in the murine thymus: the effect at later stages. Arch Histol Cytol 60:65–78

Jacob J, Kelsoe G, Rajewsky K, Weiss U (1991) Intraclonal generation of antibody mutants in germinal centres. Nature 354:389–392

Jacobson EB, Caporale LH, Thorbecke G (1974) Effect of thymus cell injections on germinal center formation in lymphoid tissues of nude (thymusless) mice. Cell Immunol 13:416–430

Jacobson MD, Weil M, Raff MC (1997) Programmed cell death in animal development. Cell 88:347–354

Kallberg E, Gray D, Teanderson T (1994) Kinetics of somatic mutation in lymph node germinal centres. Scand J Immunol 40:469–480

Kishimoto H, Surh CD, Sprent J (1995) Upregulation of surface markers on dying thymocytes. J Exp Med 181:649–655

Kizaki H, Tadakuma T, Odaka C, Muramatsu J, Ishimura Y (1989) Activation of a suicide process of thymocytes through DNA fragmentation by calcium ionophores and phorbol esters. J Immunol 143:1790–1794

Kosco MH, Pflugfelder E, Gray D (1992) Follicular dendritic cell-dependent adhesion and proliferation of B cells in vitro. J Immunol 148:2331–2339

Kuida K, Lippke JA, Ku G, Harding MW, Livingston DJ, Su MS-S, Flavell RA (1995) Altered Cytokine Export and Apoptosis in Mice Deficient in Interleukin-1β Converting Enzyme. Science 267:2000–2003

Kuida K, Zheng TS, Na S, Kuan C, Yang D, Karasuyama H, Rakic P, Flavell RA (1996) Decreased apoptosis in the brain and premature lethality in CPP32-deficient mice. Nature 384:368–372

Liu Y-J, Arpin C (1997) Germinal center development. Immunol Rev 156:111–126

Liu Y-J, Johnson GD, Gordon J, MacLennan ICM (1992) Germinal centres in T-cell-dependent antibody responses. Immunology Today 13:17–21

Mangeney M, Richard Y, Coulaud D, Tursz T, Wiels J (1991) CD77:an antigen of germinal center B cells entering apoptosis. Eur J Immunol 21:1131–1140

McConkey DJ, Hartzell P, Amador-Perez JF, Orrenius S, Jondal M (1989) Calcium-dependent killing of immature thymocytes by stimulation via the CD3/T cell receptor complex. J Immunol 143:1801–1806

Miura M, Zhu H, Rotello R, Hartwieg EA, Yuan J (1993) Induction of apoptosis in fibroblasts by IL-1β-converting enzyme, a mammalian homolog of the *C. elegans* cell death gene ced-3. Cell 75:653–660

Nakamura M, Yagi H, Ishii T, Kayaba S, Soga H, Gotoh T, Ohtsu S, Ogata M, Itoh T (1997) DNA fragmentation is not the primary event in glucocorticoid-induced thymocyte death in vivo. Eur J Immunol 27:999–1004

Nakamura M, Yagi H, Kayaba S, Ishii T, Gotoh T, Ohtsu S, Itoh T (1996) Death of germinal center B cells without DNA fragmentation. Eur J Immunol 26:1211–1216

Nakamura M, Yagi H, Kayaba S, Ishii T, Ohtsu S, Gotoh T, Itoh T (1995) Most thymocytes die in the absence of DNA fragmentation. Arch Histol Cytol 58:249–256

Ojeda F, Guarda MI, Maldonado C, Folch H (1992) A flow-cytometric method to study DNA fragmentation. J Immunol Methods 152:171–176

Pascual V, Liu Y-J, Magalski A, de Bouteiller O, Banchereau J, Capra JD (1994) Analysis of Somatic Mutation in Five B Cell Subsets of Human Tonsil. J Exp Med 180:329–339

Peitsch MC, Polzar B, Stephan H, Crompton T, MacDonald HR, Mannherz HG, Tschopp J (1993) Characterization of the endogenous deoxyribonuclease involved in nuclear DNA degradation during apoptosis (programmed cell death). EMBO J 12:371–377

Perandones CE, Illera VA, Peckham D, Stunz LL, Ashman RF (1993) Regulation of Apoptosis In Vitro in Mature Murine Spleen T Cells. J Immunol 151:3521–3529

Raff M (1992) Social controls on cell survival and cell death. Nature 356:397–400

Ritter MA, Crispe IN (1994) The thymus. IRL Press, Oxford

Surh CD, Sprent J (1994) T-cell apoptosis detected in situ during positive and negative selection in the thymus. Nature 372:100–103

Ueda N, Shah SV (1994) Apoptosis. J Lab Clin Med 124:169–177

Vaux DL, Strasser A (1996) The molecular biology of apoptosis. Proc Natl Acad Sci USA 93:2239–2244

von Boehmer H (1988) The developmental biology of T lymphocytes. Ann Rev Immunol 6:309–326

von Boehmer H (1992) Thymic selection: a matter of life and death. Immunology Today 13:454–458

Walker PR, Weaver VM, Lach B, LeBlanc J, Sikorska M (1994) Endonuclease activities associated with high molecular weight and internucleosomal DNA fragmentation in apoptosis. Exp Cell Res 213:100–106

Williams GT (1994) Apoptosis in the immune system. J Pathol 173:1–4

Wyllie AH (1980) Glucocorticoid-induced thymocyte apoptosis is associated with endogenous endonuclease activation. Nature 284:555–556

Wyllie AH (1993) Apoptosis (The 1992 Frank Rose Memorial Lecture). Br J Cancer 67:205–208

Wyllie AH, Kerr JFR, Currie AR (1980) Cell death: the significance of apoptosis. Int Roy Cytol 68:251–306

Yuan J, Shaham S, Ledoux S, Ellis HM, Horvitz HR (1993) The *C. elegans* cell death gene ced-3 encodes a protein similar to mammalian interleukin-1 β-converting enzyme. Cell 75:641–652

CHAPTER 16
Antigen Receptor-Induced Death of Mature B Lymphocytes

T. DEFRANCE, M. BERARD, and M. CASAMAYOR-PALLEJA

A. Introduction

It is established that the antigen (Ag) receptors on lymphocytes can elicit either positive or negative responses. Elucidation of the molecular parameters which govern this bi-potentiality of the Ag receptor is obviously of critical importance for the development of future immunomodulatory-based therapies in several diseases including autoimmune disorders and cancer. The concept which has long prevailed is that the switch of the B cell Ag receptor (BCR) from a negative to a positive signaling function is developmentally regulated and irreversible. This notion was essentially based on the contrasting responses elicited by Ag in immature (anergy or deletion) vs mature B lymphocytes (activation and differentiation). This rather Manichean view of the BCR signaling has been challenged by several lines of evidence over the years, some originating from early studies, others coming from more recent work in which the fate of monoclonal B cells exposed to various forms of Ag at different stages of their development has been explored (see GOODNOW et al. 1995 for review).

To start with, the assumption that the BCR is definitely wired to a positive (i.e., stimulatory) signaling pathway in mature B cells has always been at odds with the multiple reports describing that triggering of the Ag receptor on certain neoplastic mature B cells could induce their apoptosis. This point was first demonstrated by GREGORY et al. (1991) who reported that group I Burkitt lymphoma cell lines, characterized by their expression of a very restricted set of the EBV latent proteins, are susceptible to BCR-induced death. The ability of neoplastic B cells to undergo apoptosis upon ligation of their surface immunoglobulins (sIgs) was later extended to B cell lines derived from follicular lymphoma patients (ERAY et al. 1994).

Next, it has long been known that most foreign Ag have the potentiality to elicit either a positive or a negative response depending on their molecular form, dose, and route and duration of administration. This is well exemplified by the fact that various protein Ag induce either a tolerogenic or an immunogenic response when administered under a soluble deaggregated or aggregated form, respectively (MITCHISON 1964; DRESSER and MITCHISON 1968; CHILLER et al. 1971).

Finally, the recent studies of Goodnow and colleagues have paved the way for understanding how a single receptor can bring about both positive and negative responses in mature B cells. These authors have modeled the fate of B cells which encounter Ag during their early development in the bone marrow by using double transgenic mice bearing both a transgene-encoded Ag and a transgene-encoded BCR of the relevant specificity. This seminal work (see GOODNOW et al. 1995; HEALY and GOODNOW 1998 for review) has highlighted that multiple intrinsic and external parameters collectively decide whether the outcome of BCR triggering will be selection, deletion, or anergy, thus emphasizing the extraordinary plasticity of the Ag receptor. This model convincingly demonstrated that the BCR signaling, in addition to being subjected to developmental regulation, is influenced by external factors related to the physical properties of the Ag and to components of the innate and acquired immunity.

In the present review, we will first document the importance of Ag-driven apoptosis for peripheral B cell tolerance. We will then explore the possibility that the Ag receptor switches from a positive to a negative signaling function during the course of a normal antibody (Ab) response to ensure that clonal expansion of Ag-specific B cells does not lead to hyperplasia. We will discuss the importance of Ag-induced apoptosis for the regulation of homeostasis in the mature B cell compartment and how this phenomenon relates to the concept of activation-induced cell death (AICD) (see GREEN and SCOTT 1994; RUSSEL 1995 for review). Finally, we will consider some of the elements which bear weight on the regulation and execution of the apoptotic program initiated by the Ag receptors in mature B cells.

B. Antigen Receptor-Induced Death and Maintenance of Peripheral B Cell Tolerance

I. BCR-Induced Apoptosis of Germinal Center B Cells

The notion that the Ag receptor can exert an inhibitory function in B cells has been accepted for more than twenty years and originates from a series of early experiments describing that the responses of immature B cells or their tumoral counterparts are inhibited by surrogate Ags (NOSSAL and PIKE 1975; METCALF and KLINMAN 1976; CAMBIER et al. 1976; NOSSAL et al. 1979; KLINMAN et al. 1981; BOYD and SCHRADER 1981). Diversity of the B cell repertoire is primarily generated through the rearrangement of germline gene segments which combine to compose DNA sequences encoding both heavy and light chain variable regions. Given the stochastic nature of this process, it might equally create Ab specificities against foreign and self components. The negative regulatory function of the BCR towards developing B cells in the bone marrow serves the purpose of eliminating these self reactive B cells and maintaining central B cell tolerance. Nevertheless, diversification of the B cell repertoire also occurs in an Ag-driven fashion in the course of T-dependent Ab responses through the random introduction of point mutations in the V_H and V_L genes. This hypermutation process which takes place in the germinal centers (GC) of secondary B cell follicles (JACOB et al. 1991; BEREK et al. 1991) can thus potentially induce the emergence of self-reactive B cells (RAY et al. 1996). Hence additional safeguard mechanisms are necessary to ensure B cell tolerance in the periphery.

Two groups have been particularly active in examining how self-reactive B cells, generated by the hypersomatic mutations process of V genes, are eliminated from the post-immune repertoire. LINTON et al. (1991) cotransferred memory B cell precursors, defined by their low expression of HSA/CD24 (LINTON et al. 1989), with hemocyanin-primed T helper cells into irradiated mice. A primary Ab response against the hapten DNP was then generated by exposing splenic fragments of the recipients to DNP-hemocyanin in vitro. These authors showed that addition of Ag coupled to a non-cognate carrier subsequently to a primary and secondary antigenic stimulation carried in a cognate system precluded their differentiation into Ab-secreting cells. This constituted the first demonstration that B cells can also be tolerized during the course of an ongoing immune response. However, this study could not define whether this "second window" of tolerance was associated or not with a peculiar stage of B cell development. This issue was clarified by PULENDRAN et al. (1994, 1995a), who showed that the tolerance to NP conjugates induced in mice by injection of a soluble deaggregated form of NP-human serum albumin was associated with a drastic reduction of the GC development. Finally, three concordant reports (PULENDRAN et al. 1995b; HAN et al. 1995; SHOKAT and GOODNOW 1995) revealed that apoptosis was underlying the tolerizing mechanism operating in the GC and thereby definitely established that the BCR can transduce negative signals in mature B lymphocytes. Basically, these observations documented that a secondary and massive injection of soluble Ag at the peak of the primary response induces dramatic B cell death in the GC. Five important features of the Ag-driven B cell apoptosis in the GC were defined:

1. It is unrelated to the carrier part of Ag and occurs both when the secondary antigenic stimulation is made with the carrier used for the primary immunization and with cross-reacting Ag lacking T cell recognition epitopes. It is therefore the direct consequence of B cell Ag receptor occupancy.
2. It is strictly restricted to the GC as the secondary injection of soluble Ag does not perturb the development of Ag-specific B cells in the extrafollicular foci.
3. It targets the high-affinity mutant B cell clones. This point was demonstrated by the underexpression of the VDJ rearrangement conferring high affinity to NP in GC B cells which survived prolonged exposure to NP-protein soluble conjugates.
4. It is independent from the Fas signaling pathway since it can be reproduced in "lpr" mice.
5. It is at least partially reversed by the transgene-encoded expression of Bcl-2.

Since GC constitute the only anatomical site where diversification and selection of the post-immune repertoire occur, it was postulated that the Ag-driven apoptotic pathway is instrumental in eliminating autoreactive B cell mutants generated incidentally by the somatic hypermutation process in GC. The unexpected susceptibility of GC B cells to BCR-mediated killing was confirmed by two in vitro studies conducted on human B cells which showed that a surrogate Ag could induce apoptosis of isolated GC B cells provided that they had received an activation signal through CD40 (GALIBERT et al. 1996; BILLIAN et al. 1997).

Collectively these findings led to the hypothesis that the central B cell tolerance mechanism which allows for deletion of self-reactive immature B cells in the bone marrow is reactivated in the GC (KELSOE 1996; PULENDRAN et al. 1997). A series of recent experimental data at first seemed to support the assumption that GC and immature B cells present functional similarities. First, it was reported that the products of the Rag 1 and Rag 2 genes are reexpressed in the GC (HAN et al. 1996). Second, it was demonstrated that the RAG proteins expressed by GC B cells are enzymatically active, thus implying that receptor editing operates in GC B cells (HAN et al. 1997; PAPAVASILIOU et al. 1997; MEFFRE et al. 1998). Third, other molecules the expression of which was previously thought to be restricted to early B cell developmental stages, namely TdT, V-preB, and the λ-like component of the human pre-B cell receptor, were also found on human GC B cells (MEFFRE et al. 1998). However the recent experiments of HERTZ et al. (1998) indicated that the comparison between GC and immature B cells has some limitation. As convincingly demonstrated by these authors, receptor editing in the GC opposed to the bone marrow is driven by low-affinity binding Ag and suppressed by high-affinity ligands. Hence, receptor editing in the GC, instead of maintaining self-tolerance can rather be envisaged as the last opportunity for low affinity mutants to improve their Ag-binding capacities.

The mechanism whereby soluble Ag drives B cell apoptosis in the GC is not entirely clear. We will consider two hypotheses.

The first contends that flooding established GC with massive doses of soluble Ag hampers the delivery of survival signals provided through physical interactions between B cells and follicular dendritic cells (FDCs). In this model, soluble Ag would merely operate by passively subtracting B cells from the influence of FDCs through its competition with the FDC-bound immune complexes. Alternatively, soluble Ag could fail to trigger efficiently the BCR and promote B cell survival because of its low degree of reticulation. Whatever the option, this hypothesis implies that elimination of GC B cells driven by injection of soluble Ag only models the fate of autoreactive mutants reacting with soluble proteins and therefore not associated with the FDC network. If this assumption is correct, self-reactive B cells would undergo apoptosis in the GC as the result of deprivation of FDC-derived trophic factors. This situation is schematically represented by the "passive deletion model" proposed in Fig. 1.

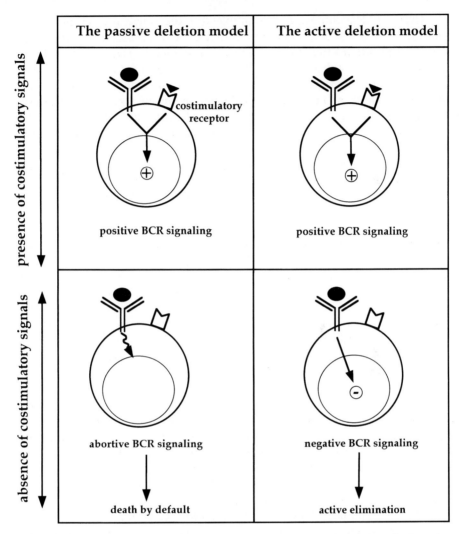

Fig. 1. Putative models for the Ag-driven apoptosis of B cells in the germinal center. The passive deletion model postulates that costimulatory signals, possibly provided by the FDCs, complement those delivered via the BCR to ensure B cell survival. These additional stimuli could modulate the strength and quality of the signals transmitted by the BCR. They will not be provided if B cells recognize an Ag which is not associated with the FDC network as it could be the case for a soluble self antigen. This model infers that GC B cells are committed to die unless they receive the appropriate antiapoptotic signals. In this case the BCR does not direct apoptosis of auto reactive B cells and their demise is comparable to that induced by trophic factor deprivation. The active deletion model postulates that the BCR is constitutively connected to an apoptotic pathway in GC B cells and that undefined costimulatory signals can overcome its negative regulatory function. Self-reactive B cells would undergo apoptosis as the consequence of BCR cross-linking because they could not benefit from the protection afforded by the putative costimulatory receptors

The second hypothesis is compatible with the concept developed above that the BCR is wired to a death pathway in GC B cells as it is in immature B cells. It postulates that soluble Ag is actively driving elimination of GC B cells by delivering a death signal. On the one hand, an Ag-mediated deletion mechanism appears superfluous, as the lack of appropriate help from FDCs or T cells could be sufficient to secure censoring of self-reactive B cells in the GC. On the other hand, it might be necessary to eradicate fully the danger of autoAb production due to bystander B cell activation. This cannot be ruled out since most of the T cell molecules exerting a B cell stimulatory function, including the CD40 ligand (CD40L), can be produced in a soluble form.

This hypothesis naturally raises the vexing question of how engagement of the same receptor can exert two opposite effects, that is to promote survival of high affinity mutants and induce apoptosis of autoreactive mutants. In a recent review dealing with this issue, HEALY and GOODNOW (1998) propose three external key parameters which might influence the nature of the response elicited by ligation of the BCR: (1) the physical properties of the Ag; (2) the duration of the antigenic stimulation; and (3) the association of Ag with costimuli. It cannot be completely excluded that self-components in the GC may structurally differ from exogenous Ag because they fail to reproduce the optimal spacing, organization, and reticulation of foreign Ag imposed by their immobilization on FDCs. However, there is no strong evidence to support the assumption that self and foreign Ag should necessarily differ by their quantity or avidity. The duration of stimulation is irrelevant to the present question of negative and positive selection of B cells in the GC. The third possibility, which is that Ag promotes death of GC B cells unless they receive anti-apoptotic signals from their environment (schematically represented by the "active deletion model" in Fig. 1), deserves further examination. In this model, maintenance of B cell tolerance in the GC would require that these costimulatory signals remain inaccessible to self-reactive B cells. The nature of these putative ancillary signals remains elusive to date. Adhesion molecules such as ICAM-1 and VCAM-1 on the FDCs can be considered since blocking these adhesive interactions inhibit the protective effect exerted by FDCs on the spontaneous apoptosis of GC B cells in vitro (KOOPMAN et al. 1994). However co-signals of that kind would spare the self-reactive mutants which recognize an Ag associated with FDCs. There is also some evidence that T cell-derived signals could fulfill this costimulatory function since activated T cells and IL4 have been demonstrated to antagonize the BCR-induced apoptosis of CD40-activated human GC B cells in vitro (GALIBERT et al. 1996; BILLIAN et al. 1997). Finally, the B cell co-receptor complex CD19/CD21/TAPA-1 stands as an interesting candidate since: (1) it includes a receptor for activated products of complement C3 (CD21); (2) it amplifies B cell responses at low Ag concentrations (for review, see TEDDER et al. 1997); and (3) foreign but not self-Ag are expected to be associated with the C3d component of complement and thus to engage this complex. This hypothesis will be discussed further in the following section.

Although the observations recalled above make a strong case for the susceptibility of GC B cells to Ag-induced apoptosis, it remains that the development of Ag-specific GC is also impeded when the tolerogenic form of Ag (i.e., soluble deaggregated) is administered 4–7 days before immunization (KARVELAS and NOSSAL 1992). This implies that the tolerizing treatment inhibits the production of high affinity Abs in at least two different ways – abortion of an established GC reaction and impairment of the process leading to the development of GC. Whether Ag-driven B cell apoptosis is also responsible for the lack of GC in the latter situation is not known. Strikingly, extrafollicular foci are preserved whatever the time schedule of injection of toleragen, indicating that the extrafollicular pathway is less sensitive to tolerization than the GC pathway (PULENDRAN et al. 1995a, b). Whether or not this observation reflects the fact that tolerization susceptibility in the mature B cell compartment is developmentally regulated remains to be determined.

II. BCR-Induced Apoptosis of Virgin and Memory B Cells

Susceptibility to Ag-driven apoptosis should not be regarded as a unique property of GC B cells inasmuch as it is also clearly documented that Ag can exert a negative regulatory function at other stages of B cell development. The first evidence that virgin (IgD^+/IgM^+) B cells can be tolerized by anti-Ig Abs was provided by studies describing that crosslinking of sIgM on mature B cells can inhibit their subsequent proliferation and Ab secretion in response to a variety of stimulatory factors (KEARNEY et al. 1978; MELCHERS et al. 1980; ISAKSON et al. 1980; MARUYAMA et al. 1985). It was next demonstrated that the tolerizing effect of anti-Ig Abs on mature B cells is independent of the negative regulatory pathway coupled to the Fcγ receptors as it can be obtained with F(ab')2 fragments of anti-Ig Abs (GAUR et al. 1993). At that time, the interpretation of these findings was that triggering of the BCR in the absence of T cell help either exerts a cytostatic effect or prevents the terminal maturation of B cells into Ig secreting cells. PARRY et al. (1994a,b) were the first to demonstrate that the negative outcome of extensive BCR ligation on murine virgin B cell responses can also be correlated with their entry into apoptosis. It was at first believed that the capacity to transduce an inhibitory signal to virgin B cells was not equally shared by IgD and IgM molecules, and that only the latter isotype could exert a tolerizing effect (KIM et al. 1992; GAUR et al. 1993). However this notion was challenged by the work of PARRY et al. (1994a) showing that anti-IgM and anti-IgD Abs equally promote apoptosis of murine virgin B cells when they are used in a highly polymerized form (immobilized on plastic or coupled to a biotin/avidin system).

Resting memory B cells can also be tolerized. This point was first demonstrated by a study of JOHNSON and JEMMERSON (1992) using a model in which memory B cells were recovered 17 days after immunization of mice with cytochrome c (cyt c) conjugated to ovalbumin and transferred into hemocyanin (Hy)-primed irradiated recipients. Memory B cells were tested for their ability to mount a recall Ab response to Hy-cyt c using the splenic focus assay. Exposition of resting memory B cells, isolated on the basis of their high density, to a tolerogenic polymerized form of cyt c before adoptive transfer strongly antagonized their subsequent response to a challenge injection made with the immunogenic form of the Ag. Efficient tolerization of memory B cells was prevented in three instances: (1) when the concentrations of cyt c polymer were lowered; (2) when monovalent cyt c was substituted for the polyvalent form of cyt c during the tolerizing phase; and (3) when T cells were artificially activated by anti-CD3 Abs. Consequently, this report confirmed that both the physical parameters linked to Ag itself (valency, concentration) and the availability of cognate T cell help strongly influence the nature of the response elicited by ligation of the BCR on memory B cells. The study described above did not address the question of the molecular mechanism underlying tolerization of resting memory B cells induced by a multivalent form of Ag. We have recently demonstrated (BERARD et al. 1999) that surrogate Ag can promote apoptosis of isolated human memory B cells in vitro. However, our experimental model differed from that described above on two points. Firstly, prior activation of human memory B cells with anti-Ig Abs or CD40L was mandatory to render them susceptible to BCR-mediated apoptosis. Secondly, extensive clustering of the Ag receptors was not required to deliver a death signal to activated human memory B cells. The implications of these contrasting observations are discussed in the following two sections.

C. Antigen Receptor-Induced Death and Homeostatic Regulation of the Mature B Cell Compartment

LIU et al. (1989) first reported that anti-Ig and anti-CD40 Abs deliver short-term and long-term survival signals to freshly isolated human GC B cells, respectively. This observation constituted the basis for a consensus model for selection of the post-immune repertoire in the GC. This model postulates that selection of high affinity centrocytes is secured first through their binding to the immune complexes on FDCs followed by cognate interaction with helper T cells in the light zone of the GC. However, anti-Ig Abs provoke a completely opposite cellular response when they are provided to GC B cells which have been stimulated by CD40L beforehand inasmuch as they induce their apoptosis (GALIBERT et al. 1996; BILLIAN et al. 1997). A 24h preculture of GC B cells with CD40L was found to be sufficient to modify the outcome of BCR signaling on GC B cells (BILLLIAN et al. 1997). These data thus evoke the possibility of an interplay between the CD40 and the BCR signaling pathways which could allow the latter to switch from a survival to a pro-apoptotic function. Nevertheless, the relation between these in vitro findings and the concept proposed by Nossal, Goodnow, and Kelsoe of an Ag-driven negative selection process operating in the GC, is not entirely clear. The in vitro data imply that Ag-mediated elimination of self-reactive B cells could only occur after GC B cells have received a CD40 signal. If one assumes that T cells are the principal source of CD40-L in the GC, then it follows that the censoring mechanism for B cell autoreactivity would operate after, or concurrently with, the delivery of T cell help. Such a hypothesis cannot be definitely ruled out but recent observations made in the laboratory (BERARD et al. 1999) question the assumption that susceptibility to BCR-mediated killing is a unique feature of activated GC B cells. As mentioned previously, we found that CD40 stimulation or triggering of the BCR can also "prime" human memory B cells and to a lesser degree virgin B cells for subsequent BCR-mediated apoptosis. Therefore we believe that, rather than being the strict illustration of a negative selection process, the susceptibility of CD40-activated B cells to BCR-induced apoptosis reflects a more generalized behavior applicable to any B cell developmental stage. Based on our in vitro findings, we propose that at least two intrinsic properties of B cells can regulate the outcome of BCR signaling – their maturational stage and their activation status.

The notion that T lymphocytes can be sensitized by Ag or another primary activation stimulus to undergo apoptosis upon rechallenge of the Ag receptor has received a great deal of support and has led to definition of the concept of AICD (see GREEN and SCOTT 1994; RUSSEL 1995 for review). The term AICD was originally coined after the description of the induction of death promoted by TCR agonists in T cell hybridomas (ASHWELL et al. 1987; UCKER et al. 1989; SHI et al. 1990). One of the most classical illustrations of this phenomenon is the biphasic development of the T cell response in mice injected with a bacterial super Ag (JONES et al. 1990; KAWABE and OCHI 1991; GONZALO et al. 1992). In this experimental model, T cells bearing the appropriate TCR (i.e., the particular $V\beta$ recognized by the super Ag) are expanded in the early phase of the response and are next deleted as the result of their apoptotic death. It is now admitted that repeated exposure to any Ag can induce death of both $CD4^+$ and $CD8^+$ T cells. Ag-induced apoptosis of activated T cells is considered to be a crucial feedback regulatory mechanism necessary for the immune system to limit the immune response within strict boundaries and prevent establishment of a pre-neoplastic stage. As such, AICD is instrumental in preserving homeostasis of the mature T cell compartment. There is compelling evidence that the death pathway activated by rechallenge of the Ag receptor on cycling T cells is not directly connected to the TCR but involves the coordinated induction and triggering of the death receptors Fas (DHEIN et al. 1995; BRUNNER et al. 1995; JU et al. 1995; ALDERSON et al. 1995) and TNF receptor type 2 (ZHENG et al. 1995). AICD may result from both cell contact-dependent and independent mechanisms. In the first situation neighboring T cells are killed by armed T cells expressing one of the death ligands (Fas-ligand/Fas-L or TNF). In the second situation which can be seen as cell suicide, the soluble form of the death ligand is produced and consumed by the same cell.

The BCR-induced apoptosis of CD40-activated human B cells therefore presents striking similarities with the phenomenon of AICD inasmuch as it is induced by surrogate Ag and is critically dependent on the activation status of the cells. We have elements suggesting that the susceptibility of mature B cell subsets to AICD is correlated with their cycling ability (BERARD et al. 1999, in press) and thus with the amplitude of the response they can generate as it has been previously demonstrated for T cells (BOEHME and LENARDO 1993; ZHU and ANASETTI 1995). In this context, the fact that activated GC and memory B cells are more prone to AICD than virgin B cells is coherent with the high proliferative potential of the former two B cell subsets. Accordingly, the feedback down-regulatory control exerted by Ag should apply with an increased force during the GC reaction and in the course of secondary Ab responses. This argues for the notion that AICD preserves homeostasis of the mature B cell compartment by preventing overexpansion of B cells when they are exposed to repeated or continuous antigenic stimulation. What function can we ascribe to AICD during GC development? One element of response comes from the observation that IL-4 protects CD40-activated human GC B cells from apoptosis induced by anti-Ig Abs (GALIBERT et al. 1996; BILLIAN et al. 1997), thus suggesting that AICD might be impeded if T cell help is available. As reported by the group of MacLennan, the numbers of Ag-specific T cells in the GC rise during approximately the first 10 days after immunization, reach a plateau, then fall by day 20 as the GC reaction vanishes (GULBRANSON-JUDGE and MACLENNAN 1996). Since the Ag stocks immobilized on the FDC are unlikely to be consumed during the GC reaction, it is conceivable that, at a certain stage of their development in the GC, B cells are exposed to Ag while T cell help gradually becomes limiting. We propose that this configuration might favor activation-induced death eventually causing the extinction of the GC reaction. The concept of B cell AICD could constitute the basis for an alternate interpretation of the massive B cell apoptosis in GC induced by administration of high doses of soluble Ag. Following this line of reasoning, it can be speculated that flooding established GC with soluble Ag might artificially render Ag accessible to a much larger fraction of the GC population than in the normal physiological situation. Under these experimental conditions, B cells which BCR has engaged might simply outnumber T cells and thus fail to receive efficient protection from AICD.

D. Positive and Negative Signaling Through the BCR

I. Biochemical Events Associated with the Alternative BCR Signaling Pathways

In the present section, we will only provide an overall perspective of the parameters which influence the nature of the response elicited by triggering of the BCR since this issue has been extensively discussed elsewhere (GOODNOW et al. 1995; HEALY and GOODNOW 1998). Basically, two distinct experimental approaches have been used to explore the biochemical modifications associated with the transduction of a negative signal through the BCR. As we will see below, none of these models truly meet the criteria required for elucidation of the transduction pathway evoked during Ag receptor-induced apoptosis of mature B cells. Proper dissection of the opposing pathways connected to the BCR still awaits the availability of an in vitro model in which B cells could be induced to mount either a positive or a negative (apoptotic) response by exposure to different forms of a BCR agonist. However, they provide important clues on the set of downstream effector molecules susceptible to constitute regulatory points for the generation of differential signaling responses.

In the first model, the murine immature B lymphoma cell line WEHI-231, which can only respond to BCR ligation by undergoing apoptosis, was used to analyze the signaling cascade coupled to the Ag receptor when it is wired to a death pathway. These studies demonstrated that certain proximal non-receptor-type protein-tyrosine kinase such as blk (YAO and SCOTT 1993) or some of their substrates such as the HS1 protein (YAMANASHI et al. 1997), are necessary for the apoptotic response of WEHI-231 to anti-IgM Abs. Moreover, disruption by gene targeting of the genes encoding Syk, Lyn (TAKATA et al. 1994) or phospholipase C $\gamma 2$ (TAKATA et al. 1995) has demonstrated that these early elements of the BCR signaling pathway are crucial for BCR-induced apoptosis in a chicken B cell line model. However, as the second messengers described above are equally recruited during positive and negative responses elicited by engagement of the BCR, they cannot account for the decision of the cells to engage either the activation or apoptosis pathway. So far, the only biochemical event which would be specifically correlated with the apoptotic outcome of BCR triggering in WEHI-231 is the activation of a sphingomyelinase and subsequent production of ceramide (GOTTSCHALK et al. 1995; WIESNER et al. 1997). Extrapolation of these findings to normal mature B lymphocytes remains to be done with caution since WEHI-231 is a transformed immature B cell line.

The second model, developed by the group of Goodnow is based on a comparative study of the biochemical events associated with BCR signaling in naive and tolerant B cells. These experiments have provided seminal information on the biochemical basis of B cell anergy but it should be stressed that they may not model exactly what is happening when Ag evokes a death signal in mature B cells. The experimental strategy used by these authors relies on the use of two types of transgenic mice carrying either a HEL-specific transgene-encoded BCR, or both the BCR anti-HEL transgene and a transgene encoding a soluble form of HEL. Naive anti-HEL B cells have never encountered Ag during their early development. Their response to immunization with HEL is representative of positive BCR signaling. Tolerant B cells have been chronically exposed to a non-deletional (i.e., soluble) form of HEL during the preimmune phase of their development. Their response to HEL immunization is considered to be exclusively negative. These surveys have revealed that both the quality and the quantity of the second messengers recruited by the BCR have an impact on the nature of the subsequent cellular response. The differences in signal quality following acute antigenic stimulation of naive and tolerant B cells have been documented in the study of HEALY et al. (1997). They can be summarized in three points: (1) the amplitude of the calcium response is high in naive B cells, low in tolerant B cells; (2) the JNK MAP kinase is activated in naive but not in tolerant B cells; and (3) NF-κB is activated in naive but not in tolerant B cells.

The importance of the "quantity" or strength of the antigenic signal for determining the outcome of BCR triggering is best exemplified by the study conducted with the "moth-eaten" mutant mice, deficient in the protein tyrosine phosphatase SHP-1, by CYSTER and GOODNOW (1995). SHP-1 deficiency was found to exaggerate the intracellular calcium elevation consecutive to Ag binding and to convert the "anergizing" signal that soluble HEL provides to developing monoclonal anti-HEL B cells into an apoptotic signal. Interestingly, it was recently demonstrated that apoptosis is responsible for the growth inhibitory signal provided by crosslinking FcγRII to sIgs in resting murine B cells (ASHMAN et al. 1996). Since SHP-1 is recruited to the cytoplasmic domain of FcγRII under these experimental conditions (D'AMBROSIO et al. 1995), this suggests that the negative signaling function of the BCR may prevail when SHP-1 is retrieved from the BCR signaling pathway. Two other membrane receptors, CD22 (DOODY et al. 1995; LAW et al. 1996) and CD5 (for B-1 cells) (BIKAH et al. 1996) can also recruit SHP-1 and as such are susceptible to tune the sensitivity threshold of the BCR. Accordingly, the phenotype of mice in which the CD22 gene has been disrupted (O'KEEFE et al. 1996; OTIPOBY et al. 1996; SATO et al. 1996) is similar to that observed for SHP-1 deficient mice. Hence, the respective ligands of CD22 and CD5, i.e., proteins containing $\alpha 2$, 6-sialylated sugars and CD72 respectively, may potentially affect the outcome of BCR signaling.

II. Parameters Affecting the Outcome of BCR Signaling

Based on the work of the group of Goodnow and others, we will now consider some of the external parameters which have an impact on the nature of the response elicited by engagement of the BCR.

1. Physical Properties of the Ag

The first parameter is the molecular form of Ag which encompasses variables such as its concentration, avidity, valency, association with Abs, and complement. A typical illustration of this notion is the observation that protein Ag behaves as efficient toleragen when provided under a soluble deaggregated form while immunogenic when administered under an aggregated form (MITCHISON 1964; DRESSER and MITCHISON 1968; CHILLER et al. 1971). However, it seems difficult to make a strict correlation between Ag valency and the induction of a particular type of response. For example, extensive cross-linking of sIgs is mandatory for the induction of apoptosis in resting murine B cells (PARRY et al. 1994a,b) while F(ab)'$_2$ fragments of anti-Ig Abs, unlikely to cause extensive clustering of the Ag receptors, induce apoptosis of activated GC and memory B cells (BILLIAN et al. 1997; BERARD et al. 1999). This assumption is also in agreement with a series of studies conducted by the group of Dintzis and documenting the immunosuppressive effect of highly reticulated forms of Ag obtained by coupling multiple hapten or peptidic groups to dextran polymers (DINTZIS and DINTZIS 1992; SYMER et al. 1995; WATSON et al. 1996). The tolerizing effect of such polymerized Ag was shown to vary greatly depending on their molecular weight and hapten density. This implies that in addition to their valency, the geometry, mass and organization of the antigenic molecules are crucial for the outcome of B cell responses.

As mentioned above, activated B cells as opposed to resting B cells, do not require extensive Ag receptor clustering to undergo apoptosis in response to surrogate Ag. Owing to the role of SHP-1 in setting the BCR signaling threshold, it could be interesting to examine whether the enhanced vulnerability of activated B cells to BCR-induced apoptosis is correlated with a decreased expression of SHP-1 or SHP-1-recruiting molecules. It is still unclear whether the level of expression of SHP-1 can be modulated upon B cell activation but there is evidence that GC B cells, which are characterized by a high susceptibility to BCR-induced apoptosis, have a strongly reduced expression of this tyrosine phosphatase (DELIBRIAS et al. 1997).

2. Costimulatory Signals

a. Activated Complement Fractions

The group of Fearon has provided compelling evidence for the potent costimulatory function of the CD19/CD21/TAPA-1 complex when the signal transmitted by the BCR is a positive one, i.e., when B cells are exposed to an immunogenic form of Ag (see FEARON and CARTER 1995 for review). By contrast, the impact of these coreceptors on Ag-driven apoptosis is much less documented and the literature on this subject is confusing. On the one hand, there is evidence that CD19 can potentiate negative signaling through the BCR. For example, concurrent engagement of CD19 has been reported to potentiate BCR-induced apoptosis both in the Burkitt lymphoma cell line Ramos and in human tonsillar B cells treated with a highly multivalent form of anti-Ig Abs (CHAOUCHI et al. 1995). Furthermore, overexpression of CD19 dramatically reduces the output of mature B cells from the bone marrow, presumably by enhancing clonal deletion (ZHOU et al. 1994; ENGEL et al. 1995).

On the other hand, convincing data also argue for a protective function exerted by the CD19/CD21/TAPA-1 co-receptor complex on BCR-induced apoptosis. In particular, exploration of the responses of CR2 (CD21/CD35)-deficient mice has provided support for this notion. FISCHER et al. (1998) have compared the responses of monoclonal anti-HEL transgenic B cells bred into either a $CR2^+$ or a $CR2^-$ genetic background, after transfer into wild-type recipients immunized with low- or high-affinity Ag variants. Although CR2⁻ B cells could be found within GC following immunization with a high affinity Ag, they failed to participate in the GC reaction. One of the possible interpretation of these findings is that binding of Ag in the absence of ancillary signals from the complement receptors is detrimental to B cell survival in the GC. In keeping with this, it is striking that both injection of a massive dose of soluble Ag (PULENDRAN et al. 1995b; HAN et al. 1995; SHOKAT and GOODNOW 1995), unlikely to be complexed with complement fragments, and that of a soluble CR2 construct (FISCHER et al. 1998) similarly cause the disruption of established GC. In addition, KOZONO et al. (1995) have shown that BCR-mediated killing of the immature B cell line WEHI-231 can be prevented by coligation of sIgs and complement receptors 1 (CD35) and 2 (CD21). Altogether these observations raise the possibility that complement receptors may direct connection of the BCR to a positive signaling pathway in GC B cells.

b. T Cells and Microbial Factors

It was documented long ago that mature B cells can be rendered tolerant if they are exposed to high Ag concentrations in the absence of T cell help (PIKE et al. 1981). The assumption that, in certain circumstances, a tolerogenic signal can be converted into an immunogenic one if cognate T cell help is available is supported by several lines of evidence. It has been demonstrated in various in vitro experimental models that the apoptotic signal provided by a surrogate Ag to resting or activated B cells can be reversed in the presence of activated T cells (BILLIAN et al. 1997), T-cell-derived soluble factors such as IL-4 (PARRY et al. 1994a; GALIBERT et al. 1996; BILLIAN et al. 1997), or membrane-bound effector molecules such as CD40-L (PARRY et al. 1994a; NOMURA et al. 1996) or CD5 (NOMURA et al. 1996). Death induced by extensive crosslinking of sIgs on mature B cells has also been shown to be prevented by thymo-independent Ag such as LPS or dextran sulfate (NOMURA et al. 1996).

E. Molecular Control of the Apoptosis Sensitivity Threshold in Mature B Cells

Apart from the external influences that we have reviewed above, the decision of the BCR to promote death can also be influenced by signals from within the cells, inherent in their activation status and maturational stage. One of the most important checkpoints on the road which leads to programmed cell death is that which decides whether the death sentence delivered by the apoptotic stimulus will be executed or not. This checkpoint is under the control of multiple cytoplasmic regulatory molecules, exerting either an anti-apoptotic or a death-inducing function. Due to the increasing numbers of identified death regulators and to the complexity of their interactions, we do not pretend to draw an extensive and definitive picture of their respective implication in the regulation of B lymphocyte survival. Therefore, we will focus on six regulatory molecules which have received particular attention from researchers interested in B cell physiology. Five of them belong to the Bcl-2 family (Bcl-2, Bcl-x, Bax, Mcl-1, and Bad), and the sixth one is the proto-oncogene c-Myc.

I. Developmental Regulation of the Survival Genes

Bcl-2 is the founding member of a family of death regulatory genes which was initially isolated from the t(14, 18) chromosomal breakpoint constituting one of the hallmarks of follicular lymphomas (TSUJIMOTO et al. 1984). Bcl-2 is considered as the prototypic survival gene since its overexpression protects cells from a variety of apoptotic signals including growth factor deprivation, glucocorticoids, γ-irradiation, among others (see CORY 1995; YANG and KORSMEYER 1996 for review). In the past five years the Bcl-2 family has expanded and now comprises both anti- and pro-apoptotic molecules. In mammals, the death antagonists include Bcl-2, the long form of Bcl-x (Bcl-x_L), Bcl-w, Mcl-1, and A1. The death inducers are: Bax, the short form of Bcl-x (Bcl-x_S), Bak, Bik, Bid, Bad (see CHAO and KORSMEYER 1998 for review), and the recently identified Bim molecule (O'CONNOR et al. 1998). The various members of the Bcl-2 family physically interact with each other to form homo- or heterodimers through conserved domains designated as Bcl-2 homology regions (BH1 to BH3) (YIN et al. 1994; CHITTENDEN et al. 1995; ZHA et al. 1996). Whether molecules such as Bax possess a pro-apoptotic effector function per se or mainly act by preventing molecules such as Bcl-2 from exerting their death inhibitory function is not entirely clear yet. However, there is general agreement on the notion that the relative cellular concentrations of the pro- and anti-apoptotic members of the Bcl-2 family are determinant for the survival of the cells. At least part of the pool of the Bcl-2, Bcl-x, and Bax polypeptides is located at the junction between the inner and outer mitochondrial membranes (KROEMER 1997). Recent evidence indicates that these three molecules interfere with some crucial elements of the mitochondrial function such as the fall in transmembrane potential, the production of reactive oxygen species, and the release of cytochrome c (KROEMER et al. 1997).

The propensity of GC B cells to undergo spontaneous apoptosis in culture was first reported by LIU et al. (1989). Since Abs directed against sIgs or CD40 were found to prevent programmed cell death of isolated GC B cells in vitro, it was postulated that these cells are committed to die unless they receive appropriate rescuing signals from Ag and T cells. It was next demonstrated that the increased death susceptibility of GC B cells was correlated with their lack of Bcl-2 expression (LIU et al. 1991). These experiments constituted the first evidence for a strong positive correlation between increased death vulnerability and the modulation of expression of a so-called survival gene. Since then, several studies have documented the distribution of other apoptosis regulatory molecules in mature human B cell subsets. A summary of these results, shown in Table 1, emphasizes that expression of these molecules is develop-

Table 1. Pattern of expression of six apoptosis regulators in human B cell subsets[a]

	Virgin	Memory	GC
Bcl-2	+	+	−
Bclx$_L$	−	+	+
Mcl-1	±	±	+
Bax	−	±	+
Bad	−	−	+
c-Myc[b]	−	+	+

+, strong expression; ±, low/intermediate levels of expression; −, undetectable.
[a] This table is exclusively based on the analysis of tonsillar B cells by: (1) immunoenzymatic staining of tissue sections (KRAJEWSKI et al. 1994a,b), (2) immunoblot performed on isolated B cell subsets (OHTA et al. 1995; GHIA et al. 1998), and (3) RT-PCR in sorted B cell subsets (MARTINEZ-VALDEZ et al. 1996).
[b] For c-Myc, distribution of the transcript only, no data available on the expression of the protein.

mentally regulated during the Ag-dependent maturation process of B cells. This assumption is exemplified by the observation that their constitutive expression in GC B cells is strikingly different from that observed in virgin and memory B cells. Although GC B cells are characterized by the extinction of the Bcl-2 molecule, they are still positive for the expression of two other death repressors (Bcl-x$_L$ and Mcl-1). However, unlike virgin and memory B cells, they also constitutively express three death-inducing molecules (Bax, Bad, and c-Myc). This finding thus points towards the notion that the vulnerability of mature B cells to apoptotic stimuli relies on the ratio between pro- and anti-apoptotic molecules. The validity of this concept was confirmed by a series of studies in which the equilibrium between death inducers and death repressors was artificially modified. These experiments involved testing the susceptibility to Ag receptor-induced apoptosis of various lymphoma cell lines in which genes encoding either anti-apoptotic (Bcl-x$_L$) or pro-apoptotic molecules (Bax) have been overexpressed. This experimental approach demonstrated that overexpression of Bcl-x$_L$ (MERINO et al. 1995; ISHIDA et al. 1995; CHOI et al. 1995; WIESNER et al. 1997) but not that of Bcl-2 (CUENDE et al. 1993; CHOI et al. 1995) protects the immature cell line WEHI-231 from cell death induced by anti-Ig Abs. Conversely, WEINMANN et al. (1997) showed that transfection of Bax could induce a Burkitt lymphoma cell line resistant to sIg-induced apoptosis to switch to a sensitive phenotype. In conclusion, these observations suggest that the Bcl-x$_L$/Bax rather than the Bcl-2/Bax ratio plays a crucial role in defining the sensitivity threshold of B cells to BCR-induced apoptosis.

II. Activation-Induced Regulation of the Survival Genes

In agreement with the data discussed above, biological stimuli, such as CD40L, which protect WEHI cells from BCR-induced apoptosis have been described to raise expression of the long form of Bcl-x (ISHIDA et al. 1995; CHOI et al. 1995; WANG et al. 1995). As expected from the transfection experiments showing that Bcl-2 fails to protect WEHI cells from BCR-induced apoptosis, expression of the Bcl-2 transcript and protein was not affected by engagement of CD40 on these cells (CHOI et al. 1995; WANG et al. 1995). In mature murine B cells, Abs to CD40 as well as other mitogenic stimuli such as LPS, soluble anti-IgM Abs, and combinations of phorbol esters and inonophores also enhance Bcl-x_L expression without affecting the constitutive expression of Bcl-2 (GRILLOT et al. 1996; CHOI et al. 1996). Hence, Bcl-x_L but not Bcl-2 is likely to be involved in regulating the apoptosis susceptibility in activated B lymphocytes.

To what extent can we extrapolate these findings to the process of BCR-induced death in mature B cells? Although the expression of the survival genes of the Bcl-2 family has not yet been found during Ag receptor-induced apoptosis of mature B cells, the study of GRILLOT et al. (1996) provides some information on the impact of Bcl-2 and Bcl-x on this process. They explored this question in an in vivo setting by injecting mice with anti-IgD Abs, thus reproducing at a polyclonal level the situation in which the Ag receptor is crosslinked in the absence of T cell help. The subsequent deletion of mature B cells was followed in the spleens of four types of mice : wild-type animals, mice carrying either a Bcl-2 or a Bcl-x transgene, and mice carrying both Bcl-2 and Bcl-x transgenes. Their results indicate that partial protection from anti-IgD-induced apoptosis of splenic B cells is afforded when mice carry both the Bcl-x and Bcl-2 transgenes but not when they express either one or the other of these transgenes alone. This suggests that full protection against the apoptotic signal delivered via the BCR in mature B cells most likely requires either collaboration between different survival molecules or the concomitant decline of proapoptotic factors.

Comparatively few studies have dealt with the expression of death-inducers following B cell activation. Activation-induced modulation of Bax was reported by OHTA et al. (1995) who showed that activation of human neonatal B cells by the T-independent Ag SAC and IL-2 increased expression of the Bax protein. This issue was also addressed by BARGOU et al. (1995) who demonstrated that sIgM-induced apoptosis of the Burkitt lymphoma cell line BL41 was preceded by a rise in the expression of the Bax protein. Finally, our own results (BERARD et al. 1999) have showed that three transcripts encoding pro-apoptotic molecules (Bax, c-Myc, and p53) are upregulated following ligation of the BCR or CD40 in human tonsillar B cells. However, these activation stimuli had a differential impact on the Bcl-x_L/Bax ratio in virgin and memory B cells. In virgin B cells, engagement of either one or the other of these membrane receptors led to a strong upregulation of the Bcl-x_L mRNA but marginally affected the Bax transcript. In contrast, activated memory B cells were characterized by a prominent expression of the Bax transcript while the levels of expression of the Bcl-x_L mRNA were only marginally affected. Altogether, these findings indicate that upregulation of pro-apoptotic molecules such as Bax can fulfill two distinct functions. First, as for anti-Ig-stimulated Burkitt lymphoma cell lines, it can be consecutive to the delivery of the apoptotic insult and directly initiate death. Second, as observed for human memory B cells, it can occur in response to stimuli which favor a mitogenic response rather than apoptosis. In this case, the rise in Bax expression and concomitant downregulation of Bcl-x_L would predispose B cells to undergo apoptosis upon reexposure to Ag by lowering their threshold of death susceptibility.

The available evidence suggests that c-Myc is involved in regulating apoptosis mediated via the Ag receptor. In WEHI cells, ligation of sIgM induces a biphasic modulation of the c-Myc transcript, that is a transient increase within the first hour of stimulation, followed by a strong downregulation of its expression (LEE et al. 1995). However, it is not yet clear whether it is the initial rise or the decline phase of c-Myc expression which is instrumental in BCR-induced apoptosis. On the one hand, blocking c-Myc function by the means of antisense oligodeoxynucleotides prevents the induction of apoptosis promoted by extensive crosslinking of sIgM on mature murine splenic B cells (SCOTT et al. 1996). On the other hand, signals which protect WEHI cells from anti-IgM-induced apoptosis (such as CD40 L) have been shown to sustain c-Myc expression (SCHAUER et al. 1996). However, since c-Myc is placed at the branching of the proliferation and apoptosis pathways it might influence cell survival in different ways. In other words, the intrinsic proapoptotic function of c-Myc (EVAN et al. 1992) and the proliferation block imposed by its downregulation might be equally detrimental to cell survival.

F. The Executioners of the BCR Apoptotic Pathway

I. Early Transduction Events

It is not the purpose of the present section to provide a detailed survey of the transduction pathway connected to the death domain (DD)-containing receptors. However, we will briefly review the current knowledge in the field because these elements are important for the understanding of the possible relationship between the Fas and BCR apoptotic pathways. The DD-containing receptors belong to the TNF receptor superfamily. They all comprise an homologous sequence of 80 amino acids in their intracytoplasmic portion which is referred to as the death domain because it is mandatory for transduction of the apoptotic signal. There are five cloned bona fide death receptors to date: the TNF receptor 1 (TNF-R1), Fas (CD95/APO-1), TRAMP (DR3/APO-3/WSL/LARD), TRAIL-R1 (DR4/APO-2), and TRAIL-R2 (DR5). The most proximal cytoplasmic element of transduction of the apoptotic signal via the DD-receptors is a so-called adapter molecule which binds to the oligomerized DD and recruits downstream mediators via a specific amino acid sequence located in its N-terminal region and designated as the death effector domain (DED). Treatment of sensitive cells with an agonistic anti-Fas Ab followed by immunoprecipitation of Fas has allowed for the identification of a group of four proteins responsible for the early steps of the death signal transduction (see SCHULZE-OSTHOFF et al. 1998 for review) via Fas. Two of them were identified as different molecular forms of the adapter molecule FADD, one of them is a caspase (see below) designated as caspase 8 (FLICE), and the fourth still awaits molecular characterization. The postulated scenario for the early biochemical events induced by Fas triggering is the following: (1) trimerization of Fas in the membrane; (2) binding of FADD to the DD of Fas; (3) recruitment of FLICE by FADD via its DED; (4) activation, i.e., processing of the proenzymatic form of FLICE; and (5) recruitment and activation by activated caspase 8 of other downstream second messengers. Adapter molecules in which this DED has been truncated can still bind to the DD of Fas but can no longer recruit the downstream caspases. Such a truncated form of FADD has been shown to function as a dominant negative mutant and to protect the cells from Fas-mediated apoptosis. It can be used to study the Fas-dependency of certain signaling pathways.

II. The Caspase Cascade

Dissection of the distal molecular events responsible for the irrevocable decision of the cell to die has been a matter of intensive research, probably because the key elements of the executor machinery of cell death are likely to be shared by most apoptotic pathways. In the nematode, the terminal irreversible effector step of cell death is controlled by the product of the Ced-3 gene. The mammalian equivalent of Ced-3 is the cytoplasmic cysteine protease interleukin-1β converting enzyme (ICE), required for processing of the IL-1β precursor to the active cytokine (YUAN et al. 1993). ICE was the first identified member of a multigene family of proteolytic enzymes designated as Caspases (*c*ysteinyl *a*spartic *a*cid *s*pecific *p*roteases) that all cleave their substrates at specific aspartate residues (see COHEN 1997; MILLER 1997 for reviews). Caspases are synthesized under an inactive proenzyme form (30–50kDa) which is processed to produce an enzymatically active complex composed of the shorter cleavage products of 10kDa and 20kDa, respectively. As caspases act in a stepwise fashion and behave as substrates for each other, their sequential activation during the apoptotic process is often referred to as the caspase cascade. Certain caspases like FLICE/Caspase 8 are proximal to the death-inducing receptors in the plasma membrane (BOLDIN et al. 1996; MUZIO et al. 1996; MEDEMA et al. 1997) while others act downstream of the core of the apoptotic pathway, i.e., the mitochondria (see SCHULZE-OSTHOFF et al. 1998 for review). At the distal end of the apoptotic pathway, caspases cleave various cellular substrates responsible for the nuclear and membrane degradations which "sign" the execution of apoptosis. These substrates include inhibitors of DNAse, enzymes involved in DNA repair and gene maintenance such as poly(ADP-ribose) polymerase/PARP, cytoskeleton proteins, cell cycle regulators, etc. Recent studies have been conducted to determine whether caspases intervene in the apoptotic pathway coupled to the Ag receptor. These experiments which have been mostly performed on immature (ANDJELIC and LIOU 1998) and mature B lymphoma cell lines (RICKERS et al. 1998; LENS et al. 1998) convincingly demonstrated that caspase 3 (CPP32/YAMA) is involved in BCR-induced killing. Since this caspase is also an element of the signaling cascade coupled to Fas, it suggests that the BCR and the Fas apoptotic pathways might at least partially converge at a certain point. Nonetheless experiments conducted by the group of Van Lier (LENS et al. 1998) on a Burkitt lymphoma cell line suggest that the apoptosis effectors acting upstream of caspase 3 in the Fas and BCR signaling pathways are distinct. Their data can be summarized as follows. First, the cleavage products of caspase 3 generated during BCR or Fas-induced apoptosis differ by their size, suggesting that the caspases responsible for the processing of procaspase 3 along these two pathways are distinct. Second, the activation of caspase 3 consecutive to BCR triggering is delayed as compared to the kinetics of caspase 3 activation following engagement of Fas. Third, transfection of a responding Burkitt lymphoma line with a dominant negative form of FADD (FADD-DN) does not affect BCR-induced apoptosis. This latter finding is coherent with previous reports documenting that Fas blocking reagents (soluble Fas, antagonistic Abs) do not affect the death signal provided through the BCR (DANIEL et al. 1997; BILLIAN et al. 1997; BERARD et al. 1999) and that activated human B cells lack detectable expression of the transcript encoding Fas-L (DANIEL et al. 1997). Altogether, these findings demonstrate that, as opposed to the mechanism underlying activation-induced death of T cells, the Fas/Fas-L system is not involved in the Ag-receptor-induced apoptosis of mature B cells. However, the possibility that BCR-mediated killing operates through indirect triggering of another death domain-containing receptor cannot be formally excluded. Indeed, there is some redundancy at the level of the proximal transducing elements involved in the apoptotic pathway coupled to the DD-containing receptors. In fact, in addition to FADD, four other adapter molecules can be recruited by the DD-containing receptors – TRADD (HSU et al. 1995), RIP (STANGER et al. 1995), RAIDD (DUAN and DIXIT 1997), and CRADD (AHMAD et al. 1997) – and promote apoptosis when overexpressed in model cell lines. Hence, the efficiency of a FADD DN protein for blocking a given apoptotic pathway will depend on the levels of endogenous expression of FADD and the other adapter molecules.

Interestingly, SCAFFIDI et al. (1998) have established that Fas can be connected to two different death pathways, depending on the cell type considered. These authors distinguish type I cells, characterized by an early activation of caspase 8 (FLICE) at the level of the death-inducing signaling complex (DISC) and type II cells for which caspase 8 processing and activation occurs later, downstream of the mitochondria. The pathway used by type I cells operates independently of the perturbations of the mitochondrial functions, while the pathway used by type II cells is fully dependent upon mitochondrial activity. Accordingly, the pathway used by type II but not by type I cells is blocked by Bcl-2 and Bcl-x which both interfere with permeability transition of the mitochondria. The transfection experiments conducted on the immature cell line WEHI-231 have also demonstrated that the BCR death pathway in these cells is sensitive to the anti-apoptotic effect of Bcl-x_L (GOTTSCHALK et al. 1994; MERINO et al. 1995). In conclusion, the BCR apoptotic pathway certainly uses some of the downstream caspases (such as caspase 3) but may or may not utilize the proximal components of the DD receptor signaling pathway. It is dependent on mitochondrial contribution and is also characterized by the late cleavage of caspase 3. Altogether, these elements raise the possibility that the death pathway connected to the BCR may present similarities with that coupled to Fas in type II cells.

G. Concluding Remarks

There are multiple pathways leading to apoptosis in B cells, and molecules such as FcγRII (ASHMAN et al. 1996), MHC class II (NEWELL et al. 1993; TRUMAN et al. 1994), and class I (GENESTIER et al. 1997) molecules, the BCR and Fas have all been reported to induce B cell death under certain circumstances. Why is there such a profusion of receptors capable of inducing death? Why can some of them, beside their long-recognized immunostimulatory function, also promote apoptosis? The precise answer to these questions is still elusive but it appears that the immune system has developed multiple strategies to prevent dysregulated expansion of the lymphoid cells which might otherwise lead to autoimmune, lymphoproliferative diseases and malignancies. This emphasizes the crucial importance of the negative control of the immune response to preserve integrity of the organism.

Still, how can we reconcile the fact that BCR ligation protects mature B cells from Fas-mediated killing with the pro-apoptotic effect of BCR agonists reviewed in this chapter? Although the issue of the regulation of Fas-induced apoptosis in the B cell compartment is beyond the scope of this review, we would like to comment briefly on these two apparently opposing functions of the BCR. Our hypothesis is that Fas- and BCR-induced apoptosis do not serve identical purposes and intervene at distinct stages of the B cell maturation process. The model presented in Fig. 2 illustrates this point. The available data are consistent with the hypothesis that Fas-induced apoptosis plays an important role during the initiation phase of B cell responses, i.e., in the T zones of secondary lymphoid organs where T and B cells physically interact. At this stage, the BCR exerts an anti-apoptotic function by protecting Ag-specific B cells from Fas-mediated apoptosis. As we and others have proposed, this mechanism could be instrumental in preventing CD40-mediated bystander B cell

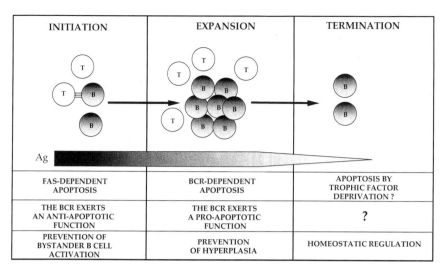

Fig. 2. Biological functions fulfilled by the BCR- and Fas-induced apoptosis during B cell responses. The B cell maturation pathway in response to antigenic stimulation is divided into three phases. The initiation phase occurs in the T zone of secondary lymphoid organs and involves close physical interaction with activated T cells which have been primed by interdigitating dendritic cells. Activated T cells are expected to express both the ligands for CD40 and Fas and are equally armed to induce activation or apoptosis of B cells. At this stage, ligation of the BCR on cells which are not yet actively cycling protects them from Fas-mediated killing. The expansion phase can take place both in the GC or in the extrafollicular foci where B cells have an intense mitogenic activity. Prolonged or repeated exposure of cycling B cells to Ag and the concomitant decline of T cell help would favor the feedback pro-apoptotic effect of Ag, thereby ensuring downsizing of the responding B cell population

activation and in recruiting, among the diverse Ag-specific B cell clones, those which display the strongest Ag-binding capacities. Later, during the expansion phase, when B cells are actively cycling and Ag is non-limiting, the signaling program coupled to the BCR is redirected towards a pro-apoptotic function. At this stage, Ag itself plays an active part in downsizing the responsive B cell population in order to prevent hyperplasia. Other apoptotic mechanisms such as growth factor deprivation might take over at the terminal stage of the response when Ag becomes limiting.

References

Ahmad M, Srinivasula SM, Wang L, Talanian RV, Litwack G, Fernandes-Alnemri T, Alnemri ES (1997) CRADD, a novel human apoptotic adapter molecule for caspase-2, and FasL/Tumor Necrosis factor receptor-interacting protein RIP. Cancer Res 57:615–619

Alderson MR, Tough TW, Davis-Smith T, Braddy S, Falk B, Schooley KA, Goodwin RG, Smith CA, Ramsdell F, Lynch DH (1995) Fas ligand mediates activation-induced cell death in human T lymphocytes. J Exp Med 181:71–77

Andjelic S, Liou H-C (1998) Antigen receptor-induced B lymphocyte apoptosis mediated via a protease of the caspase family. Eur J Immunol 28:570–581

Ashman RF, Peckham D, Stunz LL (1996) Fc receptor off-signal in the B cell involves apoptosis. J Immunol 157:5–11

Ashwell JD, Cunningham RE, Noguchi PD, Hernandez D (1987) Cell growth cycle block of T cell hybridomas upon activation with antigen. J Exp Med 165:173–194

Bargou RC, Bommert K, Weimann P, Daniel PT, Wagener C, Mapara MY, Dörken B (1995) Induction of Bax-α precedes apoptosis in a human B lymphoma cell line: potential role of the Bcl-2 gene family in surface IgM-mediated apoptosis. Eur J Immunol 25:770–775

Berek C, Berger A, Apel M (1991) Maturation of the immune response in germinal centres. Cell 67:1121–1129

Berard M, Casamayor-Pallejà M, Billian G, Bella C, Mondiere P, Defrance T (1999) Activation sensitizes human memory B cells to BCR-induced apoptosis. Immunology 98:47–54

Berard M, Mondière P, Casamayor-Pallejà M, Hennino A, Bella C, Defrance T (1999) Mitochonchia connects the antigen receptor to effector casjases during B cell receptor-induced apoptosis in normal human B cells. J. Immunol. In press

Bikah G, Carey J, Ciallella JR, Tarakhovsky A, Bondada S (1996) CD5-mediated negative regulation of antigen-receptor-induced growth signals in B-1 B cells. Science 274:1906–1909

Billian G, Mondière P, Berard M, Bella C, Defrance T (1997) Antigen receptor-induced apoptosis of human germinal center B cells is targeted to a centrocytic subset. Eur J Immunol 27:405–414

Boehme SA, Lenardo MJ (1993) Propiocidal apoptosis of mature T lymphocytes occurs at S phase of the cell cycle. Eur J Immunol 23:1552–1560

Boldin MP, Goncharov TM, Goltsev YV, Wallach D (1996) Involvement of MACH, a novel MORT1/FADD-interacting protease, in Fas/APO-1 and TNF receptor-induced cell death. Cell 85:803–815

Boyd AW, Schrader JW (1981) The regulation of growth and differentiation of a murine B cell lymphoma. II. The inhibition of WEHI 231 by anti-immunoglobulin antibodies. J Immunol 126:2466–2469

Brunner T, Mogil RJ, LaFace D, Yoo NJ, Mahboudi A, Echeverri F, Martin SJ, Force WR, Lynch DH, Ware CF, Green DR (1995) Cell-autonomous Fas (CD95)/Fas-ligand interaction mediates activation-induced apoptosis in T-cell hybridomas. Nature 373:441–444

Cambier JC, Kettman JR, Vitetta ES, Uhr JW (1976) Differential susceptibility of neonatal and adult murine spleen cells to in vitro induction of B-cell tolerance. J Exp Med 144:293–297

Chao DT, Korsmeyer SJ (1998) Bcl-2 family: regulators of cell death. Annu Rev Immunol 16:395–419

Chaouchi N, Vazquez A, Galanaud P, Leprince C (1995) B cell antigen receptor-mediated apoptosis. Importance of accessory molecules CD19 and CD22, and of surface IgM cross-linking. J Immunol 154:3096–3104

Chiller JM, Habicht GS, Weigle WO (1971) Kinetic differences in unresponsiveness of thymus and bone marrow cells. Science 171:813–815

Chittenden T, Flemington C, Houghton AB, Ebb RG, Gallo GJ, Elangovan B, Chinnadurai G, Lutz RJ (1995) A conserved domain in Bak, distinct from BH1 and BH2, mediates cell death and protein binding functions. EMBO J 14:5589–5596

Choi MSK, Boise LH, Gottschalk AR, Quintans J, Thompson CB, Klaus GGB (1995) The role of Bcl-xL in CD40-mediated rescue from anti-μ-induced apoptosis in WEHI-231 B lymphoma cells. Eur J Immunol 25:1352–1357

Choi MSK, Holman M, Atkins CJ, Klaus GGB (1996) Expression of Bcl-x during mouse B cell differentiation and following activation by various stimuli. Eur J Immunol 26:676–682

Cohen GM (1997) Caspases: the executioners of apoptosis. Biochem J 326:1–16
Cory S (1995) Regulation of lymphocyte survival by the Bcl-2 gene family. Annu Rev Immunol 13:5413–543
Cuende E, Ales-Martinez JE, Ding L, Gonzalez-Garcia M, Martinez C, Nunez G (1993) Programmed cell death by Bcl-2-dependent mechanisms in B lymphoma cells. EMBO J 12:1555–1560
Cyster JG, Goodnow CC (1995) Protein tyrosine phosphatase 1 C negatively regulates antigen receptor signaling in B lymphocytes and determines thresholds for negative selection. Immunity 2:13–24
D'Ambrosio D, Hippen KL, Minskoff SA, Mellman I, Pani G, Siminovitch KA, Cambier JC (1995) Recruitment and activation of PTP1 C in negative regulation of antigen receptor signaling by FcγRIIB1. Science 268:293–297
Daniel PT, Oettinger U, Mapara MY, Bommert K, Bargou R, Dörken B (1997) Activation and activation-induced death of human tonsillar B cells and Burkitt lymphoma cells: lack of CD95 (Fas/APO-1) ligand expression and function. Eur J Immunol 27:1029–1034
Delibrias CC, Floettmann JE, Rowe M, Fearon DT (1997) Downregulated expression of SHP-1 in Burkitt lymphomas and germinal center B lymphocytes. J Exp Med 186:1575–1583
Dhein J, Walczak H, Bäumler C, Debatin K-M, Krammer PH (1995) Autocrine T-cell suicide mediated by APO-1(Fas/CD95). Nature 373:438–441
Dintzis HM, Dintzis RZ (1992) Profound specific suppression by antigen of persistent IgM, IgG, and IgE antibody production. Proc Natl Acad Sci USA 89:1113–1117
Doody GM, Justement LB, Delibrias CC, Matthews RJ, Lin J, Thomas ML, Fearon DT (1995) A role in B cell activation for CD22 and the protein tyrosine phosphatase SHP. Science 269:242–244
Dresser DW, Mitchison NA (1968) The mechanism of immunological paralysis. Adv Immunol 8:129–181
Duan H, Dixit VM (1997) RAIDD is a new "death" adapter molecule. Nature 385:86–89
Engel P, Zhou L-J, Ord DC, Sato S, Koller B, Tedder TF (1995) Abnormal B lymphocyte development, activation and differentiation in mice that lack or overexpress the CD19 signal transduction molecule. Immunity 3:39–50
Eray M, Tuomikoski T, Wu H, Nordström T, Andersson LC, Knuutila S, Kaartinen M (1994) Cross-linking of surface IgG induces apoptosis in a Bcl-2 expressing human follicular lymphoma line of mature B cell phenotype. Intern Immunol 6:1817–1827
Evan G, Wyllie AH, Gilbert CS, Littlewood TD, Land H, Brooks M, Waters CM, Penn LZ, Hancock DC (1992) Induction of apoptosis in fibroblasts by c-myc protein. Cell 69:119–128
Fearon DT, Carter RH (1995) The CD19/CR2/TAPA-1 complex of B lymphocytes: linking natural to acquired immunity. Annu Rev Immunol 13:127–149
Fischer MB, Goerg S, Shen L, Prodeus AP, Goodnow CC, Kelsoe G, Carroll MC (1998) Dependence of germinal center B cells on expression of CD21/CD35 for survival. Science 280:582–585
Galibert L, Burdin N, Barthélémy C, Meffre G, Durand I, Garcia E, Garrone P, Rousset F, Banchereau J, Liu Y-J (1996) Negative selection of human germinal center B cells by prolonged BCR cross-linking. J Exp Med 183:2075–2085
Gaur A, Yao X, Scott DW (1993) B cell tolerance induction by cross-linking of membrane IgM, but not IgD, and synergy by cross-linking of both isotypes. J Immunol 150:1663–1669
Genestier L, Meffre G, Garrone P, Pin JJ, Liu Y-J, Banchereau J, Revillard JP (1997) Antibodies to HLA class I alpha 1 domain trigger apoptosis of CD40-activated human B lymphocytes. Blood 90:726–735
Ghia P, Boussiotis VA, Schultze JL, Cardoso AA, Dorfman DM, Gribben JG, Freedman AS, Nadler LM (1998) Unbalanced expression of Bcl-2 family proteins in fol-

licular lymphoma: contribution of CD40 signaling in promoting survival. Blood 91:244–251
Gonzalo JA, Alboran IMD, Ales-Martinez JE, Martinez AC, Kroemer G (1992) Expansion and clonal deletion of peripheral T cells induced by bacterial superantigen is independent of the interleukin-2 pathway. Eur J Immunol 22:1007–1011
Goodnow CC, Cyster JG, Hartley SB, Bell SE, Cooke MP, Healy JI, Akkaraju S, Rathmell JC, Pogue SL, Shokat KP (1995) Self-tolerance checkpoints in B lymphocyte development. Adv Immunol 59:279–363
Gottschalk AR, Boise LH, Thompson CDB, Quintans J (1994) Identification of immunosuppressant-induced apoptosis in a murine B-cell line and its prevention by Bcl-x but not Bcl-2. Proc Natl Acad Sci USA. 91:7350–7354
Gottschalk AR, McShan CL, Kilkus J, Dawson G, Quintans J (1995) Resistance to anti-IgM-induced apoptosis in a WEHI-231 subline is due to insufficient production of ceramide. Eur J Immunol 25:1032–1038
Green DR, Scott DW (1994) Activation-induced apoptosis in lymphocytes. Curr Opin Immunol 6:476–487
Gregory CD, Dive C, Henderson S, Smith CA, Williams GT, Gordon J, Rickinson AB (1991) Activation of Epstein-Barr virus latent genes protects human B cells from death by apoptosis. Nature 349:612–614
Grillot DAM, Merino R, Pena JC, Fanslow WC, Finkelman FD, Thompson CB, Nunez G (1996) Bcl-x exhibits regulated expression during B cell development and activation and modulates lymphocyte survival in transgenic mice. J Exp Med 183:381–391
Gulbranson-Judge A, MacLennan ICM (1996) Sequential antigen-specific growth of T cells in the T zones and follicles in response to pigeon cytochrome c. Eur J Immunol 26:1830–1837
Han S, Zheng B, Dal Porto J, Kelsoe G (1995) In situ studies of the primary immune response to (4-hydroxy-3-nitrophenyl) acetyl. IV. Affinity-dependent, antigen driven B cell apoptosis in germinal centers as a mechanism for maintaining self-tolerance. J Exp Med 182:1635–1644
Han S, Zheng B, Schatz DG, Spanopolou E, Kelsoe G (1996) Neoteny in lymphocytes: Rag 1 and Rag 2 expression in germinal center B cells. Science 274:2094–2097
Han S, Dillon SR, Zheng B, Shimoda M, Schlissel MS, Kelsoe G (1997) V(D)J recombinase activity in a subset of germinal center B lymphocytes. Science 278:301–305
Healy JI, Dolmetsch RE, Timmerman LA, Cyster JG, Thomas ML, Crabtree GR, Lewis RS, Goodnow CC (1997) Different nuclear signals are activated by the B cell receptor during positive versus negative signaling. Immunity 6:419–428
Healy JI, Goodnow CC (1998) Positive versus negative signaling by lymphocyte antigen receptors. Annu Rev Immunol 16:645–670
Hertz M, Kouskoff V, Nakamura T, Nemazee D (1998) V(D)J recombinase induction in splenic B lymphocytes is inhibited by antigen-receptor signalling. Nature 394:292–295
Hsu H, Xiong J, Goeddel DV (1995) The TNF receptor 1-associated protein TRADD signals cell death and NF-κB activation. Cell 81:495–504
Isakson PC, Krolick KA, Uhr JW, Vitetta ES (1980) The effect of anti-immunoglobulin antibodies on the in vitro proliferation and differentiation of normal and neoplastic murine B cells. J Immunol 125:886–892
Ishida T, Kobayashi N, Tojo T, Ishida S, Yamamoto T, Inoue J-I (1995) CD40 signaling-mediated induction of Bcl-xL, Cdk4, and Cdk6. J Immunol 155:5527–5535
Jacob J, Kelsoe G, Rajewsky K, Weiss U (1991) Intraclonal generation of antibody mutants in germinal centres. Nature 354:389–392
Johnson JG, Jemmerson R (1992) Tolerance induction in resting memory B cells specific for a protein antigen. J Immunol 148:2682–2689
Jones LA, Chin LT, Longo DL, Kruisbeek AM (1990) Peripheral clonal elimination of functional T cells. Science 250:1726–1729

Ju ST, Panka DJ, Cui H, Ettinger R, El-Khatib M, Sherr DH, Stanger BZ, Marshak-Rothstein A (1995) Fas (CD95)/FasL interactions required for programmed cell death after T cell activation. Nature 373:444–448

Karvelas M, Nossal GJV (1992) Memory cell generation ablated by soluble protein antigen by means of effect on T- and B-lymphocyte compartments. Proc Natl Acad Sci USA 89:3150–3154

Kawabe Y, Ochi A (1991) Programmed cell death and extrathymic reduction of Vβ8 + CD4+ T cells in mice tolerant to Staphylococcus Aureus enterotoxin B. Nature 349:245–248

Kearney JF, Klein J, Bockman DE, Cooper MD (1978) B cell differentiation induced by lipopolysaccharide. V. Suppression of plasma cell maturation by anti-μ: mode of action and characteristics of suppressed cells. J Immunol 120:158–166

Kelsoe G (1996) Life and death in germinal centers (redux). Immunity 4:107–111

Kim K-M, Ishigami T, Hata D, Higaki Y, Morita M, Yamaoka K, Mayumi M, Mikawa H (1992) Anti-IgM but not anti-IgD antibodies inhibit cell division of normal human mature B cells. J Immunol 148:29–34

Klinman NR, Schrater AF, Katz DH (1981) Immature B cells as the target for in vivo tolerance induction J Immunol 126:1970–1973

Koopman G, Keehnen RMJ, Lindhout E, Newman W, Shimizu Y, Van Seventer GA, De Groot C, Pals ST (1994) Adhesion through the LFA-1 (CD11a/CD18)-ICAM-1 (CD154) and the VLA-4 (CD49d)-VCAM-1 (CD106) pathways prevents apoptosis of germinal center B cells. J Immunol 152:3760–3767

Kozono Y, Duke RC, Schleicher MS, Holers VM (1995) Co-ligation of mouse complement receptors 1 and 2 with surface IgM rescues splenic B cells and WEHI-231 cells from anti-surface IgM-induced apoptosis. Eur J Immunol 25:1013–1017

Krajewski S, Krajewska M, Shabaik A, Miyashita T, Wang H-G, Reed JC (1994a) Immunohistochemical determination of in vivo distribution of Bax, a dominant inhibitor of Bcl-2. Am J Pathol 145:1323–1336

Krajewski S, Bodrug S, Gascoyne R, Berean K, Krajewska M, Reed JC (1994b) Immunohistochemical analysis of Mcl-1 and Bcl-2 proteins in normal and neoplastic lymph nodes. Am J Pathol 145:515–525

Kroemer G, Zamzami N, Susin SA (1997) Mitochondrial control of apoptosis. Immunol Today 18:44–51

Kroemer G (1997) The proto-oncogene Bcl-2 and its role in regulating apoptosis. Nature Med 3:614–620

Law CL, Sidorenko SP, Chandran KA, Zhao Z, Shen SH, Fischer EH, Clark EA (1996) CD22 associates with protein tyrosine phosphatase 1 C, Syk, and phospholipase C-gamma (1) upon B cell activation. J Exp Med 183:547–560

Lee H, Wu M, La Rosa FA, Duyao MP, Buckler AJ, Sonenshein GE (1995) Role of the Rel-family of transcription factors in the regulation of c-myc gene transcription and apoptosis of WEHI 231 murine B-cells. Curr Top Microbiol Immunol 194:247–255

Lens SMA, Den Drijver BFA, Pötgens AJG, Tesselaar K, Van Oers MHJ, Van Lier RAW (1998) Dissection of the pathways leading to antigen receptor-induced and Fas/CD95-induced apoptosis in human B cells. J Immunol 160:6083–6092

Linton PJ, Decker DJ, Klinman NR (1989) Primary antibody forming cells and secondary B cells are generated from separate precursor cell subpopulations. Cell 59:1049–1059

Linton P-J, Rudie A, Klinman NR (1991) Tolerance susceptibility of newly generating memory B cells. J Immunol 146:4099–4104

Liu Y-J, Joshua DE, Williams GT, Smith CA, Gordon J, MacLennan ICM (1989) Mechanism of antigen-driven selection in germinal centres. Nature 342:929–931

Liu Y-J, Mason DY, Johnson GD, Abbot S, Gregory CD, Hardie DL, Gordon J, MacLennan ICM (1991) Germinal center cells express Bcl-2 protein after activation by signals which prevent their entry into apoptosis. Eur J Immunol. 21:1905–1910

Martinez-Valdez H, Guret C, De Bouteiller O, Fugier I, Banchereau J, Liu Y-J (1996) Human germinal center B cells express the apoptosis-inducing genes Fas, c-Myc, p53, and Bax but not the survival gene Bcl-2. J Exp Med 183:971–977

Maruyama S, Kubagawa H, Cooper MD (1985) Activation of human B cells and inhibition of their terminal differentiation by monoclonal anti-μ antibodies. J Immunol 135:192–199

Medema JP, Scaffidi C, Kischkel FC, Shevchenko A, Mann M, Krammer PH, Peter ME (1997) FLICE is activated by association with the CD95 Death-Inducing Signaling Complex (DISC). EMBO J 16:2794–2804

Meffre E, Papavasiliou F, Cohen P, De Bouteiller O, Bell D, Karasuyam H, Schiff C, Banchereau J, Liu Y-J, Nussenzweig MC (1998) Antigen receptor engagement turns off the V(D)J recombination machinery in human tonsil B cells. J Exp Med 188:765–772

Melchers F, Anderson J, Lernhardt W, Schreier MH (1980) Roles of surface-bound immunoglobulin molecules in regulating the replication and maturation to immunoglobulin secretion of B lymphocytes. Immunol Rev 52:89–114

Merino R, Grillot DAM, Simonian PL, Muthukkumar S, Fanslow WC, Bondada S, Nunez G (1995) Modulation of anti-IgM-induced B cell apoptosis by Bcl-x_L and CD40 in WEHI-231 cells. Dissociation from cell cycle arrest and dependence on the avidity of the antibody-IgM receptor interaction. J Immunol 155:3830–3838

Metcalf ES, Klinman NR (1976) In vitro tolerance induction of neonatal murine B cells. J Exp Med 143:1327–1340

Miller DK (1997) The role of the caspase family of cysteine proteases in apoptosis. Semin Immunol 9:35–49

Mitchison MA (1964) Induction of immunological paralysis in two zones of dosage. Proc R Soc Lon (Biol) 161:275–280

Muzio M, Chinnaiyan AM, Kischkel FC, O'Rourke K, Shevchenko A, Ni J, Scaffidi C, Bretz JD, Zhang M, Gentz R, Mann M, Krammer PH, Peter ME, Dixit VM (1996) FLICE, a novel FADD-homologous ICE/CED-3-like protease is recruited to the CD95(Fas/APO-1) death-inducing signaling complex. Cell 85:817–827

Newell KM, VanderWall J, Beard KS, Freed JH (1993) Ligation of major histocompatibility complex class II molecules mediates apoptotic cell death in resting B lymphocytes. Proc Natl Acad Sci USA 90:10459–10463

Nomura T, Han H, Howard MC, Yagita H, Yakura H, Honjo T, Tsubata T (1996) Antigen receptor-mediated B cell death is blocked by signaling via CD72 or treatment with dextran sulfate and is defective in autoimmunity-prone mice. Intern Immunol 8:867–875

Nossal GJV, Pike BL (1975) Evidence for the clonal abortion theory of B lymphocyte tolerance. J Exp Med 141:904–917

Nossal GJV, Pike BL, Battye FL (1979) Mechanisms of clonal abortion tolerogenesis. II Clonal behaviour of immature B cells following exposure to anti-μ chain antibody. Immunology 37:203–215

O'Connor L, Strasser A, O'Reilly LA, Hausmann G, Adams JM, Cory S, Huang DCS (1998) Bim: a novel member of the Bcl-2 family that promotes apoptosis. EMBO J 17:384–395

Ohta K, Iwai K, Kasahara Y, Taniguchi N, Krajewski S, Reed JC, Miyawaki T (1995) Immunoblot analysis of cellular expression of Bcl-2 family proteins, Bcl-2, Bax, Bcl-x and Mcl-1, in human peripheral blood and lymphoid tissues. Intern Immunol 7:1817–1825

O'Keefe TL, Williams GT, Davies SL, Neuberger MS (1996) Hyperresponsive B cells in CD22-deficient mice. Science 274:798–801

Otipoby KL, Andersson KB, Draves KE, Klaus SJ, Garr AG, Kerner JD, Perlmutter RM, Law C-L, Clark EA (1996) CD22 regulates thymus independent responses and the lifespan of B cells. Nature 384:634–637

Papavasiliou F, Casellas R, Suh H, Qin X-F, Besmer E, Pelanda R, Nemazee D, Rajewsky K, Nussenzweig MC (1997) V(D)J recombination in mature B cells: a mechanism for altering antibody responses. Science 278:298–301

Parry SL, Hasbold J, Holman M, Klaus GGB (1994a) Hypercross-linking surface IgM or IgD receptors on mature B cells induces apoptosis that is reversed by costimulation with IL-4 and anti-CD40. J Immunol 152:2821–2829

Parry SL, Holman MJ, Hasbold J, Klaus GG (1994b) Plastic-immobilized anti-mu or anti-delta antibodies induce apoptosis in mature murine B lymphocytes. Eur J Immunol 24:974–979

Pike BL, Battye FL, Nossal GJV (1981) Effect of hapten valency and carrier composition on the tolerogenic potential of hapten-protein conjugates. J Immunol 126: 89–94

Pulendran B, Karvelas M, Nossal GJV (1994) A form of immunologic tolerance through impairment of germinal center development. Proc Natl Acad Sci USA 91:2639–2643

Pulendran B, Smith KGC, Nossal GJV (1995a) Soluble Ag can impede affinity maturation and the germinal center reaction but enhance extrafollicular immunoglobulin production. J Immunol 155:1141–1150

Pulendran B, Kannourakis G, Nouri S, Smith KGC, Nossal GJV (1995b) Soluble antigen can cause enhanced apoptosis of germinal-centre B cells. Nature 375: 331–334

Pulendran B, Van Driel R, Nossal GJV (1997) Immunological tolerance in germinal centres. Immunol Today 18:27–32

Ray SK, Putterman C, Diamond B (1996) Pathogenic autoantibodies are routinely generated during the response to foreign antigen: a paradigm for autoimmune disease. Proc Natl Acad Sci USA 93:2019–2024

Rickers A, Brockstedt E, Mapara MY, Otto A, Dörken B, Bommert K (1998) Inhibition of CPP32 blocks surface IgM-mediated apoptosis and D4-GDI cleavage in human BL60 Burkitt lymphoma cells. Eur J Immunol 28:296–304

Russel JH (1995) Activation-induced death of mature T cells in the regulation of immune responses. Current Biol 7:382–388

Sato S, Miller AS, Inaoki M, Bock CB, Jansen PJ, Tang ML, Tedder TF (1996) CD22 is both a positive and negative regulator of B lymphocyte antigen receptor signal transduction: altered signaling in CD22-deficient mice. Immunity 5:551–562

Scaffidi C, Fulda S, Srinivasan A, Friesen C, Li F, Tomaselli KJ, Debatin K-M, Krammer PH, Peter ME (1998) Two CD95(APO-1/Fas) signaling pathways. EMBO J 17: 1675–1687

Schauer SL, Wang Z, Sonenshein GE, Rothstein TL (1996) Maintenance of nuclear factor-κB/Rel and c-myc expression during CD40 ligand rescue of WEHI 231 early B cells from receptor-mediated apoptosis through modulation of IκB proteins. J Immunol 157:81–86

Schulze-Osthoff K, Ferrari D, Los M, Wesselborg S, Peter ME (1998) Apoptosis signaling by death receptors. Eur J Biochem 254:439–459

Scott DW, Lamers M, Köhler G, Sidman CL, Maddox B, Carsetti R (1996) Role of c-Myc and CD45 in spontaneous and anti-receptor-induced apoptosis in adult murine B cells. Intern Immunol 8:1375–1385

Shi Y, Szalay MG, Paskar L, Boyer M, Singh B Green DR (1990) Activation-induced cell death in T cell hybridomas is due to apoptosis: morphological aspects and DNA fragmentation. J Immunol 144:3326–3333

Shokat KM, Goodnow CC (1995) Antigen-induced B-cell death and elimination during germinal-centre immune responses. Nature 375:334–338

Stanger BZ, Leder P, Lee T-H, Kim E, Seed B (1995) RIP: a novel protein containing a death domain that interacts with Fas/APO-1 (CD95) in yeast and causes cell death. Cell 81:513–523

Symer DE, Reim J, Dintzis RZ, Voss EW Jr, Dintzis HM (1995) Durable elimination of high affinity, T cell-dependent antibodies by low molecular weight antigen arrays in vivo. J Immunol 155:5608–5616

Takata M, Sabe H, Hata A, Inazu T, Homma Y, Nukada T, Yamamura H, Kurosaki T (1994) Tyrosine kinase Lyn and Syk regulate B cell receptor-coupled Ca^{2+} mobilization through distinct pathways. EMBO J 13:1341–1349

Takata M, Homma Y, Kurosaki T (1995) Requirement of phospholipase C-gamma 2 activation in surface immunoglobulin M-induced B cell apoptosis. J Exp Med 182:907–914

Tedder TF, Inaoki M, Sato S (1997) The CD19-CD21 complex regulates signal transduction thresholds governing humoral immunity and autoimmunity. Immunity 6:107–118

Truman JP, Ericson ML, Choqueux-Seebold CJ, Charron DJ, Mooney NA (1994) Lymphocyte programmed cell death is mediated via HLA class II DR. Int Immunol 6:887–896

Tsujimoto Y, Finger LR, Yunis J, Nowell PC, Croce CM (1984) Cloning of the chromosome breakpoint of neoplastic B cells with the t(14;18) translocation. Science 226:1097–1099

Ucker DS, Ashwell JD, Nickas G (1989) Activation-driven T cell death. I. Requirements for de novo transcription and translation and association with genome fragmentation. J Immunol 143:3461–3469

Wang Z, Karrras JG, Howard RG, Rothstein TL (1995) Induction of Bcl-x by CD40 engagement rescues sIg-induced apoptosis in murine B cells. J Immunol 155:3722–3725

Watson DC, Reim J, Dintzis HM (1996) Suppression of the antibody response to a polymorphic peptide from the platelet alloantigen integrin β3 with low molecular weight antigen arrays. J Immunol 156:2443–2450

Weinmann P, Bommert K, Mapara MY, Dörken B, Bargou RC (1997) Overexpression of the death-promoting gene Bax-α sensitizes human BL-41 Burkitt lymphoma cells for surface IgM-mediated apoptosis. Eur J Immunol 27:2466–2468

Wiesner DA, Kilkus JP, Gottschalk AR, Quintans J, Dawson G (1997) Anti-immunoglobulin-induced apoptosis in WEHI 231 cells involves the slow formation of ceramide from sphingomyelin and is blocked by Bcl-x. J Biol Chem 272:9868–9846

Yamanashi Y, Fukuda T, Nishizumi H, Inazu T, Higashi K, Kitamura D, Ishida T, Yamamura H, Watanabe T, Yamamoto T (1997) Role of tyrosine phosphorylation of HS1 in B cell antigen receptor-mediated apoptosis. J Exp Med 185:1387–1392

Yang E, Korsmeyer SJ (1996) Molecular thanatopsis: a discourse on the Bcl2 family and cell death. Blood 88:386–401

Yao XR, Scott DW (1993) Antisense oligodeoxynucleotides to the blk tyrosine kinase prevent anti-μ chain-mediated growth inhibition and apoptosis in a B-cell lymphoma. Proc Natl Acad Sci USA 90:7946–7950

Yin X-M, Oltvai ZN, Korsmeyer SJ (1994) BH1 and BH2 domains of Bcl-2 are required for inhibition of apoptosis and heterodimerization with Bax. Nature 369:321–323

Yuan J, Shaham S, Ledoux S, Ellis HM, Horvitz HR (1993) The C elegans cell death gene ced-3 encodes a protein similar to mammalian interleukin-1β-converting enzyme. Cell 75:641–652

Zha H, Aimé-Sempé C, Sato T, Reed JC (1996) Proapoptotic protein Bax heterodimerizes with Bcl-2 and homodimerizes with Bax via a novel domain (BH3) distinct from BH1 and BH2. J Biol Chem 271:7440–7444

Zheng L, Fisher G, Miller RE, Peschon J, Lynch DH, Lenardo MJ (1995) Induction of apoptosis in mature T cells by tumour necrosis factor. Nature 377:348–351

Zhou L-J, Smith HM, Waldscmidt TJ, Schwarting R, Daley J, Tedder TF (1994) Tissue-specific expression of the human CD19 gene in transgenic mice inhibits antigen-independent B lymphocytes development. Mol Cell Biol 14:3884–3894

Zhu L, Anasetti C (1995) Cell cycle control apoptosis in human leukemic T cells. J Immunol 154:192–200

CHAPTER 17
Modulation of Apoptosis and Maturation of the B-Cell Immune Response

G. KOOPMAN

A. Introduction

Apoptosis plays a central role in shaping both the T and B-cell immune repertoire. Apoptosis is involved both in the positive selection of immunocompetent lymphocytes, via deletion of noncompetent cells, as well as in the negative selection of lymphocytes that, for instance, have an undesired reactivity against auto-antigens. The natural history of a B lymphocyte can be divided into two subsequent phases, with apoptosis playing a role in each of them (Fig. 1). The first phase, which takes place in the bone marrow, consists of the development of mature, immunocompetent B lymphocytes from the pluripotent stem cell. This phase is presumed to be largely independent of T lymphocytes and antigen and is guided by the recombination of immunoglobulin genes into a functional membrane immunoglobulin (Ig) receptor. During this phase positive selection results in expansion of B cells that have successfully rearranged their Ig genes, while B cells with a faulty Ig receptor are deleted through apoptosis. The second phase of B-cell development is antigen dependent and takes place in the secondary lymphoid organs such as lymph nodes, spleen, and mucosa associated lymphoid tissues. It is initiated by specific recognition of antigen by the B-cell Ig receptor and results in activation and proliferation, thereby enlarging the pool of B cells specific for a given antigen. Some of the B cells differentiate into soluble Ig producing plasma cells, others develop into memory B cells (Fig. 1). In contrast to mature B cells, which express IgM and IgD, memory B cells have undergone isotype switching and express IgA, IgG, or IgE receptors. In addition, the Ig receptor is modified through somatic mutation (see Sect. B.III) and the B cells are re-selected on the basis of the changed affinity of their Ig receptor, which results in generation of memory B cells with an increased binding affinity for antigen. Apoptosis plays a major role in this selection process. This chapter treats the regulation of B-cell apoptosis in this antigen dependent phase of the B-cell immune response. Discussed are: (1) the antigen driven B-cell maturational process in detail; (2) the molecules involved in this process; (3) the regulation of B-cell maturation; and (4) the regulation of B-cell survival. Finally, results are summarized within the framework of a recently proposed triple check model of B-cell maturation.

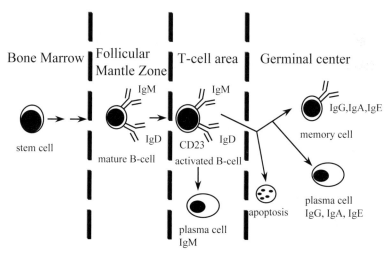

Fig. 1. The natural history of B-cell development. This figure represents an overview of the process of B-cell development schematically divided into four maturation zones. In the bone marrow (*zone 1*) stem cells differentiate into mature B cells that migrate to the follicular mantle zone of the secondary lymphoid organs (*zone 2*). After antigen encounter these B cells interact with activated T cell in the T-cell area of the lymphoid tissues (*zone 3*) after which part of the B cells mature into IgM producing plasma cells while other B cells proliferate and form a germinal center (*zone 4*). In the germinal center B cells are further selected on the basis of the antigen binding affinity of their Ig receptors. The large majority of them die by apoptosis, while others undergo isotype class switching and differentiate into memory or plasma cells

B. Antigen Dependent B-Cell Maturation in Secondary Lymphoid Organs

I. Anatomical Organization of the B-Cell Immune Response

The initiation of B-cell immune responses by thymus dependent antigens is a complex event requiring the close collaboration between antigen presenting dendritic cells, T cells, and B cells. The initial activation of antigen specific B cells is thought to take place in the T-cell area of secondary lymphoid organs, where antigen stimulated T cells provide help to the B cell (Fig. 1) (MACLENNAN and GRAY 1986; TEW et al. 1990). T cells stimulate B-cell proliferation and differentiation by release of cytokines like IL-2, IL-4, or IL-10, and by direct cell–cell contact involving both adhesion molecules and cross linking of CD40 on the B cells by interaction with CD154 expressed on the T cell (see DURIE et al. 1994 for review). Subsequently part of the stimulated B cells proliferate and form the germinal center, which presents a specialized B-cell compartment where further maturation takes place. Other B cells mature into plasma cells, producing predominantly IgM antibodies. Binding of these antibodies to antigen leads to the formation of immune complexes, some of which are trapped by follicular dendritic cells (FDC), a cell type that is present in the germinal center which is essential to B-cell maturation. The B-cell maturation in the germinal center microenvironment is discussed below in detail.

II. The Germinal Center Microenvironment

1. Cellular Composition of the Germinal Center

The main cellular constituents of the germinal center are activated B lymphocytes, follicular dendritic cells (FDC), tingible body macrophages, and T lymphocytes (STEIN et al. 1982; BUTCHER et al. 1982; ROUSE et al. 1982). FDC are large cells with elongated cytoplasmic extensions that form the framework of the germinal center (NOSSAL et al. 1968; SZAKAL and HANNA 1968). FDC express Fc receptors as well as complement receptors, through which they can bind antigen-antibody complexes (GERDES et al. 1983; PETRASCH et al. 1990; SCHRIEVER et al. 1989). These complexes can remain bound to the FDC for long periods of time, in undegraded form, thereby forming an antigen reservoir. Antigen on the FDC can be presented to the B cells either directly, or in the form of immune complex coated bodies, so-called iccosomes, that are released by the FDC (SZAKAL et al. 1988).

T cells are essential to germinal center formation; in their absence germinal center formation, isotype switching, and B-memory cell generation do not take place (see NIEUWENHUIS et al. 1992 for review). In contrast to their nongerminal center counterparts they are L-selectin (CD62L) negative and most of the cells express CD57.

2. B-Cell Subpopulations

As stated earlier, the germinal center is essential to the antigen-dependent maturation and differentiation of B cells. During this maturational process B cells go through a sequence of phenotypic and functional alterations, ultimately resulting in the formation of B-memory cells as well as plasma cells. As a consequence, B cells in the secondary lymphoid tissues can be subdivided into a number of phenotypically distinct subpopulations that occupy different zones within the follicular B-cell compartment (Fig. 2) (HARDIE et al. 1993; LIU et al. 1992; GRAY 1993). Thus upon antigenic challenge secondary follicles are formed that have a mantle zone comprised of mature IgM, IgD positive B cells and a germinal center containing activated IgD negative B cells (Fig. 2) (HARDIE et al. 1993; LIU et al. 1992; GRAY 1993). Mantle zone B cells can be further subdivided into resting cells that have not encountered antigen and B cells that have undergone their first activation and have upregulated CD23 and HLA-II expression (Fig. 2) (DEFRANCO et al. 1984; CAMBIER and

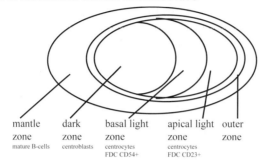

Fig. 2A,B. Schematic representation of germinal center architecture and B-cell maturational steps during an immune response: **A** B cells in the secondary lymphoid organs largely reside in follicles, where mature resting B cells are found in the mantle zone, while the activated B cells in part form a germinal center that consists of subcompartments containing centroblasts or centrocytes and different subsets of FDCs; **B** during an immune response B cells go through a series of phenotypic changes which in part are associated with their transition through a germinal center reaction. Here they also undergo isotype class switching and somatic hypermutation

CAMPBELL 1992; KLAUS et al. 1987; NOELLE et al. 1984). These cells then form centroblasts, which are rapidly proliferating cells that divide every 6–7 h. This high proliferation rate is required for the formation of a germinal center, as it is estimated that each germinal center is formed out of only one to three antigen stimulated B cells (KROESE et al. 1987). Centroblasts are $CD77^+CD10^+CD38^+CD27^+PNA^+$, while CD44, CD39, CD62L, Ig receptors, and Bcl-2 are downregulated (BUTCHER et al. 1982; KRAAL et al. 1982; MANGENEY et al. 1991; FEUILLARD et al. 1995; MAURER et al. 1990; MAURER et al. 1992). Following the centroblast stage, B cells develop into centrocytes that show reexpression of surface Ig receptors (Fig. 2) (LAGRESLE et al. 1993; LIU et al. 1989; KREMMIDIOTIS and ZOLA 1995; FEUILLARD et al. 1995). The majority of these cells have undergone Ig isotype switching and express IgG, IgA, or IgE. They have lost expression of CD77, but are otherwise, with regard to their surface markers, similar to centroblast, i.e., $CD10^+CD38^+CD27^+PNA^+CD44^-CD39^-CD62L^-Bcl-2^-$. These cells do not proliferate further and readily undergo apoptosis (LIU et al. 1989). Finally, plasma cells producing either IgA, IgG, or IgE class antibodies and B-memory cells, expressing CD44, CD39, CD27, and negative for CD77, CD10, and CD38 are formed (LAGRESLE et al. 1993; LIU et al. 1995). The B-cell maturational stages described above are found in distinct subregions of the germinal center that also differ in FDC composition (Fig. 2) (LIU et al. 1992; HARDIE et al. 1993). The centroblasts reside in the dark zone of the germinal center. This region contains only few and relatively small FDC. It is assumed that in this region the B-cell Ig repertoire is further diversified through somatic mutation (see below). Subsequently the centrocytes enter the light zone that is densely populated with FDC. The light zone is further subdivided into a basal light zone, where the FDC are strongly ICAM-1 (CD54) positive, and an apical light zone containing FDC with strong CD23 and moderate ICAM-1 expression (LIU et al. 1992; HARDIE et al. 1993). In the light zone the B cells, that now express Ig receptors that are modified by somatic hypermutation, are either selected for further differentiation or deleted through apoptosis. Apoptotic cells are degraded by local macrophages, described as tingible body macrophages. Finally, an outer zone has been postulated that contains CD75 positive B cells that might be traveling between the diverse follicular compartments (LIU et al. 1992; HARDIE et al. 1993).

Several subsets have been described that, on the basis of their phenotype, are thought to span the gap between the "activated" $CD23^+IgD^+IgM^+$ positive B cell and the Ig negative germinal center B cell, i.e., the so-called "germinal center founder cell." Thus occasionally transitional, IgD positive, germinal centers are found (LIU et al. 1996a; LENS et al. 1996a). In LIU et al. (1996a) these cells were found to express the germinal center marker CD38. They could be further subdivided into an IgM positive and IgM negative subset (Fig. 2) (LIU et al. 1996a). The IgM positive cells were found to be partly small noncycling cells and partly blastoid KI67 positive, proliferating cells (LEBECQUE et al. 1997). These cells were also shown to be extremely sensitive for apoptosis induction. Recently an additional IgD positive germinal center subset was described, characterized by expression of CD70 (LENS et al. 1996a). These cells were found to carry the naive B-cell markers CD44 and CD39 as well as the germinal center/memory cell marker CD27, while they were negative for CD10 or CD38 (Fig. 2). These cells therefore seem to represent a mantle zone/germinal center intermediate preceding the $IgD^+,CD38^+$ cell type described above. All B-cell subpopulations described thus far are listed in Fig. 2 in a sequential maturation order. As depicted also in Fig. 2, B cells with a "founder" cell and centroblast phenotype are strongly proliferating, while propensity to undergo apoptosis is seen from the "founder" cell to the centrocyte stage of development.

III. Ig Switching and Somatic Hypermutation

The three most prominent changes that take place in B cells that go through a germinal center maturation phase are isotype switching, affinity maturation, and memory cell formation (Fig. 2) (GRAY 1993; MACLENNAN and GRAY 1986; NIEUWENHUIS et al. 1992; TEW et al. 1990; BEREK et al. 1991; BEREK 1992; BEREK et al. 1985; GRIFFITHS et al. 1984). As described above, naive B cells express IgM and IgD class surface Ig receptors, while memory cells typically express IgA, IgG, or IgE. Recently the switching process was studied in detail by LIU et al. (1996b), who investigated the appearance of sterile transcripts, which are transcripts containing an I exon upstream of the S region and are found only during the first phase of the switching process, in tonsil B-cell subpopulations. These transcript were found to be present in centrocytes only, indicating that Ig class switching starts at the transition from the centroblast to the centrocyte stage (Fig. 2). Interestingly, some of the centroblasts were found to have deleted their IgM locus and to express IgG or IgA transcripts, despite the absence of sterile transcript. This may indicate that some of these B cells have undergone Ig switching earlier during a previous germinal center cycle and are in fact representing reactivated memory cells going through an additional germinal center reaction.

During an immune response there is an increase in the antigen binding affinity, a process called affinity maturation. Affinity maturation is the result of two distinct processes, somatic hypermutation and immune selection (GRIFFITHS et al. 1984; BEREK et al. 1985, 1991; BEREK 1992). Somatic hypermutation is a unique process through which random mutations are generated in the Ig heavy and light chains. Through this process further diversity, besides Ig gene rearrangement and junctional diversity occurring during B-cell lymphopoiesis in the bone marrow, is added to the B-cell Ig receptor. Somatic hypermutations are absent in IgD^+, $CD77^-$, $CD38^-$ mantle zone B cells and are detected at a low level in the IgD^+, IgM^+, $CD38^+$ germinal center "founder" cell and at high levels in all further maturated B-cell subpopulations (LIU et al. 1996a; PASCUAL et al. 1994). The majority of these mutations will result in defective V genes or in V genes that encode variable domains with a decreased affinity for the antigen. Only in a few cases will these mutations lead to an increase in affinity. High affinity B cells are then positively selected for further maturation into memory B cells or plasma cells, while the B cells with low affinity Ig receptors die through apoptosis (see Sects. D and E).

C. Cell Surface Molecules Involved in Regulation of B-Cell Maturation and Apoptosis

As already stated, the immune repertoire is shaped by selection. Cells that are positively selected will mature into functional immune cells, while the superfluous nonselected cells are deleted through apoptosis. Often the processes of maturation and apoptosis are interrelated and in fact regulated by the same molecules. In this section, two groups of cell surface molecules, the TNF/NGF receptor family and adhesion molecules, that have been implicated in maturation/apoptosis regulation will be described. Because these families contain a large number of different molecules, a description of which is beyond the scope of this chapter, only those molecules that have specifically been implicated in the regulation of apoptosis of B cells will be described in detail. Although cytokines are also important in the regulation of maturation and apoptosis they will not be described separately and are only referred to in Sects. D and E as these processes are described in detail.

I. The TNF/NGF Receptor Family

The TNF/NGF receptor family, with the two exceptions T2 and A53R, are all type I membrane proteins with sequence homology confined to the extracellular region (SMITH et al. 1994; BAZZONI and BEUTLER 1996). Several cysteine-rich pseudorepeats are present in the extracellular region, each containing about 6 cysteines and 40 amino acids. These molecules are expressed on the cell surface, although many receptors are also released in soluble form by proteolysis. Molecules belonging to this family are, amongst others; TNFR I, TNFR II, NGFR, CD27, CD30, CD40, CD95, OX40, 4–1BB, TRAIL, TRANCE-L/RANK (BAZZONI and BEUTLER 1996; ANDERSON et al. 1997; PAN et al. 1997; SHERIDAN et al. 1997; WONG et al. 1997a,b). Their ligands are also structurally related to each other and belong to the TNF family. All TNF family members, except LTα which appears to be a secreted protein, are type II membrane proteins, with a sequence homology in the C-terminus extracellular region, which folds into a β-plated sheet sandwich. Typically these molecules form trimeric molecular complexes. Binding between a TNF receptor family member and its trimeric ligand generally leads to trimerization of the TNF receptor molecule, which results in functional activation of the receptor. Broadly, TNF receptor family molecules can be subdivided into molecules that induce cell activation and proliferation and molecules that carry a so-called "death domain" and induce apoptosis. The most prominent TNF receptor molecule involved in B-cell activation is CD40. CD40 is strongly expressed on B cells and dendritic cells, while its ligand CD154 is mainly expressed on activated CD4 T cells (Table 1) (ARMITAGE et al. 1992; LEDERMAN et al. 1992b; LIU et al. 1989; ROY et al. 1993). Triggering of CD40 has been implicated in

Table 1. Cell surface molecules involved in the regulation of apoptosis or maturation of B cells. The receptor/ligand pairs, family designation, function and expression pattern, within the lymphoid system, from a limited set of molecules that are referred to in this chapter are shown

Receptor/Ligand	Family	Function	Expression
CD40/CD154	TNF-R/TNF	B-cell proliferation, B-cell and DC maturation	CD40: B, DC, FDC
			CD154: activated T
CD27/CD70	TNF-R/TNF	T- and B-cell proliferation	CD27: T cell, activated B
			CD70: activated T and B
CD95/CD95Ligand	TNF-R/TNF	T- and B-cell apoptosis	CD95: activated T and B
			CD95L: CTL, Th1, NK cells
CD134/OX40L	TNF-R/TNF	T-cell proliferation B-cell proliferation and Ig secretion	CD134: activated T OX40L: activated T and B, DC
TNF-R/TNFα	TNF-R/TNF	Apoptosis and DC maturation	TNF-R: broad
			TNFα: macrophages, lymphocytes
CD11aCD18/CD54 (LFA1/ICAM-1)	Integrin/Ig family	Adhesion, proliferation	CD11aCD18: broad
			CD54: B, activated T, DC, FDC
CD49dCD29/CD106 (VLA-4/VCAM-1)	Integrin/Ig family	Adhesion, proliferation	CD49dCD29: B and T
			CD106: DC, FDC
CD44/hyaluronic acid		Adhesion, proliferation	CD44: broad

DC, dendritic cell; FDC, follicular dendritic cell.

B-cell proliferation induction, Ig class switching, and memory cell formation (GRAY et al. 1994; GALIBERT et al. 1996b; ALLEN et al. 1993; CALLARD et al. 1993; LEDERMAN et al. 1992a, 1994; LIU et al. 1992; LANE et al. 1992; ARMITAGE et al. 1992; NOELLE et al. 1992; SPRIGGS et al. 1992; ROUSSET et al. 1991; JABARA et al. 1990). CD27 is present on T cells, activated B cells, germinal centers, and memory B cells, while its ligand CD70 is found on activated B cells, activated T cells, and some stromal cells in the thymus (HINTZEN et al. 1994, 1995). Triggering of CD27 has been shown to provide a co-stimulatory signal to T cells, enhancing proliferation, while more recently co-stimulation of B-cell proliferation was also documented (AGEMATSU et al. 1994, 1995; GOODWIN et al. 1993; HINTZEN et al. 1995). Recently, cross linking of OX40Ligand, which is expressed on anti-IgD or anti-CD40 stimulated B cells, was shown to enhance B-cell proliferation and Ig production, while it had no effect on Ig class switching (STUBER et al. 1995).

Both CD95 and TNFR I and II carry a so-called death domain sequence within their cytoplasmic domain (SMITH et al. 1994; BAZZONI and BEUTLER

1996). Cross linking of these molecules through interaction with their counter receptors results in activation of an intracellular signal transduction cascade ultimately leading to initiation of a cell death program. Triggering of CD95 has been implicated in initiation of both T- and B-cell death (KRAMMER et al. 1994; NAGATA and GOLSTEIN 1995; TRAUTH et al. 1989). CD95 is present on activated T-and B cells and strongly expressed on germinal center B cells (KRAMMER et al. 1994; NAGATA and GOLSTEIN 1995; TRAUTH et al. 1989; MIYAWAKI et al. 1992; MOLLER et al. 1993; ROTHSTEIN et al. 1995; LAGRESLE et al. 1995; DEBATIN et al. 1990). Expression of CD95L is much more limited and only documented on cytotoxic T lymphocytes, T helper 1 cells, and NK cells (OSHIMI et al. 1996; RAMSDELL et al. 1994; NAGATA and GOLSTEIN 1995; SUDA et al. 1993). The role of TNFα, lymphotoxins, and the 55kD and 75kD TNF receptors in the regulation of B-cell proliferation and differentiation is less clear. However, the fact that germinal centers are absent in lymphotoxin α and in 55kD TNF receptor deficient mice highlights their importance in the generation of the secondary B-cell immune response (MATSUMOTO et al. 1997). Interestingly, instead of inducing apoptosis, TNFα was recently shown to inhibit anti-Ig induced apoptosis in a Burkitt lymphoma cell line (LENS et al. 1996b).

II. Adhesion Molecules

Adhesion molecules are cell surface receptors that mediate the binding of cells to other cells or to the extracellular matrix (for reviews see SPRINGER 1990; HEMLER 1990; HEMLER and LOBB 1995; SPRINGER 1994). In this chapter only those adhesion molecules that have been shown to play a role in cell activation, besides adhesion, will be discussed further. The role of the integrin family of adhesion molecules in cell activation is especially well documented. Integrins are heterodimeric membrane proteins that interact either with extracellular matrix proteins like collagen, laminin, or fibronectin, or cell surface bound counter receptors which generally belong to the Ig superfamily. The most widely studied of them, with regard to cell activation, are LFA-1 (CD11a/CD18) that interacts with ICAM-1 (CD54), and VLA-4 (CD49d/CD29) that binds to VCAM-1 (CD106) and fibronectin. LFA-1 has a broad tissue distribution, but is negative on FDC, while VLA-4 is expressed on T cells and B cells only (for reviews see SPRINGER 1990; HEMLER 1990; HEMLER and LOBB 1995; SPRINGER 1994). The LFA-1 counter receptor ICAM-1 is found on B cells, activated T cells, dendritic cells, FDC, and endothelium, while the VLA-4 counter receptor VCAM-1 is positive on dendritic cells, FDC, and endothelium. Cross linking of LFA-1 or VLA-4 provides a co-stimulatory signal, enhancing anti-CD2 and anti-CD3 induced T-cell proliferation (SHIMIZU et al. 1990; VAN SEVENTER et al. 1990). Besides integrins, CD44, which forms a heterogeneous group of molecules that are all derived from a single transcript through alternative splicing, has been implicated in cell activation as it enhances both cell proliferation and cell adhesion (DENNING et al. 1990;

HUET et al. 1989; KOOPMAN et al. 1990; SHIMIZU et al. 1989). CD44 can interact with extracellular matrix molecules like hyaluronic acid, collagen, fibronectin, and laminin (JALKANEN and JALKANEN 1992; LESLEY et al. 1990; MIYAKE et al. 1990; UNDERHILL et al. 1987).

D. Regulation of B-Cell Maturation

The recognition of distinct B-cell subsets, that arise during an immune response, and the development of magnetic cell separation techniques, which has made it possible to purify these subsets in sufficient numbers from, for instance, inflamed human tonsils, has spurred research on the factors involved in the regulating B-cell maturation. Specifically the differentiation of mature resting B cells into centroblasts, of centroblasts into centrocytes, and of centrocytes into either memory cells or plasma cells has been studied (Fig. 3). In vitro studies showed that CD40 in combination with cytokines including IL-2, IL-4, and IL-10 is involved in many of these maturation steps (ARPIN et al. 1995; LAGRESLE et al. 1995; LIU et al. 1989, 1991; GALIBERT et al. 1996b; CASAMAYOR-PALLEJA et al. 1996). Moreover, blocking of CD40-CD154 interactions in vivo with anti CD154 mAb was found to inhibit germinal center and memory B-cell formation (FOY et al. 1994; HAN et al. 1995a). However, in another study, blocking the same pathway with a CD40 construct did not prevent germinal center formation, although memory B-cell formation and Ig class switching were impaired (GRAY et al. 1994). The difference between these models lies in the fact that the anti-CD154 mAb may cross link CD154, besides blocking the CD40-CD154 interaction, and that it may in fact be this CD154 cross linking that inhibits T cell activation and the release of soluble cytokines and thus thereby the signal necessary for germinal center formation (VAN ESSEN et al. 1995). However this may be, the CD40-CD154 pathway is central to many features associated with the B-cell immune response, as also evidenced by the severe immune perturbation in X-linked hyper IgM syndrome patients, who lack a functional CD40 molecule (ALLEN et al. 1993; CALLARD et al. 1993) and as a consequence cannot switch their Ig genes. In vitro studies using purified follicular dendritic cells or a follicular dendritic-like cell line suggest that, besides their involvement in rescue from apoptosis, these cells are also important in the regulation of B-cell maturation (CHOE et al. 1996, 1997; CLARK et al. 1995; KIM et al. 1995; LINDHOUT et al. 1994; GROUARD et al. 1995).

I. The Initiation of the B-Cell Immune Response, Formation of Centroblasts

Binding of antigen to the membrane Ig receptor results in B-cell activation. Some multivalent antigens that contain multiple identical epitopes can extensively cross link the Ig receptor and directly induce B-cell proliferation. These antigens are called thymus independent (TI). For the thymus dependent

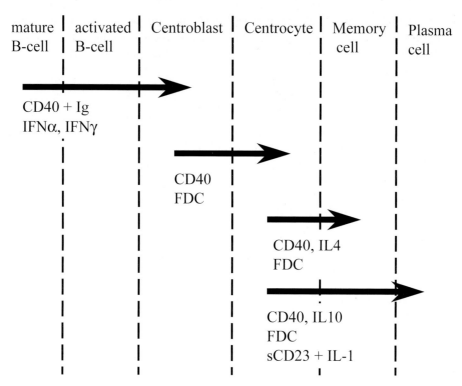

Fig. 3. Regulation of B-cell maturation. Studies on purified B-cell subpopulations have identified the factors involved in the regulation of four subsequent steps in B-cell maturation during an immune response

(TD) antigens, however, the induction of B-cell proliferation and maturation requires co-stimulatory signals (help) provided by the T cell. The initial activation of B cells therefore usually takes place in the T-cell areas of the lymphoid tissues, where antigen is presented in the context of MHC molecules by dendritic cells to the T cell (MACLENNAN and GRAY 1986; TEW et al. 1990; INABA and STEINMAN 1987; KING and KATZ 1989). Following their activation, T cells upregulate expression of the CD154 and CD54 cell surface molecules that interact with CD40 and CD11a/CD18 expressed on the B cell, resulting in B-cell activation (TOHMA et al. 1991; NOELLE et al. 1992; LANE et al. 1992; LEDERMAN et al. 1992b). In addition, T cells secrete B-cell stimulatory cytokines like IL-2, IL-4, and IL-10. These cytokines in combination with the antigenic stimulation and interaction of co-stimulatory cell surface molecules already mentioned drive B cells into cell cycle. In addition, triggering of CD40 on the B cell induces CD23 and upregulates MHC class II expression, thereby generating an activated B-cell phenotype (DEFRANCO et al. 1984; CAMBIER and CAMPBELL 1992; KLAUS et al. 1987; NOELLE et al. 1984). How a germinal center is formed from these activated B cells is still largely unclear. However, recently a possible role for migration inducing factors produced by FDC was proposed

(VAN DER VOORT et al. 1997). Thus, FDC were found to produce the growth and motility factor hepatocyte growth factor/scatter factor. Its ligand c-met was found to be expressed on $CD38^+CD77^+$ centroblast phenotype B cells (VAN DER VOORT et al. 1997). Another factor possibly involved in the regulation of B-cell migration to the germinal center is BCA-1, a CXC chemokine that is expressed in follicles and interacts with CXCR5 on the B cell (GUNN et al. 1998; LEGLER et al. 1998). Recently, GALIBERT et al. (1996b) studied the induction of centroblast markers on nongerminal center, CD38 negative, tonsillar B cells. Stimulation of these B cells via CD40 and the Ig receptor was shown to induce expression of CD95, carboxy peptidase M, and CD38. In addition the cells became sensitive to apoptosis. CD38 expression could also be induced by $IFN\alpha$ or $IFN\gamma$. However, the cells remained negative for CD10 and positive for CD44, indicating that only a partial germinal center phenotype was generated. Moreover, the fact that these experiments were performed with CD38 negative cells, which comprise both naive IgD positive and memory IgD negative B cells, calls for caution in their interpretation.

II. Differentiation of Centroblasts into Centrocytes

As stated earlier, centroblasts are rapidly proliferating, apoptosis sensitive cells whose Ig V genes undergo somatic hypermutation (FEUILLARD et al. 1995; MANGENEY et al. 1991; PASCUAL et al. 1994; BUTCHER et al. 1982; KRAAL et al. 1982). During this phase the few initially activated B cells are greatly expanded, while through somatic hypermutation a pool of related B cells with diversified Ig receptors is generated. These cells then differentiate into nonproliferating but still apoptosis sensitive centrocytes (LAGRESLE et al. 1993; LIU et al. 1989; KREMMIDIOTIS and ZOLA 1995). Recently CHOE et al. (1997) showed that culturing centroblasts in the presence of CD154, an FDC like cell line, IL-2, and IL-10 downregulated CD77 expression, which was further decreased by addition of anti-Ig, implicating these molecules in the regulation of centroblast to centrocyte differentiation. However, the same culture conditions resulted in up-regulation of CD44, which is not expressed on centrocytes but on further differentiated memory B cells. Therefore these molecules are not specifically involved in centrocyte formation but drive the entire differentiation route from centroblasts to memory cell.

III. Differentiation of Centrocytes into Memory and Plasma Cells

Centrocytes form the focal point in the germinal center reaction. The large majority of them die by apoptosis and only those cells whose mutated Ig receptor have a high affinity for the antigen are selected for further maturation. This selection process and the molecules involved in apoptosis regulation will be dealt with in Sect. E. Subsequently the selected B cells mature further into either memory or plasma cells. In vitro studies indicate that CD40 also plays an important role in this phase of the B-cell maturation. Thus, ARPIN et al.

(1995) found that continuous stimulation of germinal center B cells via CD40 in the presence of IL-4 and IL-10 resulted in the formation of memory phenotype cells, while a two-phase culture system with initial activation via CD40 plus IL-4 and IL-10 followed by culture with IL-4 and IL-10 in the absence of CD40 led to the formation of plasma cells. CASAMAYOR-PALLEJA et al. (1996) and LAGRESLE et al. (1995), however, found that triggering of CD40 induced only a partial memory phenotype and that instead addition of CD45RO$^+$ T cells was required for downmodulation of, for instance, CD77 and CD23. Studies by CHOE et al. (1996) stressed the role of IL-10 in driving the differentiation of centrocytes into plasma cells. Triggering of CD21 on the B cell, via interaction with its counter receptor CD23, plus addition of IL-1α have also been shown to promote plasma cell formation (BJORCK et al. 1993; LIU et al. 1991). Thus, in conclusion, despite mutual differences in these reports, T cells through expression of CD154 and secretion of cytokines are important in driving B-cell maturation in the germinal center. This point is further supported in studies (Kosco et al. 1988; Kosco 1991) which showed that in vitro proliferation of germinal center B cells isolated from immunized mice required help by autologous primed T cells. In a mixed culture system containing FDC, B cells, and T cells from immunized mice it was shown that antigen sequestered on the FDC is subsequently taken up by the B cells and presented by them to the T cells, which then provide B-cell help.

CD154, the ligand of CD40, is expressed on activated CD4 T cells (ARMITAGE et al. 1992; LEDERMAN et al. 1992b). Importantly, preformed CD154 was shown to be present in the cytoplasm of CD45RO memory T cells, and to be expressed on the cell surface within 5 min after TCR cross linking (CASAMAYOR-PALLEJA et al. 1995). Immunohistochemical studies have shown CD154 expressing T cells to be present in the T-cell area, where the immune response is started, and in the outer zone of the germinal center (CASAMAYOR-PALLEJA et al. 1995). The lack of CD154 expression in the light zone of the germinal center may indicate that triggering of CD40 only becomes important in the regulation of B-cell survival and maturation after an initial phase of B-cell selection in the germinal center light zone, thus at a point that the selected cells start migrating out of the germinal center. RT-PCR analysis of germinal center T cells, defined as CD57 expressing CD4 T cells, showed strong expression of IL-4 mRNA, while IL-2, IL-6, IL-10, TNFα, or IFNγ mRNA was detected in the germinal center T cells of some tonsils only (BOWEN et al. 1991; BUTCH et al. 1993), stressing the importance of IL-4 in the germinal center reaction.

Another cell type, besides the T cell, that plays a central role in B-cell maturation is the follicular dendritic cell. As explained in Sect. B.II, FDC can bind antigen in its native form for long periods of time. Studies using anti-Ig as a surrogate for antigen have shown that cross linking of the Ig receptor on centrocytes mainly affects B-cell survival, while specific effects on centrocyte maturation have not been documented (LIU et al. 1989). However, anti-CD40 plus cytokine induced germinal center B-cell proliferation and Ig production can be further enhanced by Ig receptor cross linking (LAGRESLE et al. 1993). Besides their function in the delivery of antigens to the B cell, FDC also have a more direct effect on B-cell maturation. Addition of purified FDC or an FDC cell line to anti-Ig or anti-CD40 stimulated B cells enhanced their proliferation and in combination with IL-10 promoted plasma cell formation and Ig secretion (CHOE et al. 1996, 1997; CLARK et al. 1995; GROUARD et al. 1995; KIM et al. 1995). Using a transwell system this stimulatory effect by FDC was shown to involve soluble factors, while direct cell–cell contact also contributed to the proliferation stimulatory effect (KIM et al. 1995). The nature of these soluble factors is still unclear, especially because RT-PCR analysis on purified FDC has shown no expression of IL-1α, IL-1β, IL-2, IL-3, IL-4, IL-5, IL-6, IL-10, TNFα, or IFNγ (BUTCH et al. 1993).

E. Regulation of B-Cell Survival

In contrast to mature B cells and memory B cells, germinal center B cells spontaneously undergo apoptosis upon in vitro culture (LIU et al. 1989). Germinal center B cells can be subdivided into centroblasts, which are rapidly proliferating cells that initially form the germinal center and reside in the germinal center dark zone, and centrocytes that are nonproliferating and reside in the light zone (LIU et al. 1992; HARDIE et al. 1993). Importantly, while centroblasts have lost surface Ig expression the centrocytes are surface Ig positive and can therefore interact with antigen (LAGRESLE et al. 1993; KREMMIDIOTIS and ZOLA 1995). The interaction between antigen and the Ig receptor is though to be the key determinant in the selection of these B cells. However, besides antigen other molecules have been described to affect germinal center B-cell survival. Antigen as well as these additional survival signals are delivered to the B cell via two cell types that are present in the germinal center – the follicular dendritic cell and the T cell.

I. Apoptosis Regulation by Antigen and Follicular Dendritic Cells

The importance of antigen in apoptosis regulation can be inferred from in vitro studies where it was found that cross linking of the Ig receptor on purified germinal center B cells inhibited their entry into apoptosis (LIU et al. 1989). However, induction of apoptosis by excessive Ig cross linking in vitro as well as by injection of antigen in vivo shortly after initiation of an immune response has also been documented (GALIBERT et al. 1996a; HAN et al. 1995b; PULENDRAN et al. 1995; SHOKAT and GOODNOW 1995). Antigen induced apoptosis was found to be independent of CD95, as similar effects were seen in lpr mice, which do not express CD95 (SMITH et al. 1995). Importantly, excessive cross linking of Ig receptors did not induce apoptosis of mature B cells or memory B cells, indicating that only at the germinal center stage of development are B cells sensitive to this effect (GALIBERT et al. 1996a). Thus, these studies indicate that the germinal center forms a site for both positive as well as negative selection and that only those B cells with an intermediate affinity for the antigen are selected for further maturation.

FDC play an important role in bringing the germinal center B cells into contact with the antigen. FDC have been reported to express several Fc receptors (CD23, CD16, and CD32) as well as complement receptors (CD35, CD21, CD11b) (GERDES et al. 1983; SCHRIEVER et al. 1989; PETRASCH et al. 1990) through which they can bind antigen-antibody complexes (TEW and MANDEL 1979; KLAUS et al. 1980). Antigen on the FDC is either presented directly to the B cells, or in the form of iccosomes, which are released by the FDC (SZAKAL et al. 1988). The interaction between FDC and germinal center B cells and T cells is regulated by adhesion molecules (FREEDMAN et al. 1990; LOUIS et al. 1989; KOOPMAN et al. 1991; KOSCO et al. 1992; RICE et al. 1991). FDC strongly express ICAM-1 (CD54) and VCAM-1 (CD106). Through these molecules they can interact with LFA-1 (CD11a/18) and VLA-4 (CD49d/CD29) on germinal center B cells. In vitro studies using purified FDC and germinal center B cells have shown that their interaction mainly involves the binding of LFA-1 on the B cell to its counter receptor, ICAM-1, on the FDC (LOUIS et al. 1989; KOOPMAN et al. 1991; KOSCO et al. 1992). However, binding between VLA-4 on the B cell and VCAM-1 on the FDC also plays a role (FREEDMAN et al. 1990; KOOPMAN et al. 1991; KOSCO et al. 1992). Binding of germinal center B cells to FDC inhibits their entry into apoptosis, while disruption of the FDC/B-cell binding by using mAb directed against these adhesion molecules promotes B-cell apoptosis (KIM et al. 1995; LINDHOUT et al. 1993, 1994). Besides the presentation of antigen, the adhesive interaction itself was also shown to contribute to inhibition of apoptosis (KOOPMAN et al. 1994). Thus, triggering of the adhesion receptors LFA-1 and VLA-4 on the B cell, though binding to their ligands ICAM-1 and VCAM-1 that were immobilized on plastic surfaces, prevented B-cell entry into apoptosis (KOOPMAN et al. 1994). Importantly, anti-Ig stimulation and adhesion receptor triggering were found to act synergistically. In vivo this synergy between signals delivered through antigen and adhesion receptors may be crucial for effective B-cell activation, as antigen levels in vivo might be too low to induce fully B-cell activation by itself. Indeed, similar to what has been described for the T cell, low levels of antigen may induce only a small initial activation that, however, may switch the LFA-1 and VLA-4 receptors from an inactive into an active binding mode (DANG et al. 1990; DUSTIN and SPRINGER 1989; VAN KOOYK et al. 1989). Subsequent binding to their adhesive counter receptors will then provide additional stimulation leading to full cell activation (DRANSFIELD and HOGG 1989; KOOPMAN et al. 1992; VAN KOOYK et al. 1989; VAN NOESEL et al. 1988). In other cell culture systems adhesion molecules have now also been implicated in apoptosis regulation. Thus binding of CHO cells via the integrin molecule CD49e/CD29 to fibronectin inhibited their entry into apoptosis under low serum culture conditions, and dexamethason or anti-CD3 induced apoptosis of a mouse T-cell hybridoma was inhibited by binding via CD44 to hyaluronic acid (ZHANG et al. 1995; AYROLDI et al. 1995).

II. Apoptosis Regulation by T Cells

Several studies have shown that triggering of CD40 on germinal center B cells prevents their entry into apoptosis (ARPIN et al. 1995; LEDERMAN et al. 1994; LENS et al. 1996b; LIU et al. 1989). As stated in Sect. D, CD154, the ligand of CD40, is found on T cells residing in the T-cell areas, where the B-cell immune response is initiated and in the outer zone of the germinal center (CASAMAYOR-PALLEJA et al. 1995). Although hardly any information is available on the outer zone it was suggested to form a transitory compartment, where B cells are either driven back into the germinal center for another round of selection or differentiate further into memory or plasma cells and leave the germinal center (HARDIE et al. 1993; LIU et al. 1992). The role of CD40 in apoptosis regulation in the germinal center may therefore be limited to this transition phase and linked to its function in the regulation of maturation of germinal center B cells (see Sect. D). Interestingly, part of the effect of CD40 on apoptosis inhibition may be mediated through adhesion molecules, as recently anti-CD40 mediated rescue from anti-IgM induced apoptosis in the B-cell line DND-39 was shown to be abolished by anti-LFA-1 and anti-ICAM-1 mAb (SUMIMOTO et al. 1994). Importantly, triggering of CD40 has been shown to increase LFA-1 dependent cell adhesion (BARRETT et al. 1991).

CD95 is strongly expressed on germinal center B cells, while mature resting B cells are CD95 negative (GALIBERT et al. 1996b; LAGRESLE et al. 1995; MIYAWAKI et al. 1992; MOLLER et al. 1993). Expression of CD95 was shown to be induced on mature B cells by stimulation via CD40, while stimulation with anti-IgM did not upregulate CD95 expression (MOLLER et al. 1993; GARRONE et al. 1995; LAGRESLE et al. 1995; ROTHSTEIN et al. 1995). These cells subsequently became sensitive to anti-CD95 induced cell death (GARRONE et al. 1995; LAGRESLE et al. 1995; ROTHSTEIN et al. 1995; NAKANISHI et al. 1996). Similar results have been reported using a Burkitt lymphoma cell line (LENS et al. 1996b). Thus, it seems that the initial CD40 mediated B-cell activation in the T-cell areas induces a state of CD95-dependent apoptosis sensitivity in their progeny, i.e., the germinal center B cells. However, purified germinal center B cells spontaneously die by apoptosis upon in vitro culture without additional triggering of the CD95 molecule (LIU et al. 1989). Conflicting data have been published regarding the effect of anti-CD95 on the level of germinal center B-cell apoptosis (CLEARY et al. 1995; KOOPMAN et al. 1997; LAGRESLE et al. 1995; LIU et al. 1995). While CLEARY et al. (1995) found increased apoptosis after 10h of culture, LIU et al. (1995) found that anti-CD95 increased the amount of apoptosis only after 4h of culture and not after 12h or 24h of culture, and LAGRESLE et al. (1995) found no change in apoptosis during a 2–12 h culture period. We previously described no increase in apoptosis by addition of anti-CD95 after 16h or 48h of culture (KOOPMAN et al. 1997). However, that anti-CD95 did have an effect on these cells was shown by the observation that using anti-CD95 in combination with anti-Ig or adhesion molecule mediated rescue signals did result in B-cell apoptosis, even when anti-Ig and adhesion molecules were used in combination (KOOPMAN et al. 1997). Thus although the CD95 molecule on germinal center B cells is functionally active, its involvement in the spontaneous apoptosis sensitivity of these B cells is still unclear. Indeed, lpr mice were shown to be able to generate a normal germinal center response following antigenic stimulation and memory B cells were formed that had gone through somatic hypermutation and selection (SMITH et al. 1995). However, studies in lpr mice have also shown CD95 to be required for elimination of auto-reactive B cells (RATHMELL et al. 1995). Possibly there are two phases in B-cell selection during a secondary immune response, i.e., (1) CD95-independent apoptosis of centrocytes in the germinal center light zone and (2) CD95-dependent apoptosis at a later phase. Although at present it is unclear what other factors may regulate apoptosis of germinal center B cells, an interesting candidate might be TRAIL (APO-2 ligand), which induces apoptosis via binding to DR4 and DR5, while binding to the decoy receptor DcR1 reduces apoptosis induction (JEREMIAS et al. 1998; PAN et al. 1997; SHERIDAN et al. 1997).

Binding of germinal center B cells to FDC, which is known to prevent their entry into apoptosis, also prevented apoptosis of these cells in the presence of anti-CD95 (KOOPMAN et al. 1997). In view of the results described above, this inhibition of CD95 mediated apoptosis could not be attributed to antigen presentation or adhesive interactions alone, and should involve other as yet unidentified factors.

Conflicting data on apoptosis of germinal center B cells have been reported regarding the effect of combined triggering of CD40 and CD95 (KOOPMAN et al. 1997; LAGRESLE et al. 1995; CLEARY et al. 1995). The outcome of these studies, apoptosis or survival, seems to be determined in part by the time course of the experiment. For instance, when germinal center cells were cultured in the presence of anti-CD40 plus anti-CD95, low amounts of apoptosis were found in a 10h experiment by CLEARY et al. (1995) and a 16h experiment performed by us (KOOPMAN et al. 1997), while high numbers of apoptotic cells were seen after 48h of culture (KOOPMAN et al. 1997; LAGRESLE et al. 1995). Thus, CD40 can only temporarily prevent CD95 mediated induction of apoptosis. This transient nature of apoptosis inhibition conforms to the short time that B cells reside in the germinal center and the reported fast up-regulation and down-regulation of CD154 expression on T cells (CASAMAYOR-PALLEJA et al. 1995; YELLIN et al. 1994). Both the CD40 mediated rescue signal and the CD95 mediated death signal could be delivered by T cells residing in the germinal center. However, while CD154 is expressed on T cells in the germinal center outer zone, expression of the ligand of CD95 on germinal center T cells has so far not been described (CASAMAYOR-PALLEJA et al. 1995).

III. Triple Check Hypothesis of B-Cell Selection

The amplification in magnitude and the increase in antigen binding affinity that occur during a B-cell immune response are potentially hazardous to the organism as they may amplify not only antigen specific responses but also reactions to the organism itself. The B-cell immune response is therefore under tight regulatory control. In a recent paper we proposed a triple check model where the appropriate reactivity of B cells to nonself molecules only is achieved by: (1) T-cell dependent stimulation by antigen specific T cells in the T-cell area during initiation of the immune response; (2) antigen dependent selection of B cells with mutated Ig genes in the germinal center light zone; and (3) T-cell dependent selection during the final stages of B-cell maturation (Fig. 4) (LINDHOUT et al. 1997).

The germinal center reaction is a potentially dangerous event in the formation of a B-cell immune response as it generates high numbers of easily activated B cells that bear Ig receptors with a high antigen binding affinity. It is probable therefore that antigen recognition by a B cell is in itself not sufficient to drive a germinal center response and that help by antigen specific T cells is required. These helper T cells are activated in an MHC class II restricted manner by professional antigen presenting cells, thereby limiting the possibility of undesired reactions against for instance auto-antigens (Fig. 4). Subsequently the B cells that go through a germinal center reaction undergo somatic hypermutation in their IgV genes, thereby altering the specificity of their Ig receptors. As a consequence a second round of B-cell selection is necessary. We have proposed that this second round of selection not only involves recognition of antigen, that is presented by the FDC, but also additional help by antigen stimulated T cells. As discussed above, triggering of Ig receptors, CD40, adhesion receptors, and CD95 have all been implicated in the regulation of survival of germinal center B cells. Therefore, both FDC, that carry antigen and provide adhesive interactions, and T cells, that carry CD154 and potentially the CD95 ligand, seem to be involved. However, in the germinal center the FDC and CD154 positive T cells seem to reside in different compartments, i.e., the light zone and outer zone respectively (CASAMAYOR-PALLEJA

Fig. 4. Triple check B-cell maturation model. The development of a B-cell immune response is under tight regulatory control. This model represents three separate phases where B cells are checked for their appropriate reactivity to the antigenic challenge. (1) During initial activation in the T-cell area antigen specific T cells stimulated by professional antigen presenting dendritic cells, provide help to antigen triggered B cells. (2) Within the light zone of the germinal center, B cells, that have mutated their Ig receptors, are selected on the basis of their antigen binding affinity. FDC provide both the antigen and co-stimulatory molecules. (3) Before leaving the germinal center, B cells undergo a final check in the outer zone, where T cells are encountered that can provide survival and differentiation signals via expression of CD154 that binds to CD40 on the B cell or a death signal via expression of CD95 ligand that interacts with CD95 on the B cell

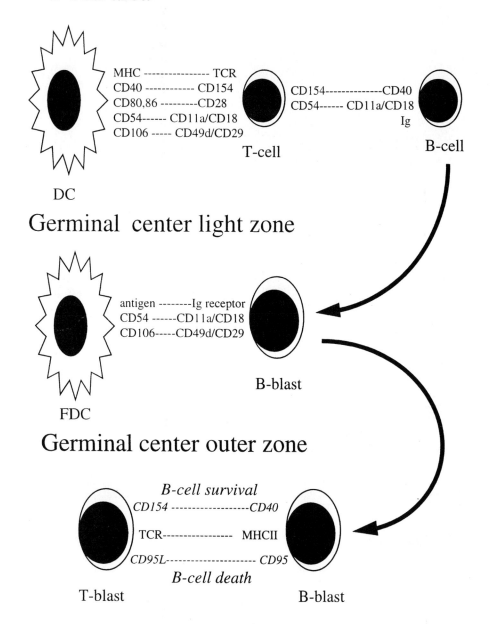

et al. 1995). We therefore think that there are two phases in the selection of germinal center B cells. First, B cells are selected by FDC on the basis of their antigen binding affinity (Fig. 4). In this phase B-cell selection is probably independent of the CD95 apoptosis pathway as germinal center formation and affinity maturation were found to be unperturbed in lpr mice (SMITH et al. 1995). Possibly, germinal center B cells harbor an active death program, which results in their spontaneous entry into apoptosis upon in vitro culture. Indeed, recently an NUC-18-like endonuclease activity was found to be present in nuclei that were extracted from germinal center B cells causing DNA fragmentation in the nuclei after incubation at 37 C (LINDHOUT et al. 1995). Importantly, it was found that, while triggering of CD40 could inhibit apoptosis formation in whole germinal center B cells, it could not prevent the DNA fragmentation in subsequently purified nuclei. In contrast, interaction with FDC was found to reverse completely endonuclease activity in the nuclei (LINDHOUT et al. 1995).

Notwithstanding their spontaneous entry into apoptosis, germinal center B cells are still sensitive to CD95 induced apoptosis (CLEARY et al. 1995; KOOPMAN et al. 1997; LAGRESLE et al. 1995; LIU et al. 1995). Their fate may thus depend on the balance between CD40 mediated rescue and CD95 mediated death signals, which can both be provided by the T cell. We speculate that it is the specific recognition of antigen by these T cells, for instance, presented by the already activated B cells that are surrounding the T cells, that determines whether rescue or death signals predominate. Indeed this would provide a second T cell, and therefore MHC restricted, check on the specificity of the via somatic hypermutation altered specificity of the B-cell Ig receptor. Importantly, while CD95 is not required for germinal center formation, it does play a role in the elimination of self-reactive B cells (RATHMELL et al. 1995; TAKAHASHI et al. 1994). Triggering of CD40 at this phase of the B-cell immune response may also be important because of its role in the differentiation of B cells into memory cells.

Acknowledgements. I would like to thank Dr. S.T. Pals and Dr. R. van der Voort (Academic Medical Center, Amsterdam, The Netherlands) for critical reading of the manuscript.

References

Agematsu K, Kobata T, Sugita K, Freeman GJ, Beckmann MP, Schlossman SF, Morimoto C (1994) Role of CD27 in T cell immune response. Analysis by recombinant soluble CD27. J Immunol 153:1421–1429

Agematsu K, Kobata T, Yang FC, Nakazawa T, Fukushima K, Kitahara M, Mori T, Sugita K, Morimoto C, Komiyama A (1995) CD27/CD70 interaction directly drives B cell IgG and IgM synthesis. Eur J Immunol 25:2825–2829

Allen RC, Armitage RJ, Conley ME, Rosenblatt H, Jenkins NA, Copeland NG, Bedell MA, Edelhoff S, Disteche CM, Simoneaux DK et al. (1993) CD40 ligand gene defects responsible for X-linked hyper-IgM syndrome [see comments]. Science 259:990–993

Anderson DM, Maraskovsky E, Billingsley WL, Dougall WC, Tometsko ME, Roux ER, Teepe MC, DuBose RF, Cosman D, Galibert L (1997) A homologue of the TNF receptor and its ligand enhance T-cell growth and dendritic-cell function. Nature 390:175–179

Armitage RJ, Fanslow WC, Strockbine L, Sato TA, Clifford KN, Macduff BM, Anderson DM, Gimpel SD, Davis-Smith T, Maliszewski CR et al. (1992) Molecular and biological characterization of a murine ligand for CD40. Nature 357:80–82

Arpin C, Dechanet J, Van Kooten C, Merville P, Grouard G, Briere F, Banchereau J, Liu YJ (1995) Generation of memory B cells and plasma cells in vitro. Science 268:720–722

Ayroldi E, Cannarile L, Migliorati G, Bartoli A, Nicoletti I, Riccardi C (1995) CD44 (Pgp-1) inhibits CD3 and dexamethasone-induced apoptosis. Blood 86:2672–2678

Barrett TB, Shu G, Clark EA (1991) CD40 signaling activates CD11a/CD18 (LFA-1)-mediated adhesion in B cells. J Immunol 146:1722–1729

Bazzoni F, Beutler B (1996) The tumor necrosis factor ligand and receptor families. N Engl J Med 334:1717–1725

Berek C (1992) The development of B cells and the B-cell repertoire in the microenvironment of the germinal center. Immunol Rev 126:5–19

Berek C, Berger A, Apel M (1991) Maturation of the immune response in germinal centers. Cell 67:1121–1129

Berek C, Griffiths GM, Milstein C (1985) Molecular events during maturation of the immune response to oxazolone. Nature 316:412–418

Bjorck P, Elenstrom-Magnusson C, Rosen A, Severinson E, Paulie S (1993) CD23 and CD21 function as adhesion molecules in homotypic aggregation of human B lymphocytes. Eur J Immunol 23:1771–1775

Bowen MB, Butch AW, Parvin CA, Levine A, Nahm MH (1991) Germinal center T cells are distinct helper-inducer T cells. Hum Immunol 31:67–75

Butch AW, Chung GH, Hoffmann JW, Nahm MH (1993) Cytokine expression by germinal center cells. J Immunol 150:39–47

Butcher EC, Rouse RV, Coffman RL, Nottenburg CN, Hardy RR, Weissman IL (1982) Surface phenotype of Peyer's patch germinal center cells: implications for the role of germinal centers in B cell differentiation. J Immunol 129:2698–2707

Callard RE, Armitage RJ, Fanslow WC, Spriggs MK (1993) CD40 ligand and its role in X-linked hyper-IgM syndrome. Immunol Today 14:559–564

Cambier JC, Campbell KS (1992) Membrane immunoglobulin and its accomplices: new lessons from an old receptor. FASEB J 6:3207–3217

Casamayor-Palleja M, Feuillard J, Ball J, Drew M, MacLennan IC (1996) Centrocytes rapidly adopt a memory B cell phenotype on co-culture with autologous germinal centre T cell-enriched preparations. Int Immunol 8:737–744

Casamayor-Palleja M, Khan M, MacLennan IC (1995) A subset of CD4+ memory T cells contains preformed CD40 ligand that is rapidly but transiently expressed on their surface after activation through the T cell receptor complex. J Exp Med 181:1293–1301

Choe J, Kim HS, Armitage RJ, Choi YS (1997) The functional role of B cell antigen receptor stimulation and IL-4 in the generation of human memory B cells from germinal center B cells. J Immunol 159:3757–3766

Choe J, Kim HS, Zhang X, Armitage RJ, Choi YS (1996) Cellular and molecular factors that regulate the differentiation and apoptosis of germinal center B cells. Anti-Ig down-regulates Fas expression of CD40 ligand-stimulated germinal center B cells and inhibits Fas-mediated apoptosis. J Immunol 157:1006–1016

Clark EA, Grabstein KH, Gown AM, Skelly M, Kaisho T, Hirano T, Shu GL (1995) Activation of B lymphocyte maturation by a human follicular dendritic cell line, FDC-1. J Immunol 155:545–555

Cleary AM, Fortune SM, Yellin MJ, Chess L, Lederman S (1995) Opposing roles of CD95 (Fas/APO-1) and CD40 in the death and rescue of human low density tonsillar B cells. J Immunol 155:3329–3337

Dang LH, Michalek MT, Takei F, Benaceraff B, Rock KL (1990) Role of ICAM-1 in antigen presentation demonstrated by ICAM-1 defective mutants. J Immunol 144:4082–4091

Debatin KM, Goldmann CK, Bamford R, Waldmann TA, Krammer PH (1990) Monoclonal-antibody-mediated apoptosis in adult T-cell leukaemia. Lancet 335:497–500

DeFranco AL, Ashwell JD, Schwartz RH, Paul WE (1984) Polyclonal stimulation of resting B lymphocytes by antigen-specific T lymphocytes. J Exp Med 159:861–880

Denning SM, Le PT, Singer KH, Haynes BF (1990) Antibodies against the CD44 p80, lymphocyte homing receptor molecule augment human peripheral blood T cell activation. J Immunol 144:7–15

Dransfield I, Hogg N (1989) Regulated expression of Mg2+ binding epitope on leukocyte integrin alpha subunits. Embo J 8:3759–3765

Durie FH, Foy TM, Masters SR, Laman JD, Noelle RJ (1994) The role of CD40 in the regulation of humoral and cell-mediated immunity. Immunol Today 15:406–411

Dustin ML, Springer TA (1989) T-cell receptor cross-linking transiently stimulates adhesiveness through LFA-1. Nature 341:619–624

Feuillard J, Taylor D, Casamayor-Palleja M, Johnson GD and MacLennan IC (1995) Isolation and characteristics of tonsil centroblasts with reference to Ig class switching. Int Immunol 7:121–130

Foy TM, Laman JD, Ledbetter JA, Aruffo A, Claassen E, Noelle RJ (1994) gp39-CD40 interactions are essential for germinal center formation and the development of B cell memory. J Exp Med 180:157–163

Freedman AS, Munro JM, Rice GE, Bevilacqua MP, Morimoto C, McIntyre BW, Rhynhart K, Pober JS, Nadler LM (1990) Adhesion of human B cells to germinal centers in vitro involves VLA-4 and INCAM-110. Science 249:1030–1033

Galibert L, Burdin N, Barthelemy C, Meffre G, Durand I, Garcia E, Garrone P, Rousset F, Banchereau J, Liu YJ (1996a) Negative selection of human germinal center B cells by prolonged BCR cross-linking. J Exp Med 183:2075–2085

Galibert L, Burdin N, de Saint-Vis B, Garrone P, Van Kooten C, Banchereau J, Rousset F (1996b) CD40 and B cell antigen receptor dual triggering of resting B lymphocytes turns on a partial germinal center phenotype. J Exp Med 183:77–85

Garrone P, Neidhardt EM, Garcia E, Galibert L, van Kooten C, Banchereau J (1995) Fas ligation induces apoptosis of CD40-activated human B lymphocytes. J Exp Med 182:1265–1273

Gerdes J, Stein H, Mason DY, Ziegler A (1983) Human dendritic reticulum cells of lymphoid follicles: their antigenic profile and their identification as multinucleated giant cells. Virchows Arch B Cell Pathol Incl Mol Pathol 42:161–172

Goodwin RG, Alderson MR, Smith CA, Armitage RJ, VandenBos T, Jerzy R, Tough TW, Schoenborn MA, Davis-Smith T, Hennen K et al. (1993) Molecular and biological characterization of a ligand for CD27 defines a new family of cytokines with homology to tumor necrosis factor. Cell 73:447–456

Gray D (1993) Immunological memory. Annu Rev Immunol 11:49–77

Gray D, Dullforce P, Jainandunsing S (1994) Memory B cell development but not germinal center formation is impaired by in vivo blockade of CD40-CD40 ligand interaction. J Exp Med 180:141–155

Griffiths GM, Berek C, Kaartinen M, Milstein C (1984) Somatic mutation and the maturation of immune response to 2-phenyl oxazolone. Nature 312:271–275

Grouard G, de Bouteiller O, Banchereau J, Liu YJ (1995) Human follicular dendritic cells enhance cytokine-dependent growth and differentiation of CD40-activated B cells. J Immunol 155:3345–3352

Gunn MD, Ngo VN, Ansel KM, Ekland EH, Cyster JG, Williams LT (1998) A B-cell-homing chemokine made in lymphoid follicles activates Burkitt's lymphoma receptor-1. Nature 391:799–803

Han S, Hathcock K, Zheng B, Kepler TB, Hodes R, Kelsoe G (1995a) Cellular interaction in germinal centers. Roles of CD40 ligand and B7–2 in established germinal centers. J Immunol 155:556–567

Han S, Zheng B, Dal Porto J, Kelsoe G (1995b) In situ studies of the primary immune response to (4-hydroxy-3- nitrophenyl)acetyl. IV. Affinity-dependent, antigen-driven B cell apoptosis in germinal centers as a mechanism for maintaining self-tolerance. J Exp Med 182:1635–1644

Hardie DL, Johnson GD, Khan M, MacLennan IC (1993) Quantitative analysis of molecules which distinguish functional compartments within germinal centers. Eur J Immunol 23:997–1004

Hemler ME (1990) VLA proteins in the integrin family: structures, functions, and their role on leukocytes. Annu Rev Immunol 8:365–400

Hemler ME, Lobb RR (1995) The leukocyte beta 1 integrins. Curr Opin Hematol 2:61–67

Hintzen RQ, de Jong R, Lens SM, van Lier RA (1994) CD27: marker and mediator of T-cell activation? Immunol Today 15:307–311

Hintzen RQ, Lens SM, Lammers K, Kuiper H, Beckmann MP, van Lier RA (1995) Engagement of CD27 with its ligand CD70 provides a second signal for T cell activation. J Immunol 154:2612–2623

Huet S, Groux H, Caillou B, Valentin H, Prieur AM, Bernard A (1989) CD44 contributes to T cell activation. J Immunol 143:798–801

Inaba K, Steinman RM (1987) Monoclonal antibodies to LFA-1 and to CD4 inhibit the mixed leukocyte reaction after the antigen-dependent clustering of dendritic cells and T lymphocytes. J Exp Med 165:1403–1417

Jabara HH, Fu SM, Geha RS, Vercelli D (1990) CD40 and IgE: synergism between anti-CD40 monoclonal antibody and interleukin 4 in the induction of IgE synthesis by highly purified human B cells. J Exp Med 172:1861–1864

Jalkanen S, Jalkanen M (1992) Lymphocyte CD44 binds the COOH-terminal heparin-binding domain of fibronectin. J Cell Biol 116:817–825

Jeremias I, Herr I, Boehler T, Debatin KM (1998) TRAIL/Apo-2-ligand-induced apoptosis in human T cells. Eur J Immunol 28:143–152

Kim HS, Zhang X, Klyushnenkova E, Choi YS (1995) Stimulation of germinal center B lymphocyte proliferation by an FDC- like cell line, HK. J Immunol 155:1101–1109

King PD, Katz DR (1989) Human tonsillar dendritic cell-induced T cell responses: analysis of molecular mechanisms using monoclonal antibodies. Eur J Immunol 19:581–587

Klaus GG, Bijsterbosch MK, O'Garra A, Harnett MM, Rigley KP (1987) Receptor signalling and crosstalk in B lymphocytes. Immunol Rev 99:19–38

Klaus GG, Humphrey JH, Kunkl A, Dongworth DW (1980) The follicular dendritic cell: its role in antigen presentation in the generation of immunological memory. Immunol Rev 53:3–28

Koopman G, de Graaff M, Huysmans AC, Meijer CJ, Pals ST (1992) Induction of homotypic T cell adhesion by triggering of leukocyte function-associated antigen-1 alpha (CD11a): differential effects on resting and activated T cells. Eur J Immunol 22:1851–1856

Koopman G, Keehnen RM, Lindhout E, Newman W, Shimizu Y, van Seventer GA, de Groot C, Pals ST (1994) Adhesion through the LFA-1 (CD11a/CD18)-ICAM-1 (CD54) and the VLA-4 (CD49d)-VCAM-1 (CD106) pathways prevents apoptosis of germinal center B cells. J Immunol 152:3760–3767

Koopman G, Keehnen RM, Lindhout E, Zhou DF, de Groot C, Pals ST (1997) Germinal center B cells rescued from apoptosis by CD40 ligation or attachment to follicular dendritic cells, but not by engagement of surface immunoglobulin or adhesion receptors, become resistant to CD95- induced apoptosis. Eur J Immunol 27:1–7

Koopman G, Parmentier HK, Schuurman HJ, Newman W, Meijer CJ, Pals ST (1991)

Adhesion of human B cells to follicular dendritic cells involves both the lymphocyte function-associated antigen 1/intercellular adhesion molecule 1 and very late antigen 4/vascular cell adhesion molecule 1 pathways. J Exp Med 173:1297–1304

Koopman G, van Kooyk Y, de Graaff M, Meyer CJ, Figdor CG, Pals ST (1990) Triggering of the CD44 antigen on T lymphocytes promotes T cell adhesion through the LFA-1 pathway. J Immunol 145:3589–3593

Kosco MH (1991) Cellular interactions during the germinal centre response. Res Immunol 142:245–248

Kosco MH, Monfalcone AP, Szakal AK, Tew JG (1988) Germinal center B cells present antigen obtained in vivo to T cells in vitro and stimulate mixed lymphocyte reactions. Adv Exp Med Biol 237:883–888

Kosco MH, Pflugfelder E, Gray D (1992) Follicular dendritic cell-dependent adhesion and proliferation of B cells in vitro. J Immunol 148:2331–2339

Kraal G, Weissman IL, Butcher EC (1982) Germinal centre B cells: antigen specificity and changes in heavy chain class expression. Nature 298:377–379

Krammer PH, Dhein J, Walczak H, Behrmann I, Mariani S, Matiba B, Fath M, Daniel PT, Knipping E, Westendorp MO et al. (1994) The role of APO-1-mediated apoptosis in the immune system. Immunol Rev 142:175–191

Kremmidiotis G, Zola H (1995) Changes in CD44 expression during B cell differentiation in the human tonsil. Cell Immunol 161:147–157

Kroese FG, Wubbena AS, Seijen HG, Nieuwenhuis P (1987) Germinal centers develop oligoclonally. Eur J Immunol 17:1069–1072

Lagresle C, Bella C, Daniel PT, Krammer PH, Defrance T (1995) Regulation of germinal center B cell differentiation. Role of the human APO-1/Fas (CD95) molecule. J Immunol 154:5746–5756

Lagresle C, Bella C, Defrance T (1993) Phenotypic and functional heterogeneity of the IgD- B cell compartment: identification of two major tonsillar B cell subsets. Int Immunol 5:1259–1268

Lane P, Traunecker A, Hubele S, Inui S, Lanzavecchia A, Gray D (1992) Activated human T cells express a ligand for the human B cell- associated antigen CD40 which participates in T cell-dependent activation of B lymphocytes. Eur J Immunol 22:2573–2578

Lebecque S, de Bouteiller O, Arpin C, Banchereau J, Liu YJ (1997) Germinal center founder cells display propensity for apoptosis before onset of somatic mutation. J Exp Med 185:563–571

Lederman S, Yellin MJ, Cleary AM, Pernis A, Inghirami G, Cohn LE, Covey LR, Lee JJ, Rothman P, Chess L (1994) T-BAM/CD40-L on helper T lymphocytes augments lymphokine-induced B cell Ig isotype switch recombination and rescues B cells from programmed cell death. J Immunol 152:2163–2171

Lederman S, Yellin MJ, Inghirami G, Lee JJ, Knowles DM, Chess L (1992a) Molecular interactions mediating T-B lymphocyte collaboration in human lymphoid follicles. Roles of T cell-B-cell-activating molecule (5c8 antigen) and CD40 in contact-dependent help. J Immunol 149:3817–3826

Lederman S, Yellin MJ, Krichevsky A, Belko J, Lee JJ, Chess L (1992b) Identification of a novel surface protein on activated CD4+ T cells that induces contact-dependent B cell differentiation (help). J Exp Med 175:1091–1101

Legler DF, Loetscher M, Roos RS, Clark-Lewis I, Baggiolini M, Moser B (1998) B cell-attracting chemokine 1, a human CXC chemokine expressed in lymphoid tissues, selectively attracts B lymphocytes via BLR1/CXCR5. J Exp Med 187:655–660

Lens SM, Keehnen RM, van Oers MH, van Lier RA, Pals ST, Koopman G (1996a) Identification of a novel subpopulation of germinal center B cells characterized by expression of IgD and CD70. Eur J Immunol 26:1007–1011

Lens SM, Tesselaar K, den Drijver BF, van Oers MH, van Lier RA (1996b) A dual role for both CD40-ligand and TNF-alpha in controlling human B cell death. J Immunol 156:507–514

Lesley J, Schulte R, Hyman R (1990) Binding of hyaluronic acid to lymphoid cell lines is inhibited by monoclonal antibodies against Pgp-1. Exp Cell Res 187:224–233

Lindhout E, Koopman G, Pals ST, de Groot C (1997) Triple check for antigen specificity of B cells during germinal centre reactions. Immunol Today 18:573–577

Lindhout E, Lakeman A, de Groot C (1995) Follicular dendritic cells inhibit apoptosis in human B lymphocytes by a rapid and irreversible blockade of preexisting endonuclease. J Exp Med 181:1985–1995

Lindhout E, Lakeman A, Mevissen ML, de Groot C (1994) Functionally active Epstein-Barr virus-transformed follicular dendritic cell-like cell lines. J Exp Med 179: 1173–1184

Lindhout E, Mevissen ML, Kwekkeboom J, Tager JM, de Groot C (1993) Direct evidence that human follicular dendritic cells (FDC) rescue germinal centre B cells from death by apoptosis. Clin Exp Immunol 91:330–336

Liu YJ, Barthelemy C, de Bouteiller O, Arpin C, Durand I, Banchereau J (1995) Memory B cells from human tonsils colonize mucosal epithelium and directly present antigen to T cells by rapid up-regulation of B7–1 and B7–2. Immunity 2:239–248

Liu YJ, Cairns JA, Holder MJ, Abbot SD, Jansen KU, Bonnefoy JY, Gordon J, MacLennan IC (1991) Recombinant 25-kDa CD23 and interleukin 1 alpha promote the survival of germinal center B cells: evidence for bifurcation in the development of centrocytes rescued from apoptosis. Eur J Immunol 21:1107–1114

Liu YJ, de Bouteiller O, Arpin C, Briere F, Galibert L, Ho S, Martinez-Valdez H, Banchereau J, Lebecque S (1996a) Normal human IgD+IgM- germinal center B cells can express up to 80 mutations in the variable region of their IgD transcripts. Immunity 4:603–613

Liu YJ, Johnson GD, Gordon J, MacLennan IC (1992) Germinal centres in T-cell-dependent antibody responses. Immunol Today 13:17–21

Liu YJ, Joshua DE, Williams GT, Smith CA, Gordon J, MacLennan IC (1989) Mechanism of antigen-driven selection in germinal centres. Nature 342:929–931

Liu YJ, Malisan F, de Bouteiller O, Guret C, Lebecque S, Banchereau J, Mills FC, Max EE, Martinez-Valdez H (1996b) Within germinal centers, isotype switching of immunoglobulin genes occurs after the onset of somatic mutation. Immunity 4:241–250

Louis E, Philippet B, Cardos B, Heinen E, Cormann N, Kinet-Denoel C, Braun M, Simar LJ (1989) Intercellular contacts between germinal center cells. Mechanisms of adhesion between lymphoid cells and follicular dendritic cells. Acta Oto-Rhino-Laryngologica Belgica 43:297–320

MacLennan IC, Gray D (1986) Antigen-driven selection of virgin and memory B cells. Immunol Rev 91:61–85

Mangeney M, Richard Y, Coulaud D, Tursz T, Wiels J (1991) CD77: an antigen of germinal center B cells entering apoptosis. Eur J Immunol 21:1131–1140

Matsumoto M, Fu YX, Molina H, Huang G, Kim J, Thomas DA, Nahm MH, Chaplin DD (1997) Distinct roles of lymphotoxin alpha and the type I tumor necrosis factor (TNF) receptor in the establishment of follicular dendritic cells from non-bone marrow-derived cells. J Exp Med 186:1997–2004

Maurer D, Fischer GF, Fae I, Majdic O, Stuhlmeier K, Von Jeney N, Holter W, Knapp W (1992) IgM and IgG but not cytokine secretion is restricted to the CD27+ B lymphocyte subset. J Immunol 148:3700–3705

Maurer D, Holter W, Majdic O, Fischer GF, Knapp W (1990) CD27 expression by a distinct subpopulation of human B lymphocytes. Eur J Immunol 20:2679–2684

Miyake K, Underhill CB, Lesley J, Kincade PW (1990) Hyaluronate can function as a cell adhesion molecule and CD44 participates in hyaluronate recognition. J Exp Med 172:69–75

Miyawaki T, Uehara T, Nibu R, Tsuji T, Yachie A, Yonehara S, Taniguchi N (1992) Differential expression of apoptosis-related Fas antigen on lymphocyte subpopulations in human peripheral blood. J Immunol 149:3753–3758

Moller P, Henne C, Leithauser F, Eichelmann A, Schmidt A, Bruderlein S, Dhein J, Krammer PH (1993) Coregulation of the APO-1 antigen with intercellular adhesion molecule- 1 (CD54) in tonsillar B cells and coordinate expression in follicular center B cells and in follicle center and mediastinal B-cell lymphomas. Blood 81:2067–2075

Nagata S, Golstein P (1995) The Fas death factor. Science 267:1449–1456

Nakanishi K, Matsui K, Kashiwamura S, Nishioka Y, Nomura J, Nishimura Y, Sakaguchi N, Yonehara S, Higashino K, Shinka S (1996) IL-4 and anti-CD40 protect against Fas-mediated B cell apoptosis and induce B cell growth and differentiation. Int Immunol 8:791–798

Nieuwenhuis P, Kroese FG, Opstelten D, Seijen HG (1992) De novo germinal center formation. Immunol Rev 126:77–98

Noelle R, Krammer PH, Ohara J, Uhr JW, Vitetta ES (1984) Increased expression of Ia antigens on resting B cells: an additional role for B-cell growth factor. Proc Natl Acad Sci USA 81:6149–6153

Noelle RJ, Ledbetter JA, Aruffo A (1992) CD40 and its ligand, an essential ligand-receptor pair for thymus- dependent B-cell activation. Immunol Today 13:431–433

Nossal GJ, Abbot A, Mitchell J, Lummus Z (1968) Antigens in immunity. XV. Ultrastructural features of antigen capture in primary and secondary lymphoid follicles. J Exp Med 127:277–290

Oshimi Y, Oda S, Honda Y, Nagata S, Miyazaki S (1996) Involvement of Fas ligand and Fas-mediated pathway in the cytotoxicity of human natural killer cells. J Immunol 157:2909–2915

Pan G, Ni J, Wei YF, Yu G, Gentz R, Dixit VM (1997) An antagonist decoy receptor and a death domain-containing receptor for TRAIL [see comments]. Science 277:815–818

Pascual V, Liu YJ, Magalski A, de Bouteiller O, Banchereau J, Capra JD (1994) Analysis of somatic mutation in five B cell subsets of human tonsil. J Exp Med 180:329–339

Petrasch S, Perez-Alvarez C, Schmitz J, Kosco M, Brittinger G (1990) Antigenic phenotyping of human follicular dendritic cells isolated from nonmalignant and malignant lymphatic tissue. Eur J Immunol 20:1013–1018

Pulendran B, Kannourakis G, Nouri S, Smith KG, Nossal GJ (1995) Soluble antigen can cause enhanced apoptosis of germinal-centre B cells [see comments]. Nature 375:331–334

Ramsdell F, Seaman MS, Miller RE, Picha KS, Kennedy MK, Lynch DH (1994) Differential ability of Th1 and Th2 T cells to express Fas ligand and to undergo activation-induced cell death. Int Immunol 6:1545–1553

Rathmell JC, Cooke MP, Ho WY, Grein J, Townsend SE, Davis MM, Goodnow CC (1995) CD95 (Fas)-dependent elimination of self-reactive B cells upon interaction with CD4+ T cells. Nature 376:181–184

Rice GE, Munro JM, Corless C, Bevilacqua MP (1991) Vascular and nonvascular expression of INCAM-110. A target for mononuclear leukocyte adhesion in normal and inflamed human tissues. Am J Pathol 138:385–393

Rothstein TL, Wang JK, Panka DJ, Foote LC, Wang Z, Stanger B, Cui H, Ju ST, Marshak-Rothstein A (1995) Protection against Fas-dependent Th1-mediated apoptosis by antigen receptor engagement in B cells. Nature 374:163–165

Rouse RV, Ledbetter JA, Weissman IL (1982) Mouse lymph node germinal centers contain a selected subset of T cells– the helper phenotype. J Immunol 128:2243–2246

Rousset F, Garcia E, Banchereau J (1991) Cytokine-induced proliferation and immunoglobulin production of human B lymphocytes triggered through their CD40 antigen. J Exp Med 173:705–710

Roy M, Waldschmidt T, Aruffo A, Ledbetter JA, Noelle RJ (1993) The regulation of the expression of gp39, the CD40 ligand, on normal and cloned CD4+ T cells. J Immunol 151:2497–2510

Schriever F, Freedman AS, Freeman G, Messner E, Lee G, Daley J, Nadler LM (1989) Isolated human follicular dendritic cells display a unique antigenic phenotype. J Exp Med 169:2043–2058

Sheridan JP, Marsters SA, Pitti RM, Gurney A, Skubatch M, Baldwin D, Ramakrishnan L, Gray CL, Baker K, Wood WI, Goddard AD, Godowski P, Ashkenazi A (1997) Control of TRAIL-induced apoptosis by a family of signaling and decoy receptors [see comments]. Science 277:818–821

Shimizu Y, van Seventer GA, Horgan KJ, Shaw S (1990) Roles of adhesion molecules in T-cell recognition: fundamental similarities between four integrins on resting human T cells (LFA-1, VLA-4, VLA-5, VLA-6) in expression, binding, and costimulation. Immunol Rev 114:109–143

Shimizu Y, Van Seventer GA, Siraganian R, Wahl L, Shaw S (1989) Dual role of the CD44 molecule in T cell adhesion and activation. J Immunol 143:2457–2463

Shokat KM, Goodnow CC (1995) Antigen-induced B-cell death and elimination during germinal-centre immune responses. Nature 375:334–338

Smith CA, Farrah T, Goodwin RG (1994) The TNF receptor superfamily of cellular and viral proteins: activation, costimulation, and death. Cell 76:959–962

Smith KG, Nossal GJ, Tarlinton DM (1995) FAS is highly expressed in the germinal center but is not required for regulation of the B-cell response to antigen. Proc Natl Acad Sci USA 92:11628–11632

Spriggs MK, Armitage RJ, Strockbine L, Clifford KN, Macduff BM, Sato TA, Maliszewski CR, Fanslow WC (1992) Recombinant human CD40 ligand stimulates B cell proliferation and immunoglobulin E secretion. J Exp Med 176:1543–1550

Springer TA (1990) Adhesion receptors of the immune system. Nature 346:425–434

Springer TA (1994) Traffic signals for lymphocyte recirculation and leukocyte emigration: the multistep paradigm. Cell 76:301–314

Stein H, Gerdes J, Mason DY (1982) The normal and malignant germinal centre. Clin Haematol 11:531–559

Stuber E, Neurath M, Calderhead D, Fell HP, Strober W (1995) Cross-linking of OX40 ligand, a member of the TNF/NGF cytokine family, induces proliferation and differentiation in murine splenic B cells. Immunity 2:507–521

Suda T, Takahashi T, Golstein P, Nagata S (1993) Molecular cloning and expression of the Fas ligand, a novel member of the tumor necrosis factor family. Cell 75:1169–1178

Sumimoto S, Heike T, Kanazashi S, Shintaku N, Jung EY, Hata D, Katamura K, Mayumi M (1994) Involvement of LFA-1/intracellular adhesion molecule-1-dependent cell adhesion in CD40-mediated inhibition of human B lymphoma cell death induced by surface IgM crosslinking. J Immunol 153:2488–2496

Szakal AK, Hanna MG Jr (1968) The ultrastructure of antigen localization and virus-like particles in mouse spleen germinal centers. Exp Mol Pathol 8:75–89

Szakal AK, Kosco MH, Tew JG (1988) A novel in vivo follicular dendritic cell-dependent iccosome-mediated mechanism for delivery of antigen to antigen-processing cells. J Immunol 140:341–353

Takahashi T, Tanaka M, Brannan CI, Jenkins NA, Copeland NG, Suda T, Nagata S (1994) Generalized lymphoproliferative disease in mice, caused by a point mutation in the Fas ligand. Cell 76:969–976

Tew JG, Kosco MH, Burton GF, Szakal AK (1990) Follicular dendritic cells as accessory cells. Immunol Rev 117:185–211

Tew JG, Mandel TE (1979) Prolonged antigen half-life in the lymphoid follicles of specifically immunized mice. Immunology 37:69–76

Tohma S, Hirohata S, Lipsky PE (1991) The role of CD11a/CD18-CD54 interactions in human T cell-dependent B cell activation. J Immunol 146:492–499

Trauth BC, Klas C, Peters AM, Matzku S, Moller P, Falk W, Debatin KM, Krammer PH (1989) Monoclonal antibody-mediated tumor regression by induction of apoptosis. Science 245:301–305

Underhill CB, Green SJ, Comoglio PM, Tarone G (1987) The hyaluronate receptor is identical to a glycoprotein of Mr 85,000 (gp85) as shown by a monoclonal antibody that interferes with binding activity. J Biol Chem 262:13142–13146

van der Voort R, Taher TE, Keehnen RM, Smit L, Groenink M, Pals ST (1997) Paracrine regulation of germinal center B cell adhesion through the c- met-hepatocyte growth factor/scatter factor pathway. J Exp Med 185:2121–2131

van Essen D, Kikutani H, Gray D (1995) CD40 ligand-transduced co-stimulation of T cells in the development of helper function. Nature 378:620–623

van Kooyk Y, van de Wiel-van Kemenade P, Weder P, Kuijpers TW, Figdor CG (1989) Enhancement of LFA-1-mediated cell adhesion by triggering through CD2 or CD3 on T lymphocytes. Nature 342:811–813

van Noesel C, Miedema F, Brouwer M, de Rie MA, Aarden LA, van Lier RA (1988) Regulatory properties of LFA-1 alpha and beta chains in human T-lymphocyte activation. Nature 333:850–852

Van Seventer GA, Shimizu Y, Horgan KJ, Shaw S (1990) The LFA-1 ligand ICAM-1 provides an important costimulatory signal for T cell receptor-mediated activation of resting T cells. J Immunol 144:4579–4586

Wong BR, Josien R, Lee SY, Sauter B, Li HL, Steinman RM, Choi Y (1997a) TRANCE (tumor necrosis factor [TNF]-related activation-induced cytokine), a new TNF family member predominantly expressed in T cells, is a dendritic cell-specific survival factor. J Exp Med 186:2075–2080

Wong BR, Rho J, Arron J, Robinson E, Orlinick J, Chao M, Kalachikov S, Cayani E, Bartlett FS, 3rd, Frankel WN, Lee SY, Choi Y (1997b) TRANCE is a novel ligand of the tumor necrosis factor receptor family that activates c-Jun N-terminal kinase in T cells. J Biol Chem 272:25190–25194

Yellin MJ, Sippel K, Inghirami G, Covey LR, Lee JJ, Sinning J, Clark EA, Chess L, Lederman S (1994) CD40 molecules induce down-modulation and endocytosis of T cell surface T cell-B cell activating molecule/CD40-L. Potential role in regulating helper effector function. J Immunol 152:598–608

Zhang Z, Vuori K, Reed JC, Ruoslahti E (1995) The alpha 5 beta 1 integrin supports survival of cells on fibronectin and up-regulates Bcl-2 expression. Proc Natl Acad Sci USA 92:6161–6165

CHAPTER 18
The Neuroprotective and Neuronal Rescue Effect of (−)-Deprenyl

K. Magyar and B. Szende

A. Summary

(−)-Deprenyl treatment is able to increase the dopaminergic tone in the CNS by several mechanisms. It inhibits the normal metabolic degradation of dopamine and the metabolites formed from the drug reduce the uptake and promote the release of the transmitter. The age-related increase in MAO-B activity can also be blocked by (−)-deprenyl administration, which can decrease the resulting oxidative damage of the CNS. (−)-Deprenyl pre-treatment can inhibit the formation of toxins from pre-toxins and their selective uptake into the nerve endings. In small doses (−)-deprenyl is also effective in post-treatment schedules, having a neuronal rescue effect partly due to the inhibition of apoptosis of the neurones by the drug. (−)-Deprenyl is still the most widely used MAO inhibitor in the treatment of Parkinson's disease (PD). It is administered alone or in combination with levodopa. The treatment can postpone the need for levodopa or potentiate its effect. The usage of (−)-deprenyl treatment in Alzheimer's disease (AD) is less frequent than in PD, but some results indicate a mild improvement in cognitive functions of the patients with AD.

B. Introduction

Deprenyl (phenyl-isopropyl-methyl-propargylamine) was synthesized in 1962 by Ecsery in the Chinoin Pharmaceutical Works, Hungary (Parnham 1993). The first paper regarding its pharmacological activity was published by Knoll et al. (1965). Deprenyl and especially its (−)-optical isomer (selegiline) is a selective irreversible inhibitor of monoamine oxidase type B (MAO-B) (Magyar et al. 1967; Knoll and Magyar 1972). Like most of the MAO inhibitors, it was developed as an antidepressant, but in a selective dose, needed to induce MAO-B inhibition, (−)-deprenyl does not provide any antidepressive activity (Sandler 1981; Magyar 1993). As a selective irreversible inhibitor of MAO-B it is free from the "cheese reaction," which was frequently reported after the administration of MAO-A blockers to patients who had consumed foods rich in indirectly acting sympathomimetic amines, e.g., tyramine (Youdim and Finberg 1987; Palfreyman et al. 1988; Jarrott and Vajda 1987).

(−)-Deprenyl has been used in the treatment in Parkinson's disease alone or in combination with levodopa, as a putative neuroprotective agent. Its mechanism of action is rather complex. It seems probable that the antioxidant and dopamine sparing activity, as well as the neuroprotective and neuronal rescue effect of the drug, cannot be explained solely by its irreversible enzyme inhibitory action (MAGYAR et al. 1996). Studies on laboratory animals indicated that (−)-deprenyl can protect dopaminergic neurones by a mechanism independent of MAO-B inhibition (TATTON et al. 1997).

C. Clinical Benefits of (−)-Deprenyl Treatment

(−)-Deprenyl combined with levodopa was first used for the treatment of PD in 1977 (BIRKMAYER et al. 1977). Birkmayer and his co-workers were also the first who reported on the basis of a retrospective study the neuroprotective effect of the combined (−)-deprenyl treatment in patients with advanced PD (1985). More recently, TETRUD and LANGSTON (1989) carried out a prospective double blind study on a small number of early Parkinsonians, comparing (−)-deprenyl with placebo. Their conclusion, drawn from these studies after a month wash-out period, was that (−)-deprenyl treatment delayed the need for levodopa therapy and apparently slowed down disease progression. The largest, and one of the most reliable, multicenter trials, analyzing the clinical benefits of (−)-deprenyl treatment in 800 patients, is known as the DATATOP study (deprenyl and tocopherol antioxidative therapy of Parkinsonism). It was a prospective, randomized, placebo controlled, double-blind trial, in which (−)-deprenyl monotherapy was analyzed in patients with early PD. The effects of (−)-deprenyl (10 mg/day), tocopherol (vitamin E; 2000 IU/day), and (−)-deprenyl plus tocopherol were compared with placebo in the mild form of PD. The end point of the trial was when the level of the disability of the patients required the introduction of levodopa therapy. In accordance with the former studies this trial has also proved that (−)-deprenyl delayed the development of disabilities necessitating levodopa therapy (PARKINSON STUDY GROUP 1989a,b, 1993). In addition to the neuroprotective effect, (−)-deprenyl treatment has a significant symptomatic activity due to its dopamine sparing effect, i.e., the inhibition of dopamine metabolism and uptake (OLANOW 1996). Concerning uptake inhibition, the metabolites of (−)-deprenyl (amphetamine and methylamphetamine) are especially effective (TEKES et al. 1988; MAGYAR 1994; MAGYAR et al. 1996). The metabolites, in spite of being (−)-isomers, can also elicit some release of dopamine. The rise of phenylethylamine (PEA) concentration in the central nervous system may also play a role in dopamine potentiation, because a high level of PEA, due to (−)-deprenyl treatment, can enhance dopaminergic activity (OLANOW and CALNE 1991). Nevertheless, findings experienced after a suitably long drug withdrawal at the end of the study support the view that (−)-deprenyl treatment can slow down disease progression. When (−)-deprenyl was administered in a combined therapy, it led to the reduction of the dose of levodopa and a decrease in response fluctuations due to levodopa treatment (on-off, end of the dose dyskinesia) (WESSEL 1993).

D. Effect of (−)-Deprenyl Against Oxidative Stress

(−)-Deprenyl treatment might protect neurones from oxidative damage and death by reducing the production of H_2O_2 due to the inhibition of the normal metabolism of dopamine by MAO-B (COHEN and SPINA 1989; OLANOW 1990). It is well known that in the presence of Fe^{++} ion, H_2O_2 can be converted to hydroxyl radicals (·OH) and hydroxyl ions (OH⁻). Reactive species, such as ·O_2^- and ·OH, can induce lipid peroxidation of the membrane and thereby may cause neuronal rupture and death (SIMONIAN and COYLE 1996). Studies on platelet MAO-B activity have shown that an age-dependent increase can be observed in the enzyme activity. It was also demonstrated that platelet MAO-B activity is higher in some neurodegenerative diseases, like PD and dementia of the Alzheimer type. The inhibition of the overproduction of H_2O_2 after a certain age or in neurodegenerative diseases might lead to neuroprotection (STROLIN-BENEDETTI and DOSTERT 1989; BERRY et al. 1994).

Some authors have shown that deprenyl in a concentration lower than needed to inhibit MAO-B can decrease the damage due to oxidative shock (WU et al. 1993, 1994; CHIUEH et al. 1994). This protection could be due to the increase of scavenger mechanisms. Long term treatment with (−)-deprenyl can enhance the synthesis of Cu/Zn dependent superoxide dismutase (SOD1) and Mn dependent superoxide dismutase (SOD2) or catalase (CARRILLO et al. 1991, 1992, 1993; KNOLL 1988) in some experimental animals. The increased scavenger activity might also protect neurons from oxidative damage. It has been reported that (−)-deprenyl treatment increased the life span of laboratory animals, rats, and mice (MILGRAM et al. 1990; YEN and KNOLL 1992; KITANI et al. 1993; FREISLEBEN et al. 1994; KNOLL et al. 1994), but contradictory data were also published (GALLAGHER et al. 1998). A recent clinical trial of (−)-deprenyl found an increased mortality at five years after treatment (LEES 1995). The methods of this study were seriously criticized by the authors of other clinical trials who found a decrease in mortality in Parkinsonian patients after (−)-deprenyl treatment (GERLACH et al. 1996; JELLINGER 1996; MAKI-IKOLA et al. 1996; OLANOW et al. 1996).

E. Selegiline Induced Neuroprotection Against Toxic Insults

The mechanism of neuroprotection was examined by using chemicals as selective toxins in animal experiments. A selective injury can be documented in nerves that inactivate their natural transmitter by means of a membrane-bound, high-affinity, energy- and sodium-dependent monoamine transporter.

Structural analogues of the transmitters among these toxins can be taken up by the same carrier transport (BAUMGARTEN and ZIMMERMANN 1992). Selective toxins for the dopaminergic, noradrenergic, serotonergic, and cholinergic nerves are available which can degenerate these nerve endings.

It has become apparent in recent years that (−)-deprenyl pre-treatment can protect neurones from a variety of toxins, which induce neurodegeneration. The neuroprotective effect of (−)-deprenyl against MPTP (1-methyl-4-phenyl-1,2,3,6-tetrahydropyridine) (LANGSTON et al. 1983, 1984), 6-hydroxydopamine (KNOLL 1987), and DSP-4 [N-(2-chloroethyl)-N-ethyl-2-bromobenzylamine] has been widely demonstrated (ROSS and RENYI 1976). Similar protection due to (−)-deprenyl pre-treatment was shown against a central cholinergic neurotoxin, AF64A (methyl-β-acetoxyethyl-2-chloroethylamine) (RICCI et al. 1992).

The mechanism of MPTP toxicity was excellently reviewed by GLOVER et al. (1986). The substance is a preferential substrate for MAO-B (SALACH et al. 1984), as its oxidation is highly sensitive to inhibition by (−)-deprenyl. It inhibits MPTP oxidation to the toxic metabolite MPP$^+$ (1-methyl-4-phenylpiridine), which is actively taken up by the dopaminergic nerve terminals via the DA re-uptake processes (JAVITCH et al. 1985). Since the formation of the neurotoxin MPP$^+$ from MPTP is MAO-B dependent, all the selective inhibitors of MAO-B can potentially prevent MPTP-induced neurodegeneration in vivo.

Inhibitors of DA uptake, like desipramine (DMI) and mazindol, are also capable of preventing MPTP-induced neurodegeneration. Since (−)-deprenyl – and mainly its metabolites (amphetamine and methylamphetamine) – are potent inhibitors of DA uptake (Table 1), all of them play a considerable role in the prevention of MPTP-induced neurotoxicity by inhibiting the re-uptake process (HÁRSING et al. 1979; MAGYAR 1991). The toxicity induced by MPTP, which is still the best primate model of Parkinsonian syndrome, is a two-step process (Fig. 1).

The 6-hydroxydopamine (6-OH-DA) treatment induces nigro-striatal degeneration which can also be prevented by pre-treatment with (−)-deprenyl (KNOLL 1988). The mechanism underlying the neural degeneration depends on the formation of 6-hydroxyquinone from 6-OH-DA, this step being fol-

Table 1. The effect of deprenyl, methylamphetamine (MA), on the synaptosomal uptake in vitro in rats

Compounds	IC$_{50}$ in mol/l[a]		
	NA Hypothalamus	DA Striatum	5-HT Hippocampus
(−)-deprenyl	5.1×10^{-5}	1.0×10^{-4}	5.0×10^{-3}
(+)-deprenyl	1.7×10^{-5}	2.4×10^{-5}	3.6×10^{-2}
(−)-MA	3.5×10^{-6}	4.2×10^{-5}	$>1.0 \times 10^{-2}$
(+)-MA	3.5×10^{-7}	6.0×10^{-7}	1.9×10^{-2}

[a] Method: SNYDER and COYLE (1969).

1. Transformation of pretoxins to toxins

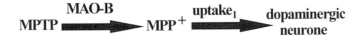

2. Selective uptake of the toxins

Fig. 1. Models of the neurotoxic mechanisms

lowed by an uptake into the dopaminergic nerve endings. 6-Hydroxyquinone initiates neural degeneration due to the generation of free radicals. The inhibition of MAO-B by (−)-deprenyl cannot play a significant role in the prevention of neurotoxicity, caused by this toxin (BERRY et al. 1994). (−)-Deprenyl is a weak inhibitor of noradrenaline (NA) and DA uptake. Nevertheless, the uptake inhibition elicited by the inhibitor, and mainly by the metabolites, can contribute to the protective effect of the drug, e.g., the inhibition of re-uptake appears to be the most probable effective mechanism against 6-OH-DA induced toxicity.

DSP-4, originally described by ROSS and RENYI (1976), is a beta-haloethylamine derivative of benzylamine which interacts with the presynaptic components of the adrenergic synapse with a one-step process (Fig. 1). The toxin is an alkylating agent that forms covalent bonds with electrophilic centers on its site of action and exerts irreversible effects. The recovery from the irreversible effect of DSP-4 is based on the synthesis of a new transport protein. The irreversible damage of the presynaptic nerve endings caused by DSP-4 can be blocked by the coadministration of uptake inhibitors, such as DMI. The selectivity of the compound to the noradrenergic synapse is based on the selective uptake into the noradrenergic nerve endings. The protection mechanism also requires a functional transporter system which can be competitively and reversibly inhibited by the protective agent.

(−)-Deprenyl, but not the MDL 72974/A, another potent selective MAO inhibitor, was capable of preventing depletion of NA in the mouse hippocampus induced by DSP-4 (FINNEGAN et al. 1990). We reported as early as 1972 that deprenyl and its optical isomers inhibit ^{3}H-NA uptake into cerebral cortex slices of mice (KNOLL and MAGYAR 1972) and in the synaptosomal fraction of the rat brain (TEKES et al. 1988). Recent experiments revealed that not only the parent compound but also it metabolites are responsible for the inhibition of the synaptosomal uptake of NA and DA (MAGYAR 1994). Neither (−)-deprenyl nor its metabolites inhibit the synaptosomal uptake of serotonin.

SKF-525A pre-treatment decreased the protective effect of (–)-deprenyl against the DSP-4 induced NA depletion, while the SKF-525A treatment in itself, did not influence NA level of the hippocampus (MAGYAR et al. 1996; MAGYAR 1997). From these data it can be concluded that the inhibition of the metabolism of (–)-deprenyl decreases the protective capacity of the inhibitor against DSP-4. When (–)-deprenyl was administered in a dose of 0.5–1 mg/kg orally, it induced a marked degree of protection against DSP-4 toxicity (MAGYAR et al. 1996), comparable to that caused by 10 mg/kg of (–)-deprenyl given intraperitoneally (FINNEGAN et al. 1990). This finding might be due to the intensive ("first pass") metabolism of (–)-deprenyl, occurring after oral administration.

Although it is widely accepted that the inhibition of the carrier mediated re-uptake process of NA plays an essential role in the prevention of DSP-4 induced neurotoxicity, some contradictory data are also cumulating in the literature (BERRY et al. 1994). GIBSON (1987) reported that clorgyline, which has similar inhibitory properties to (–)-deprenyl on NA re-uptake, does not protect against DSP-4 toxicity. It has also been published that some relatively short chain aliphatic compounds, such as N-2-hexyl-N-methyl-propargylamine (2-HxMP), are potent selective inhibitors of MAO-B, with the lack of uptake inhibitory potency, and able to protect DSP-4 toxicity (YU et al. 1994). It is apparent that the toxicity induced by DSP-4 is more complex than had been thought, but the role of the uptake inhibition in the protection cannot be ruled out.

Exogenous neurotoxins are good models to elicit selective neurodegeneration, but an endogenous neurotoxin which could be responsible for a common neurodegenerative disease, like PD or AD, has not been found yet in spite of extensive studies.

In addition to the neuroprotective activity, (–)-deprenyl is also effective in post-treatment schedule in small doses. It elicits neuronal rescue effects in a dose too low to inhibit MOA-B activity.

F. Apoptosis in Neurodegenerative Diseases

Apoptosis or programmed active cell death is a basic feature of ontogenesis and also occurs in adult tissues such as bone marrow, intestinal mucosa, thymus, skin, etc. (WYLLIE et al. 1986). A series of physiological signals and also damaging agents (viruses, toxins, ionizing radiation, etc.) can induce or stimulate apoptosis. Apoptosis has been thought to occur in cells which have entered either the G_1 or G_2 phase. Neurons in the adult central nervous system do not undergo renewal. In spite of this, increasing evidence shows that neuronal apoptosis can result from a wide range of insults like trophic insufficiency, excitatory amino acids, metamphetamine, and others (TATTON and CHALMERS-REDMAN 1996). Apoptosis can be identified in neuronal cells, in culture, or in tissue by demonstration of nuclear and cytoplasmic shrinkage (fluorescent dye methods; flow cytometry) and by showing cleavage of nuclear DNA (polyacrylamide gel-electrophoresis to identify the "ladder" pattern of oligonucleosomal DNA fragmentation; terminal deoxynucleotidyl-transferase-mediated dUTP-x nick-end labeling, also called the TUNEL method).

Apoptosis contributes to neuronal loss in human neurodegenerative diseases, such as PD, AD, and amyotrophic lateral sclerosis (ALS). ANGLADE et al. (1995) reported apoptotic cell death of nigral dopaminergic neurons in PD. COTMAN and ANDERSON (1995) suggested a potential role for apoptosis in neurodegeneration and AD. SU et al. (1994) produced immunohistochemical evidence of apoptosis in AD. This finding correlates with the DNA damage and apoptosis described by ANDERSON et al. (1996), who also showed co-localization of apoptosis with c-Jun, using immunohistochemistry.

LASSMANN et al. (1995) evaluated apoptotic cell death in AD by in situ end-labeling of fragmented DNA. The same was done by DRAGUNOW et al. (1995) in AD temporal lobes and Huntington's disease striatum. YOSHIYAMA et al. (1994) found that apoptosis-related antigen Le (Y) and nick-end labeling are positive in spinal motor neurons in amyotrophic lateral sclerosis. Moreover, according to the work of MULLER et al. (1992), the AIDS protein gp120 of HIV-1 induces apoptosis in rat cortical cell cultures.

G. Effect of Deprenyl on Neuronal Apoptosis

The pharmacological effects of (−)-deprenyl are numerous and varied in their nature and the neuroprotective as well as neuronal rescue effect cannot be explained solely by the MAO-B inhibitory action of this compound.

A series of both in vitro and in vivo studies has shown that (−)-deprenyl can reduce neuronal apoptosis caused by a variety of agents, without inhibiting MAO-B.

The first data on the anti-apoptotic effect of (−)-deprenyl was published by TATTON et al. (1994a), who used serum and nerve growth factor withdrawal to induce apoptosis in cultured PC12 human pheochromocytoma cells. (−)-Deprenyl reduced both cell death and internucleosomal DNA degradation in a concentration-dependent manner and was effective at concentrations below 10^{-9} mol/l. These concentrations are too low to inhibit MAO-B, and a mode of action other than MAO-B inhibition should be implied regarding the anti-apoptotic effect of (−)-deprenyl.

At the same time, (+)-deprenyl did not increase survival of PC12 cells after serum and nerve growth factor withdrawal, neither did other MAO-B inhibitors. The apoptosis-reducing effect of (−)-deprenyl could be suspended by addition of cycloheximide or actinomycin D, i.e., transcriptional or translational inhibitors of protein synthesis, pointing to the fact that new protein synthesis was required for the above-mentioned action of (−)-deprenyl.

Regarding apoptosis induced by serum deprivation of PC12 cells, LINDENBOIM et al. (1995) reported that cells from all phases of the cell cycle are damaged upon serum deprivation and the apoptotic cell death of non-synchronized PC12 cells may occur from each phase of the cell cycle. This finding also points to the possibility that the apoptosis-preventing action of (−)-deprenyl is also cell cycle independent.

Decrease or complete loss of trophic support induces apoptotic death of most types of cells, and thus also of nerve cells. Crush or transection of nerves represent a fairly single model to induce neuronal apoptosis caused by deprivation of target-derived trophic support.

The studies of ANSARI et al. (1993a), JU et al. (1994), SALO and TATTON (1992), and OH et al. (1993) revealed that (−)-deprenyl reduces the death of facial motoneurons of immature as well as adult rats, caused by axotomy. TATTON et al. (1994b) made the statement that reduction of nerve cell death by (−)-deprenyl occurs without monoamine oxidase inhibition.

According to BUYS et al. (1995), retinal ganglion cells which die by apoptosis after damage to their axons caused by optic nerve crush can be at least partially rescued by administration of (−)-deprenyl, in vivo. The increased survival of retinal ganglion cells is possibly caused through a transcriptionally-dependent blockade of apoptosis.

The protective effects of (−)-deprenyl were examined by NAOI et al. (1998) on apoptotic DNA damage induced by an endogenous neurotoxin in human dopaminergic neuroblastoma (SH-SY5Y) cells. The DNA damage was quantitatively measured by a single cell electrophoresis (COMET) assay. Pretreatment of the cells with (−)-deprenyl protected the cells from apoptosis, and the effects could be detected even after the washing out of (−)-deprenyl, suggesting that the intracellular process, such as synthesis of anti-oxidative proteins, may be induced by (−)-deprenyl.

The anti-apoptotic effect of (−)-deprenyl has been investigated by our group (MAGYAR et al. 1996, 1998a,b; SZENDE and MAGYAR 1998) using two human melanoma cell lines (M-1 and A-2058). Melanocytes are of neuroectodermal origin like pheochromocytomas, and therefore the neuronal rescue exerted by (−)-deprenyl was assumed. According to the in vitro studies performed on melanoma cells, serum deprivation for five days resulted in an excessive number of apoptotic cells of the cell cultures. Apoptosis was verified by morphology of the cells, as well as by flow cytometry and by TUNEL assay. Very low doses – similar to those applied by TATTON et al. (1994a) in the case of PC12 cells – of (−)-deprenyl (10^{-7} mol/l to 10^{-13} mol/l) caused an approximately two-day delay in the onset of apoptosis. At the same time, (+)-deprenyl was ineffective (Figs. 2 and 3). This latter finding is also in accordance with the results of TATTON et al. (1994a) obtained on PC12 cells.

In further experiments, (−)-deprenyl was administered in higher doses (10^{-2} mol/l to 10^{-4} mol/l) to A-2058 melanoma and HT-1080 fibrosarcoma cells in culture. In these experiments no serum deprivation was applied and the treatment was started 2 h after plating. Total eradication of the A-2058 melanoma cells was caused by 10^{-2} mol/l (−)-deprenyl. The type of cell death proved to be apoptosis. Subsequently 10^{-3} mol/l (−)-deprenyl resulted in 50%

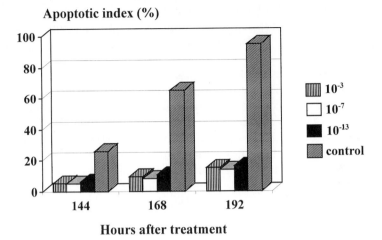

Fig. 2. The effect of (−)-deprenyl on apoptosis of M1 cell cultures. Apoptotic index = apoptotic cells in % compared to control

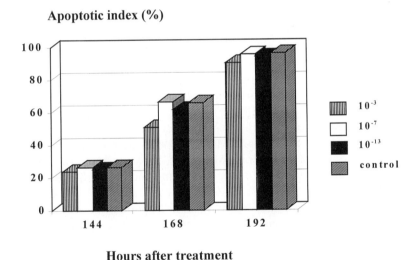

Fig. 3. The effect of (+)-deprenyl on apoptosis of M1 cell cultures. Apoptotic index = apoptotic cells in % compared to control

apoptosis 72 h after treatment. It should be mentioned that TATTON et al. (1994a) also found significant increase of apoptotic cell death of PC12 cells when (−)-deprenyl was applied at 10^{-3} mol/l in MEM.

In the case of HT-1080 fibrosarcoma cells, 10^{-3} mol/l and 10^{-4} mol/l (−)-deprenyl also caused apoptotic cell death in a dose-dependent manner. However, 10^{-2} mol/l (−)-deprenyl administration resulted in non-apoptotic, so-called cytoplasmic, vacuolar, non-lysosomal active cell death (CLARKE 1990) 24h, 48h and 72h after (–)-deprenyl administration. These results indicate that (–)-deprenyl may influence apoptosis and other types of cell death in a dose-dependent manner.

Most of the experimental work on the anti-apoptotic action of (–)-deprenyl has been carried out using cells of neuronal origin. FANG et al. (1995) investigated the effect of (–)-deprenyl in a non-neuronal cell model, namely apoptosis of mouse thymocytes induced by dexamethasone. (–)-Deprenyl did not exhibit any detectable protective effect on the thymocytes from apoptosis. This important finding shows that (–)-deprenyl can selectively prevent apoptosis depending on cell types and the mechanism of apoptosis which may also depend on cell type. It is important that an anti-apoptotic agent that blocks neurodegeneration should not disturb programmed cell death in other tissues essential for maintaining normal physiology.

H. Possible Mode of Action of (–)-Deprenyl on Apoptosis

The fact that (–)-deprenyl can increase neuronal survival without inhibiting MAO-B leads to the conclusion that this compound or one of its metabolites interrupts a still unrecognized process that leads to the death of catecholaminergic neurons (TATTON and CHALMERS-REDMAN 1996).

One of the main metabolites of (–)-deprenyl is (–)-desmethyl-deprenyl. This compound has been reported to reduce neuronal cell death by TATTON and CHALMERS-REDMAN (1996) and recently by NAOI et al. (1998) using the endogenous neurotoxin model. On the other hand, in the serum-deprived melanoma cell model SZENDE and MAGYAR (1998) did not find any protective effect of (–)-desmethyl-deprenyl. Moreover, in high doses (–)-desmethyl-deprenyl caused apoptotic cell death of both melanoma and fibrosarcoma cell cultures.

In vivo, 20 mg/kg daily subcutaneous treatment of A-2058 human melanoma xenografts growing in immune-deprived mice with (–)-desmethyl-deprenyl resulted in a significant growth-retardation of the melanoma.

In their earlier studies, Tatton's group also found that (–)-deprenyl and not its major metabolites, rescued axotomized immature facial motoneurons (ANSARI et al. 1993b). Recently TATTON and CHALMERS-REDMAN (1996) claimed that the way of administration (oral or subcutaneous) may influence the metabolism of (–)-deprenyl, i.e., subcutaneously administered (–)-deprenyl is converted into (–)-desmethyl-deprenyl in higher amounts.

With the accumulating data on the mechanism of apoptosis, a number of genes have been shown to be involved in the promotion or inhibition of apoptosis.

The early events in apoptosis are controlled by the BAX/BCL family, which are positioned in the nucleus, endoplasmic reticulum, but mainly in the mitochondrial membranes. Among these genes Bcl_2, Bcl_{xL} are anti-apoptotic and Bax as well as Bcl_{x3} have pro-apoptotic properties (OLTVAI and KORSMEYER 1994). Similarly, the interleukin-converting enzyme family (ICE) has pro-apoptotic (ICH-1_L) and anti-apoptotic (ICH-1_S) members (TAKAHASHI and EARNSHAW 1996). It has also been shown that c-Jun when overexpressed, increases neuronal apoptosis and antisense oligonucleotides against c-Jun reduced neuronal apoptosis (SCHLINGENSIEPEN et al. 1994; HAM et al. 1995).

The recent comprehensive studies of TATTON and CHALMERS-REDMAN (1996) provided evidence that mitochondria contribute to the initiation of neuronal apoptosis. When the permeability transition pore of the mitochondrial membrane is opened, molecules like Bcl_{xS} and ICH-1_L, and also holocytochrome c can escape the mitochondria and initiate apoptosis.

Decrease in mitochondrial membrane potential is also a very early event of apoptosis in neurodegenerative models. It has also been shown by TATTON and CHALMERS-REDMAN (1996) and TATTON (1998) that (−)-deprenyl as well as (−)-desmethyl-deprenyl modifies apoptosis through a mitochondrial mechanism. (−)-Deprenyl alters the expression of genes that influence cell viability mainly by its capacity of maintaining mitochondrial membrane potential. The maintenance of mitochondrial membrane potential is at least partially caused by the increase in Bcl_2 and Bcl_{xL}. The increased activity of these two genes, as well as decreased Bax synthesis, can be achieved by the administration of (−)-deprenyl. Furthermore, WU et al. (1993) showed the antioxidant effect of (−)-deprenyl on hydroxyl radical formation, in concentrations too low to inhibit MAO-B, indicating that (−)-deprenyl might act as a hydroxyl radical scavenger. This finding is important because reduction in apoptosis induced by Bcl_2 overexpression is associated with decrease in oxidative radical levels and reduced peroxidation of membrane lipids (HOCKENBERY et al. 1993).

The ongoing studies may elucidate more details with respect to the action of (−)-deprenyl and its metabolites on the apoptotic cascade, first of all in relation to caspase activity.

References

Anderson AJ, Su JH, Cotman CW (1996) DNA damage and apoptosis in Alzheimer's disease: colocalization with c-Jun immuno-reactivity, relationship to brain area and the effect of post-mortem delay. J Neurosci 16:1710–1719

Anglade P, Michel P, Marquez J, Mouatt-Prient A, Ruberg M, Hirsh EC et al. (1995) Apoptotic degeneration of nigral dopaminergic neurons in Parkinson's disease. Proc Nat Acad Sci 21:489–493

Ansari KS, Yu PH, Kruck TX, Tatton WG (1993a) Rescue of axotomized immature rat-facial motoneurons by R(−)-deprenyl: stereospecificity and independence from monoamine oxidase inhibition. J Neurosci 13:4042–4053

Ansari KS, Zhang F, Holland DH, Yu PH, Tatton WG (1993b) R(−)-deprenyl, not its major metabolites, rescue axotomized immature facial motoneurons. (Abstract) Soc Neurosci 19:243

Baumgarten HG, Zimmermann G (1992) In: Herken H, Hucho F (eds) Selective neurotoxicity (Handbook of Exp Pharm 102). Springer, Berlin Heidelberg New York, pp 225–292

Berry MD, Juorio AV, Paterson IA (1994) Possible mechanism of action of (−)-deprenyl and other MAO-B inhibitors in some neurologic and psychiatric disorders. Neurobiology 44:141–161

Birkmayer W, Riederer P, Ambrozi L, Youdim MB (1977) Implications of combined treatment with "Madopar" and L-deprenyl in Parkinson's disease. A long-term study. Lancet 1(8009):439–443

Birkmayer W, Knoll J, Riederer P, Youdim MB, Hars V, Marton J (1985) Increased life expectancy resulting from addition of L-deprenyl to Madopar treatment in Parkinson's disease: a longterm study. J Neural Transm 64(2):113–127

BuysYM, Trope GE, Tatton WG (1995) (−)-Deprenyl increases the survival of retinal ganglion cells after optic nerve crush. Curr Eye Res 14:119–126

Carrillo MC, Kanai S, Nokubo M, Kitani K (1991) (−)-Deprenyl induces activities of both superoxide dismutase and catalase but not of glutathione peroxidase in the striatum of young male rats. Life Sci 48:517–521

Carrillo MC, Kitani K, Kanai S, Sato Y, Ivy GO (1992) The ability of (−)-deprenyl to increase superoxide dismutase activities in the rat tissue and brain region selectively. Life Sci 50:1985–1992

Carrillo MC, Kanai S, Sato Y, Nokubo M, Ivy GO, Kitani K (1993) The optimal dosage of (−)-deprenyl for increasing superoxide dismutase activity in several brain regions decreases with age in male Fischer 344 rats. Life Sci 52:1925–1934

Chiueh CC, Huang SJ, Murphy DL (1994) Suppression of hydroxyl radical formation by MAO inhibitors: a novel possible neuroprotective mechanism in dopaminergic neurotoxicity. J Neural Transm [Suppl] 41:189–196

Clarke PGH (1990) Developmental cell death: morphological diversity and multiple mechanisms. Anat Embryol 181:195–213

Cohen G, Spina MB (1989) Deprenyl suppresses the oxidant stress associated with increased dopamine turn-over. Ann Neurol 26:689–690

Cotman CW Anderson AJ (1995) A potential role for apoptosis in neurodegeneration and Alzheimer's disease. Mol Neurobiol 10:19–45

Dragunow M, Faull RL, Lawlor P, Beilharz EJ, Singleton K, Walker BB, Mee E (1995) In situ evidence for DNA fragmentation in Huntington's disease striatum and Alzheimer's disease temporal lobes. Neuroreport 6:1053–1057

Fang J, Zuo DM, Yu PH (1995) Lack of protective effect of R(−)-deprenyl on programmed cell death of mouse thymocytes induced by dexamethasone. Life Sci 57:15–22

Finnegan KT, Skratt JS, Irwin I, DeLanney LE, Langston JW (1990) Protection against DSP-4-induced neurotoxicity by deprenyl is not related to its inhibition of MAO-B. Eur J Pharmacol 184:119–126

Freisleben HJ, Lehr F, Fuchs J. (1994) Lifespan of immunosuppressed NMRI-mice is increased by deprenyl. J Neural Transm [Suppl] 41:231–236

Gallagher IM, Clow A, Glover V (1998) Long term administration of (−)-deprenyl increases mortality in male Wistar rats. J Neural Transm [Suppl] 52:315–320

Gerlach M, Riederer P, Vogt H (1996) Effect of adding selegiline to levodopa in early, mild Parkinson's disease: "on treatment" rather than intention to treat analysis should have been used. [letter] BMJ 312:704

Gibson CJ (1987) Inhibition of MAO-B, but not MAO-A blocks DSP-4 toxicity on central NE neurones. Eur J Pharmacol 141:135–138

Glover V, Gibb C, Sandler M (1986) The role of MAO in MPTP toxicity – A review. J Neural Transm [Suppl] XX 65–76

Ham J, Babij C, Whitfield J, Pfarr CM, Lallemand D, Yaniv M, Rubin LL (1995) A c-Jun dominant negative mutant protects sympathetic neurons against programmed cell death. Neuron 14:927–939

Hársing LG, Magyar K, Tekes K, Vizi ES, Knoll J (1979) Inhibition by deprenyl of dopamine uptake in rat striatum: a possible correlation between dopamine uptake and acetylcholine release inhibition. Pol J Pharmacol Pharm 31:297–307

Hockenbery DM, Oltvai ZN, Xiao-Ming Y, Korsmeyer SJ (1993) Bcl-2 functions in an antioxidant pathway to prevent apoptosis. Cell 75:241–251

Jarrott B, Vajda FJE. (1987) The current status of monoamine oxidase and its inhibitors. Med J Aust 146:634–8

Javitch JA, d'Amato RJ, Strittmatter SM, Snyder SH (1985) Parkinsonism-inducing neurotoxin, N-methyl-4-phenyl-1,2,3,6-tetrahydropyridine: uptake of the metabolite N-methyl-4-phenylpyridine by dopamine neurons explains selective toxicity. Proc Natl Acad Sci USA 82:2173–2177

Jellinger KA (1996) Effect of adding selegiline to levodopa in early, mild Parkinson's disease: causes of death need confirmation. BMJ [Lett] 312:704–705

Ju WJH, Holland DP, Tatton WG (1994) (–)-Deprenyl alters the time course of death of axotomized facial motoneurons and the hypertrophy of neighboring astrocytes in immature rats. Exp Neurol 126:233–246

Kitani K, Kanai S, Sato Y, Ohta M, Ivy GO, Carrillo MC (1993) Chronic treatment of (–)-deprenyl prolongs the life span of male Fischer 344 rats: further evidence. Life Sci 52:281–288

Knoll J (1987) R-(–)-Deprenyl ((–)-deprenyl, Mogervan) facilitates the activity of the nigro-striatal dopaminergic neuron. J Neural Transm 25:45–66

Knoll J (1988) The striatal dopamine dependence of life span in male rats: longevity study with (–)-deprenyl. Mech Ageing Dev 46:237–262

Knoll J, Magyar K (1972) Some puzzling pharmacological effects of monoamine oxidase inhibitors. In: Costa E, Sandler M (eds) Monoamine oxidases (New Vistas Adv in Biochem Psychopharmacol, vol 5). Raven Press, New York, pp 393–408

Knoll J, Ecseri Z, Kelemen K, Nivel J, Knoll B (1965) Phenylisopropylmethylpropinylamine (E-250), a new spectrum psychic energizer. Arch Int Pharmacodyn Ther 155:154–164

Knoll J, Yen TT, Kiklya I. (1994) Sexually low performing male rats die earlier than their high performing peers and (–)-deprenyl treatment eliminates this difference. Life Sci 54:1047–1057

Langston JW, Ballard P, Tetrud JW, Irwin I (1983) Chronic parkinsonism in humans due to a product of meperidine-analog synthesis. Science 219:979–980

Langston JW, Langston EB, Irwin I. (1984) MPTP-induced parkinsonism in human and non-human primates – clinical and experimental aspects. Acta Neurol Scand 100 [Suppl]:49–54

Lassmann H, Bancher C, Breitschopf H, Wegiel J, Bobinski M, Jeelinger K, Wisniewski HM (1995) Cell death in Alzheimer's disease evaluated by DNA fragmentation in situ. Acta Neuropathol Ber 89:35–41

Lees AL and the Parkinson's Disease Research Group of the United Kingdom. (1995) Comparison of therapeutic effects and mortality data of levodopa combined with selegiline in patients with early, mild Parkinson's disease. BMJ 311:1602–1607

Lindenboim L, Diamond R, Rothenberg E, Stein R (1995) Apoptosis induced by serum deprivation of PC12 cells is not preceded by growth arrest and can occur at each phase of the cell cycle. Cancer Res 55:1242–1247

Magyar K (1991) Neuroprotective effect of deprenyl and p-fluor-deprenyl. Paneuropean Society of Neurology, Second Congress, Vienna, 26

Magyar K (1993) Pharmacology of monoamine oxidase type B inhibitors. In: Szelenyi I (ed) Inhibitors of monoamine oxidase B. Birkhauser Verlag, Basel, pp 125–143

Magyar K (1994) Behaviour of (–)-deprenyl and its analogues. J Neural Transm 41:167–175

Magyar K (1997) The role of the metabolism of (–)-deprenyl in neuroprotection. In: Proceedings of the 11th ESN Meeting. Teelken AW, Korf J (eds) Neurochem 303–308

Magyar K, Vizi ES, Ecseri Z, Knoll J (1967) Comparative pharmacological analysis of the optical isomers of phenyl-isopropyl-methyl-propinylamine (E-250). Acta Physiol Acad Sci Hung 32(4):377–387

Magyar K, Szende B, Lengyel J, Tekes K (1996) The pharmacology of B-type selective monoamine oxidase inhibitors; milestones in (−)-deprenyl research. J Neural Transm (Suppl) 48:29–43

Magyar K, Szende B, Haberle D, Gaál J, Tarcali J (1998a) The neuroprotective and neuronal rescue effect of selegiline. 18th European Winter Conference on Brain Research. Arc 2000 (France). 7–14 March, p 42

Magyar K, Szende B, Lengyel J, Tarcali J, Szatmári I (1998b) The neuroprotective and neuronal rescue effects of (−)-deprenyl. J Neural Transm (Suppl) 52:109–123

Maki-Ikola O, Kilkku O, Heinonen E (1996) Effect of adding selegiline to levodopa in early, mild Parkinson's disease: other studies have not shown increased mortality. [letter] BMJ 312:704–705

Milgram NW, Racine RJ. Nellis P. Mendonca A, Ivy GO. (1990) Maintenance on L-deprenyl prolongs life in aged male rats. Life Sci 47:415–420

Muller WE, Schroder HC, Ushijima H, Drapper J, Bormann J (1992) gp120 of HIV-1 induces apoptosis in rat cortical cell cultures: prevention by memantine. Eur J Pharmacol 226:209–214

Naoi M, Maruyama W, Yagi K, Youdim M (1998) Anti-apoptotic function of (−)-deprenyl and related compounds. 8th Amine Oxidase Workshop. International Workshop on Monoamine Oxidases, Trace Amines, Neuroprotection and Neuronal Rescue. Balatonöszöd, Lake Balaton, Hungary, 6–10 September p 16

Oh C, Murray B, Bhattacharya N, Holland D, Tatton WG (1993) (−)-Deprenyl alters the survival of adult facial motoneurons after axotomy: increases in vulnerable C57BL strain but decreases in Mnd mutants. J Neurosci Res 38:64–74

Olanow CW (1990) Oxidation reactions in Parkinson's disease. Neurology 40 (10 Suppl 3):32–39

Olanow CW (1996) Deprenyl in the treatment of Parkinson's disease: clinical effects and speculations on mechanism of action. J Neural Transm [Suppl] 48:75–84

Olanow CW, Calne D (1991) Does selegiline monotherapy in Parkinson's disease act by symptomatic or protective mechanisms? Neurology 42:13–26

Olanow CW, Godbold JH, Koller W (1996) Effect of adding selegiline to levodopa in early, mild Parkinson's disease: patients taking selegiline may have received more levodopa than necessary. [letter] BMJ 312:702–703

Oltvai ZN, Korsmeyer SJ (1994) Checkpoints of dueling dimers foil death wishes. Cell 79:189–192

Palfreyman MG, McDonald IA, Bey P, Schechter PJ, Sjoerdsma A. (1988) Design and early clinical evaluation of selective inhibitors of monoamine oxidase. Prog Neuropsychopharmacol Biol Psychiatry 12:967–87

Parkinson Study Group (1989a) DATATOP: a multicenter controlled clinical trial in early Parkinson's disease. Arch Neurol 46:1052–1060

Parkinson's Study Group (1989b) Effect of deprenyl on the progression of disability in early Parkinson's disease. NEJM 321:1364–1371

Parkinson's Study Group (1993) Effects of Tocopherol and deprenyl on the progression of disability in early Parkinson's disease. NEJN. 328:176–183

Parnham MJ (1993) The history of l-deprenyl. In: Szelenyi I (ed) Inhibitors of monoamine oxidase B. Birkhauser, Basel, pp 237–251

Ricci A, Mancini M, Strocchi P, Bongrani, Bronzetti E (1992) Deficits in cholinergic neurotransmission markers induced by ethylcholine mustard aziridinium (AF64 A) in the rat hippocampus: sensitivity to treatment with the monoamine oxidase-B inhibitor l-deprenyl. Drugs Exp Clin Res VIII(5):163–171

Ross SB, Renyi AL (1976) On the long-lasting inhibitory effect of N-(2-chloroethyl)-N-ethyl-2-bromobenzylamine (DSP-4) on the active uptake of adrenaline. J Pharm Pharmacol 28:458–459

Salach JI, Singer TP, Castagnoli N, Trevor A (1984) Oxidation of the neurotoxic amine 1-methyl-4-phenyl-1,2,3,6-tetrahydropyridine (MPTP) by monoamine oxidases A and B and suicide inactivation of the enzymes by MPTP. Biochem Biophys Res Comm 125:831–835

Salo PT, Tatton WG (1992) Deprenyl reduces the death of motoneurons caused by axotomy. J Neurosci Res 31:394–400

Sandler M (1981) Monoamine oxidase inhibitor efficacy in depression and the "cheese effect". Psychol Med 11:455–458

Schlingensiepen KH, Wollnik F, Kunst M, Schlingensiepen R, Herdegen T, Brysch W (1994) The role of Jun transcription factor expression and phosphorylation in neuronal differentiation, neuronal cell death, and plastic adaptations in vivo. Cell Mol Neurobiol 14:487–505

Simonian NA, Coyle JT (1996) Oxidative stress in neurodegenerative diseases Annu Rev Pharmacol Toxicol 36:83–106

Snyder SH, Coyle JT (1969) Regional differences in ^{3}H-norepinephrine and ^{3}H-dopamine uptake into rat brain homogenates. J Pharmacol Exp Ther 165:78–86

Strolin-Benedetti M, Dostert P. (1989) Monoamine oxidase, brain ageing and degenerative diseases. Biochem Pharmacol 38:-561

Su JH, Anderson AJ, Cummings BJ, Cotman CW (1994) Immunohistochemical evidence for apoptosis in Alzheimer's disease. Clin Neurosci Neuropathol 5:2529–2533

Szende B, Magyar K (1998) Apoptotic and anti-apoptotic effect of deprenyl and desmetyl-deprenyl on human cell lines. 8th Amine Oxidase Workshop. International Workshop on Monoamine Oxidases, Trace Amines, Neuroprotection and Neuronal Rescue. Balatonöszöd, Hungary, 6–10 September p 17

Takahashi A, Earnshaw WC (1996) ICE-related proteases in apoptosis. Curr Opin Genet Dev 6:50–55

Tatton WG (1998) Agents that block mitochondrial initiation of apoptosis: a new opportunity for neuroprotection. Satellite Symposium to the 3rd Congress of the European Federation of Neurological Societies. Selegiline in the Treatment of Neurodegenerative Diseases. Sevilla, Italy, 19 September, p 4

Tatton WG, Chalmers-Redman RME (1996) Modulation of gene expression rather than monoamine oxidase inhibition: (–)-Deprenyl-related compounds in controlling neuro-degeneration. Neurology 47 (Suppl 3) S171–183

Tatton WG, Ju WYL, Holland DP, Tai CE, Kwan MM (1994a) (–)-Deprenyl reduces PC12 cell apoptosis by inducing new protein synthesis. J Neurochem 63:1572–1574

Tatton WG, Seniuk NA, Ju WYH, Ansari KS (1994b) Reduction of nerve cell death by deprenyl without monoamine oxidase inhibition. In: Monoamine Oxidase Inhibitors in Neurological Diseases. Lieberman A, Olanow O, Youdim MBH, Tipton K (eds), Raven Press, New York, pp 217–248

Tatton WG, Chalmers-Redman RME, Ju WYH, Waida J, Tatton NA (1997) Apoptosis in neurodegenerative disorders: potential for therapy by modifying gene transcription. J Neural Transm [Suppl] 49:245–268

Tekes K, Tóthfalusi L, Gaál J, Magyar K (1988) Effect of MAO inhibitors on the uptake and metabolism of dopamine in rat and human brain. Pol J Pharmac Pharm 40:653–658

Tetrud JW, Langston JW (1989) The effect of deprenyl (selegiline) on the natural history of Parkinson's disease. Science 245:519–122

Tetrud JW, Langston JW (1992) Protective and preventive therapeutic strategies: monoamine oxidase inhibitors. Neurol-Clin 10 (2):541–552

Wessel K (1993) MAO-B inhibitors in neurological disorders with special reference to selegiline. In: Szelenyi I (ed) Inhibitors of monoamine oxidase B. Birkhauser Verlag Basel, pp 253–275

Wu RM, Chiueh CC, Pert A, Murphy DL (1993) Apparent antioxidant effect of L-deprenyl on hydroxyl radical formation and nigral injury elicited by MPP$^+$ in vivo. Eur J Pharmacol 243:241–248

Wu RM, Mohanakumar KP, Murohy DL, Chiueh CC (1994) Antioxidant mechanism and protection of nigral neurones against MPP$^+$ toxicity by deprenyl (selegiline). Ann NY Acad Sci 738:214–221

Wyllie AH, Kerr JFR, Currie AR (1986) Cell death: the significance of apoptosis. Int Rev Cytol 68:251–306
Yen TT, Knoll J (1992) Extension of lifespan in mice treated with Dinh lang (Policias fruticosum L.) and (−)-deprenyl. Acta Physiol Hung 79:119–124
Yoshiyama Y, Yamada T, Asanuma K, Asahi T (1994) Apoptosis related antigen, Le(Y) and nick-end labeling are positive in spinal motor neurons in amyotrophic lateral sclerosis. Acta Neuropathol Berl 88:207–211
Youdim MBH, Finberg JPM (1987) Monoamine oxidase B inhibition and the "cheese effect". J Neural Transm [Suppl] 25:27–33
Yu PH, Davis BA, Fang J, Boulton AA (1994) Neuroprotective effects of some monoamine oxidase-B inhibitors against DSP-4 induced noradrenaline depletion in the mouse hippocampus. J Neurochem 63:1820–1828

Subject Index

A
ABAH (4-aminobenzoic acid hydrazide) 279, 297
acetaminophen 343, 353
– apoptosis 348–350
acetaminophen 4, 345
N-acetylcysteine 68, 264, 266, 284, 293, 349
acitretin 82
actinomycin D 86, 463
actractylis gummifera L. 86
actractyloside 86
ADAM 17 (a desintegrin and metalloprotease 17) 74
adenosylmethionine 85
adenovirus 50, 243
– adeno-associated viruses (AAV) 248
adhesion molecules 437, 438
adrenocorticotropic hormone 14
AF64A (methyl-β-acetoxyethyl-2-chloroethylamine) 460
AIF (apoptosis inducing factor) 66
alcohol 65, 72
alkoxyl radicals 276
ALS (amyotropic lateral sclerosis) 202, 463
Alzheimer's disease 199, 200, 457
α-amanitin 76
amphetamine 457
β-amyloid 262
anticancer drug bleomycin 65, 73
antigen presenting cells 38
antimycin 160
antioxidants 86, 257–267, 289
– thiol antioxidants 263–265
antiphospholipid autoantibodies 170
antitumor drugs 333
AP-1 (activator protein-1) 74
APO-1 61, 365
APO-2 ligand 445
apoptin 87
apoptosis
– activation 9–12
– biochemistry 7–9
– characteristics/scopes 1, 2
– incidence 12–23
– inducers, *table* 15
– inhibitors, *table* 19
– molecular mediators 37–44
– morphology 6, 7
– occurence 6
apoptotic bodies 5
arachidonate metabolites 322
ascorbic acid 264
atrophy, pathological 13, 14
aurintricarboxylic acid 88
autoimmunity 187, 188
– autoimmune disorders 52
AZT 130

B
B cells 45, 46
– antigen receptor-induced death 399–420
– death 390–392
– immune response, modulation of apoptosis and maturation 429–448
– maturation 386–389
– – regulation 438–442
– receptor (BCR) 390, 399
– selection 390, 446–448
– survival, regulation 442–448
B7 45, 182, 183
baculovirus 50
Bad 413
Bak 39, 49
Bax 39, 49, 67, 413
4-1BB 435
Bcl-2 family of genes 8, 11, 16, 39–41, 80, 116, 181, 212–215, 401, 413, 467
– expression 49, 59, 60, 133
Bcl-2 transgene 18
Bcl-x 16
Bcl-xL 39, 60, 71, 347, 413, 467
BCNU 52
BCR (B cell Ag receptor) 390, 399

betubinic acid   182
bile acids   83–86
bleomycin   52, 73, 343
blood cell disorders   20, 21
t-BOOH   281
bryostatin   119
BSO (buthionine sulfoxymine)   68, 260

C

C1q deficiency   169, 170
CAD (caspase activated desoxyribonuclease)   63
caenorhabditis elegans   163, 167, 199, 394
calmidazolium   88
camptothecin   112
cancer   51
captopril   263
carcinogens, chemical   15, 16, 296
carcinoma   48, 49
carotenoids   258
caspase (cystinoaspartic acid specific protease)   8, 41, 42, 45, 49, 61, 68, 75, 84, 112, 122, 182, 208–211, 295, 343, 394, 417–419
– caspase 1   50
– caspase 3   62
– caspase 8   66
– pro-caspase 8   62
catalase   258, 262–264, 277, 283, 459
CD8+ cytotoxic T cells   38
CD14   165, 166
CD19   404
CD21   404
CD27   38, 435
CD28   182, 183
CD28   45
CD30   38, 435
CD40   38, 404, 412, 435
CD95   16, 18, 38, 45, 50, 61, 133, 135, 294, 364, 365, 435
– mediated apoptosis   136, 137
– receptor   343
CD154   430
CED-4 (cenorhabditis elegans death-4)   66
CED-5   167, 168
CED-6   167, 168
CED-7   167
cell cycle regulators   204, 215–218
cell turnover   12
centroblast   438–440
centrocyte   440
ceramide   81, 82, 122, 291–294, 299, 367
chelerythrine   84, 111

chemokines   153
chemotherapeutic drugs   15, 16, 52, 237
cisplatin   52, 73, 244, 343
clonal anergy   185
clotrimazole   88
c-Myc   413
cocaine   87
colchicine   14, 49, 152
complement receptors   166, 167
concanavalin A   71, 78, 90
copper   14, 72, 73
corticosteroids   111, 382
coryneobacterium parvum   77
co-stimulatory molecules   42, 43
cowpox virus   50
CPP32   41, 62
Creutzfeld-Jacob disease   202
crmA   11
CTLA-4   182, 183
curcuma longa   87
curcumin   87
cyanide   4
cyclins   112, 217, 218
cycloheximide   43, 70, 77, 86, 122, 130, 332, 463
cyclosporin A   130, 190
cyproterone acetate   80
cysteine   263, 266, 289
cystine   263
cytokines   42, 132–134, 138, 182, 186, 257, 280, 296, 297, 320, 322, 343, 345, 430
– immunosuppressive   153, 157
– stimulation   329
– type I/II   139
cytosine arabinoside   343
cytotoxins   2

D

daclizumab   190
daunorubicin   52
death receptor   416
DED (death effector domain)   417
dendritic cells   46
– follicular   442–444
deoxyrubicine   43
(-)-deprenyl   457
(+)-deprenyl   463
– (-)-desmethyl-deprenyl   466
dexamethasone   60, 74, 88, 259, 260, 262, 466
diamide   259
dimethylsulfoxide   77
diphenyleneidonium (DPI)   279
DISC (death-inducing signaling complex)   418

Subject Index

diseases (*see* syndromes)
DMI (desipramine)   460
DMSO   279, 284
doxorubicin   52, 260, 330, 343
drosophila   164
DSP-4 (N-(2-chloroethyl)-N-ethyl-
  2-bromobenzylamine)   460
DTT (dithiothreitol)   258, 264, 265

E
EB virus   50
encephalitis, HIV   202
eosinophils   46, 47
– apoptosis   357–370
erythroxylon coca   87
ethanol   74, 78, 81, 111, 344, 345
N-ethylmaleimide   330
etoposide   87, 112, 182
etretinate   82

F
FADD (Fas-associated protein with
  death domain)   61, 76
– adaptor protein   76
Fas ligand/antigen   22, 257, 401, 407,
  416, 418
– anti-Fas antibody   71
– cell surface Fas protein   73
– mediated apoptosis   60–74
– mRNA   73
Fas/FADD complex   68
FGF   297
FLICE (FADD-like interleukin-1
  converting enzym)   62
FLIPPs (Fas-associated death-domain-
  like interleukin-1 converting enzyme
  inhibitory proteins)   66
fluorocitrate   4
forskolin   85
Fos   204
fraticidal killing   72, 73
fumonisin B   87
fusarium moniliforme   87

G
gamma ray   16
gene therapy   235–250
genistein   70
germander   87, 88
glucocorticoids   121, 171, 190, 259,
  333, 343, 363, 364
glutamate   40, 51, 203
glutathione   44, 261, 275, 277, 283,
  289, 290, 293, 294, 351, 353b
glutathione peroxidase   262
glycochenodeoxycholate   83, 84

glycodeoxycholate   83
GM-CSF (granulocyte macrophage
  colony stimulating factor)   320, 329
granzyme B   43
GSH   264

H
H7   70
heat-shock   12, 48, 332, 333
heavy metals   51, 203
hepatocytes   47, 48
– apoptosis
– – methotrexate   344–347
– – triggered by natural substances
  59–91
herbimycin A   70
herpes virus   50, 249
HIV disease   127–144, 249
HIV-encephalitis   202
hornet, oriental   89
15-HPETE (15-hydroperoxyeicosa-
  tetraenoic acid)   259
Huntington's disease   202
2-HxMP (N-2-hexyl-N-methyl-
  propargylamine)   462
hydrocortisone sodium phosphate   385
hydrogen peroxide (H_2O_2)   40, 43, 51,
  203, 275, 277, 278, 283, 300
t-butyl hydroperoxide   261
hydroperoxides, organic   276
6-hydroxydopamine   460
hydroxyl radicals   275, 283, 298
7-hydroxystaurosporine   112, 119
hypermutation, somatic   434
hyperplasia, regression   13
hypersensitivity   49, 50
hyperthermia   16, 284, 330
hypochlorous acid   275–278, 297
hypoxia   122, 151, 257

I
IAPs (inhibitor of apoptosis)   66
– c-IAPs (cellular inhibitor of
  apoptosis)   76
ICAD (inhibitor of caspase activated
  desoxyribonuclease)   63
ICAM-3 (intercellular adhesion
  molecule)   162
ICE (interleukin-1β converting enzyme)
  63
– ICE/Ced-3 family, cysteine protease   9
Ig switching   434
immune activation/hyperactivation   132,
  133
immunity, cell-mediated   16–18
immunosuppressive drugs   52, 191

inflammation 49, 50
inhibitor polypeptides 44
integrin β_2 162
interferon 72, 122
– γ-interferon 121, 139, 182, 188, 329
interleukin-1β 320, 329
– converting enzyme 41
– converting enzyme/Ced-3 family 8, 394
interleukin-2 8, 182, 188, 430
interleukin-3 9
interleukin-4 182, 412, 430
interleukin-5 357
interleukin-8 320
interleukin-10 153, 320, 430
interleukin-converting enzyme family 467
involution 13
iodoacetate 4
ionizing radiation 121, 151, 237
ionomycin 183, 265
ischemia 18, 19
– cerebral 201, 205
3-isobutyl-1-methylxanthine 85
isotretinoin (13-cis-retinoic acid) 82

J
Jun 204

K
kainate 40
Kupffer cells 48
– clearance of apoptotic lymphocytes 319–336

L
LAP (latency associated peptide) 78
lazaroid 83
lead nitrate 323, 329
lectin-like receptor, hepatic 322–325
lentivirus 249
leucointegrin α/β_2 162
leukemia 48
LFA-1 162
α-lipoic acid 266
lipopolysaccharides, bacterial 90
liver apoptosis 321, 322
lupus erythematosus, systemic (SLE) 169
lymphocytes, activated 71, 72
lymphoma 48

M
MACH-1 (Mort-1-associated CED-3) 62
macrophages 46

mazindol 460
Mcl-1 413
memory cells 440–442
menidione 40, 261
2-mercaptoethanol 263, 266
mercaptopropionic acid 264
mercury 14
metalloproteinase 74
methotrexate 52, 73, 343, 353
– induced apoptosis 343–348
methylamphetamine 457
methylcholanthrene 22
microcystin-LR 89
microcystis aeroginosa 89
mitochondria 290, 291
mitochondrial membranes, permeabilization 63–66
mitogens 183
mitomycin 52
Mn-SOD (manganese-containing superoxide dismutase) 76, 288, 293
MPP+ (1-methyl-4-phenylpiridine) 460
MPTP (1-methyl-4-phenyl-1,2,3,6-tetrahydropyridine) 200, 203, 460
murine ABC1 167

N
NAC (N-acetyl-L-cysteine) 258
nafenopin 51, 80, 81
L-NAME (N-omega-nitro-L-arginine methyl ester) 279
necrosis
– biochemistry 5, 6
– morphology 5
– occurence 4, 5
neoplasms, malignant 21
nerve cell death 197–218
– degenerative 203, 204
neurodegenerative disorders 19, 20, 51
neuronal cells 47
neutrophils 47
NF-κB 39, 68, 72, 76, 293
NFκB/ReL factor 80
NGF (neuronal growth factor) 202
– receptor family 435–437
nitric oxide (NO) 86, 275, 276, 278, 285–287, 365, 366
nitrogen mustard 343
3-nitropropionic acid 202
nitroxyl radicals 276
NK cells 38
L-NMMA (N(G)-monomethyl-L-arginine) 279
nordihydroguaiaretic acid 266

O

6-OHDA (6-hydroxydopamine) 200, 203, 460
olomoucine 114
oncogene 296
ONYX-015 virus 247, 248
oral tolerance 187, 188
ouabain 4
OX40 435
oxidative stress 12, 65, 330, 333
oxygen metabolites, reactive 4, 16

P

p35 11
p53 tumor suppressor gene
– alteration in human cancers 240, 241
– and gene therapy 241–250
– homologues 241
– induction of apoptosis 239, 240
– structure, function 236–239
paclitaxel 89
Parkinson's disease 200, 201, 457
PDGF (platelet-derived growth factor) 329
D-penicillamine 263
perforin 43, 72
perillyl alcohol 89
peroxyl radicals 276
peroxynitrite 200, 275, 278, 284, 285, 287–290, 299, 300
phagocytosis, liver 321–330
phenylethylamine (PEA) 458
phorbol 12-myristate-13-acetate 84, 116
phorbol esters 111, 116
phorbol myristate acetate 265
phosphatidylserine 59, 81, 82
– PSRs (phosphatidylserine receptors) 166
PKC isozymes (C-type protein kinases) 109–123
plasma cells 440–442
platelet activating factor 153
podophyllotoxin 87
prednisone 14
progeny virus 50
progesterone 13, 121
propionibacterium acnes 71
prostaglandins 74
– prostaglandin E_2 153, 320
protooncogene 413
pseudomonas aeroginosa exotoxin A 77, 90
PSRs (phosphatidylserine receptors) 166
pyrrolidine dithiocarbamate 265

Q

quinolinic acid 202

R

radiation 8, 12, 16, 111
rapomycin 190
redox status, cellular 258–261
regulatory molecules 413
retinoid acid 82, 83
– 9-cis-retinoid acid 190
– receptors (RAR$\alpha/\beta/\gamma$) 82
retrovirus 243, 248
RIP (receptor interacting protein) 76
ROS (reactive oxygen species) 258ff.
– and apoptosis 275–302
roscovitine 114
RXRα/β (retinoid X receptor) 82

S

scavenger receptors 165
selegiline 457
SERCA (sarcoplasmatic/endoplasmic reticulum Ca^{2+}-ATPase) 67
serine/threonine kinase inhibitor 70
serum deprivation 257
SIN-1 (3-morpholinosydnonimine hydrochloride) 279, 287
SKF-525A 462
SLE (systemic lupus erythematosus) 169
SOD (superoxide dismutase) 258, 264, 277, 300
– Cu/Zn SOD 267, 459
– Mn-SOD (manganese-containing superoxide dismutase) 76, 288, 293
– SOD2 459
sodium nitroprusside 279
solamargine 89
solanum incanum 89
sphingomyelinase 366–368
sphingosine 43
– sphingosine-1-phosphate 81, 82
status epilepticus 201, 202
staurosporine 89, 111, 114, 116, 119
– 7-hydroxystaurosporine 112, 119
storosporine 43
superantigens 183
superoxide anions 275, 278, 280, 285, 290, 294, 297, 298, 300
SV40 virus 50
syndromes/diseases (names only)
– *Alzheimer's* disease 199, 200, 457
– *Creutzfeld-Jacob* disease 202
– *Huntington's* disease 202
– *Parkinson's* disease 200, 201, 457

T

T cells 44, 45
– apoptosis 179–191
– – regulation 444, 445
– CD8+ cytotoxic 38
TACE (TNF-α converting enzyme) 74
TAPA-1 404
taurochenodeoxycholic acid 83
tauroursodeoxycholic acid 85
taxol 89, 115, 116
taxotere 89
taxus brevifolia 89
TCDD 10
temperature, extreme 151
tertiary butylhydroxy peroxide 299
testosterone 14
teucrium chamaedrys L. 87
TGF-β (transforming growth factor-β) 78–81, 153, 157, 295–297, 300, 320, 329
– receptor type I/II 79
thapsia garganica L. 89
thapsigargin 89, 116–119
thiol antioxidants 263–265
thiols, oxidized 264
thioredoxin 260, 266
thymocyte
– death 377–386
– differentiation 375–377
– selection 377
TNF-α (tumor necrosis factor α) 74–78, 86, 139, 153, 257, 293, 320, 329, 345, 407
– TACE (TNF-α converting enzyme) 74
TNFR (tumor necrosis factor receptor superfamily) 37–39, 135, 180, 181, 435–437
– TNF-R1 416
α-tocopherol 258, 260, 458
TRADD (TNFR1-associated death domain) 76
TRAF protein 39
TRAF-2 (TNF receptor-associated factor-2) 76
TRAIL 39, 435, 445
TRAIL-R1/R2 416
TRAMP 416
TRANCE-L/RANK 435
transcription factors, inducible 204
tributylin 10
2,4,6-trinitrobenzenesulfonic acid 287
troleandomycin 88
tumor development, genetic changes 236
tyrosine kinase 368, 369

U

U83836E 83
ultraviolet light 16, 121, 237, 257, 259, 296
– UV-C ray 12
urano neo-clerodane diterpenoids 88
ursodeoxycholic acid 85

V

4-en-valproate 351
valproic acid 343, 345, 353
– apoptosis 350–353
vespa orientalis 89
vincristin 343
viral infections 21, 22
viral proteins 11
vitamin C 258, 352
vitamin E 262, 263, 266, 352, 457

X

xanthine 280
x-ray 16, 259

Z

Z-VAD-fluorometylketone 77